Quantum Communication, Computing, and Measurement 3

Quantum Communication, Computing, and Measurement 3

Edited by

Paolo Tombesi
University of Camerino
Camerino, Italy

and

Osamu Hirota
Tamagawa University
Machida, Tokyo, Japan

SPRINGER SCIENCE+BUSINESS MEDIA, LLC

Proceedings of the 5th International Conference on Quantum Communication, Measurement, and Computing, held July 3–8, 2000, in Capri, Italy

DOI 10.1007/978-0-306-47114-8

Originally published by Kluwer Academic / Plenum Publishers in 2001
MyCopy version of the original edition 2001

http://www.wkap.nl/

10 9 8 7 6 5 4 3 2 1

A C.I.P. record for this book is available from the Library of Congress.

Proceedings of the

Fifth International Conference on

Quantum Communication, Measurement, & Computing

Capri, Italy, July 3–8 2000

Principal Organizers

Local Organizing Committee

2000 International Quantum Communication Award

The International Award on Quantum Communications has been established by the Research Institute of Tamagawa University to acknowledge an individual or a group of individuals for their pioneering contributions to the development of quantum communications. The award carries a plaque and a cash prize.

The third Quantum Communication Award was awarded to:

PAUL BENIOFF
of Argonne National Laboratory, USA

for fundamental contributions to quantum computing, especially his seminal work that first established quantum computation as a theoretical possibility.

DAVID WINELAND and CHRIS MONROE
of National Institute for Standards & Technology, USA

for pioneering experiments demonstrating quantum information processing and quantum entanglement in a system of trapped ions.

The awards were presented on July 5th, 2000 at the Awards Banquet of QCM&C-Y2K held in the Hotel La Residenza in Capri, Italy. A check in the amount of USD 4,400 was given to each award recipient.

The members of Award committee in 2000: P.Tombesi, C.M.Caves, O.Hirota, and Y.Obara: Chairman (President of Tamagawa University)

The first International Award on Quantum Communications in 1996 was awarded to the following four individuals:

C.H.BENNETT of IBM Corporation, USA
C.W.HELSTROM of University of California, San Diego, USA
A.S.HOLEVO of Steklov Mathematical Institute, Russia
H.P.YUEN of Northwestern University, USA

and the second International Award on Quantum Communications in 1998 was awarded to:

H.J.KIMBLE of California Institute of Technology, USA
P.W.SHOR of AT&T Shannon Laboratory, USA

PREFACE

This volume contains contributions based on the lectures delivered and posters presented at the Fifth International Conference on Quantum Communication, Measurement and Computing (QCM&C-Y2K). This Conference is the fifth of a successful series hosted this time in Italy, was held in Capri, 3-7 July 2000. The conference was attended by more than 200 participants from all over the world. There was also a high level of participation from graduate students, who greatly benefited from the opportunity to attend world-class conferences. The Conference Hall was hosted in La Residenza Hotel in Capri, where part of participants where housed, while others where housed in various cozy nearby hotels. All enjoyed the pleasant atmosphere offered by the island of Capri. There were 59 invited lectures given as oral presentations of 30 minutes and 94 poster papers. The major topics covered at the Conference where new experimental and theoretical results in quantum information. They were divided in five parts; i) Quantum Information and Communication, ii) Quantum Measurement, Decoherence, and Tomography, iii) Quantum Computing, iv) Cryptography, v) Entanglement and Teleportation. We were lucky in that almost all major experimental groups in the world working in this area were represented, as were the major theoreticians. There was very active audience participation. A number of graduate students and post-docs were able to present their contributions in four after dinner poster sessions. Sessions were conducted in the morning and late afternoon and this left the midday free for informal discussions, beach excursions, and sights of the island. This format was greatly appreciated. The editors would like to thank the participants who provided detailed notes for publication within the deadline for not delay too much the printing of this volume. We especially thank the conference secretaries Gabriella Bucci, Giovanna La Pietra, and in the name of all participants we graciously thank Marcella Mastrofini who organized almost every thing. Last but certainly not the least, a special thank is for Professor Enzo Valente, Director of the GARR-B, who helped the organizing committee in obtaining the fast internet connection during the conference. We wish to thank the various organizations that have provided either financial or technical support. These include: Tamagawa University, University of Camerino, NEC, SCAT, MagiQ Technologies, Elsag Informatica, Yamaha Motor Europe, Telecom Italia, Istituto Nazionale di Fisica Nucleare, Gruppo Nazionale di Struttura della Materia-CNR, Istituto

Nazionale per la Fisca della Materia, Day Office 2000. Especially, we would like to emphasize that this conference could not have been organized without the generous support of Professor Yoshiaki Obara, President of Tamagawa University. Finally, editors greatly acknowledge the precious collaboration of Mauro Fortunato, Irene Marzoli, David Vitali at the University of Camerino and Kentaro Kato at Tamagawa University. Without their help this volume would never appear.

PAOLO TOMBESI
OSAMU HIROTA

PARTICIPANTS OF THE FIFTH INTERNATIONAL CONFERENCE ON QUANTUM COMMUNICATION, COMPUTING, AND MEASUREMENT

PHOTO BY C.H. BENNETT

List of Participants

0. Stefano Olivares
1. Christian Marchetti
2. Cameron Wellard
3. Mario Ziman
4. Sergei Ulyanov
5. Steven Peil
6. Bill Wootters
7. Sergey Panfilov
8. Natalia Korolkova
9. Seth Lloyd
10. Giovanna La Pietra
11. Marco Genovese
12. Ivan Sartini
13. Gennaro Auletta
14. Bob Gelfond
15. Bruno Di Stefano
16. Peter Shor
17. Masahiro Kitagawa
18. Alexander Sergienko
19. unidentified
20. Giovanni A. della Rossa
21. Norbert Lutkenhaus
22. Zdenek Hradil
23. Anna Lawniczak
24. Duger Ulam-Orgikh
25. Howard Wiseman
26. Daniel Gottesman
27. David Poulin
28. Gerald Gabrielse
29. Christine Silberhorn
30. Harold Ollivier
31. Jan De Neve
32. Leah Henderson
33. Dietmar Fischer
34. Aldo Delgado
35. Per W. Hansen
36. Holger Mack
37. Dan Greenberger
38. Alexander Shumovsky
39. unidentified
40. Philippe Grangier
41. Jonas Soderholm
42. unidentified
43. unidentified
44. Viacheslav P. Belavkin
45. Wojciech Zurek
46. Alberto Commino
47. Osamu Hirota
48. Paolo Tombesi
49. Michael Mussinger
50. Philippe Jorrand
51. Paul Kwiat
52. Daniel Ljunggren
53. unidentified
54. Patrick Hayden
55. Mark Bowdrey
56. unidentified
57. David Vitali
58. Yasser Revez Omar
59. unidentified
60. unidentified
61. Mikhail Kolobov
62. Alexander Lvovsky
63. Stefan Weigert
64. Yoshi Yamamoto
65. David DiVincenzo
66. Giacomo Mauro D'Ariano
67. Alexander Holevo
68. Shiro Kawabata
69. unidentified
70. Tohya Hiroshima
71. Jörg Schmiedmayer
72. John G. Rarity
73. Dominik Janzing
74. Paul Benioff
75. unidentified

76. Stefan Scheel
77. Daniel Collins
78. Chris Westbrook
79. Peter Smith
80. Eduard Schmidt
81. Masanao Ozawa
82. Matteo Paris
83. Yooh-Ho Kim
84. Tatjana Curcic
85. Vittorio Giovannetti
86. unidentified
87. unidentified
88. unidentified
89. unidentified
90. unidentified
91. Lorenzo Maccone
92. unidentified
93. Igor Jex
94. Cristiano Fidani
95. Andreas Winter
96. Axel Kuhn
97. unidentified
98. unidentified
99. unidentified
100. Bassano Vacchini
101. unidentified
102. Anatoly S. Chirkin
103. Ranjit Singh
104. Sara Gasparoni
105. Anton Zeilinger
106. Rainer Blatt
107. Peter Knight
108. unidentified
109. Paolo Zanardi
110. Fabio Sciarrino
111. Ludovico Lanz
112. Gernot Alber
113. Rodney Polkinghorne
114. unidentified
115. Stephen Barnett
116. Alessandra Gatti
117. Vasily Klimov
118. Alberto Porzio
119. Alberto Barchielli
120. Petra Scudo
121. Jingbo Wang
122. Margarita Man'ko
123. Sergey Mayburov
124. Gerald Gilbert
125. John Jeffers
126. unidentified
127. Massimiliano Sacchi
128. Stefano Mancini
129. Bryan Dalton
130. Paoloplacido Lo Presti
131. Patrick Zarda
132. Jaewan Kim
133. Shao-Ming Fei
134. Marcella Mastrofini
135. David Pegg
136. Hugo Zbinden
138. Masaki Sohma
139. Mauro Fortunato
140. Gabriella Bucci
141. Ernesto F. Galvao
142. unidentified
143. Franco Wong
144. Daniel Oi
145. unidentified
146. Jeffrey Shapiro
147. Morton Rubin
148. Peter Zoller
149. Lu-Ming Duan
150. Hoi-Kwong Lo
151. Irene Marzoli
152. Stephan Schiller
153. Eugene Polzik
154. unidentified
155. Robert Garisto
156. unidentified
157. unidentified
158. Frau Zeilinger
159. Ms. Greenberger

Contents

Part I Quantum Information and Quantum Communication

1
Additivity/Multiplicativity Problems for Quantum Communication Channels 3
G. G. Amosov, A. S. Holevo, R. F. Werner

2
Universal Copying of Coherent States: A Gaussian Cloning Machine 11
N. J. Cerf, S. Iblisdir

3
Quantum Random Coding Exponent of a Symmetric State Alphabet 15
Kentaro Kato, Osamu Hirota

4
Information and Distance in Hilbert Space 19
Lev B. Levitin, Tom Toffoli, Zac Walton

5
Superadditivity with Mixed Letter States 27
Masao Osaki

6
Quantum State Recognition 31
Masahide Sasaki, Alberto Carlini

7
Superadditivity in Capacity of Quantum Channel by Classical Pseudo-Cyclic Codes 35
Shogo Usami, Tsuyoshi Sasaki Usuda, Ichi Takumi, Ryohei Nakano, Masayasu Hata

8
Property of Mutual Information for M-ary Quantum-State Signals 39

Tsuyoshi Sasaki Usuda, Ichi Takumi, Ryohei Nakano, Masao Osaki, Masayasu Hata

9
Performances of Binary Block Codes Used on Binary Classical-Quantum Channels 43
Pawel Wocjan, Dejan E. Lazic, Thomas Beth

Part II Quantum Measurement, Decoherence, and Tomography

10
Entropy and Information Gain in Quantum Continual Measurements 49
A. Barchielli

11
Experimental Quantum State Discrimination 59
Stephen M. Barnett, Roger B. M. Clarke, Vivien M. Kendon, Erling Riis, Anthony Chefles, Masahide Sasaki

12
Spontaneous Intrinsic Decoherence in Rabi Oscillations Experiments 69
Rodolfo Bonifacio, Stefano Olivares

13
Quantum Tomography, Teleportation, and Cloning 79
Giacomo Mauro D'Ariano

14
Decoherence versus the Idealization of Microsystems as Correlation Carriers between Macrosystems 87
Ludovico Lanz, Bassano Vacchini, Olaf Melsheimer

15
Quantum Measurement, Information, and Completely Positive Maps 97
Masanao Ozawa

16
On the Number of Elements Needed in a POVM Attaining the Accessible Information 107
Peter W. Shor

17
Einselection and Decoherence from an Information Theory Perspective 115
W. H. Zurek

18
Reconstructing the Discrete Wigner Function through Complementary Measurements 127
Roberth Asplund, Gunnar Björk

19
Statistical Noise in Measuring Correlated Photon Beams 131
Stefania Castelletto, Ivo Pietro Degiovanni, Maria Luisa Rastello

20
Reconstruction Technique for a Trapped Electron 135
M. Fortunato, M. Massini, S. Mancini, D. Vitali, P. Tombesi

21
Quantum Mechanics without Statistical Postulates 139
H. Geiger, G. Obermair, Ch. Helm

22
Quantum Retrodiction 143
J. Jeffers, S. M. Barnett, D. Pegg, O. Jedrkiewicz, R. Loudon

23
Quantum-Tomography Method in Information Processing 147
M. A. Man'ko

24
Quantum Measurement Problem and State Dual Representations 151
S. N. Mayburov

25
Homodyne Characterization of Active Optical Media 155
G. Mauro D'Ariano, Matteo G. A. Paris, Massimiliano F. Sacchi

26
Quantum Cloning Optimal for Joint Measurements 159
Giacomo Mauro D'Ariano, Massimiliano Federico Sacchi

27
How Many Projections Are Needed in Quantum Tomography of Spin States? 163
Z. S. Sazonova, R. Singh

Part III Quantum Computing

28
The Representation of Numbers by States in Quantum Mechanics 171
Paul Benioff

29
Towards Quantum Computation with Trapped Calcium Ions 179
D. Leibfried, C. Roos, P. Barton, H. Rohde, S. Gulde, A. B. Mundt, F. Schmidt-Kaler, J. Eschner, R. Blatt

30
Physical Limits to Computation 189
Seth Lloyd

31
Realising Quantum Computing: Physical Systems and Robustness 199
Andrew M. Steane

32
Information Analysis of Quantum Gates for Simulation of Quantum Algorithms on Classical Computers 207
S. V. Ulyanov, S. A. Panfilov, I. Kurawaki, A. V. Yazenin

33
Quantum Probabilistic Subroutines and Problems in Number Theory 215
A. Carlini, A. Hosoya

34
Theory of the Quantum Speed Up 219
Giuseppe Castagnoli, David Ritz Finkelstein

35
Quantum Gates Using Motional States in an Optical Lattice 227
Eric Charron, Eite Tiesinga, Frederick Mies, Carl Williams

36
Quantum Error-Correcting Code for Burst Error 231
Shiro Kawabata

37
Non-Dissipative Decoherence in Ion-Trap Quantum Computers 235
Stefano Mancini, Rodolfo Bonifacio

38
Optical Qubit Using Linear Elements 239
Matteo G. A. Paris

39
Atom Chips 243
Jörg Schmiedmayer, Ron Folman

40
Decoherence and Fidelity of Single Qubit Operations in a Solid State Quantum Computer 247
C. J. Wellard, L. C. L. Hollenberg

Part IV Cryptography

41
Long Distance Entangled State Quantum Key Distribution 253
Grégoire Ribordy, Nicolas Gisin, Hugo Zbinden

42
Bunching and Antibunching from Single NV Color Centers in Diamond 261
A. Beveratos, R. Brouri, J.-P. Poizat, P. Grangier

43
Violation of Locality and Self-Checking Source: A Brief Account 269
Dominic Mayers, Christian Tourenne

44
The Unconditional Security of Quantum Key Distribution 277
Tal Mor, Vwani Roychowdhury

45
Anonymous-Key Quantum Cryptography and Unconditionally Secure Quantum Bit Commitment 285
Horace P. Yuen

46
Quantum Key Distribution Using Multilevel Encoding 295
Mohamed Bourennane, Anders Karlsson, Gunnar Björk

47
Authority-Based User Authentication and Quantum Key Distribution 299
Daniel Ljunggren, Mohamed Bourennane, Anders Karlsson

48
Improvement of Key Rate for Yuen-Kim Cryptoscheme 303
Kouichi Yamazaki

49
Stable Solid-State Source of Single Photons 307
Patrick Zarda, Christian Kurtsiefer, Sonja Mayer, Harald Weinfurter

Part V Entanglement and Teleportation

50
Non-Locality and Quantum Theory: New Experimental Evidence 313
Luigi Accardi, Massimo Regoli

51
On Entangled Quantum Capacity 325
Viacheslav P. Belavkin

52
Control of Squeezed Light Pulse Spectrum in the Kerr Medium with an Inertial Nonlinearity 335
A. S. Chirkin, F. Popescu

53
Macroscopic Quantum Superposition by Amplification of Entangled States 343
Giovanni Di Giuseppe, Francesco De Martini

54
Quantum Teleportation with Atomic Ensembles and Coherent Light 351

Lu-Ming Duan, J. I. Cirac, P. Zoller, E. S. Polzik

55
Entangled State Based on Nonorthogonal State 359
Osamu Hirota, Masahide Sasaki

56
Long-Distance High-Fidelity Teleportation Using Singlet States 367
Jeffrey H. Shapiro

57
Quantum Teleportation with Complete Bell State Measurement 375
Yoon-Ho Kim, Sergei P. Kulik, Yanhua Shih

58
Complete Quantum Teleportation with a Crossed-Kerr Nonlinearity 383
D. Vitali, M. Fortunato, P. Tombesi

59
Quantum Lithography 391
Pieter Kok, Samuel L. Braunstein, Agedi N. Boto, Daniel S. Abrams, Colin P. Williams, Jonathan P. Dowling

60
Experimental Test of Local Realism Using Non-Maximally Entangled States 399
M. Genovese, G. Brida, C. Novero, E. Predazzi

61
New Schemes for Manipulating Quantum States with a Kerr Cell 403
M. Genovese, C. Novero

62
Local and Nonlocal Properties of Werner States 407
Tohya Hiroshima, Satoshi Ishizaka

63
Maximally Entangled Mixed States in Two Qubits 411
Satoshi Ishizaka, Tohya Hiroshima

64
Engineering Bell-like States of Two High-Q Cavity Fields 415
A. Napoli, A. Messina, S. Maniscalco

65
Nondissipative Decoherence and Entanglement in the Dynamics of a Trapped Ion 419
S. Maniscalco, A. Messina, A. Napoli, D. Vitali

66
Entanglement Manipulation and Concentration in Mixed States 423
R. T. Thew, K. Nemoto, W. J. Munro

67
Ramsey Interferometry with a Single Photon Field in Cavity QED 427
S. Osnaghi, A. Rauschenbeutel, P. Bertet, G. Nogues, M. Brune, J. M. Raimond, S. Haroche

68
Entanglement Transformation at Dielectric Four-Port Devices 433
S. Scheel, L. Knöll, T. Opatrný, D.-G. Welsch

69
Quantum Noise in Polarization Measurement and Polarization Entanglement 439
Alexander S. Shumovsky

70
Bright EPR-Entangled Beams for Quantum Communication 443
Ch. Silberhorn, P. K. Lam, N. Korolkova, G. Leuchs

71
Polarization in Quantum Optics: A New Formalism and Two Experiments 449
T. Tsegaye, P. Usachev, J. Söderholm, A. Trifonov, G. Björk, M. Atatüre, M. C. Teich, A. V. Sergienko, B. E. A. Saleh

72
Spin Squeezing and Decoherence Limit in Ramsey Spectroscopy 453
Duger Ulam-Orgikh, Masahiro Kitagawa

73
Teleportation of Entanglement for Continuous Variables 457
Anatoly I. Zhiliba, Valery N. Gorbachev, Andrew I. Trubilko

74
Generation and Detection of Fock States of the Radiation Field 463
Herbert Walther

Index 475

This volume is dedicated to
the memory of
Daniel F. Walls who gave
impetus to the modern
quantum science.

I

QUANTUM INFORMATION AND QUANTUM COMMUNICATION

ADDITIVITY/MULTIPLICATIVITY PROBLEMS FOR QUANTUM COMMUNICATION CHANNELS

G. G. Amosov
Moscow Institute for Physics and Technology
gramos@mail.sitek.ru

A. S. Holevo
Steklov Mathematical Institute, Moscow
holevo@mi.ras.ru

R. F. Werner
Institut für Mathematische Physik, TU Braunschweig
R.Werner@tu-bs.de

Abstract A class of problems in quantum information theory, having an elementary formulation but still resisting solution, concerns the additivity properties of various quantities characterizing quantum channels, notably the "classical capacity", and the "maximal output purity". All known results, including extensive numerical work, are consistent with the conjecture that these quantities are indeed additive (resp. multiplicative) with respect to tensor products of channels. A proof of this conjecture would have important consequences in quantum information theory. In particular, according to this conjecture, the classical capacity or the maximal purity of outputs cannot be increased by using entangled inputs of the channel. In this paper we state the additivity/multiplicativity problems, give some relations between them, and prove some new partial results, which also support the conjecture.

Introduction

Quantum information theory [1], which makes a theoreticalbasis for physics of information and computation,is a source of challenging problems, often having elementary formulation but still resisting solution. One group of such problems concerns the additivity properties of various quantities characterizing quantum channels, notably the capacity for classical information, and the "maximal output purity", defined below. All known results, including extensive numerical work in the IBM group [2], the Quantum Information group in the Technical University of Braunschweig, and elsewhere, are consistent with the conjecture that these quantities are indeed additive (resp. multiplicative) with respect to tensor products of channels. A proof of this conjecture would have important consequences in quantum information theory: In particular, according to this conjecture, the classical capacity or the maximal purity of outputs cannot be increased by using entangled inputs of the channel.

In this paper we state the additivity/multiplicativity problems, give relations between them, and formulate some new results, which also support the conjecture[1].

1. STATEMENT OF THE PROBLEM

Let us give precise formulation of the additivity problem for the classical capacity (see [1, 3]). Let $\mathcal{B}(\mathcal{H})$ be the $*$ -algebra of all operators in a finite dimensional unitary space $\mathcal{H}$. We denote the set of states, i.e., positive unit trace operators in $\mathcal{B}(\mathcal{H})$ by $\mathcal{S}(\mathcal{H})$, the set of all m-dimensional projections by $\mathcal{P}_m(\mathcal{H})$ and the set of all projections by $\mathcal{P}(\mathcal{H})$. A quantum channel Φ is a completely positive trace preserving linear map of $\mathcal{B}(\mathcal{H})$ (we are in the finite dimensional case and we use the Schrödinger picture). These are the maps admitting the Kraus decomposition (see e.g. [3, 4])

$$\Phi(\rho) = \sum_k A_k \rho A_k^*, \tag{1}$$

where A_k are operators satisfying $\sum_k A_k^* A_k = I$.

Let $H(\rho) = -\mathrm{Tr}\rho \log \rho$ denote the von Neumann entropy of the state ρ and define

$$C(\Phi) = \max_{p_i, \rho_i} [H(\sum_i p_i \Phi(\rho_i)) - \sum_i p_i H(\Phi(\rho_i))],$$

where the maximum is taken over all finite probability distributions $\{p_i\}$ on $\mathcal{S}(\mathcal{H})$, ascribing probabilities p_i to (arbitrary) states ρ_i. The quantity $C(\Phi)$ appears as the "one-step classical capacity" of the quantum channel Φ or the capacity with unentangled input states (we refer to [3] for a detailed information-theoretic discussion and the proof of the corresponding coding theorem). A thorough discussion of the properties of $C(\Phi)$ is given in [5].

The additivity problem can be formulated as follows: let $\Phi_1, \ldots, \Phi_n$ be channels in the algebras $\mathcal{B}(\mathcal{H}_1), \ldots, \mathcal{B}(\mathcal{H}_n)$ and let $\Phi_1 \otimes \ldots \otimes \Phi_n$ be their tensor product in $\mathcal{B}(\mathcal{H}_1 \otimes \ldots \otimes \mathcal{H}_n)$. Is it true that

$$C(\Phi_1 \otimes \ldots \otimes \Phi_n) = \sum_{i=1}^n C(\Phi_i) \quad ? \tag{2}$$

This obviously holds for reversible unitary channels; in [3] the additivity was established for the so called classical-quantum and quantum-classical channels, which map from or into an Abelian subalgebra of $\mathcal{B}(\mathcal{H})$.

Another closely related problem is the additivity of a quantity, which can be read as the "maximal output purity" of a channel. In fact, there are several quantities of this kind, depending on the way we measure "purity". If we just take the von Neumann entropy as a measure of purity, we arrive at the question [6, 7] whether or not

$$\min_{\rho\in\mathcal{S}(\mathcal{H}_1\otimes\ldots\otimes\mathcal{H}_n)} H((\Phi_1\otimes\ldots\otimes\Phi_n)(\rho)) = \sum_{i=1}^{n} \min_{\rho\in\mathcal{S}(\mathcal{H}_i)} H(\Phi_i(\rho))\ ? \tag{3}$$

For a particular class of channels this property implies (2) (see the Lemma in Section 4 below).

We will also consider this problem for other measures of purity, based on the noncommutative ℓ_p-norms

$$\|A\|_p = (\mathrm{Tr}|A|^p)^{\frac{1}{p}}\ ,$$

defined for $p \geq 1$, and $A \in \mathcal{B}(\mathcal{H})$, with the operator norm $\|A\|$ corresponding naturally to the case $p = \infty$. For an arbitrary quantum channel Φ let us introduce the following notations for the "highest purity" of outputs of a channel

$$\begin{aligned} \nu_H(\Phi) &= \min_\rho H(\Phi(\rho)), & (4)\\ \nu_p(\Phi) &= \max_\rho \|\Phi(\rho)\|_p, & (5)\\ \nu_{-\infty}(\Phi) &= \min_\rho \|\Phi(\rho)^{-1}\|^{-1}, & (6) \end{aligned}$$

where the extrema are taken with respect to all input density matrices ρ. By convexity of the norms, the extrema in the above definitions are attained on pure states [in the first (resp. last) case the operator convexity of the function $x \mapsto x\log x$ (resp. $x \mapsto x^{-1}$) is also relevant].

Then the additivity/multiplicativity inequalities

$$\begin{aligned} \nu_p(\Phi_1\otimes\Phi_2) &\geq \nu_p(\Phi_1)\nu_p(\Phi_2) & (7)\\ \nu_H(\Phi_1\otimes\Phi_2) &\leq \nu_H(\Phi_1)+\nu_H(\Phi_1) & (8) \end{aligned}$$

are clear from inserting product density operators into the defining variational expressions. The standing **conjecture** is that equality always holds in these inequalities, i.e., that choosing entangled input states is never helpful for getting purer output states.

2. TENSORING WITH AN IDEAL CHANNEL

The first natural step is to establish the multiplicativity property when one factor is the identity channel.

Lemma. For $* = p, H, -\infty$

$$\nu_*(\Phi \otimes \mathrm{Id}) = \nu_*(\Phi) \ . \tag{9}$$

Since $\nu_p(\mathrm{Id}) = 1$, and $\nu_H(\mathrm{Id}) = 0$, this is indeed an instance of the additivity/multiplicativity conjecture.

Proof. We shall restrict to the case $* = p, 1 \leq p \leq \infty$. The argument in the case $* = -\infty$ is similar and the case $* = H$ follows by the argument given above.

Let us denote by $\mathcal{H}_1, \mathcal{H}_2$ the Hilbert spaces of the first and the second system, respectively. Let ϕ_{12} be a unit vector in $\mathcal{H}_1 \otimes \mathcal{H}_2$, and write $\rho_{12} = |\phi_{12}\rangle\langle\phi_{12}|$ and $\rho_1 = \mathrm{Tr}_2\rho_{12}$ for the partial state in $\mathcal{H}_1$. If Φ is the channel in $\mathcal{H}_1$, we denote $\rho'_{12} = (\Phi \otimes \mathrm{Id})(\rho_{12})$. Let us dilate the channel Φ to a unitary evolution U_{13} with the environment $\mathcal{H}_3$, initially in a pure state $\rho_3 = |\phi_3\rangle\langle\phi_3|$. The the final state of the environment is

$$\rho'_3 = \mathrm{Tr}_1 U_{13}(\rho_1 \otimes |\phi_3\rangle\langle\phi_3|)U^*_{13} \equiv \Psi(\rho_1).$$

Since the state of the composite system $\mathcal{H}_1 \otimes \mathcal{H}_2 \otimes \mathcal{H}_3$ remains pure after the unitary evolution, its partial states ρ'_{12}, ρ'_3 are isometric [4]. Therefore

$$||(\Phi \otimes \mathrm{Id})(\rho_{12})||_p = ||\rho'_{12}||_p = ||\rho'_3||_p = ||\Psi(\rho_1)||_p.$$

Now the map $\rho_1 \to \Psi(\rho_1)$ is affine and the norm is convex, therefore the maximum of the quantity above is attained on pure ρ_1, whence $\rho_{12} = \rho_1 \otimes \rho_2$, and the statement follows. $\triangle$

3. WEAK NOISE

One testing ground for the multiplicativity/additivity conjecture are channels close to the identity. For such channels the purity parameters can be evaluated in lowest order in the deviation from the identity. Doing this for each subchannel and for their tensor product, one can explicitly check the conjecture. As the following result shows, this test supports the conjecture.

Consider a channel with weak noise, i.e., choose some channel Φ on $\mathcal{B}(\mathcal{H})$, and set

$$\Phi^{(\epsilon)} = (1 - \epsilon)\mathrm{Id} + \epsilon\Phi \ . \tag{10}$$

For small ϵ this is a weak noise channel, which has the property that for *any* pure input the output will be nearly pure.

Theorem. The multiplicativity hypothesis for the quantities $\nu_p(\Phi)$ with $1 \leq p \leq \infty$ and the additivity hypothesis for the quantity $\nu_H(\Phi)$ hold true approximately in the leading order in ϵ.

Proof. In order to estimate these quantities for the weak noise channels, we need to estimate entropy, and the p-norms near a pure state. Let ρ be a density operator on a d-dimensional Hilbert space, and suppose that $\|\rho\| = 1-\epsilon+\mathbf{o}(\epsilon)$. Then the leading order of the other norms is determined completely by ϵ:

$$\begin{aligned} \|\rho\|_p &= 1-\epsilon+\mathbf{o}(\epsilon) \quad \text{for } p>1 &(11)\\ H(\rho) &= -\epsilon\log\epsilon+\mathbf{o}(\epsilon\log\epsilon)\ , &(12)\end{aligned}$$

where as usual $\mathbf{o}(\epsilon)$ stands for terms going to zero faster than ϵ as $\epsilon \to 0$. In this case we can say more: in first line we have $0 \leq$ remainder $\leq C\ \epsilon^p$, for $\epsilon < 1/2$, where C is a constant depending only on the dimension. Similarly, the estimates in the second line are independent of the details of ρ. Hence in leading order all the variational expressions are equivalent: each one amounts to maximizing ϵ.

Let us go back to the weak noise channel (10). To get high fidelity we need to maximize the leading term, so we can take $\eta = \xi$ in the following computation:

$$\begin{aligned} \nu_\infty(\Phi^{(\epsilon)}) &= \sup_{\xi,\eta}\langle\xi,\Phi^{(\epsilon)}(|\eta\rangle\langle\eta|)\xi\rangle \\ &= \sup_{\xi,\eta}\Big((1-\epsilon)|\langle\xi,\eta\rangle|^2+\epsilon\langle\xi,\Phi(|\eta\rangle\langle\eta|)\xi\rangle\Big) \\ &= 1-\epsilon+\epsilon\sup_{\xi}\langle\xi,\Phi(|\xi\rangle\langle\xi|)\xi\rangle\ +\mathbf{o}(\epsilon) \\ &= 1-\epsilon+\epsilon\nu_\flat(\Phi)+\mathbf{o}(\epsilon)\ . &(13)\end{aligned}$$

Note that in all these estimates the remainder estimates can be done uniformly for all channels, depending only on dimension.

A tensor product of weak noise channels (10) is again of the same form:

$$\begin{aligned} \Phi^{(\epsilon)} &= \Phi_1^{(\epsilon)}\otimes\cdots\Phi_n^{(\epsilon)} \\ &= (1-\epsilon)^n\mathrm{Id}+\epsilon(1-\epsilon)^{n-1}\left(\Phi_1\otimes\mathrm{Id}_{2\ldots n}+\cdots+\mathrm{Id}_{1\ldots n-1}\otimes\Phi_n\right) \\ &\quad +\mathbf{o}(\epsilon) \\ &= (1-n\epsilon)\mathrm{Id}+n\epsilon\ \delta\Phi+\mathbf{o}(\epsilon)\ ,\end{aligned}$$

where $\delta\Phi$ is the average of the n channels $\mathrm{Id}_{1\ldots k-1}\otimes\Phi_k\otimes\mathrm{Id}_{k+1\ldots n}$. Hence, in order to compute the leading order of $\nu_\infty(\Phi^{(\epsilon)})$ by formula (13) we have to determine $\nu_\flat(\delta\Phi)$. We have

$$\begin{aligned} \frac{1}{n}\sum_{k=1}^{m}\nu_\flat(\Phi_k) &\leq \nu_\flat(\delta\Phi) \\ &\leq \frac{1}{n}\sum_{k=1}^{m}\nu_\flat(\Phi_k\otimes\mathrm{Id})\end{aligned}$$

$$= \frac{1}{n}\sum_{k=1}^{m} \nu_\flat(\Phi_k).$$

Hence equality holds, which means that in the leading order in ϵ all the variational expressions for the purity quantities $\nu_*()$, with $* = p, H, \flat$ are attained at product states. $\triangle$

4. DEPOLARIZING CHANNELS

A channel is called *bistochastic* if $\Phi(I) = I$, where I is the unit operator in $\mathcal{B}(\mathcal{H})$. An important example is the *depolarizing* channel [1]

$$\Phi(\rho) = (1-p)\rho + \frac{p}{d}(\mathrm{Tr}\rho)I, \; \rho \in \mathcal{B}(\mathcal{H}), \; 0 < p < 1,$$

where $d = \dim\mathcal{H}$. A channel is called *binary* if $d = 2$.

Lemma. Let Φ be binary bistochastic channel, then

$$C(\Phi) = \log 2 - \min_{\rho\in\mathcal{S}(\mathcal{H})} H(\Phi(\rho)). \tag{14}$$

If Φ_i are binary bistochastic channels, then (3) implies (2).

In the paper [8] the relation(2) was proven for the two binary depolarizing channels Φ_1, Φ_2. The proof heavily uses Schmidt decomposition and as such does not generalizes to the case $n > 2$. The main difficulty is evaluating the entropy of the product channel. However, it appears to be possible to check the additivity in the limiting cases of "weak" and "strong" depolarization in the leading order. Let us consider a collection of depolarizing channels $\{\Phi_i\}$ in the Hilbert spaces $\mathcal{H}_i$, with parameters p_i, d_i, $i = 1, 2, \ldots, n$, and denote $\Phi = \otimes_{i=1}^n \Phi_i$, $\mathcal{H} = \otimes_{i=1}^n \mathcal{H}_i$, $d = \prod_{i=1}^n d_i$. In the following we shall use symbols I_i and $I = \otimes_{i=1}^n I_i$ for the identity operators in $\mathcal{H}_i$ and $\mathcal{H}$ respectively. Let ϵ_L be the tensor product $\phi_1 \otimes \ldots \otimes \phi_n$, where $\phi_i(\rho) = \frac{1}{d_i}\mathrm{Tr}(\rho)I_i$, $i \in L \subset \{1, 2, \ldots, n\}$, and $\phi_i(\rho) = \rho$ otherwise, $\rho \in \mathcal{S}(\mathcal{H}_i)$. Then ϵ_L is a conditional expectation onto the subalgebra $\mathcal{M}_L$, generated by operators of the form $A_1 \otimes \ldots \otimes A_n$, where $A_i = I_i$ for $i \in L$, and is normalized partial trace with respect to its commutant. It has the property $\min_{P\in\mathcal{P}(\mathcal{M}_L)} \dim P = \prod_{i=1}^n d_i^{\theta_L(i)}$, where $\theta_L(i) = 1$ if $i \in L$ and $\theta_L(i) = 0$ otherwise. Here we denoted by $\mathcal{P}(\mathcal{M}_L)$ the set of all orthogonal projections in $\mathcal{M}_L$. Notice that the inclusion $\mathcal{M}_{L_1} \subset \mathcal{M}_{L_2}$ holds if $L_2 \subset L_1$. So $\epsilon_{L_1}\epsilon_{L_2} = \epsilon_{L_1 \vee L_2}$.

We use the expansion

$$\Phi = \sum_L \prod_{i=1}^n p_i^{\theta_L(i)}(1-p_i)^{1-\theta_L(i)}\epsilon_L. \tag{15}$$

Weak (strong) depolarization corresponds to the case where all p_i (respectively, $1 - p_i$) are small parameters, which we assume to be of the same order.

Proposition The relation (2) holds in the cases of weak and strong depolarization approximately in the leading order.

Partial answers to the multiplicativity hypothesis are given by the following Theorem. In fact, multiplicativity of $\nu_\infty(\Phi)$ for binary bistochastic maps follows from a more general result in [7].

Theorem.

$$
\begin{array}{llll}
(i) & \nu_2(\Phi) & = \prod_{i=1}^{n} \nu_2(\Phi_i) & = \prod_{i=1}^{n} \left(\frac{d_i - 1}{d_i}(1 - p_i)^2 + \frac{1}{d_i} \right)^{1/2}, \\
(ii) & \nu_\infty(\Phi) & = \prod_{i=1}^{n} \nu_\infty(\Phi_i) & = \prod_{i=1}^{n} \left(1 - \frac{p_i(d_i - 1)}{d_i} \right), \\
(iii) & \nu_{-\infty}(\Phi) & = \prod_{i=1}^{n} \nu_{-\infty}(\Phi_i) & = \prod_{i=1}^{n} \frac{p_i}{d_i}.
\end{array}
$$

Acknowledgments

The work of the second author (ASH) was supported by the Research Award of A. von Humboldt Foundation. He also gratefully acknowledges financial support of the Research Institute, Tamagawa University, and of the Organizing Committee, making possible his attendance at this Conference.

Notes

1. The full presentation can be found in quant-ph/0003002 and will appear in "Problems of Information Transmission."

References

[1] C. H. Bennett, P. W. Shor, Quantum information theory, IEEE Trans. on Inform. Theory, **IT-44**, 2724-2742, 1998.

[2] C. H. Bennett, C. Fuchs, J. A. Smolin, Entanglement enhanced classical communication on a noisy quantum channel, in: Proc. 3d Int. Conf. on Quantum Communication and Measurement, ed. by C. M. Caves, O. Hirota, A. S. Holevo, Plenum, NY 1997. LANL e-print quant-ph/9611006.

[3] A. S. Holevo, Quantum coding theorems, Russian Math. Surveys **53:6**, 1295-1331, 1998. LANL e-print quant-ph/9808023.

[4] G. Lindblad, Quantum entropy and quantum measurements, in: Proc. Int. Conf. on Quantum Communication and Measurement, ed. by C. Benjaballah, O. Hirota, S. Reynaud, Lect. Notes Phys. **378**, 71-80, Springer-Verlag, Berlin 1991.

[5] B. Schumacher, M. D. Westmoreland, Optimal signal ensembles. LANL e-print quant-ph/9912122.

[6] C. Fuchs, private communication.

[7] C. King, M. B. Ruskai, Minimal entropy of states emerging from noisy quantum channels. LANL e-print quant-ph/9911079.

[8] D. Bruss, L. Faoro, C. Macchiavello, M. Palma, Quantum entanglement and classical communication through a depolarizing channel. J. Mod. Opt. **47**, 325-332, 2000. LANL e-print quant-ph/9903033.

UNIVERSAL COPYING OF COHERENT STATES: A GAUSSIAN CLONING MACHINE

N. J. Cerf, S. Iblisdir
Ecole Polytechnique, CP 165, Université Libre de Bruxelles, B-1050 Bruxelles, Belgium

Keywords: Quantum cloning, continuous quantum information, coherent states.

Abstract A Gaussian cloning machine is presented that duplicates with a same fidelity two canonically conjugate continuous variables such as the quadratures of a mode of the electromagnetic field. This cloner is displacement and rotation covariant in phase space, hence it duplicates all coherent states with a fidelity of $2/3$. The optimality of this cloner is discussed, as well as the extension to $N \rightarrow M$ continuous cloners.

Mostof the concepts of quantum computation have been initially developed for discrete variables, in particular for quantum bits. However, recent progress has suggested that *continuous* quantum variables might be experimentally easier to manipulate than their discrete counterparts [1, 2]. The present article deals with the issue of *cloning* continuous variables. In full generality, a continuous variable denotes a system described in terms of two canonically conjugate observables with continuous spectra (referred to as $\hat{x}$ and $\hat{p}$, with eigenvalues x and p, respectively) such as, for example, the two quadratures of a mode of the electromagnetic field. As a result of the no-cloning theorem [3] and $[\hat{x}, \hat{p}] = i$ (assuming $\hbar = 1$), these observables cannot both be cloned with arbitrary accuracy. Nevertheless, one might be interested in producing *imperfect* clones from one or several originals. A possible application, for example, is the security assessment of continuous quantum key distribution schemes [4]. In this context, the following questions arise: "To what extent can the copies resemble the original in accordance with quantum mechanics?", or "What unitary transformation would achieve the best possible cloning and how to experimentally implement it?" These questions have now been extensively studied for quantum bits [5, 6, 7, 8, 9] and, more generally, for d-level systems [10]. Here, we discuss these questions for continuous variables. First, the notion of continuous cloners is defined. Then, the problem of duplication is treated, and, finally, the extension to general $N \to M$ continuous cloning is discussed.

Cloning is defined as a particular unitary transformation that meets the two following symmetry requirements. Firstly, we impose it be rotation covariant in the (x,p) phase space, in the sense that the eigenstates of any linear combination of $\hat{x}$ and $\hat{p}$ should be copied equally well. Secondly, we require it to be displacement covariant, that is, if two input states are identical up to a displacement, then their respective copies should be identical up to the same displacement. A simple way to meet these two conditions is to seek for a cloning machine that, being provided with N replicas of an original state $|\psi\rangle$ as input, yields M output clones whose individual states are given each by the Gaussian mixture:

$$\rho(|\psi\rangle) = \frac{1}{2\pi\sigma^2(N,M)} \int dx\, dp\, e^{-\frac{x^2+p^2}{2\sigma^2(N,M)}} \hat{D}(x,p)|\psi\rangle\langle\psi|\hat{D}^\dagger(x,p), \quad (1)$$

where $\hat{D}(x,p)$ is the displacement operator, and $\sigma^2(N,M)$ is the cloning error variance. In the case of duplication ($N=1, M=2$), the cloning transformation can be written as [11]:

$$\hat{U}_{1,2,3} = e^{-i(\hat{x}_3-\hat{x}_2)\hat{p}_1} e^{-i\hat{x}_1(\hat{p}_2+\hat{p}_3)} e^{-i\hat{x}_2\hat{p}_3}, \quad (2)$$

where modes 1, 2 and 3 refer respectively to the original, the additional copy, and the ancilla, the two latter modes being initially prepared in the vacuum state. This operation yields Gaussian-distributed output states and effects identical x- and p-error variances regardless the input state: $\sigma^2(1,2) = 1/2$. In particular, it copies all coherent states $|\alpha\rangle$ with the same fidelity $f_{1,2} = \langle\alpha|\rho(|\alpha\rangle)|\alpha\rangle = 2/3$. The corresponding network in terms of continuous C-NOT gates is shown in Fig. 1. This machine is optimal in the sense that it is impossible to have $\sigma^2(1,2) < 1/2$. The proof of optimality, which is not presented here (see [12] for details), is based on the fact that quantum cloning should not allow to measure $\hat{x}$ and $\hat{p}$ separately on each of the two copies of state $|\psi\rangle$ with more accuracy than by performing a simultaneous optimal measurement of $\hat{x}$ and $\hat{p}$ on $|\psi\rangle$.

It is worth noting that a possible implementation of this machine can be achieved using a linear amplifier and a beam splitter (see Fig. 2), where the linear amplifier obeys the following equations [13]:

$$\hat{a}_{out} = \sqrt{2}\hat{a}_{in} + \hat{a}_z^\dagger, \qquad \hat{a}_{z'} = \hat{a}_{in}^\dagger + \sqrt{2}\hat{a}_z, \quad (3)$$

leading to an equal x- and p-error variance of $1/2$ for both clones. This implementation emphasizes that the impossibility of cloning results in an additional noise on the two output clones that originates physically from the vacuum fluctuations of the ancillae.

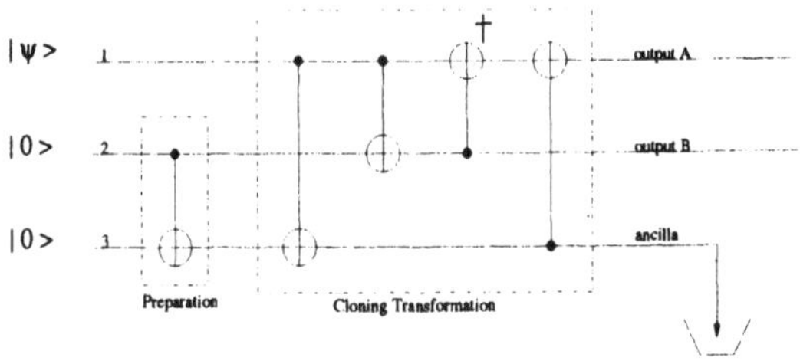

Figure 1 Quantum circuit for the $1 \rightarrow 2$ cloner.

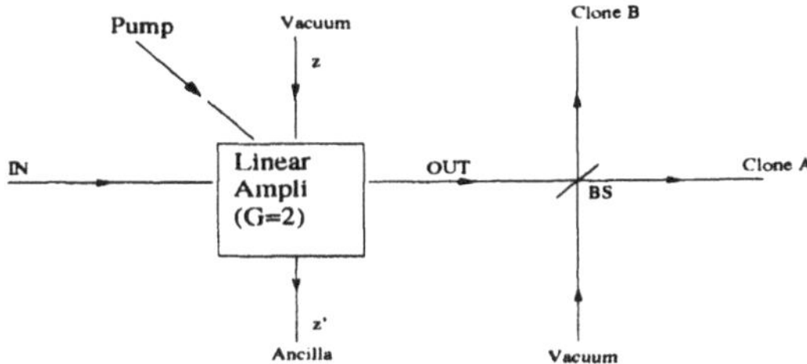

Figure 2 Implementation of a $1 \rightarrow 2$ cloner.

For the general case of $N \rightarrow M$ cloning, the optimal cloning transformation is unknown to date but we have found a lower bound on $\sigma^2_{N,M}$ when coherent states are copied [12]. This lower bound can be derived by exploiting the fact that cascading an $N \rightarrow M$ cloner with an $M \rightarrow L$ cloner results in an $N \rightarrow L$ cloner that cannot be better than the optimal $N \rightarrow L$ cloner. Using the property that the variances of two cascaded cloners add, we have $\sigma^2_{opt}(N, L) \leq \sigma^2(N, M) + \sigma^2(M, L)$. For $L \rightarrow \infty$, the cloner becomes a measurement device for which we know that ${\sigma_{opt}}^2(N, \infty) = 1/N$ when coherent states are copied. Hence,

$$\sigma^2(N, M) \geq \frac{1}{N} - \frac{1}{M}. \tag{4}$$

and the corresponding fidelity for coherent states is

$$f_{N,M} \leq \frac{MN}{MN + M - N} \tag{5}$$

In conclusion, the problem of cloning a continuous quantum variable has been investigated. A continuous cloning machine, which is translation and rotation covariant in phase space, has been found. Moreover, this machine has been shown to be optimal and achievable using only an amplifier and a beam splitter. Finally, a lower bound on the cloning-induced noise of an $N \rightarrow M$ cloner has been derived.

Acknowledgments

S. I. acknowledges support from the Fondation Universitaire Van Buuren at the Université Libre de Bruxelles.

References

[1] Braunstein, S. L. (1998). Quantum error correction for communication with linear optics. *Nature*, **394**, 47.

[2] Braunstein, S. L. and Kimble, H. (1998). Teleportation of continuous quantum variables. *Phys. Rev. Lett.*, **80**, 869.

[3] Wootters, W. K. and Zurek, W. H. (1982). A single quantum cannot be cloned. *Nature*, **299**, 802.

[4] Cerf, N., Lévy, M., and Van Assche, G. (2000). Quantum distribution of Gaussian keys with squeezed states. e-print quant-ph/0008058.

[5] Mandel, L. (1983). Is a photon amplifier always polarisation dependent? *Nature*, **304**, 188.

[6] Buzek, V. and Hillery, M. (1996). Quantum copying: Beyond the no-cloning theorem. *Phys. Rev. A*, **54**, 1844.

[7] Gisin, N. and Massar, S. (1997). Optimal quantum cloning machines. *Phys. Rev. Lett.*, **79**, 2153.

[8] Niu, C.-S. and Griffiths, R. (1998). Optimal cloning of one quantum bit. *Phys. Rev. A*, **58**, 4377.

[9] Cerf, N. J. (2000). Pauli cloning of a quantum bit. *Phys. Rev. Lett.*, **84**, 4497.

[10] Werner, R. (1998). Optimal cloning of pure states. *Phys. Rev. A*, **58**, 1827.

[11] Cerf, N. J., Ipe, A., and Rottenberg, X. (2000). Cloning of continuous quantum variables. *Phys. Rev. Lett.*, **85**, 1754.

[12] Cerf, N. J. and Iblisdir, S. (2000). Optimal N-to-M cloning of conjugate quantum variables. *Phys. Rev. A*, **62**, 040301.

[13] Caves, C. M. (1982). Quantum limits on noise in linear amplifiers. *Phys. Rev. D*, **26**, 1817.

QUANTUM RANDOM CODING EXPONENT OF A SYMMETRIC STATE ALPHABET

Kentaro Kato, Osamu Hirota
Research Center for Quantum Communications, Tamagawa University, JAPAN
{kkato,hirota}@lab.tamagawa.ac.jp

INTRODUCTION

Quantum channel coding theorem is one of the most important theorems in quantum information theory. According to the theorem, we can say that if information rate R is less than the quantum channel capacity C, then there exists the coding scheme such that the average probability of decoding error $P_{\mathrm{e}}(N, R)$ tends to zero with codeword length $N \to \infty$.

Quantum channel coding theorem was first proved for the case of pure state alphabets by Hausladen *et al.* in 1996 and has been generalized for the case of mixed state alphabets by Holevo in 1996, and by Schumacher and Westmoreland in 1997, independently [1, 2, 3]. Holevo-Schumacher-Hausladen's proof is based on the concept of typical subspaces of the Hilbert space spanned by codeword states. This proof corresponds to the proof of classical channel coding theorem based on the concept of the typical sequences of codewords.

In classical information theory another proof of channel coding theorem using a random coding exponent has been developed by Gallager[4]. In Gallager's proof, a random coding exponent is formulated as the doubly maximization of the function defined by channel matrix, *a priori* distribution on the input alphabet, and information rate with respect to all *a priori* distributions and with respect to the parameter s over $0 \leq s \leq 1$. By using the random coding exponent, we can obtain the upper bound of the average probability of decoding error in finite codeword length. Likewise another proof of quantum channel coding theorem using a quantum random coding exponent has been developed by Holevo[5]. The quantum random coding exponent is formulated as follows.

$$E_Q(R) = \max_{\xi} \max_{1 \leq s \leq 0} \left[E_{QG}(\xi; s) - sR \right] \tag{1}$$

where the quantum Gallager function is given by

$$E_{QG}(\xi;s) = -\log \mathrm{Tr}\left(\sum_{k=1}^{K} \xi_k |\psi_k\rangle\langle\psi_k|\right)^{1+s}, \tag{2}$$

and where $\mathcal{A} = \{|\psi_1\rangle, \cdots, |\psi_K\rangle\}$ and $\xi = (\xi_1, \cdots, \xi_K)$ are the input state alphabet and the *a priori* distribution on $\mathcal{A}$, *respectively*. By using the quantum random coding exponent, we can also obtain the upper bound of the average probability of quantum decoding error in finite codeword length such as

$$P_e(N,R) \leq 2\exp[-NE_Q(R)]. \tag{3}$$

In general it is difficult to derive the random coding exponent analytically both in quantum information theory and in classical one. It is however that the numerical calculation algorithm of the classical random coding exponent is well known in classical information theory. Unfortunately, there are no such algorithms in quantum information theory.

In this letter we consider the case of the symmetric state alphabet and try to derive the optimum distribution on the input alphabet analytically.

QUANTUM RANDOM CODING EXPONENT OF A SYMMETRIC STATE ALPHABET

Definition: When the states of a input alphabet satisfy the next conditions, the input alphabet is called the symmetric state alphabet:

$$|\psi_k\rangle = \hat{V}^{k-1}|\psi_1\rangle,\ k = 1, \cdots, K; \quad \hat{V}\hat{V}^\dagger = \hat{V}^\dagger\hat{V} = \hat{1}; \quad \hat{V}^K = \pm\hat{1},$$

where $\hat{1}$ is the identity operator.

For the symmetric state alphabet, we can get the next proposition.

Proposition: If the input state alphabet is a symmetric state alphabet, then, for arbitrary fixed parameter s within $0 \leq s \leq 1$, the maximum of the quantum Gallager function over all *a priori* distributions on the input alphabet is given by the uniform distribution.

Proof: (By using the convexity of $\mathrm{Tr}(\hat{A})^{1+s}$ and the symmetry of the input states, one can easily prove this. Also see Ref.[6]) ∎

According to the proposition, we can say that the quantum random coding exponent of a symmetric state alphabet is given by the uniform distribution on the input alphabet. In other words, we can reduce the calculation of the quantum random coding exponent to the maximization problem with respect only to the parameter s in this case.

EXAMPLE: Lifted Trine States

Here we consider the case of the ternary symmetric state alphabet, called lifted trine states. Lifted trine states are given as follows [7]:

$$\begin{cases} |\psi_k\rangle = \hat{V}^{k-1}|\psi_1\rangle, \quad k = 1,2,3. \\ \hat{V} = \begin{bmatrix} -\frac{1}{2} & -\frac{\sqrt{3}}{2} & 0 \\ \frac{\sqrt{3}}{2} & -\frac{1}{2} & 0 \\ 0 & 0 & 1 \end{bmatrix} \quad |\psi_1\rangle = \begin{bmatrix} \sqrt{1-\alpha} \\ 0 \\ \sqrt{\alpha} \end{bmatrix} \end{cases} \tag{4}$$

where $0 \leq \alpha \leq 1$. From the proposition, the quantum random coding exponent of lifted trine states is given by

$$E_Q(R) = \max_{0 \leq s \leq 1} \left[E_{QG}(\frac{1}{3}, \frac{1}{3}, \frac{1}{3}; s) - sR \right], \tag{5}$$

where quantum Gallager function is

$$E_{QG}(\frac{1}{3}, \frac{1}{3}, \frac{1}{3}; s) = -\log_2 \left[\alpha^{1+s} + 2\left(\frac{1}{2}(1-\alpha) \right)^{1+s} \right] \tag{6}$$

On the other hand, the quantum cutoff rate R_Q [8] is obtained as follows.

$$R_Q = \log_2 3 - \log_2 \left[1 + \frac{1}{2}(1-3\alpha)^2 \right] \quad \text{[bits/letter]} \tag{7}$$

Furthermore, by using the result of Ref.[6], the quantum channel capacity of the lifted trine states is given as

$$C = (1-\alpha) - \alpha \log_2 \alpha - (1-\alpha)\log_2(1-\alpha) \quad \text{[bits/letter]} \tag{8}$$

The quantum random coding exponent for the cases of $\alpha = 0.1$ and $\alpha = 0.7$ are drawn in *Figure 1*. The quantum cutoff rate and the quantum channel capacity are in *Figure 2*.

CONCLUSION

It was shown that if the input state alphabet is a symmetric state alphabet then the optimum distribution achieving the quantum random coding exponent is the uniform distribution. This result simplifies the calculation procedure of the quantum random coding exponent. As a concrete example, we showed the random coding exponent and the cutoff rate for the lifted trine states.

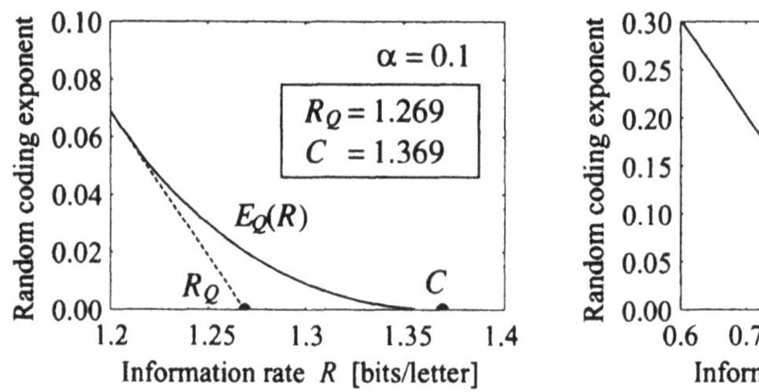

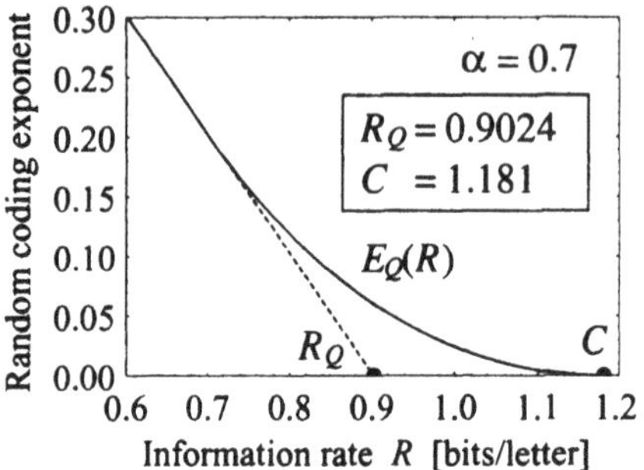

Figure 1 Quantum random coding exponent of the lifted trine states.

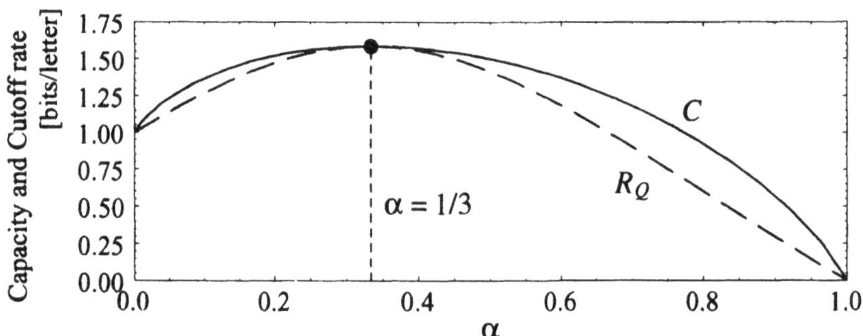

Figure 2 Quantum channel capacity and quantum cutoff rate of the lifted trine states.

References

[1] P. Hausladen, R. Jozsa, B. Schumacher, M.D. Westmoreland, W.K. Wootters, Phys. Rev. A 54 (1996) 1869.

[2] A.S. Holevo, LANL quant-ph/9611023 (1996); IEEE Trans. Inform. Theory 44 (1998) 269.

[3] B. Schumacher, M.D. Westmoreland, Phys. Rev. A 56 (1997) 131.

[4] R.G. Gallager, Information theory and reliable communication (Wiley, New York, 1968).

[5] M.V. Burnashev, A.S. Holevo, LANL quant-ph/9703013 v4 (1997).

[6] K. Kato, M. Osaki, O. Hirota, Phys. Lett. A 251 (1999) 157.

[7] P. Shor, Presentation at QCM&C-Y2K, Capri, Italy, Jul.3-8, 2000; Private communication.

[8] M. Ban, K. Kurokawa, O. Hirota, J. Opt. B 1 (1999) 206.

INFORMATION AND DISTANCE IN HILBERT SPACE

Lev B. Levitin, Tom Toffoli, Zac Walton
Department of Electrical and Computer Engineering, Boston University
8 Saint Mary's Street, Boston, Massachusetts 02215
levitin@bu.edu, tt@bu.edu, walton@bu.edu

Keywords: Distinguishable quantum states, information in quantum measurements

1. INTRODUCTION

In his remarkable 1981 paper, "Statistical Distance and Hilbert Space" [1], Wootters showed that the statistical distance between two vectors in Hilbert space is proportional to the angle between these two vectors and does not depend on the position of the vectors. He defines statistical distance as the number of distinguishable intermediate states between the two vectors. However, his notion of distinguishibility relies on the apparently arbitrary criterion that two states are distinguishable if measurements performed on n identical copies of each state yield two distributions whose means are separated by a constant factor times the sum of the standard deviations of these distributions. We use a more rigorous notion of distinguishibility based on Shannon's 12th theorem [2] and arrive at an expression for the number of distinguishable states that is consistent with Wootters' result; however, unlike that result, our expression does not depend on an arbitrary choice of the distinguishability criterion. Rather, our notion of distinguishibility is predicated on the guarantee that the measurer be able to distinguish between the quantum states with probability approaching 1 as the number n of copies of identical states in a sample tends to infinity. Wootters shows that for large n the number of distinguishable states between the vectors α_1 and α_2 is proportional to $|\alpha_1 - \alpha_2|\sqrt{n}$, where α is the angle of the vector from some reference direction in the plane spanned by the two vectors. We show here that the actual number of distinguishable states is

$$\mathrm{W}(\alpha_1, \alpha_2, n) = e^{\mathrm{I}_{\mathrm{sup}}(P;K)} = |\alpha_1 - \alpha_2|\sqrt{\frac{2n}{\pi e}}$$

where $\mathrm{I_{sup}}(P;K)$ is the maximum mutual information between the (random) quantum state and the results of measurements. We prove that this maximum is achieved for an ensemble of quantum states with the uniform distribution of the angle α for any interval $[\alpha_2, \alpha_1]$. The independence of the number of distinguishable states of the position of the interval $[\alpha_2, \alpha_1]$ is a remarkable asymptotic property that does not hold for small values of n (cf. [3]).

The result that the number of distinguishable states is proportional to the geometric distance as measured by angle in Hilbert space is quite nontrivial and noteworthy. Indeed, it suggests that the metric of Hilbert space may result not from a physical principle, but rather as a consequence of an optimal statistical inference procedure.

2. FORMULATION OF THE PROBLEM

Consider a quantum physical system whose states are unit vectors in a 2-dimensional complex Hilbert space $\mathbf{C}^2$ (the so-called "qubit"). Denote the state vector by $\mathbf{v}$ and let (Φ, Ψ) be an orthonormal basis in the Hilbert space, so that $\mathbf{v} = a\Phi + b\Psi$, where $a = \langle \mathbf{v}|\Phi\rangle$, $b = \langle \mathbf{v}|\Psi\rangle$ are inner products and $|a|^2 + |b|^2 = 1$. Then $|a|^2 = p$ and $|b|^2 = 1-p$ are probabilities of two possible outcomes of the measurement performed over the state $\mathbf{v}$ in the (Φ, Ψ) basis. Obviously, these probabilities do not depend on the phases of the coefficients a and b, and, therefore, all quantum states with the same magnitudes $|a| = x$ and $|b| = y$ are indistinguishable by this measurement. Hence, the state space S^3 can be reduced to the non-negative quadrant of a circle in a real 2-dimensional Euclidean space (Fig. 1), spanned by Φ and Ψ. Now let $\mathbf{v}_1$ and $\mathbf{v}_2$ be two distinct state vectors, such that

$$\mathbf{v}_i = x_i\Phi + y_i\Psi \text{ where } x_i = \sqrt{p_i} \text{ and } y_i = \sqrt{1-p_i}, \text{ for } i = 1,2. \quad (1)$$

Denote by α_i the angle between Φ and $\mathbf{v}_i$, so that

$$p_i = \cos^2\alpha_i, \;\; 1 - p_i = \sin^2\alpha_i, \;\; i = 1,2. \quad (2)$$

Suppose, we want to distinguish between various quantum states chosen from the interval of angles $[\alpha_2, \alpha_1]$ by performing measurements in the (Φ, Ψ) basis. Further, assume that we are allowed to perform the measurement over n identical copies of each quantum state.

Problem: What determines the number of distinguishable states, and what is the asymptotic expression for the number of states in the interval $[\alpha_2, \alpha_1]$ that can be distinguished with probability approaching 1 when n tends to infinity?

As shown in the next section, the problem can be rigorously analyzed by applying concepts and results of Shannon's information theory.

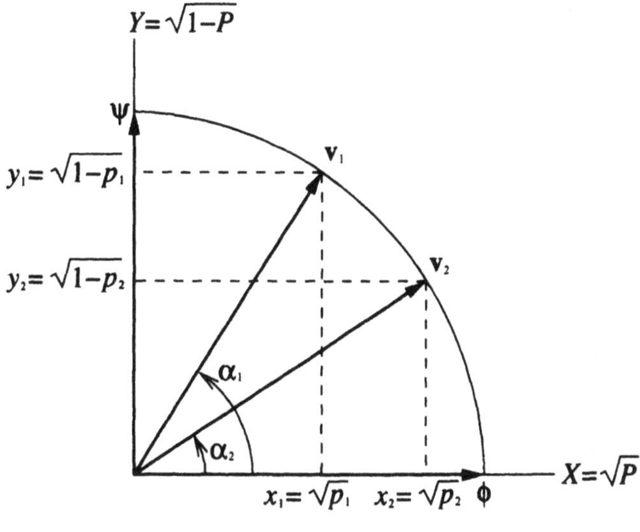

Figure 1

3. INFORMATION-THEORETICAL DESCRIPTION

Suppose the state vectors are chosen from the angle interval $[\alpha_2, \alpha_1]$ with certain probability density function (pdf) $\mathrm{P}_{\mathbf{A}}(\alpha)$, where $\mathbf{A}$ is a random variable that takes on values from $[\alpha_2, \alpha_1]$, $\alpha \in [\alpha_2, \alpha_1]$. Let $\mathrm{P}_P(p)$ be the pdf of the random variable P that takes on values p, where p is the probability of the state vector to be projected as the result of the measurement onto basis vector Φ. Obviously, $P = \cos^2 \mathbf{A}$, and the value of P (or of $\mathbf{A}$) characterizes uniquely the chosen quantum state. In a series of n measurements, let K be the (random) number of measurements which have resulted in projectios onto Φ. The conditional probability distribution of K given P is binomial:

$$\mathrm{P}_{K/P}(k/p) = \binom{n}{k} p^k (1-p)^{n-k}, \quad \text{where } k = 0, 1, \ldots, n. \tag{3}$$

The values of K obtained in the measurement are the only data available from which one can infer about the value of P, i.e., about the choice of a quantum state.

Let $\mathrm{P}_K(k)$ be the marginal probability distribution of K. The information $\mathrm{I}(K; P)$ in K about P is given by

$$\mathrm{I}(K;P) = \int_{p_1}^{p_2} \sum_{k=0}^{n} \mathrm{P}_P(p) \mathrm{P}_{K/P}(k/p) \ln \frac{\mathrm{P}_{K/P}(k/p}{\mathrm{P}_K(k)} \, dp. \tag{4}$$

The importance of considering information $\mathrm{I}(K, P)$ stems from Shannon's 12th theorem [2] which, for our setting of the problem, can be rephrased in the following way.

Let $\mathrm{S} = \{\mathbf{p}\}$, where $\mathbf{p}$ is an n-dimensional vector $\mathbf{p} = (p, p, \ldots, p)$ and $p \in [p_1, p_2]$ be the set of all possible input signals and $\mathbf{Z}_n = \{0, 1, \ldots, n\}$ be the set of all output signals in a communication channel with a conditional probability distribution given by (3). Then for any $\varepsilon > 0$ and $n \geq n(\varepsilon)$ the maximum number $\mathrm{W}(n, \varepsilon)$ of input signals that can be chosen from S in such a way that the probability of error (incorrect decision about $\mathbf{p}$ based on the value of the output signal $k \in \mathcal{Z}_n$) does not exceed ε satisfies the asymptotic property:

$$\lim_{n \to \infty} \frac{\log \mathrm{W}(n, \varepsilon)}{\mathrm{I}_{\mathrm{sup}}(K, P)} = 1, \tag{5}$$

where $\mathrm{I}_{\mathrm{sup}}$ is the least upper bound of $\mathrm{I}(K, P)$ given by (4) over all possible probability distributions $\mathrm{P}_P(p)$ of the input parameter P.

Note that the asymptotic expression for $\mathrm{W}(n, \varepsilon)$ in fact does not depend on ε. If means that for $n \to \infty$ the number of distinct input signals (different values of P) that can be distinguished with probability arbitrarily close to 1 is $e^{\mathrm{I}_{\mathrm{sup}}(K,P)}$. The problem is reduced now to the computation of $\mathrm{I}_{\mathrm{sup}}(K, P)$ under the condition that P takes on values in $[p_1, p_2]$. This problem is very difficult, in general. However, the following important theorem will be helpful.

Define individual information in $P = p$ about K as

$$\mathrm{I}(K; p) = \sum_{k=0}^{n} \mathrm{P}_{K/P}(k/p) \ln \frac{\mathrm{P}_{K/P}(k/p)}{\mathrm{P}_K(k)}. \tag{6}$$

As is well known (e.g. [4]), $\mathrm{I}(K; P)$ achieves the maximum value $\mathrm{I}_{\mathrm{sup}}(K; P)$ for such a distribution $\mathrm{P}_P(p)$ that there exists a constant I such that

$$\mathrm{I}(K; p) = \mathrm{I} \ \text{ for all } \ p \ \text{ such that } \ \mathrm{P}_p(p) > 0 \tag{7}$$

and

$$\mathrm{I}(K; p) < \mathrm{I} \ \text{ for all } \ p \ \text{ such that } \ \mathrm{P}_p(p) = 0\,. \tag{8}$$

Then $\mathrm{I}_{\mathrm{sup}}(K; P) = \mathrm{I}$.

4. THE NUMBER OF DISTINGUISHABLE STATES

When n is large, the binomial distribution (3) can be well-approximated by a Gaussian distribution:

$$\mathrm{P}_{K/P}(k/p) = \binom{n}{k} p^k (1-p)^{n-k} \approx \frac{1}{\sqrt{2\mathbf{p}ip(1-p)n}} e^{-\frac{(k-pn)^2}{2p(1-p)n}}\,. \tag{9}$$

For large n, distribution (9) has a very sharp maximum at $k = pn$, so that the Laplace method [5] can be used for evaluation of integrals involving (9).

Consider a uniform distribution over the angle interval $[\alpha_2, \alpha_1]$,

$$\mathrm{P}_{\mathsf{A}}(\alpha) = \frac{1}{|\alpha_1 - \alpha_2|} \quad \text{for } \alpha \in [\alpha_2, \alpha_1]\,. \tag{10}$$

The corresponding distribution of the probability P is

$$\mathrm{P}_P(p) = \mathrm{P}_{\mathsf{A}}(\alpha) \left| \frac{d\alpha}{dp} \right| = \frac{1}{2|\alpha_1 - \alpha_2|} \quad \text{where } \alpha_i = \cos^{-1}\sqrt{p_i}\,,\ i = 1, 2\,. \tag{11}$$

We will prove that for large n this distribution yields the maximum of $\mathrm{I}(K; P)$. The marginal probability distribution $\mathrm{P}_K(k)$ can be evaluated as follows (assuming $p_2 > p_1$):

$$\begin{aligned} \mathrm{P}_K(k) &= \int_{p_1}^{p_2} \mathrm{P}_P(p) \mathrm{P}_{K/P}(k/p)\, dp \\ &\approx \frac{1}{2|\alpha_1 - \alpha_2|} \int_{p_1}^{p_2} \frac{1}{p(1-p)\sqrt{2\pi n}} e^{-\frac{(k-np)^2}{2p(1-p)n}}\, dp\,. \end{aligned} \tag{12}$$

If the point of maximum $p = \frac{k}{n}$ of the exponential in the integrand is within the interval $[p_1, p_2]$, the integration interval can be extended to $(-\infty, \infty)$. Otherwise, the value of the integral approaches zero when n tends to infinity. Thus, for large n we obtain:

$$\mathrm{P}_K(k) \approx \begin{cases} \frac{1}{2|\alpha_1 - \alpha_2|\sqrt{k(n-k)}} & \text{if } np_1 \le k \le np_2 \\ 0 & \text{otherwise.} \end{cases} \tag{13}$$

Note that, as could be expected, the distribution of K for large n is the discrete counterpart of the distribution of P. Now we can evaluate the individual information $\mathrm{I}(K; p)$.

$$\begin{aligned} \mathrm{I}(K; p) &\approx \sum_{k=\lceil np_1 \rceil}^{\lfloor np_1 \rfloor} \mathrm{P}_{K/P}(k/p) \ln \frac{\mathrm{P}_{K/P}(k/p)}{\mathrm{P}_K(k)} \\ &\approx \int_{np_1}^{np_2} \frac{dk}{p(1-p)\sqrt{2\pi n}} e^{-\frac{(k-np)^2}{2p(1-p)n}} \left[\ln \mathrm{P}_{K/P}(k/p) - \ln \mathrm{P}_K(k) \right] \end{aligned} \tag{14}$$

The first term in (14) is the differential entropy of a Gaussian distribution (with the opposite sign), the second one can be evaluated by the Laplace method. Hence, asymptotically,

$$\begin{aligned} \mathrm{I}(K;p) &= -\frac{1}{2}\ln[2\pi e p(1-p)n] + \frac{1}{2}\ln[p(1-p)n^2] + \ln 2|\alpha_1 - \alpha_2| \\ &= \frac{1}{2}\ln\frac{2n}{\pi e} + \ln|\alpha_1 - \alpha_2| \end{aligned} \tag{15}$$

Note that $\mathrm{I}(K;p)$ takes on the same value for any $p \in [p_1, p_2]$. Hence, distribution (10) (or (11)) is the optimal one for large n, and the maximum information $\mathrm{I}_{\mathrm{sup}}(K;P)$ is expressed asymptotically as given below.

$$\mathrm{I}_{\mathrm{sup}}(K;P) = \frac{1}{2}\ln\frac{2n}{\pi e} + \ln|\alpha_1 - \alpha_2| \tag{16}$$

Thus, the number of distinguishable quantum states in the interval of angles $[\alpha_2, \alpha_1]$ is proportional to the length of the interval and to $\sqrt{n}$. It does not depend on the position of the interval in the circle.

$$\mathrm{W}(n, \alpha_1, \alpha_2) = e^{\mathrm{I}_{\mathrm{sup}}(K;P)} = |\alpha_1 - \alpha_2|\sqrt{\frac{2n}{\pi e}} \tag{17}$$

Of course, the range of **A** may consist of several separated intervals. Then (17) remains valid, as long as n is sufficiently large, so that each interval has many distinguishable states; also, $|\alpha_1 - \alpha_2|$ should be replaced by the total length of the intervals.

5. CONCLUSION

The main result of the paper can be summarized as follows. The number of distinguishable quantum states in a 2-dimensional Hilbert space is proportional to the square root of the number of identical copies of each state measured and to the total length of the angle intervals occupied by the state vectors. Surprisingly, it does not depend on the position of the arcs comprising the range on the unit circle. These results can be generalized for the N-dimensional Hilbert space of states of a quantum system. As in the 2-dimensional case, the unit sphere can be reduced to the non-negative orthant of the unit sphere in the real N-dimensional Euclidean space. It turns out [6] that the number of distinguishable states depends only on the area Ω of the domain on the unit sphere from which the states can be chosen, but does not depend on the shape and position of this domain. The optimal distribution is uniform over the domain in angular (polar) coordinates, and the number of distinguishable states is $\mathrm{W}(n, \Omega) = \Omega(cn)^{\frac{N-1}{2}}$ where c is a constant.

Acknowledgments

T.T. was partially supported by the Department of Energy under grant DE-FG02-99ER25414. Z.W. acknowledges financial support by the National Science Foundation and the Boston University Photonics Center.

References

[1] Wootters, W.K. Statistical Distance and Hilbert Space, *Phys. Rev. D*, 23:357–362, 1981.

[2] Shannon, C.E. and Weaver, W. *The Mathematical Theory of Communication*. University of Illinois Press, 1949, p. 76.

[3] Levitin, L. B., Entropy Defect and Information for Two Quantum States, *Open Systems and Information Dynamics* 2:319–329, 1994.

[4] Mansuripur, M. *Introduction to Information Theory*. Prentice-Hall, 1987, p. 66.

[5] De Bruijn, N. G. *Asymptotic Methods in Analysis*. North Holland Publishing Co., 1958, Ch. 4.

[6] Levitin, L. B., Toffoli, T., Walton, Z. Information and Distinguishability of Quantum States, to be published.

SUPERADDITIVITY WITH MIXED LETTER STATES

Masao Osaki
Research Center for Quantum Communications, Tamagawa University, Japan
osaki@lab.tamagawa.ac.jp

1. INTRODUCTION

We consider the situation sending a conventional (Shannon) information through some quantum medium [1]. One of purposes is to find some codewords which approach to the quantum channel capacity C [2, 3]. To accomplish this, the superadditivity is indispensable. The maximum mutual information without coding C_1 is necessary to judge the superadditivity. However C_1 is analytically derived only in the case of the binary pure state signal [4]. Hence the codewords with superadditivity are found only for this signal. For the other signals, further investigations into C_1 and appropriate codewords still remain to be done. Here we numerically derive C_1 for the binary mixed state signals. With the derived C_1, the amount of superadditivity is analyzed for the codeword length 2 and 3.

2. C_1 FOR BINARY MIXED STATES

We consider the binary mixed states in spin 1/2 system represented as follows:

$$\hat{\rho}_i = \begin{bmatrix} \frac{1}{2} + \left(\frac{1}{2} - f_i\right)\cos\theta & \pm\left(\frac{1}{2} - f_i\right)\sin\theta \\ \pm\left(\frac{1}{2} - f_i\right)\sin\theta & \frac{1}{2} - \left(\frac{1}{2} - f_i\right)\cos\theta \end{bmatrix},$$

where $i = 1, 2$, and '$\pm$' $\Rightarrow$ '$+$' for $i = 1$, '$\pm$' $\Rightarrow$ '$-$' for $i = 2$. f_i $(0 \leq f_i \leq \frac{1}{2})$ and θ $(0 \leq \theta \leq \frac{\pi}{2})$ represent the degree of mixture and an angle from z-axis, respectively. In order to identify C_1, 3 detection operators $\Pi_j, j = 1, 2, 3$ are enough to be considered [5].

$$\hat{\Pi}_1 = |\omega_1\rangle\langle\omega_1|, \quad \hat{\Pi}_2 = |\omega_2\rangle\langle\omega_2|, \quad \hat{\Pi}_3 = \hat{I} - \hat{\Pi}_1 - \hat{\Pi}_2,$$

where $\hat{I}$ represents the identity operator and $|\omega_j\rangle = m_j \begin{bmatrix} \cos\phi_j \\ \sin\phi_j \end{bmatrix}$ is a measurement state with domains of parameters, $0 < m_j \leq 1, 0 \leq \phi_j \leq 2\pi$. The parameters m_j, ϕ_j which give $\hat{\Pi}_3 < 0$ are excluded. Since the conditional

Quantum Communication, Computing, and Measurement 3
Edited by P. Tombesi and O. Hirota, Kluwer Academic/Plenum Publishers, 2001

probabilities $P(j|i)$ are given by $P(j|i) = \mathrm{Tr}\hat{\Pi}_j\hat{\rho}_i$, the mutual information is calculated with *a priori* probability ξ_1. In Table 1, the domains and steps of parameters used in the numerical derivation of C_1 are summarized.

Table 1 Domains and steps of parameters in derivation of C_1.

param.	domain	step	param.	domain	step
θ	$0 \le \theta \le \pi/2$	$\pi/12$	ϕ_j	$0 \le \phi_j \le 2\pi$	$\pi/12$
m_j	$0 \le m_j \le 1$	$1/12$	ξ_1	$0.4 \le \xi_1 \le 0.6$	0.01

In the case of $f_1 = f_2$, C_1 is attained by two detection operators ($m_1 = m_2 = 1$), fixed $(\phi_1, \phi_2) = (\frac{\pi}{4}, -\frac{\pi}{4}), \forall f$, and equal *a priori* probabilities $\xi_1 = \xi_2 = 1/2$. Hence C_1 can be represented analytically.

$$C_1 = (1 - P_e)\log[2(1 - P_e)] + P_e \log[2P_e], \text{ with } P_e = \frac{1 - (1 - 2f)\sin[\theta]}{2}.$$

On the other hand, C is analytically derived as follows:

$$C = (1 - f)\log[1 - f] + f\log[f] - \lambda_+ \log[\lambda_+] - \lambda_- \log[\lambda_-],$$

where $\lambda_\pm$ are eigen-values of $\hat{\rho} = \frac{1}{2}(\hat{\rho}_1 + \hat{\rho}_2)$, $\lambda_\pm = \frac{1 \pm (1-2f)\cos[\theta]}{2}$.

When $f_1 \neq f_2$, C_1 is attained by two detection operators, unsymmetrical ϕ_i, and unbalanced *a priori* probabilities.

3. SUPERADDITIVITY

For the codeword length 2, there is a set of codewords by which the superadditivity is found in pure letter states [6]. The larger amount of superadditivity is obtained by codeword states $\hat{\rho}_1^{(2)} = \hat{\rho}_1 \otimes \hat{\rho}_1$, $\hat{\rho}_2^{(2)} = \hat{\rho}_2 \otimes \hat{\rho}_2$, $\hat{\rho}_3^{(2)} = \hat{\rho}_1 \otimes \hat{\rho}_2$ with *a priori* probabilities $\xi_1^{(2)} = \xi_2^{(2)} = 0.4664$, $\theta = \frac{\pi}{14}$ in letter states, and $\eta = 1.398$[rad.] for measurement states in Ref. [6]. In calculation of the mutual information I_2 with mixed letter states, $f_1 = f_2 = f$ is assumed. The same codewords and measurement states are also employed with the fourth detection operator $\hat{\Pi}_4^{(2)} = \hat{I}^{(2)} - \sum_{j=1}^{3} \hat{\Pi}_j^{(2)}$ to compensate the influence of mixture. I_2 is depicted in Fig. 1 and 2. The superadditivity can be found when $f < 0.0005$.

For codeword length 3, we also use the codewords by which the superadditivity is found [7]. The larger amount of superadditivity is obtained by $\hat{\rho}_1^{(3)} = \hat{\rho}_1 \otimes \hat{\rho}_1 \otimes \hat{\rho}_1$, $\hat{\rho}_2^{(3)} = \hat{\rho}_1 \otimes \hat{\rho}_2 \otimes \hat{\rho}_2$, $\hat{\rho}_3^{(3)} = \hat{\rho}_2 \otimes \hat{\rho}_1 \otimes \hat{\rho}_2$, $\hat{\rho}_4^{(3)} = \hat{\rho}_2 \otimes \hat{\rho}_2 \otimes \hat{\rho}_1$, with $\xi_i^{(3)} = 1/4, \forall i$, $\theta = \frac{\pi}{6}$ in letter states, and square-root-detection (SRD).

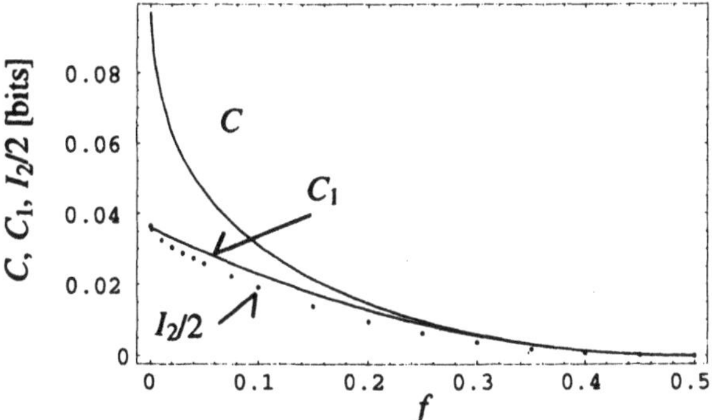

Figure 1 Comparison of mutual information.

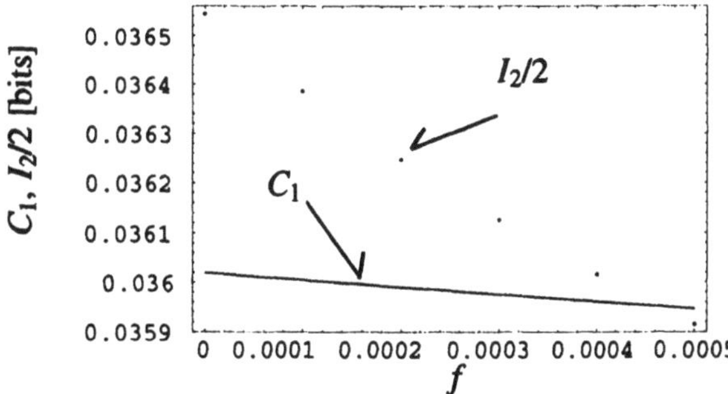

Figure 2 Detailed comparison of mutual information.

To derive I_3 for mixed letter states, $f_1 = f_2 = f$ is assumed and the same codewords are employed. The detection operators of SRD are defined as follows:

$$\hat{\Pi}_j = \hat{G}^{-\frac{1}{2}}\hat{\rho}_j^{(3)}\hat{G}^{\dagger-\frac{1}{2}}, \quad \hat{G} = \sum_k \hat{\rho}_k^{(3)}.$$

The operators $\hat{\Pi}'_j \equiv \hat{\Pi}_j\Big|_{f\to 0}$ are also the resolution of identity. For each detection operators, I_3 is depicted in Fig. 3 and 4. The superadditivity can be found only in the case of $f \leq 0.015$ with $\left\{\hat{\Pi}'_j\right\}$.

4. CONCLUSION

In this paper we numerically derived C_1 of the binary mixed state signal. The mutual informations of codeword states, I_2 and I_3, were calculated and were compared with C_1 and C. The expected amount of the superadditivity was given when the pure state preparation is too difficult in experimental set up.

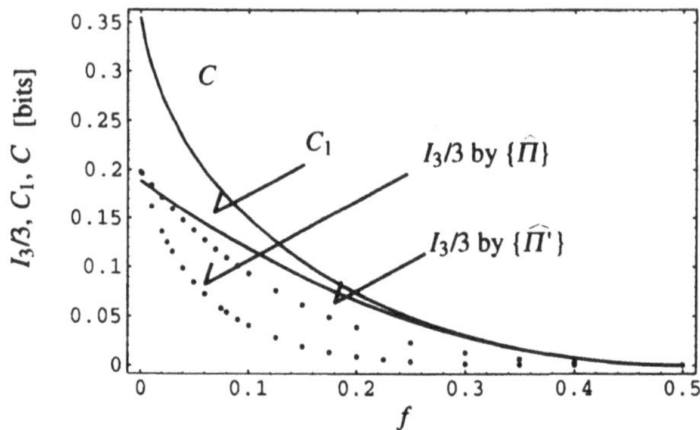

Figure 3 Comparison of mutual information.

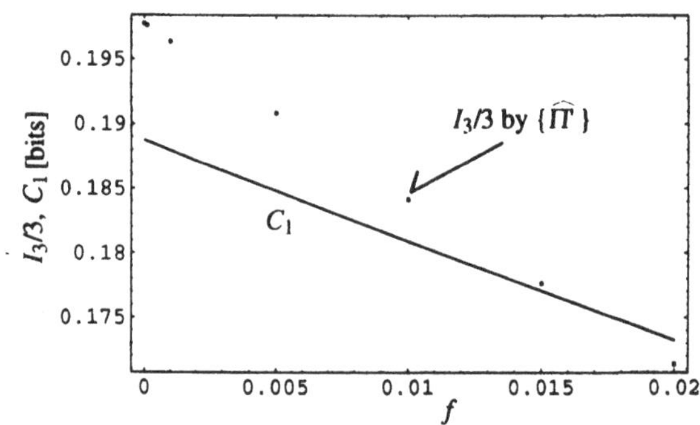

Figure 4 Detailed comparison of mutual information.

References

[1] A.S.Holevo, J. Multivar. Anal., **3**, 337, 1973.

[2] A.S.Holevo, IEEE, Trans. Inform. Theory, **IT-44**, 269, 1998.

[3] B.Schumacher and M.D.Westmoreland, Phys. Rev. A **56**, 131, 1997.

[4] M.Ban, K.Kurokawa, and O.Hirota, J. Opt. B: Quantum Semiclass. Opt., **1**, 206, 1999.

[5] E.B.Davies, IEEE, Trans. Inform. Theory, **IT-24**, 596, 1978.

[6] J.R.Buck, S.J.van Enk, and C.A.Fuchs, quant-ph/9903039 ver.2, 1999.

[7] M.Sasaki, K.Kato, M.Izutsu, and O.Hirota, Phys. Lett. A, **236**, 1, 1997.

QUANTUM STATE RECOGNITION

Masahide Sasaki, Alberto Carlini
Communications Research Laboratory, Ministry of Posts and Telecommunications
Koganei, Tokyo 184-8795, Japan
psasaki@crl.go.jp

1. INTRODUCTION

Let us consider the following *pattern matching* problem [1]. We have at our disposal a database of recorded persons' pictures, classified in different classes according to certain defined features. Now we are given somebody's picture and we want to know to which class the person should belong. We scan the database by comparing the defined features of the given sample with those of the classes. If the patterns of the sample have a good enough matching with the elements of a certain class, we can then say that the person is *recognized*, i.e., that the person belongs to that pattern class. We would like to consider a similar problem in quantum mechanical context.

We are given a *feature state* $|f\rangle$ which is usually unknown. We want then to classify this $|f\rangle$ into a proper pattern class among a set of classes $\{C_i\}$ that we have. For simplicity, we assume that each class C_i is represented by a typical pattern called *template state* $|g_i\rangle$. The problem is to pick up the template which best matches with $|f\rangle$. We also assume that N identical samples of $|f\rangle$ are available. We shall consider this problem in the Bayesian formalation of quantum mearsurement theory [2, 3].

2. BAYESIAN FORMULATION

Let us consider the simplest case where there are only two classes and the input feature state and the template states are all pure state of a two state system. They are confined in the same plane of a great circle of the Bloch sphere.

$$|g_0\rangle = \cos\frac{\theta}{2}\,|\uparrow\rangle + \sin\frac{\theta}{2}\,|\downarrow\rangle, \quad |g_1\rangle = \sin\frac{\theta}{2}\,|\uparrow\rangle + \cos\frac{\theta}{2}\,|\downarrow\rangle, \tag{1}$$

where θ is a given parameter. The input feature state is

$$|f\rangle = \cos\frac{\phi}{2}\,|\uparrow\rangle + \sin\frac{\phi}{2}\,|\downarrow\rangle, \tag{2}$$

Quantum Communication, Computing, and Measurement 3
Edited by P. Tombesi and O. Hirota, Kluwer Academic/Plenum Publishers, 2001

where *a priori* distribution of ϕ is assumed to be uniform $P(\phi) = \frac{1}{2\pi}$.

We want to decide to which template this is closer in the sense of higher state overlap, that is, we are to maximize the average *score* defined by

$$\bar{S}(N) = \sum_{j=0}^{1} \frac{1}{2\pi} \int_0^{2\pi} d\phi \, \mathrm{Tr}\, (\hat{\Pi}_j \hat{F}) | \langle f | g_j \rangle |^2, \quad \hat{F} \equiv |f\rangle \langle f|^{\otimes N} \tag{3}$$

where $\{\hat{\Pi}_0, \hat{\Pi}_1\}$ is a probability operator measure (POM) representing a recognition process. The N product state $|F\rangle$ is expressed by

$$|F\rangle \equiv |f\rangle^{\otimes N} = \sum_{k=0}^{N} \sqrt{\begin{pmatrix} N \\ k \end{pmatrix}} \left(\cos\frac{\phi}{2}\right)^{N-k} \left(\sin\frac{\phi}{2}\right)^{k} |k\rangle\,, \tag{4}$$

where $\{|k\rangle\}$ is the symmetric bosonic basis with k the number of $|\downarrow\rangle$. A POM $\{\hat{\Pi}_0, \hat{\Pi}_1\}$ is defined on the $N+1$-dimensional subspace spanned by $\{|k\rangle\}$. By introducing the score operator

$$\hat{W}_j \equiv \frac{1}{2\pi} \int_0^{2\pi} d\phi \hat{F} | \langle f | g_j \rangle |^2, \tag{5}$$

Eq. (3) is then written as

$$\bar{S}(N) = \mathrm{Tr}\, (\hat{W}_0 \hat{\Pi}_0) + \mathrm{Tr}\, (\hat{W}_1 \hat{\Pi}_1) = \frac{1}{2} + \mathrm{Tr}\left[(\hat{W}_0 - \hat{W}_1) \hat{\Pi}_0 \right]. \tag{6}$$

To maximize $\mathrm{Tr}\,[(\hat{W}_0 - \hat{W}_1)\hat{\Pi}_0]$ $\hat{\Pi}_0$ should be the projection onto the subspace corresponding to the positive eigenvalues of the operator $\hat{W}_0 - \hat{W}_1$. Let the diagonalizing operator be $\hat{P}$ such that

$$\hat{P}(\hat{W}_0 - \hat{W}_1)\hat{P}^{\dagger} = \left(\sum_{k=0}^{N} \lambda_k |k\rangle \langle k| \right) \cos\theta, \tag{7}$$

where $\lambda_0 > \lambda_1 > .... > \lambda_N$. Due to the symmetry of $\hat{W}_0 - \hat{W}_1$, $\lambda_N = -\lambda_0$, $\lambda_{N-1} = -\lambda_1$, and so on. The optimal strategy can then be constructed as $\hat{\Pi}_i = \hat{P}^{\dagger} \hat{\Pi}_i^{\mathrm{MV}} \hat{P}$ $(i = 0, 1)$ where

$$\hat{\Pi}_0^{\mathrm{MV}} = |0\rangle \langle 0| + \cdots + \left|\tfrac{N-1}{2}\right\rangle \left\langle \tfrac{N-1}{2}\right|, \tag{8}$$

$$\hat{\Pi}_1^{\mathrm{MV}} = \left|\tfrac{N+1}{2}\right\rangle \left\langle \tfrac{N+1}{2}\right| + \cdots + |N\rangle \langle N|, \tag{9}$$

where N is assumed to be odd. The maximum average score is then

$$\bar{S}_{\mathrm{OPT}}(N) = \frac{1}{2} + \sum_{k=0}^{\frac{N+1}{2}} \lambda_k \cos\theta. \tag{10}$$

This strategy consists of two steps: the first step is the unitary operation $\hat{P}$ on the N-product input pattern $\hat{F}$ and the second step is the measurement of $\hat{P}\hat{F}\hat{P}^{\dagger}$ by the POM $\{\hat{\Pi}_0^{MV}, \hat{\Pi}_1^{MV}\}$ corresponding to the separable measurement on each copy and majority voting afterwards. One possible circuit structure is shown in figure 1 (a) for the case of $N = 3$. After performing $\hat{P}$, each

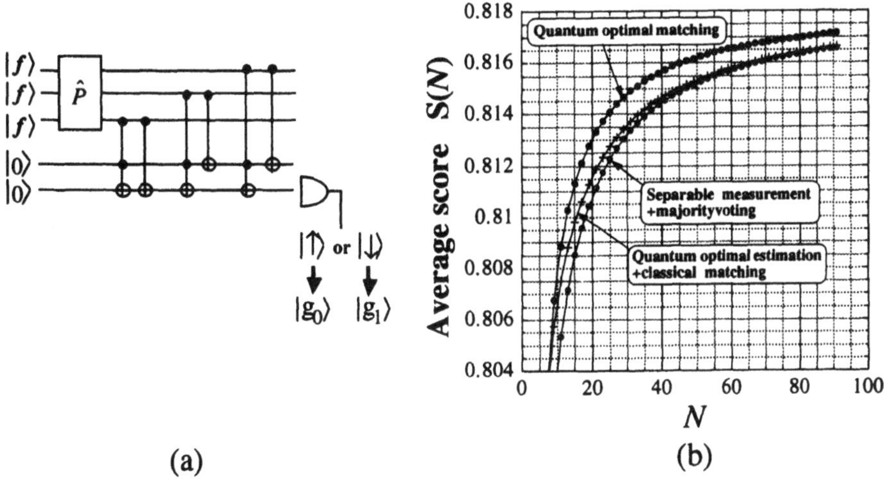

Figure 1 (a) A quantum circuit for the optimal recognizer for $N = 3$. (b) The average score as a function of N.

input is processed inteructively with two ancillary qubits via controlled NOT gates. By measureing the second ancillary qubit with the basis $\{|\uparrow\rangle, |\downarrow\rangle\}$, we can decide at the minimum classification error the best matched template.

In figure 1 (b), the average scores obtained by several strategies are compared as a function of N. (θ is taken as 0.) The black circles represent the performance of the optimal recognizer. The white circles correspond to the separable measurement + majority voting strategy which is just $\{\hat{\Pi}_0^{MV}, \hat{\Pi}_1^{MV}\}$. The effect of $\hat{P}$ is seen as the reduction of required number of copies to attain the same level of the average score. The curve denoted by "+" corresponds to the strategy consisting of quantum state estimation and classical matching between a reconstructed feature state and templates. As for the optimal state estimation, we can use the POM derived by Derka et al. [4]. The detail will be reported elsewhere. As seen, the strategy of the optimal state estimation + classical matching can be the near optimal for small N, while as N increases it deviate from the optimal and becomes closer to the separable measurement + majority voting strategy. Thus a whole process of recognition has to stay entirely in the quantum domain until the very final measurement is performed. Any intermediate classical process degrade the total performance, that is, waste of input samples.

3. CONCLUDING REMARK

We have only considered the simplest case of quantum state recognition: binary classification of a two state system with templates in pure state. The efficiency was compared in terms of required numbers of input samples, say N_{Q} for the optimal strategy and N_{C} for a semiclassical one. It would then be worth considering a computational complexity $\chi(N)$ of recognizers for those numbers and comparing $\chi(N_{\mathrm{Q}})$ and $\chi(N_{\mathrm{C}})$ as the total efficiency. This might give an insight into an exploration of good quantum algorithms along with Bayesian optimization.

More practical models may include multiple classes each of which contains several templates corresponding to some possible features. This leads to a full mixed-state problem and an appropriate classification criterion should be carefully chosen such as quantum relative entropy other than fidelity. Further discussions will be reported elsewhere.

Acknowledgments

The authors would like to thank T. Hattori for introducing us to pattern matching problem. They also thank R. Jozsa for giving valuable suggestions. They would also like to thank A. S. Holevo, C. A. Fuchs, O. Hirota, S. M. Barnett, and A. Chefles for helpful discussions.

References

[1] *Digital Pattern Recognition*, ed. by K. S. Fu, (Springer-Verlag, Berlin, Heidelberg, New York, 1976)

[2] C. W. Helstrom : *Quantum Detection and Estimation Theory* (Academic Press, New York, 1976).

[3] A. S. Holevo : *Probabilistic and Statistical Aspects of Quantum Theory* (North-Holland, Amsterdam, 1982).

[4] R. Derka, V. Buzek, and A. K. Ekert, Phys. Rev. Lett. **80**, 1571 (1998).

SUPERADDITIVITY IN CAPACITY OF QUANTUM CHANNEL BY CLASSICAL PSEUDO-CYCLIC CODES

Shogo Usami, Tsuyoshi Sasaki Usuda, Ichi Takumi, Ryohei Nakano
Dept. A. I. and Computer Science, Nagoya Institute of Technology,
Gokiso-cho, Showa-ku, Nagoya, 466-8555, Japan
sho@ics.nitech.ac.jp

Masayasu Hata
Dept. Applied Information Technology, Aichi Prefectural University, Japan

Keywords: Channel capacity, superadditivity, mutual information, square-root measurement

Abstract Analytical solution of mutual information by square-root measurement for binary linear code is given. Applying the solution, we show properties of superadditivity in capacity by BCH codes with codeword length 15 and 31. Furthermore, we refer to codes almost achieving the quantum channel capacity.

Introduction

It is known that there is superadditivity in classical capacity of quantum channel [1] :

$$C_n + C_m \leq C_{n+m}, \tag{1}$$

where C_i is channel capacity or maximum mutual information with codeword length i. Recently, clear examples of strict superadditivity for binary pure-state signals with finite codeword length were demonstrated [2] by showing first inequality of

$$C_1 < I_n(X;Y)/n \leq C_n/n, \tag{2}$$

where $I_n(\cdot;\cdot)$ is mutual information of a code with codeword length n.

However, to calculate mutual information with large number of codewords is very difficult because of computational cost.

In this paper, first we show the analytical solution of mutual information when we use classical linear codes for coding scheme and use square-root measurement as decoding scheme. Second, we show properties of superadditivity by classical pseudo-cyclic codes by applying the analytical solution. In particular, detailed properties for primitive BCH-codes with codeword length 15 and 31 are clarified. We will simplify the analytical solution for specific codes and show the properties of superadditivity for the codes with long codeword length (more than 10^{10}). Furthermore, we will refer to a code almost achieving quantum channel capacity $C = \lim_{n\to\infty} C_n/n$.

1. ANALYTICAL SOLUTION OF MUTUAL INFORMATION

We assume *a priori* probabilities of all codewords are equal. By using group covariancy of linear codes [3], we obtain the analytical solution of mutual information as

$$I_n(X;Y) = k + \sum_{j=0}^{2^k-1} P(j|0) \log P(j|0), \tag{3}$$

$$\begin{aligned} P(j|0) &= \left\{ (\Gamma_{2^k})^{\frac{1}{2}}_{0,j} \right\}^2 \\ &= \left\{ \frac{1}{2^k} \sum_{k'=0}^{2^k-1} (-1)^{w_{\mathrm{H}}(j\cdot k')} \sqrt{\sum_{l=0}^{2^k-1} (-1)^{w_{\mathrm{H}}(k'\cdot l)} (\Gamma_{2^k})_{0,l}} \right\}^2, \end{aligned} \tag{4}$$

where n is codeword length, Γ_{2^k} is the Gram matrix, 2^k is the number of codewords, $w_{\mathrm{H}}(i)$ denotes the Hamming weight of i in binary notation, and $j \cdot k'(k' \cdot l)$ means "AND" operation for each bit when $j(k')$ and $k'(l)$ are represented as binary numbers of n-digits.

By using the above formula, we can compute mutual information with large number of codewords (up to about 1,000,000 codewords)

2. PROPERTIES OF SUPERADDITIVITY FOR LARGE SCALE CODES

In this section, we show properties of superadditivity by classical pseudo-cyclic codes by applying Eqs.(3) and (4).

Fig.1 shows mutual information per letter $I_n(X;Y)/n$ and a quantity AF of primitive BCH codes of codeword length 15 and 31 with respect to inner product between letter states κ. Here, AF is defined as

$$AF = \frac{I_n(X;Y)/n - C_1}{C - C_1}, \tag{5}$$

which represents how much is superadditivity achieved to C. We call it the achievement factor. If $AF = 0$, $I_n(X;Y)/n = C_1$ while if $AF = 1$, one has $I_n(X;Y)/n = C$, i.e., the quantum channel capacity is achieved. We also define the maximum value of AF with respect to $\kappa : AF_{\max} = \max_\kappa AF$. In Fig.1, $(n, n-m)$ means $I_n(X;Y)/n$ or AF for the pseudo-cyclic code with codeword length n, the order of generator polynomial m, and 2^{n-m} codewords. From Figs.1(c) and (d), it seems that $AF_{\max}$ for primitive BCH code with the same codeword length have almost the same values. The longer codeword length is, the wider range of κ in which superadditivity is achieved.

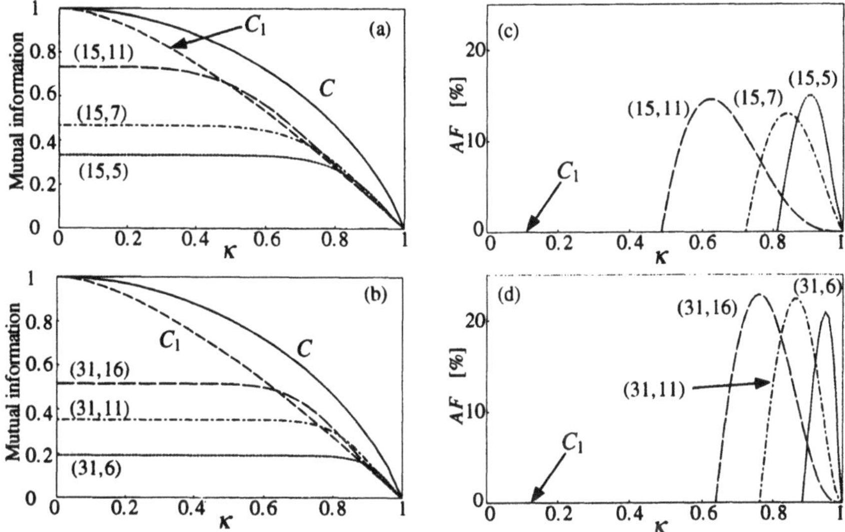

Figure 1 (a) and (b) Mutual information per letter, (c) and (d) achievement factor with respect to κ for primitive BCH codes with codeword length 15 ((a) and (c)) and 31 ((b) and (d)).

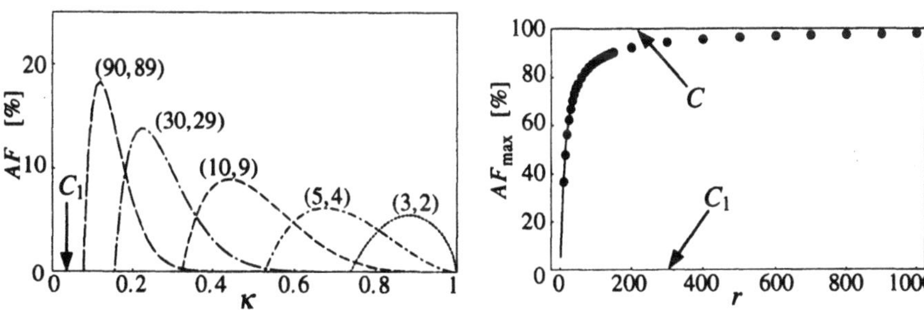

Figure 2 Achievement factor with respect to κ for parity check codes.

Figure 3 Maximum value of achievement factor for simplex codes with respect to r. $AF_{\max}$ with exact solution (solid line) and with approximation (dots).

3. CODES ALMOST ACHIEVING CAPACITY

Here we try to calculate mutual information for specific codes with longer codeword length. We can simplify the analytical solutions for parity check codes and $(2^r - 1, r + 1)$ primitive BCH codes. As an example, we show AF for parity check codes with codeword length up to 90 in Fig.2. We also check properties of mutual information for $(2^r - 1, r + 1)$ primitive BCH codes and simplex codes. As a result, we find that $AF_{\max}$ for $(2^r - 1, r+1)$ BCH code and $(2^r - 1, r)$ simplex code have almost the same values when the codeword length is long. In the following we show the property of $AF_{\max}$ for $(2^r - 1, r)$ simplex code with extremely long codeword length. We compute exact $AF_{\max}$ and approximate $AF_{\max}$ by Maclaurin expansion with the condition that codeword length is extremely ong and κ is closed to 1. Fig.3 show the property. The longer codeword length is, the larger $AF_{\max}$ is. We can see from Fig.3, the quantum channel capacity C is almost achieved by simplex codes ($AF_{\max} > 97\%$ when $r > 700$).

4. CONCLUSION

In this paper, we gave the analytical solution of mutual information for any binary linear code with square-root measurement as decoding process. By applying the formula, mutual information for any binary linear code with up to about 1,000,000 codewords can be computed. Furthermore, we have shown properties of mutual information for extremely long codeword length by approximation of mutual information for simplex code. As a result, we find that the quantum channel capacity C is almost achieved by simplex codes. We conjecture that C is almost achieved by BCH code for almost all range of κ since mutual information for $(2^r - 1, r + 1)$ BCH and $(2^r - 1, r)$ simplex codes has similar property and $AF_{\max}$ for primitive BCH code with the same codeword length have almost the same values.

Acknowledgments

S. Usami is supported by the Hori Information Science Promotion Foundation.

References

[1] A. S. Kholevo, Problemy Peredachi Informatsii, 15, 3, 1979.

[2] M. Sasaki *et al.*, Phys. Rev. A **58**, 146, 1998.

[3] T. S. Usuda and I. Takumi, in *Quantum Communication, Computing and Measurement* 2, P. Kumar, et. al. (Eds.), pp.37-42 (2000).

PROPERTY OF MUTUAL INFORMATION FOR M-ARY QUANTUM-STATE SIGNALS

Tsuyoshi Sasaki Usuda, Ichi Takumi, Ryohei Nakano
Dept. A. I. and Computer Science, Nagoya Institute of Technology,
Gokiso-cho, Showa-ku, Nagoya, 466-8555, Japan
usuda@ics.nitech.ac.jp

Masao Osaki
Research Center for Quantum Communications, Tamagawa University, Japan

Masayasu Hata
Dept. Applied Information Technology, Aichi Prefectural University, Japan

Keywords: Channel capacity, mutual information, square-root measurement

Abstract Numerical solutions of maximum mutual information for 3-ary phase-shift keyed coherent-state signals and 3-ary real-symmetric (lifted trine) signals, which are linearly independent signals, are shown. A condition for achieving maximum mutual information for M-ary linearly independent signals is also discussed.

Introduction

The maximization problem of mutual information is one of important topics in quantum information theory for sending classical information over a quantum channel. However, solutions for specific signals are given only for a few cases [1, 2, 3], e.g., maximum mutual information without coding C_1 for binary pure-state signals, accessible information for M-ary linearly dependent signals in two-state systems, etc. In this paper, we compute numerical solutions of maximum mutual information for 3-ary phase-shift-keyed coherent-state signals and 3-ary real-symmetric (or lifted trine) signals. Then we discuss a condition for achieving maximum mutual information for M-ary linearly independent signals.

1. PRELIMINARY

In quantum communication systems for sending classical information, a transmitter transmits a quantum state $|\psi_i\rangle \in \mathcal{H}$ which corresponds to a classical input signal x_i with *a priori* probability ξ_i and a receiver detects the signal. A quantum detection process is described by a set of detection operators $\{\hat{\Pi}_j\}$ which is a positive operator-valued measure (POM). A detection operator $\hat{\Pi}_j$ corresponds to an output signal y_j. We can express the conditional probability that the signal y_j is detected when the signal x_i was sent as $P(j|i) = \langle\psi_i|\hat{\Pi}_j|\psi_i\rangle$. The mutual information $I_1(\{\xi_i\};\{\hat{\Pi}_j\})$ is defined by $P(j|i)$ as in the classical information theory. Then we define the channel capacity without coding as follows.

$$C_1 = \max_{\{\xi_i\}} \max_{\{\hat{\Pi}_j\}} I_1(\{\xi_i\};\{\hat{\Pi}_j\}). \tag{1}$$

In the same way, the channel capacity with codeword length n is defined as maximum mutual information for n-th extended signals. Furthermore the quantum channel capacity is defined as $C = \lim_{n\to\infty} C_n/n$.

2. COMPUTATION OF C_1

Here we try to compute C_1 for 3-ary phase-shift-keyed (3PSK) coherent-state signals and 3-ary real-symmetric (or lifted trine) signals.

The M-ary PSK coherent-state signals are defined as

$$|\psi_k\rangle = |\alpha e^{\frac{2k\pi i}{M}}\rangle, \quad k = 0, 1, \cdots, M-1, \tag{2}$$

where $\alpha e^{\frac{2k\pi i}{M}}$ is the complex amplitude of a coherent state. The M-ary real-symmetric signals $\{|\phi_k\rangle\}$ are symmetric signals in which inner products between different two signals are equal. That is,

$$\langle\phi_i|\phi_j\rangle = \kappa, \quad i \neq j, \tag{3}$$

where $-1/(M-1) < \kappa \leq 1$. To compute C_1 is difficult since we must maximize mutual information with respect to up to 9 detection operators and *a priori* probabilities. Here we assume *a priori* probabilities of 2 of 3 signals are equal ($\xi_0 = \xi_1$) because C_1 for 3PSK signals are attained when $\xi_0 = \xi_1$ in cases of very weak or intense light [2]. And we assume the detection process is represented by von Neumann measurement because all examples of C_1 up to now are attained by von Neumann measurements [1, 2, 3]. We denote the maximum mutual information under the above condition by $\tilde{C}_1(\leq C_1)$.

Figs. 1(a) and (c) show $\tilde{C}_1$ and Figs. (b) and (d) show *a priori* probabilities ξ_0 with which $\tilde{C}_1$ are attained. We can see from Figs. 1(b) and (d) that for 3PSK (3-ary real-symmetric) signals, $\tilde{C}_1$ is attained by using 2 of 3 signals,

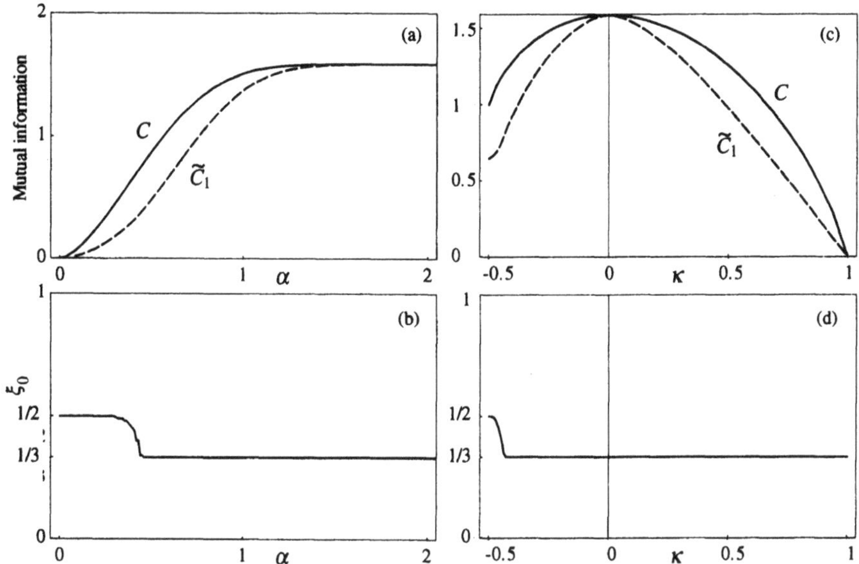

Figure 1 $\tilde{C}_1$ (dashed line) and C (solid line) for (a) 3PSK and (c) 3-ary real-symmetric signals. *A priori* probabilities ξ_0 with which $\tilde{C}_1$ are attained for (b) 3PSK and (d) 3-ary real-symmetric signals.

i.e. $\xi_0 = 1/2$, if $\alpha \to 0$ ($\kappa \to -0.5$) and is attained by using 3 of 3 signals equiprobably, i.e. $\xi_0 = 1/3$, if $\alpha > 0.5$ ($\kappa > -0.4$). The signals are linearly dependent when $\kappa = -0.5$ for 3-ary real-symmetric signals. Therefore the signals are *almost linearly dependent* when $\kappa \approx -0.5$. Moreover by comparing $\tilde{C}_1$, I_{SRM}, and I_{BIN}, we find $\tilde{C}_1 \approx \max\{I_{\mathrm{SRM}}, I_{\mathrm{BIN}}\}$. Here I_{SRM} (I_{BIN}) is mutual information by using 3 (2) of 3 signals equiprobably and applying square-root measurement (SRM) as the detection process.

3. PROPERTY OF MUTUAL INFORMATION

The above result means that for M-ary signals, d' and N are closely related and $\tilde{C}_1$ is attained by abandoning $M - N$ states and making the signals be linearly independent. Here, d' is dimension of space almost spanned by M-ary signals and $N(\leq M)$ is number of signals should be used. In the following, we compare the properties of mutual information by using $N(< M)$ of M signals with dimension d' of space almost spanned by M-ary signals. d' is evaluated by the eigenvalues λ_j of Gram matrix $\Gamma = [\sqrt{\xi_i \xi_j}\langle\psi_i|\psi_j\rangle]$: Let $\lambda_j (j = 0, 1, \cdots, M-1)$ be eigenvalues of Gram matrix. If $\lambda_0 + \cdots \lambda_{d'-1} \approx 1$ and $\lambda_{d'} + \cdots \lambda_{M-1} \approx 0$, then approximate dimension is regarded as d'. Here SRM is applied for the calculation of the mutual information.

Fig. 2(a) shows mutual information for 8-ary real-symmetric signals. From Fig. 2(a), using $M - 1$ of M states is better when $\kappa \approx -1/(M-1)$. Fig. 2(b) shows property of eigenvalues. From Fig. 2(b), the eigenvalue λ_0 is larger than

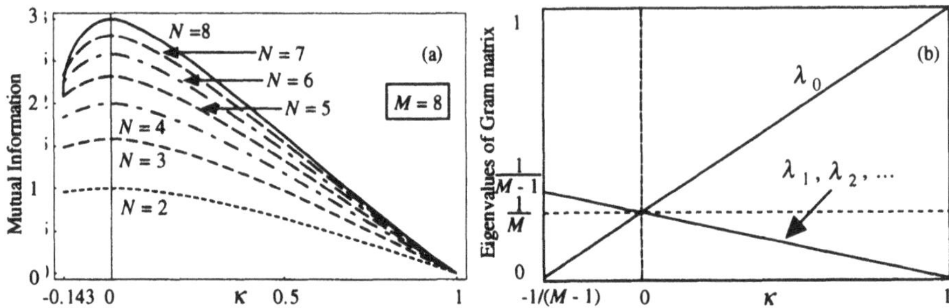

Figure 2 (a) Mutual information for M-ary real-symmetric signals ($M = 8$). (b) The eigenvalues for M-ary real-symmetric signals.

$M - 1$ degenerate eigenvalues $\lambda_1, \cdots, \lambda_{M-1}$ when $\kappa > 0$ but $\lambda_1, \cdots, \lambda_{M-1}$ are larger than λ_0 when $\kappa < 0$. Therefore, $\lambda_1 + \cdots + \lambda_{M-1} \approx 1$ (so that $d' = M - 1$) when $\kappa \approx -1/(M-1)$. Similar discussion is possible for M-ary PSK signals.

4. CONCLUSION

In this paper, we computed maximum mutual information for 3PSK and 3-ary real-symmetric signals. Then we discussed a condition for achieving maximum mutual information for M-ary linearly independent signals. As a result, the number of signals $N(< M)$ achieving maximum mutual information and the approximate dimension d' of the signal space are related. As P. Shor pointed out in Ref. [4], mutual information has interesting behavior when the signals are *almost linearly dependent*. Developing the above discussion, we mention in Ref. [5] the condition for achieving higher superadditivity in classical capacity [6, 7].

Acknowledgments

This research is supported in part by Grant-in-Aid for Encouragement of Young Scientists (No.11750319) from Japan Society for the Promotion of Science.

References

[1] M. Ban, K. Kurokawa, and O. Hirota, J. Opt. B, **1**, 206, 1999.

[2] M. Osaki, M. Ban, and O. Hirota, QCM&C98.

[3] M. Osaki, QCM&C-Y2K.

[4] P. Shor, QCM&C-Y2K, LANL quant-ph/0009077.

[5] S. Usami, et. al., Proc. of ISITA2000, in press.

[6] M. Sasaki, K. Kato, M. Izutsu, and O. Hirota, Phys. Rev. A **58**, 146 (1998).

[7] J. R. Buck, S. J. van Enk, and C. A. Fuchs, Phys. Rev. A **61**, 032309 (2000).

PERFORMANCES OF BINARY BLOCK CODES USED ON BINARY CLASSICAL-QUANTUM CHANNELS

Binary block codes attaining the expurgated and cutoff rate lower bound on the error exponent

Pawel Wocjan, Dejan E. Lazic, Thomas Beth
Institut für Algorithmen und Kognitive Systeme
Am Fasanengarten 3a, D–76 131 Karlsruhe, Germany
wocjan@ira.uka.de

Keywords: Classical-quantum channels, binary block codes, error exponent

Abstract A conceptually simple method for derivation of lower bounds on the error exponent of specific families of block codes used on classical-quantum channels with arbitrary signal states over a finite dimensional Hilbert space is presented. It is shown that families of binary block codes with appropriately rescaled binomial multiplicity enumerators used on binary classical-quantum channels and decoded by the suboptimal decision rule introduced by Holevo attain the expurgated and cutoff rate lower bounds.

Let C denote a classical-quantum (c-q) channel [1] with the finite input alphabet A $= \{a_1, a_2, \ldots, a_q\}$ and the corresponding channel output signal states $S_1, S_2, \ldots, S_q$ given by density matrices over a finite dimensional Hilbert space. A codeword $\mathbf{x}_m = (x_{m1}, x_{m2}, \ldots, x_{mN})$, $x_{mn} \in$ A, $m = 1, \ldots, M$, transmitted over C induces the corresponding product state $\mathbf{S}_m = S_{m1} \otimes S_{m2} \otimes \ldots \otimes S_{mN}$ as channel output. A *block code* B of length N and size $M \leq q^N$ is a collection $\{\mathbf{x}_1, \mathbf{x}_2, \ldots, \mathbf{x}_M\}$ of M codewords from A^N, where $R = \log_2(M)/N$ is the *code rate*. The quantum decoder for B used on C is statistically characterized by a *quantum decision rule* D $= (\mathbf{D}_1, \mathbf{D}_2, \ldots, \mathbf{D}_M)$ which is a collection of M *decision operators* satisfying $\sum_{m=1}^{M} \mathbf{D}_m \leq I$. The *error effect* $e_m(\mathbf{x}_j)$ is the conditional probability that the quantum decoder makes the decision in favor of the codeword $\mathbf{x}_j$ when the codeword $\mathbf{x}_m$ ($m \neq j$) is transmitted over C and is given by $e_m(\mathbf{x}_j) = \mathrm{Tr}(\mathbf{S}_m \mathbf{D}_j)$. The *word error probability* is given by $P_{em} = P[\hat{\mathbf{x}} \neq \mathbf{x}_m \mid \mathbf{x}_m] = \sum_{j \neq m} e_m(\mathbf{x}_j)$ where $\hat{\mathbf{x}}$ is the decision made by the rule D. The *overall block decoding error probability* is given by $P_e = \frac{1}{M}\sum_{m=1}^{M} P_{em}$. Once a block code B has been chosen, the *optimal decision rule* D_{opt} minimizes P_e for this code when used on C. For a given length N and fixed rate R there exists at least one *optimal block code* B_{opt} producing the minimal overall block decoding error probability P_{eopt} on the channel C, if decoded by the corresponding optimal decision rule D_{opt}. Asymptotic reliability performances of C are characterized by the *error exponent* given by $E(R) = \lim_{N\to\infty} \sup\{-\log_2[P_{\mathrm{eopt}}(R,N)/N]\}$, so that $P_e \geq P_{\mathrm{eopt}}(R,N) = 2^{-N\cdot E(R)+o(N)}$. Lower bounds on the error exponent are usually obtained by the *ensemble averaging*, or *random coding*, technique which consists of calculating the average overall block decoding error probability over the ensemble of all codes that are possible over the encoding space. However, as shown in [2] for classical channels, it is possible to define error exponents and capacity measures for specific code families. These methods are now generalized for c-q channels.

The binary c-q channel C with input alphabet $\{0,1\}$ and signal states S_1 and S_2 is considered. Only binary codes $[N,R]$ (of length N and rate $R = \log_2(M)/N \leq 1$) can be used on C. The following lemma can be proved for Holevo's suboptimal decision rule [1] $\mathrm{D_H}$ whose decision operators are given by $\mathbf{D}_m = \Phi^{-1/2}\,\mathbf{S}_m^{1/2}\,\Phi^{-1/2}$ where $\Phi = \sum_{j=1}^{M} \mathbf{S}_j^{1/2}$.

Lemma 9.1 (Union bound) *The overall block decoding error probability P_e of $[N,R]$ used on C and decoded by the suboptimal decision rule $\mathrm{D_H}$ is upper-bounded by the union bound*

$$P_{\mathrm{e}} \leq \tilde{P}_{\mathrm{e}} = \frac{1}{M}\sum_{m=1}^{M}\sum_{\substack{j=1\\ j\neq m}}^{M} 2^{-N[-\underline{\mathrm{d}}_{\mathrm{H}}(\mathbf{x}_m,\mathbf{x}_j)\log_2(c)]}$$

where $c = \mathrm{Tr}\sqrt{S_1}\sqrt{S_2}$ and $\underline{\mathrm{d}}_{\mathrm{H}}(\mathbf{x}_m,\mathbf{x}_j) = \mathrm{d_H}(\mathbf{x}_m,\mathbf{x}_j)/N$ is the normalized Hamming distance between the codewords $\mathbf{x}_m$ and $\mathbf{x}_j$.

The union bound depends on Hamming distances among the codewords of $[N,R]$ and the channel parameter c. This permits to use a new method, derived from [2], in order to estimate the error exponent and capacity performances of specific code families used on the c-q channel C. For this purpose some necessary definitions and terms are introduced.

The *multiplicity enumerators* of a codeword $\mathbf{x}_m$ of the binary linear or non-linear code $[N,R]$ are given by $\mathbf{M}_{\mathrm{H}}(\mathbf{x}_m) = (M_{m0}, M_{m1}, \ldots, M_{mN})$ where M_{mn} represents the *multiplicity* (number) of codewords on the Hamming distance n from $\mathbf{x}_m$. M_{m0} is by definition equal to one and $\sum_{n=0}^{N} M_{mn} = M$. The *average multiplicity enumerators* of $[N,R]$ are given by $\underline{\mathbf{M}}_{\mathrm{H}}[N,R] = (\underline{M}_0, \underline{M}_1, \ldots, \underline{M}_N)$ satisfying $\sum_{n=0}^{N} \underline{M}_n = M$ where $\underline{M}_n = \frac{1}{M}\sum_{m=1}^{M} M_{mn}$ is the average multiplicity of codewords on the Hamming distance n. The *weight enumerators* of $[N,R]$ are given by $\mathbf{W}_{\mathrm{H}}[N,R] = (A_0, A_1, \ldots, A_N)$ where A_n is the number of codewords of Hamming weight equal to n. A_0 and A_N can take the values 0 or 1 only and $\sum_{n=0}^{N} A_n = M$. For linear codes multiplicity enumerators of all codewords are equal and coincide with the corresponding weight enumerators. For the non-redundant binary linear code of rate $R = 1$ ($M = 2^N$) the weight enumerators (and the multiplicity enumerators) are given by the corresponding binomial coefficients $\binom{N}{n}$, $n = 0, \ldots, N$. Since $\sum_{n=0}^{N} \binom{N}{n} = 2^N$, it is obvious that there is no redundant binary block code $[N,R]$ ($R < 1$) with average multiplicity enumerators or weight enumerators given by all binomial coefficients $\binom{N}{n}$, $n = 0, \ldots, N$. However, appropriately *rescaled binomial coefficients* $b\binom{N}{n}$ with $b = 2^{N(R-1)}$ satisfy the necessary condition $\sum_{n=0}^{N} b\binom{N}{n} = 2^{RN} = M$ for average multiplicity enumerators of redundant block codes.

For an infinite family of block codes *asymptotic performances* can be considered if the length of its members tends to infinity. This is the case for *fixed rate sequences* $\mathrm{FRS}(R) = \{[N_i, R]\}_{i=1}^{\infty}$ of block codes with the same rate $R = \log_2(M_i)/N_i$ and increasing length N_i. In the asymptotic analysis it is convenient to express the average multiplicity enumerators of $[N_i, R]$ by their *average multiplicity exponents* $\mathrm{M}_{\mathrm{H}}[N_i, R] = (\underline{\mathrm{M}}_0^{(i)}, \underline{\mathrm{M}}_1^{(i)}, \ldots, \underline{\mathrm{M}}_{N_i}^{(i)})$ where $\underline{\mathrm{M}}_n^{(i)} = \frac{1}{N_i}\log_2(\underline{M}_n^{(i)})$ so that $\sum_{n=0}^{N_i} 2^{N_i \underline{\mathrm{M}}_n^{(i)}} = M_i$. The corresponding asymptotic values (if they exist) form an infinite sequence representing the *asymptotic average multiplicity exponents* (AAME), which are given by $\mathrm{M}_{\mathrm{FRS}} = (\underline{\mathrm{M}}_0, \underline{\mathrm{M}}_1, \ldots, \underline{\mathrm{M}}_n, \ldots)$, where $\underline{\mathrm{M}}_n = \lim_{i\to\infty} \frac{1}{N_i}\log_2(\underline{M}_n^{(i)})$. All normalized Hamming distances $\underline{\mathrm{d}}_{\mathrm{H}} = n/N_i$ are in the interval $[0,1]$ and the asymptotic values, given by $\underline{\delta}_{\mathrm{H}} = \lim_{i\to\infty} n/N_i$, are dense enough. Consequently, the AAME can be replaced by a continuous function $\mathrm{M}_{\mathrm{FSR}}(\underline{\delta}_{\mathrm{H}})$ of the argument $\underline{\delta}_{\mathrm{H}}$ interpolating the discrete values of $\mathrm{M}_{\mathrm{FRS}}$. This function is called the *interpolated asymptotic average multiplicity exponent* (IAAME) of FRS(R).

Lemma 9.2 *If the average multiplicity enumerators of the codes of* $\mathrm{FRS}(R)$ *are given by rescaled binomial coefficients, i.e.*

$$\left\{\underline{M}_n^{(i)} = \frac{o(2^n)}{2^{N_i(1-R)}}\binom{N_i}{n}\right\}_{n=0}^{N_i}, \quad \sum_{n=0}^{N_i} \underline{M}_n^{(i)} = M_i\,,$$

than the IAAME is given by $\mathrm{M}_{\mathrm{FRS}}(\underline{\delta}_{\mathrm{H}}) = \mathrm{H}_2(\underline{\delta}_{\mathrm{H}}) - (1-R)$, *where* $\mathrm{H}_2(x) = -x\log_2(x) - (1-x)\log_2(1-x)$ *is the binary entropy function.*

Theorem 1 (Cutoff rate lower bound) *FRS with rescaled binomial average multiplicity enumerators used on the binary c-q channel* C *and decoded by the suboptimal decision rule* D_{H} *have an error exponent whose lower bound is the* *cutoff rate bound* $\underset{\sim}{E}_{cut}(R) = R - R_0$ *where* $R_0 = 1 - \log_2(1+c)$ *is the cutoff rate.*

In the proof of Theorem 1 one sees that *only* codewords on the effective Hamming distance $\underline{\delta}_{\mathrm{Heff}} = c/(1+c)$ determine the cutoff rate bound. This distance does not depend on the rate and is generally different from the asymptotic *minimal Hamming distance* of the FSR. The *Gilbert-Varshamov lower bound* (GVB) on the asymptotic minimal normalized distance of binary block codes $[N, R]$ is given by $\underline{\delta}_{\mathrm{GV}}(R) = \mathrm{H}_2^{-1}(1-R)$, where H_2^{-1} denotes the inverse of the binary entropy function in $[0, \frac{1}{2}]$. Note that all FRS with IAAMEs having negative values can be expurgated: all codewords corresponding to the negative values can be removed without changing the rate in the asymptotic case. Expurgated FSR(R) with rescaled binomial multiplicity enumerators satisfy the GVB and their *expurgated IAAME* is given by $\mathrm{M_{FSR}}(\underline{\delta}_{\mathrm{H}}) = -\infty$ for $\underline{\delta}_{\mathrm{H}} \leq \underline{\delta}_{\mathrm{GV}}$ and $\mathrm{H}_2(\underline{\delta}_{\mathrm{H}}) - (1-R)$ otherwise.

Lemma 9.3 *Fixed rate sequences of linear block codes with rescaled binomial multiplicity enumerators satisfy the GVB.*

Theorem 2 (Expurgated lower bound) *Fixed rate sequences with rescaled binomial multiplicity enumerators satisfying the GVB used on the binary c-q channel* C *and decoded by the suboptimal decision rule* $\mathrm{D_H}$ *have an error exponent whose lower bound is the expurgated one* $\underset{\sim}{E}_{\mathrm{ex}}(R) = -\log_2(c)\,\underline{\delta}_{\mathrm{GV}}(R)$ *for rates below* $R_{\mathrm{ex}} = 1 - \mathrm{H}_2\left(\underline{\delta}_{\mathrm{Heff}}\right)$ *and the cutoff rate bound above* R_{ex}.

This representation $\underset{\sim}{E}_{\mathrm{ex}}(R) = -\log_2(c)\,\underline{\delta}_{\mathrm{GV}}(R)$ is the quantum analog of the *Gilbert-Varshamov Bhattacharyya distance* [2, 3].

References

[1] A. S. Holevo, *Reliability function of general classical-quantum channel*, LANL preprint no. quant-ph/9907087.

[2] D. E. Lazic, V. Senk, *A Direct Geometrical Method for Bounding the Error Exponent for Any Specific Family of Channel Codes - Part I: Cutoff Rate Lower Bound for Block Codes*, IEEE Trans. Inform. Th., Vol. 38, No. 4, pp. 1548–1559, September 1992.

[3] J. K. Omura, *On general Gilbert bounds*, IEEE Trans. Inform. Th., 15, pp. 661–665, 1973.

II

QUANTUM MEASUREMENT, DECOHERENCE, AND TOMOGRAPHY

ENTROPY AND INFORMATION GAIN IN QUANTUM CONTINUAL MEASUREMENTS

A. Barchielli
Politecnico di Milano, Dipartimento di Matematica,
Piazza Leonardo da Vinci 32, I-20133 Milano, Italy;
also Istituto Nazionale di Fisica Nucleare, Sezione di Milano.
barchielli@mate.polimi.it

1. INTRODUCTION

The theory of measurements continuous in time in quantum mechanics(the theory of quantum continual measurements) has been formulated by using the notions of instrument, positive operator valued (POV) measure, etc. [1, 2], by using quantum stochastic differential equations [3, 4] and by using classical stochastic differential equations (SDE's) for vectors in Hilbert spaces or for trace-class operators [5, 6, 7, 8]. In the same times Ozawa made developments in the theory of instruments [9, 10] and introduced the related notions of *a posteriori* states [11] and of information gain [12].

In Section 2 we introduce a simple class of SDE's relevant to the theory of continual measurements and we recall how they are related to instruments and *a posteriori* states and, so, to the general formulation of quantum mechanics [13]. In Section 3 we shall introduce and use the notion of information gain and the other results of paper [12] inside the theory of continual measurements.

2. STOCHASTIC DIFFERENTIAL EQUATIONS AND INSTRUMENTS

Let $\mathcal{H}$ be a separable complex Hilbert space, associated to the quantum system of interest. Let us denote by $\mathcal{B}(\mathcal{H})$ the space of bounded linear operators on $\mathcal{H}$ and by $\mathcal{T}(\mathcal{H})$ the trace-class on $\mathcal{H}$, i.e. $\mathcal{T}(\mathcal{H}) = \{\rho \in \mathcal{B}(\mathcal{H}) : \|\rho\| \equiv \operatorname{Tr}\{\sqrt{\rho^*\rho}\} < \infty\}$. Let $\mathcal{S}(\mathcal{H}) \subset \mathcal{T}(\mathcal{H})$ be the set of all statistical operators (*states*) on $\mathcal{H}$. Commutators and anticommutators are denoted by $[\ ,\ ]$ and $\{\ ,\ \}$, respectively.

Let H, L_j, S_h, $j, h = 1, 2, \ldots$, be bounded operators on $\mathcal{H}$ such that $H = H^\dagger$, $\sum_{j=1}^{\infty} L_j^\dagger L_j$ and $\sum_{h=1}^{\infty} S_h^\dagger S_h$ are strongly convergent in $\mathcal{B}(\mathcal{H})$. Let J_k be

Quantum Communication, Computing, and Measurement 3
Edited by P. Tombesi and O. Hirota, Kluwer Academic/Plenum Publishers, 2001

a bounded linear map on $\mathcal{T}(\mathcal{H})$ such that its adjoint J_k^* is a normal, completely positive map on $\mathcal{B}(\mathcal{H})$ and $\sum_{k=1}^{\infty} J_k^*[\mathbb{1}]$ is strongly convergent to a bounded operator. Then, we introduce the following operators on $\mathcal{T}(\mathcal{H})$:

$$\begin{aligned} \mathcal{L}_0[\rho] &= -\mathrm{i}[H,\rho] + \sum_{j=1}^{\infty}\left(L_j\rho L_j^\dagger - \frac{1}{2}\left\{L_j^\dagger L_j, \rho\right\}\right) \\ &\quad + \sum_{k=1}^{\infty}\left(J_k[\rho] - \frac{1}{2}\left\{J_k^*[\mathbb{1}], \rho\right\}\right), \qquad (1) \\ \mathcal{L}_1[\rho] &= \sum_{h=1}^{\infty}\left(S_h\rho S_h^\dagger - \frac{1}{2}\left\{S_h^\dagger S_h, \rho\right\}\right), \qquad (2) \\ \mathcal{L} &= \mathcal{L}_0 + \mathcal{L}_1. \qquad (3) \end{aligned}$$

The adjoint operators of $\mathcal{L}$, $\mathcal{L}_0$, $\mathcal{L}_1$ are generators of norm-continuous quantum dynamical semigroups [14, 15].

Let us now consider the following linear SDE (in the sense of Itô) for trace-class operators:

$$\begin{aligned} \mathrm{d}\sigma_t &= \mathcal{L}[\sigma_{t^-}]\,\mathrm{d}t + \sum_{j=1}^{\infty}\left(\widetilde{L}_j(t)\sigma_{t^-} + \sigma_{t^-}\widetilde{L}_j(t)^\dagger\right)\mathrm{d}\widetilde{W}_j(t) + \\ &\quad + \sum_{k=1}^{\infty}\left(\frac{1}{\lambda_k}J_k[\sigma_{t^-}] - \sigma_{t^-}\right)(\mathrm{d}N_k(t) - \lambda_k\,\mathrm{d}t); \qquad (4) \end{aligned}$$

the initial condition is $\sigma_0 = \rho \in \mathcal{S}(\mathcal{H})$ (a non-random state) and we have set

$$\widetilde{L}_j(t) = \mathrm{e}^{\mathrm{i}\omega_j t}L_j\,, \qquad \omega_j \in \mathbb{R}. \qquad (5)$$

The processes $\widetilde{W}_j(t)$ are independent standard Wiener processes, the $N_k(t)$ are independent Poisson processes of intensity $\lambda_k > 0$, which are also independent of the Wiener processes; we assume $\sum_k \lambda_k < +\infty$.

These processes are realized in a probability space $(\Omega, \mathcal{F}, Q)$; the sample space Ω is, roughly speaking, the set of possible trajectories for the processes $\widetilde{W}_j$, N_k, the event space $\mathcal{F}$ is the σ-algebra of sets of trajectories to which a probability can be given and Q is the probability law under which $\widetilde{W}_j$, N_k are independent Wiener and Poisson processes. Moreover, let $\mathcal{F}_t$ be the collection of events which are specified by giving conditions involving times only in the interval $[0,t]$. We also ask $\mathcal{F} = \mathcal{F}_\infty$. In mathematical terms the $\widetilde{W}_j$, N_k are canonical Wiener and Poisson processes, $\{\mathcal{F}_t,\ t \geq 0\}$ is their natural filtration and $\mathcal{F} = \bigvee_{t\geq 0}\mathcal{F}_t$. Finally, let us denote by $\mathbb{E}_Q$ the expectation with respect to the probability Q, i.e. $\mathbb{E}_Q[A] = \int_\Omega A(\omega)Q(\mathrm{d}\omega)$.

For every $F \in \mathcal{F}_t$ and every initial condition $\rho \in \mathcal{S}(\mathcal{H})$, let us set

$$\mathcal{I}_t(F)[\rho] = \mathbb{E}_Q[1_F \sigma_t] \equiv \int_F \sigma_t(\omega) Q(\mathrm{d}\omega); \tag{6}$$

1_F is the indicator function of the set F, i.e. $1_F(\omega) = 1$ if $\omega \in F$ and $1_F(\omega) = 0$ if $\omega \notin F$. The map $\mathcal{I}_t$ turns out to be a (completely positive) instrument [9] with value space $(\Omega, \mathcal{F}_t)$ and $\mathcal{I}_t(\cdot)^*[\mathbb{1}]$ is the associated POV measure. Then we set, $\forall F \in \mathcal{F}_t$,

$$P_\rho(F) = \mathrm{Tr}\left\{\mathcal{I}_t(F)^*[\mathbb{1}]\rho\right\} = \mathbb{E}_Q\left[\|\sigma_t\|\, 1_F\right]. \tag{7}$$

The important point in this formula is that $\|\sigma_t\|$ is a Q-martingale and this implies that the time dependent probability measures on the r.h.s. are consistent and define a unique probability P_ρ on $(\Omega, \mathcal{F})$.

The interpretation of eqs. (6) and (7) is that $\{\mathcal{I}_t,\, t \geq 0\}$ is the family of instruments describing the continual measurement, the processes $\widetilde{W}_j$, N_k represent the output of this measurement and P_ρ is the physical probability law of the output.

From eq. (6) it follows that

$$\eta_t = \mathcal{I}_t(\Omega)[\rho] = \mathbb{E}_Q[\sigma_t] \tag{8}$$

is the state to be attributed to the system at time t if the output of the measurement is not taken into account or not known; it can be called the *a priori* state at time t. It turns out that the *a priori* states satisfy the master equation

$$\frac{\mathrm{d}}{\mathrm{d}t}\eta_t = \mathcal{L}[\eta_t], \qquad \eta_0 = \rho. \tag{9}$$

If we introduce the random states

$$\rho_t = \frac{\sigma_t}{\|\sigma_t\|}, \tag{10}$$

then we have, $\forall F \in \mathcal{F}_t$,

$$\mathcal{I}_t(F)[\rho] = \mathbb{E}_Q[1_F \sigma_t] = \mathbb{E}_{P_\rho}\left[1_F \frac{\sigma_t}{\|\sigma_t\|}\right] = \int_F \rho_t(\omega) P_\rho(\mathrm{d}\omega). \tag{11}$$

According to [11], $\rho_t(\omega)$ is a family of *a posteriori* states for the instrument $\mathcal{I}_t$ and the initial state ρ, i.e. $\rho_t(\omega)$ is the state to be attributed to the system at time t when the trajectory ω of the output is known, up to time t. Note that $\eta_t = \mathbb{E}_Q[\sigma_t] = \mathbb{E}_{P_\rho}[\rho_t]$.

By using Itô's calculus, we find that the *a posteriori* states satisfy the non-linear SDE

$$\begin{aligned} \mathrm{d}\rho_t &= \mathcal{L}[\rho_{t^-}]\,\mathrm{d}t + \sum_{j=1}^{\infty}\left[\widetilde{L}_j(t)\rho_{t^-} + \rho_{t^-}\widetilde{L}_j(t)^\dagger - m_j(t)\rho_{t^-}\right]\mathrm{d}W_j(t) + \\ &\quad + \sum_{k=1}^{\infty}\left[\frac{1}{\nu_k(t)}\mathcal{J}_k[\rho_{t^-}] - \rho_{t^-}\right](\mathrm{d}N_k(t) - \nu_k(t)\,\mathrm{d}t), \end{aligned} \tag{12}$$

where

$$W_j(t) = \widetilde{W}_j(t) - \int_0^t m_j(s)\,\mathrm{d}s\,, \tag{13}$$

$$m_j(t) = \mathrm{Tr}\left\{\rho_{t^-}\left(\widetilde{L}_j(t) + \widetilde{L}_j(t)^\dagger\right)\right\}, \qquad \nu_k(t) = \mathrm{Tr}\left\{\rho_{t^-}\mathcal{J}_k^*[\mathbb{1}]\right\}. \tag{14}$$

Under the physical probability law P_ρ, the processes $W_j(t)$ are independent standard Wiener processes and the $N_k(t)$ are counting processes with stochastic intensity $\nu_k(t)$. In eq. (12) the sum in the jump term is only on the set where the stochastic intensity $\nu_k(t)$ is different from zero.

Formulae for the moments of the output can be obtained by the technique of the characteristic operator [2, 3, 4]. Let $h_{k\alpha}$ be real test functions in a suitable space; we define the characteristic operator $\mathcal{G}$ by

$$\begin{aligned} \mathcal{G}_t(h)[\rho] &= \mathbb{E}_{P_\rho}\Bigg[\exp\Bigg\{\mathrm{i}\sum_j\int_0^t h_{j1}(s)\,\mathrm{d}\widetilde{W}_j(s) \\ &\quad + \mathrm{i}\sum_k\int_0^t h_{k2}(s)\,\mathrm{d}N_k(s)\Bigg\}\rho_t\Bigg]\,; \end{aligned} \tag{15}$$

then, $\mathrm{Tr}\{\mathcal{G}_t(h)[\rho]\}$ is the characteristic functional of the output up to time t (the Fourier transform of P_ρ restricted to $\mathcal{F}_t$). By Itô's calculus we obtain

$$\frac{\mathrm{d}}{\mathrm{d}t}\mathcal{G}_t(h)[\rho] = \mathcal{K}_t(h)\circ\mathcal{G}_t(h)[\rho]\,, \tag{16}$$

$$\begin{aligned} \mathcal{K}_t(h)[\rho] &= \mathcal{L}[\rho] + \mathrm{i}\sum_j h_{j1}(t)\left[\widetilde{L}_j(t)\rho + \rho\widetilde{L}_j(t)^\dagger\right] \\ &\quad -\frac{1}{2}\sum_j h_{j1}(t)^2\rho + \sum_k\{\exp[\mathrm{i}h_{k2}(t)] - 1\}\,\mathcal{J}_k[\rho]\,. \end{aligned} \tag{17}$$

All the moments can be obtained by functional differentiation of the characteristic functional. In particular, the mean values are expressed in terms of the *a priori* states as

$$\mathbb{E}_{P_\rho}\left[\widetilde{W}_j(t)\right] = \int_0^t \mathbb{E}_{P_\rho}\left[m_j(s)\right] \mathrm{d}s\,, \quad \mathbb{E}_{P_\rho}\left[N_k(t)\right] = \int_0^t \mathbb{E}_{P_\rho}\left[\nu_k(s)\right] \mathrm{d}s\,,$$

$$\mathbb{E}_{P_\rho}\left[m_j(s)\right] = \mathrm{Tr}\left\{\eta_s\left(\widetilde{L}_j(s) + \widetilde{L}_j(s)^\dagger\right)\right\}, \quad \mathbb{E}_{P_\rho}\left[\nu_k(s)\right] = \mathrm{Tr}\left\{J_k[\eta_s]\right\},$$

and the second moments are given by

$$\mathbb{E}_{P_\rho}[X_{j\alpha}(t)X_{i\beta}(s)] = \delta_{ij}\delta_{\alpha\beta}\int_0^{\min\{t,s\}} \mathrm{d}\tau\left(\delta_{\alpha 1} + \delta_{\alpha 2}\,\mathrm{Tr}\left\{J_i[\eta_\tau]\right\}\right)$$

$$+ \int_0^t \mathrm{d}\tau_1 \int_0^{\min\{s,\tau_1\}} \mathrm{d}\tau_2\,\mathrm{Tr}\left\{\mathcal{A}_{j\alpha}(\tau_1) \circ \mathrm{e}^{\mathcal{L}(\tau_1-\tau_2)} \circ \mathcal{A}_{i\beta}(\tau_2)[\eta_{\tau_2}]\right\}$$

$$+ \int_0^s \mathrm{d}\tau_2 \int_0^{\min\{t,\tau_2\}} \mathrm{d}\tau_1\,\mathrm{Tr}\left\{\mathcal{A}_{i\beta}(\tau_2) \circ \mathrm{e}^{\mathcal{L}(\tau_2-\tau_1)} \circ \mathcal{A}_{j\alpha}(\tau_1)[\eta_{\tau_1}]\right\},$$

where $X_{j1}(t) = \widetilde{W}_j(t)$, $X_{j2}(t) = N_j(t)$, $\mathcal{A}_{j1}(t)[\rho] = \widetilde{L}_j(t)\rho + \rho\widetilde{L}_j(t)^\dagger$, $\mathcal{A}_{j2}(t) = J_j$.

The class of SDE's presented here is a particular case of the one studied in [16] and, while not so general, it contains the main detection schemes found in quantum optics [17]; also the chosen time-dependence is natural for some systems typical of quantum optics under the so called heterodyne/homodyne detection scheme.

3. ENTROPY AND INFORMATION GAIN

In [12] a measurement is called quasi-complete if the *a posteriori* states are pure for every pure initial state and it is called complete if the *a posteriori* states are pure for every (pure or mixed) initial state. So, we call *quasi-complete* the continual measurement of Section 2 if the *a posteriori* states ρ_t are *pure* (P_ρ-almost surely) *for all t and for all pure initial conditions* ρ. In [18] we proved that

Theorem 1 *The continual measurement of Section 2 is quasi-complete if and only if* $\mathcal{L}_1 = 0$ *and* $\dfrac{J_k[\rho]}{\mathrm{Tr}\left\{J_k[\rho]\right\}}$ *is a pure state for every* k *and for every pure state* ρ. *In this case there exists a partition* A_1, A_2 *of the integer numbers such that for some* $R_k \in \mathcal{B}(\mathcal{H})$ *and for some monodimensional projection* P_k *we can write* $J_k[\rho] = R_k\rho R_k^\dagger$, *for* $k \in A_1$, $J_k[\rho] = \mathrm{Tr}\left\{\rho J_k^*[\mathbb{1}]\right\} P_k$, *for* $k \in A_2$.

Our continual measurement can not be complete in the sense of [12] for a fixed time; however, it can be "asymptotically complete". Examples of this behaviour in the case of linear systems are given in [19]. In [18], we proved that

Theorem 2 *Let the continual measurement of Section 2 be quasi-complete and let $\mathcal{H}$ be finite-dimensional. If for every time t it does not exist a bidimensional projection P_t such that, $\forall j,k$, $P_t\left(\widetilde{L}_j(t)+\widetilde{L}_j(t)^\dagger\right)P_t = z_j(t)P_t$, $P_t J_k^*[\mathbb{1}]P_t = q_k(t)P_t$ for some complex numbers $z_j(t)$ and $q_k(t)$, then eq. (12) maps asymptotically, for $t\to\infty$, mixed states into pure ones, in the sense that, for every initial condition ρ, P_ρ-almost surely* $\lim_{t\to\infty}\operatorname{Tr}\{\rho_t(\mathbb{1}-\rho_t)\}=0$.

The proof of the theorems above is based on the study of the *a posteriori linear entropy* (or *purity*) $\operatorname{Tr}\{\rho_t(\mathbb{1}-\rho_t)\}$ and of its mean value. However, physically more interesting quantities are the von Neumann entropy and the relative entropy : for $x,y\in\mathcal{S}(\mathcal{H})$, $S[x]=-\operatorname{Tr}\{x\ln x\}\geq 0$, $S[x|y]=\operatorname{Tr}\{x\ln x - x\ln y\}\geq 0$ (they can also diverge) [15]. In our case we have the initial state $\rho=\rho_0=\sigma_0=\eta_0$ and the *initial entropy* $S[\rho]$, the *a priori* state η_t and the *a priori entropy* $S[\eta_t]$, the *a posteriori* states ρ_t and the *mean a posteriori entropy*

$$\mathbb{E}_{P_\rho}[S[\rho_t]] = \mathbb{E}_Q[\|\sigma_t\|\ln\|\sigma_t\| - \operatorname{Tr}\{\sigma_t\ln\sigma_t\}]\,. \tag{18}$$

By some direct computations, we obtain a first relation among these quantities:

$$S[\eta_t] - \mathbb{E}_{P_\rho}[S[\rho_t]] = \mathbb{E}_{P_\rho}[S[\rho_t|\eta_t]] \geq 0\,. \tag{19}$$

Following [12], we can also introduce the *amount of information* of the continual measurement

$$I[\rho;t] = S[\rho] - \mathbb{E}_{P_\rho}[S[\rho_t]] \tag{20}$$

and the classical amount of information. To introduce this last quantity we need some notations. Let us set $P_\rho(\mathrm{d}\omega;t)=\|\sigma_t(\omega)\|Q(\mathrm{d}\omega)$, let $\rho=\sum_\alpha w_\alpha\rho_\alpha$ be the orthogonal decomposition of ρ into pure states and P_{ρ_α}, σ_t^α, ρ_t^α, η_t^α, $m_j^\alpha(t)$, $\nu_k^\alpha(t)$ be defined starting from ρ_α as P_ρ, σ_t, ρ_t, η_t, $m_j(t)$, $\nu_k(t)$ are defined starting from ρ. Then, the *classical amount of information* of the continual measurement is defined by

$$\begin{aligned}
\text{c-}I[\rho;t] &= \sum_\alpha w_\alpha\int_\Omega \ln\left(\frac{P_{\rho_\alpha}(\mathrm{d}\omega;t)}{P_\rho(\mathrm{d}\omega;t)}\right)P_{\rho_\alpha}(\mathrm{d}\omega;t)\\
&= \sum_\alpha w_\alpha\mathbb{E}_{P_{\rho_\alpha}}\left[\ln\frac{\|\sigma_t^\alpha\|}{\|\sigma_t\|}\right]\\
&= \mathbb{E}_Q\left[\sum_\alpha w_\alpha\|\sigma_t^\alpha\|\ln\|\sigma_t^\alpha\| - \|\sigma_t\|\ln\|\sigma_t\|\right]\,.
\end{aligned} \tag{21}$$

By classical arguments, c-$I[\rho;t]$ is always positive [12]: c-$I[\rho;t] \geq 0, \forall t \geq 0, \forall \rho \in \mathcal{S}(\mathcal{H})$. Obviously, we have $I[\rho;t] \leq S[\rho]$, $I[\rho;0] = 0$, c-$I[\rho;0] = 0$. If it exists an equilibrium state η_{eq} ($\mathcal{L}[\eta_{\text{eq}}] = 0$), by (19) we have also $I[\eta_{\text{eq}};t] \geq 0$.

Theorem 3 *The classical amount of information of the continual measurement of Section 2 is non-decreasing in time and*

$$\begin{aligned}
\frac{\mathrm{d}}{dt}\,\text{c-}I[\rho;t] &= \sum_\alpha w_\alpha \mathbb{E}_{P_{\rho_\alpha}}\left[\frac{1}{2}\sum_j m_j^\alpha(t)^2 + \sum_k \nu_k^\alpha(t)\ln \nu_k^\alpha(t)\right] \\
&- \mathbb{E}_{P_\rho}\left[\frac{1}{2}\sum_j m_j(t)^2 + \sum_k \nu_k(t)\ln \nu_k(t)\right] \\
&= \sum_\alpha w_\alpha \mathbb{E}_{P_{\rho_\alpha}}\left[\frac{1}{2}\sum_j \left(m_j^\alpha(t) - m_j(t)\right)^2\right. \\
&+ \left.\sum_k \nu_k(t)\left(1 - \frac{\nu_k^\alpha(t)}{\nu_k(t)} + \frac{\nu_k^\alpha(t)}{\nu_k(t)}\ln\frac{\nu_k^\alpha(t)}{\nu_k(t)}\right)\right] \geq 0\,.
\end{aligned} \tag{22}$$

To prove this theorem one has to differentiate the last expression in (21) and to use the relationships among Q, P_ρ, P_{ρ_α}.

For quasi-complete measurements the information gain $I[\rho;t]$ has a nice behaviour.

Theorem 4 *The continual measurement of Section 2 is quasi-complete if and only if the amount of information* $I[\rho;t]$ *is non-negative for any* $\rho \in \mathcal{S}(\mathcal{H})$ *with* $S[\rho] < +\infty$ *and any* $t \geq 0$. *Moreover, if it is quasi-complete, we have* $I[\rho;t] \geq$ c-$I[\rho;t] \geq 0$, $I[\rho;t] \geq I[\rho;s]$ *for any* t, *any* $s < t$ *and any state* ρ *with* $S[\rho] < +\infty$.

Proof. All the statements but the last one are a particularization of Theorems 1 and 2 of [12] to our case. The last statement needs the use of conditional expectations. We have $I[\rho;t] - I[\rho;s] = \mathbb{E}_{P_\rho}[S[\rho_s] - \mathbb{E}_{P_\rho}[S[\rho_t]|\mathcal{F}_s]]$; by (12) $S[\rho_s] - \mathbb{E}_{P_\rho}[S[\rho_t]|\mathcal{F}_s]$ is the amount of information at time t when the initial time is s and the initial state is ρ_s and, so, it is non-negative for a quasi-complete measurement. □

Finally, if $\mathcal{H}$ is finite-dimensional, the vanishing of the purity implies the vanishing of the entropy; therefore, we have the asymptotic completeness also in the sense of the vanishing of the entropy:

> *The hypotheses of Theorem 2 imply also that* $\lim_{t\to+\infty} S[\rho_t] = 0$, P_ρ-*almost surely, and* $\lim_{t\to+\infty} I[\rho;t] = S[\rho]$.

References

[1] E. B. Davies, *Quantum Theory of Open Systems* (Academic Press, London, 1976).

[2] A. Barchielli, L. Lanz, G. M. Prosperi, *Statistics of continuous trajectories in quantum mechanics: Operation valued stochastic processes*, Found. Phys. **13** (1983) 779–812.

[3] A. Barchielli, G. Lupieri, *Quantum stochastic calculus, operation valued stochastic processes and continual measurements in quantum mechanics*, J. Math. Phys. **26** (1985) 2222–2230.

[4] A. Barchielli, *Direct and heterodyne detection and other applications of quantum stochastic calculus to quantum optics*, Quantum Opt. **2** (1990) 423–441.

[5] V. P. Belavkin, *Nondemolition measurements, nonlinear filtering and dynamic programming of quantum stochastic processes*. In A. Blaquière (Ed.), *Modelling and Control of Systems*, Lecture Notes in Control and Information Sciences **121** (Springer, Berlin, 1988) pp. 245–265.

[6] V. P. Belavkin, *A new wave equation for a continuous nondemolition measurement*, Phys. Lett. A **140** (1989) 355–358.

[7] V. P. Belavkin, *A continuous counting observation and posterior quantum dynamics*, J. Phys. A: Math. Gen. **22** (1989) L1109–L1114.

[8] A. Barchielli, V. P. Belavkin, *Measurements continuous in time and a posteriori states in quantum mechanics*, J. Phys. A: Math. Gen. **24** (1991) 1495–1514.

[9] M. Ozawa, *Quantum measuring processes of continuous observables*, J. Math. Phys. **25** (1984) 79–87.

[10] M. Ozawa, *Mathematical characterizations of measurement statistics*. In V. P. Belavkin, O. Hirota, R. L. Hudson (Eds.), *Quantum Communications and Measurement (Nottingham, 1994)* (Plenum, New York, 1995) pp. 109–117.

[11] M. Ozawa, *Conditional probability and a posteriori states in quantum mechanics*, Publ. Res. Inst. Math. Sc. Kyoto Univ. **21** (1985) 279-295.

[12] M. Ozawa, *On information gain by quantum measurements of continuous observables*, J. Math. Phys. **27** (1986) 759–763.

[13] K. Kraus, *States, Effects and Operations*, Lecture Notes in Physics **190** (Springer, Berlin, 1980).

[14] G. Lindblad, *On the generators of quantum dynamical semigroups*, Commun. Math. Phys. **48** (1976) 119–130.

[15] R. Alicki, *General theory and applications to unstable particles*. In: R. Alicki, K. Lendi, *Quantum Dynamical Semigroups and Applications*, Lecture Notes in Physics **286** (Springer, Berlin, 1987) pp. 1–94.

[16] A. Barchielli, A. S. Holevo, *Constructing quantum processes via classical stochastic calculus*, Stochastic Process. Appl. **58** (1995) 293–317.

[17] A. Barchielli, A. M. Paganoni, *Detection theory in quantum optics: stochastic representation*, Quantum Semiclass. Opt. **8** (1996) 133–156.

[18] A. Barchielli, A. M. Paganoni, *On the asymptotic behaviour of some stochastic differential equations for quantum states*, preprint of the Mathematical Department n. 420/P, July 2000.

[19] A. C. Doherty, S. M. Tan, A. S. Parkins, D. F. Walls, *State determination in continuous measurements*, Phys. Rev. A **60** (1999) 2380–2392.

EXPERIMENTAL QUANTUM STATE DISCRIMINATION

Stephen M. Barnett, Roger B. M. Clarke
Department of Physics and Applied Physics, University of Strathclyde
Glasgow, G4 0NG, UK

Vivien M. Kendon
Optics Section, Blackett Laboratory, Imperial College, London, SW7 2BW, UK

Erling Riis
Department of Physics and Applied Physics, University of Strathclyde
Glasgow, G4 0NG, UK

Anthony Chefles
Department of Physical Sciences, University of Hertfordshire
Hatfield, AL10 9AB, UK

Masahide Sasaki
Communications Research Laboratory, Ministry of Posts and Telecommunications,
Koganei, Tokyo 184-8795, Japan

Abstract We present the results of generalized measurements of optical polarization designed to provide one of three or four distinct outcomes. This has allowed us to discriminate between non-orthogonal polarization states with an error probability that is close to the minimum allowed by quantum theory. The sets of states we have prepared and measured are (i) two non-orthogonal states of linear polarisation, (ii) the trine, consisting of three states of linear polarization separated by 60° and, (iii) the tetrad, consisting of four polarization states (two linear and two elliptical) arranged in a tetrahedron on the Poincaré sphere. This has enabled us to realize the generalized measurements required for (i) optimal unambiguous discrimination between two non-orthogonal polarization states, (ii) discrimination between the three trine states and four tetrad states with minimum probability of error and, (iii) maximizing the mutual information associated with our

Quantum Communication, Computing, and Measurement 3
Edited by P. Tombesi and O. Hirota. Kluwer Academic/Plenum Publishers, 2001

trine and tetrad measurements and showing that it exceeds the value assocated with the best possible von Neumann measurement.

1. INTRODUCTION

The basic building block in quantum communications and informationis the qubit, a physical system with two orthogonal quantum states. A simple example is the horizontal and vertical states of polarization associated with a single photon, $|h\rangle$ and $|v\rangle$, from which it is possible to prepare any superposition of $|h\rangle$ and $|v\rangle$ corresponding to other states of linear, circular and elliptical polarization. In quantum communications a transmitting party (Alice) might select from a number of possible non-orthogonal polarization states to send to the receiving party (Bob). This idea is the basis of the emerging technology of quantum key distribution.

The message encoded in the transmitted photons must be retrieved by measurement, either using a conventional (von Neumann) measurement or by means of a generalized measurement [1, 2, 3]. The von Neumann measurement gives answers corresponding to one of a pair of orthogonal polarization states. Generalized measurements give one of three or more possible outcomes corresponding to the elements of a probability operator measure, (POM) [2, 3].

If there are only two possible states then it is known that a single von Neumann measurement with two possible outcomes will minimise the probability for error in identifying the state [2]. State discrimination with this minimum allowed error probability has been demonstrated for weak pulses of polarized light [4]. With only two possible states, it is also possible to discriminate between the two states without error if we allow for a third outcome indicating an inconclusive result, [5]. Near error-free discrimination between two non-orthogonal polarization states was first demonstrated by Huttner et al. [6], but they did not explicitly record their inconclusive results. We have performed an experiment in which such a generalized measurement was applied to two non-orthogonal polarization states and all outcomes were recorded [7]. Our results show that it is possible to discriminate between non-orthogonal polarization states with the probability for an inconclusive outcome at the quantum limit determined by Ivanovic, Dieks and Peres.

Unambiguous discrimination between two non-orthogonal states of a qubit is one of the simplest examples of a generalized measurement. We have also carried out experiments on polarized light prepared in one of three or four non-orthogonal states [8]. We performed two types of optimal measurements on these states. The first gives the minimum probability of error in identifying the state [2, 9, 10, 11]. The second provides the knowledge that the qubit was definitely not prepared in one out of the three or four possibilities, but does not discriminate further between the remaining possibilities [11]. We calculate the

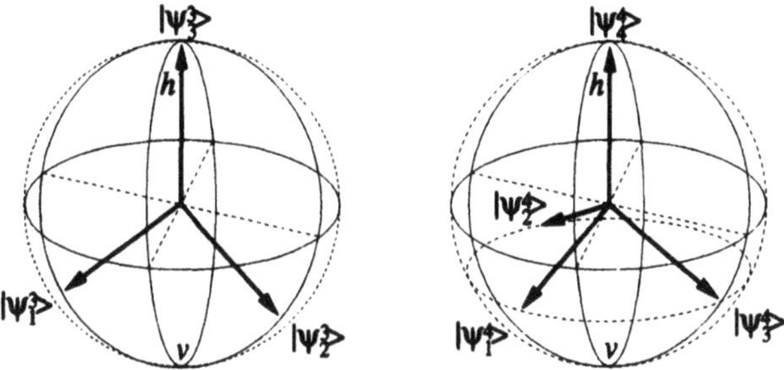

Figure 1 The trine (left) and tetrad (right) states shown on the Poincaré sphere.

mutual information attainable from our experiments and show that it exceeds that which can be obtained by the best von Neumann measurement.

2. GENERALIZED MEASUREMENTS

Our signal states are realizations of either two non-orthogonal states,

$$|\psi_{\pm}\rangle = \cos\alpha|h\rangle \pm \sin\alpha|v\rangle, \tag{1}$$

the trine ensemble [1, 12].

$$\begin{aligned} |\psi_1^3\rangle &= -\frac{1}{2}\left(|h\rangle + \sqrt{3}|v\rangle\right), \\ |\psi_2^3\rangle &= -\frac{1}{2}\left(|h\rangle - \sqrt{3}|v\rangle\right), \\ |\psi_3^3\rangle &= |h\rangle, \end{aligned} \tag{2}$$

corresponding to states of linear polarization separated by 60° (see Fig. 1, left), or tetrad ensemble [13], see Fig. 1 (right),

$$\begin{aligned} |\psi_1^4\rangle &= \frac{1}{\sqrt{3}}\left(-|h\rangle + \sqrt{2}e^{-2\pi i/3}|v\rangle\right), \\ |\psi_2^4\rangle &= \frac{1}{\sqrt{3}}\left(-|h\rangle + \sqrt{2}e^{+2\pi i/3}|v\rangle\right), \\ |\psi_3^4\rangle &= \frac{1}{\sqrt{3}}\left(-|h\rangle + \sqrt{2}|v\rangle\right), \\ |\psi_4^4\rangle &= |h\rangle, \end{aligned} \tag{3}$$

The superscript label denotes the fact that there are three members of the trine ensemble and four members of the tetrad ensemble. We also make use of the antitrine and antitetrad ensembles consisting of the polarization states that are orthogonal to the trine and tetrad states, denoted by an overbar (e.g. $|\bar{\psi}_1^4\rangle$).

The optimum detection strategies require generalized measurements which may be described in terms of the elements of a POM [2]. Each of the possible results (y_j) of the measurement corresponds to a Hermitian operator, Π_j, such that the probability of obtaining this result given that the polarization was prepared in the state $|\psi_k^N\rangle$ is

$$P(y_j|\psi_k^N) = \langle\psi_k^N|\Pi_j|\psi_k^N\rangle, \tag{4}$$

where N is the number of states making up the ensemble. For a von Neumann measurement the POM elements are projectors onto the orthonormal eigenstates of the measured observable, however, the POM elements generally will be neither normalized nor orthogonal.

If the states making up the trine and tetrad ensembles occur with equal prior probabilities then the POM elements that minimise the error probability are just these projectors [10, 14],

$$\Pi_k = \frac{2}{N}|\psi_k^N\rangle\langle\psi_k^N|. \tag{5}$$

This gives for the minimum error probability the value

$$P_e(\text{min}) = 1 - \frac{2}{N}. \tag{6}$$

This is $\frac{2}{3}$ for the trine ensemble and $\frac{1}{2}$ for the tetrad (and easily generalized to $1 - (D/N)$ for D–dimensional quantum systems).

The trine and tetrad ensembles comprise non-orthogonal states and hence errors in determining the state prepared are inevitable. It is possible, however, to achieve asymptotically error-free transmission by employing block coding schemes [15]. The required amount of redundancy associated with these blocks is characterized by the mutual information. If Alice chooses from the set of N states $\{|\psi_k^N\rangle\}$ with prior probabilities $\{p_k\}$ and Bob chooses a measurement with POM elements $\{\Pi_j\}$ and M possible outcomes $\{y_j\}$, then the mutual information is

$$\begin{aligned} I(\{\psi_k^N\} : \{y_j\}) = & \sum_{j=1}^{M}\sum_{k=1}^{N}\Bigg[p_k P(y_j|\psi_k^N) \\ \times & \log_2\left(\frac{P(y_j|\psi_k^N)}{\sum_{k'=1}^{N} p_{k'} P(y_j|\psi_{k'}^N)}\right)\Bigg], \end{aligned} \tag{7}$$

with the conditional probabilities $P(y_j|\psi_k^N)$ given by Eq. (4), and the logarithm in base two so that the information is expressed in bits.

If p_k is fixed then the problem is simplified and the resulting maximum mutual information is known as the accessible information, known exactly for a few special cases [13, 16, 17]. For the trine states with equal prior probabilities, the accessible information is attained for a measurement with POM elements based on the antitrine states

$$\Pi_j = \frac{2}{3}|\bar{\psi}_j^3\rangle\langle\bar{\psi}_j^3|. \tag{8}$$

The optimality of this measurement strategy was first conjectured in 1973 [9, 12] and finally proven by Sasaki et al. [16] in 1999. The corresponding accessible information is $\log_2 \frac{3}{2} = 0.585$ bits. This clearly exceeds the value $(-\frac{1}{3} + \frac{1}{2}\log_2 3) = 0.459$ bits which is the maximum mutual information attainable for a von Neumann measurement corresponding to finding one of two orthogonal polarizations.

For the tetrad ensemble, Davies [13] has conjectured that the accessible information corresponds to the POM elements

$$\Pi_j = \frac{1}{2}|\bar{\psi}_j^4\rangle\langle\bar{\psi}_j^4|. \tag{9}$$

The mutual information associated with this measurement strategy is $\log_2 \frac{4}{3} = 0.415$ bits. This again exceeds the value of $\frac{3}{2}(1 - \frac{1}{2}\log_2 3) = 0.311$ bits attained for the best possible von Neumann measurement.

The measurement strategy based on the antitrine and antitetrad states only assigns the correct state with probability $1/(N-1)$. It does, however, identify for certain one of the states that was not sent. In the antitrine case, this measurement strategy is related to the optimal discrimination between two non-orthogonal states. If the state to be identified is known to be one of only two of the three possible antitrine states, either $|\bar{\psi}_j^3\rangle$ or $|\bar{\psi}_k^3\rangle$, then the three possible outcomes correspond to "not-$|\bar{\psi}_j^3\rangle$" (i.e. "is $|\bar{\psi}_k^3\rangle$"), "not-$|\bar{\psi}_k^3\rangle$", and "inconclusive".

3. OPTICAL IMPLEMENTATION

The generalized measurements described in the previous section have to be implemented in practice as von Neumann measurements with simple yes/no results, but in an enlarged state space, [2, 3, 9, 18]. In our experiments, the state space is enlarged by incorporating an interferometer into the measurement apparatus, thus introducing vacuum modes through the unused input port. A single interferometer allows up to four mutually exclusive (orthogonal) possible results from a single photon input state. Figure 2 shows our experimental arrangement. Details of the experimental apparatus and procedures are given in [7, 8].

Unambiguous discrimination between two non-orthogonal states can be obtained with detectors PD1 and PD2 corresponding to certain identification of input states $|\psi_+\rangle$ and $|\psi_-\rangle$, and PD3 indicating an inconclusive result. When

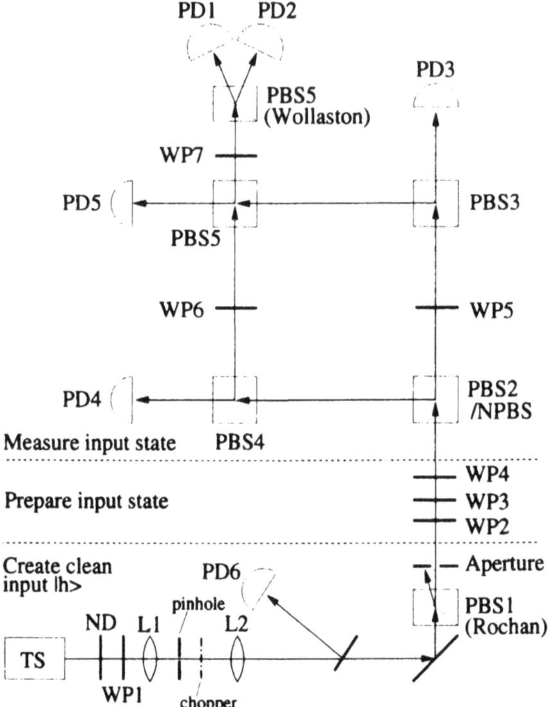

Figure 2 Experimental setup for trine and tetrad state measurements. See text for full description. L = lens, ND = neutral density filter to attenuate the light, NPBS = non-polarizing beamsplitter, PBS = polarizing beam splitter, PD = photodetector, TS = Titanium-Sapphire laser operating at 780nm, WP = waveplate, either half or quarter.

waveplate WP5 is rotated to minimise the signal in PD3, theory predicts that the count rate in PD3 should be proportional to $\cos 2\alpha$, and this was achieved experimentally to within 2.8% for $14° < \alpha < 45°$. For smaller α (nearly identical input states) the error rates rise significantly with the performance of the polarizing beam splitters being the limiting factor.

Exactly the same apparatus, with waveplate WP5 set with its axes at an angle of 17.63° anticlockwise to the horizontal, can also be used to carry out an optimal measurement to discriminate between the three trine states with minimum error. In this case the results do not identify the states for certain. If PD1 detects the photon, there is a probability of $\frac{2}{3}$ that the input state was $|\psi_1^3\rangle$, and a probability of $\frac{1}{6}$ each that it was $|\psi_2^3\rangle$ or $|\psi_3^3\rangle$, similarly for PD2 with states $|\psi_1^3\rangle$ and $|\psi_2^3\rangle$ interchanged, and PD3 identifying state $|\psi_3^3\rangle$.

In order to implement the POM corresponding to the maximum mutual information, instead of constructing the optical network that corresponds to POM elements $|\bar{\psi}_j^3\rangle\langle\bar{\psi}_j^3|$, and applying it to the trine states, we use the optical net-

States	Experiment	Ideal Theory	Noisy Theory	von Neumann
trine	$0.312\ ^{+0.017}_{-0.08}$	0.333	0.302	0.459
antitrine	$0.491\ ^{+0.011}_{-0.027}$	0.585	0.486	0.459
tetrad	$0.209\ ^{+0.013}_{-0.010}$	0.208	0.194	0.311
antitetrad	$0.363\ ^{+0.09}_{-0.024}$	0.415	0.355	0.311

Table 1 Mutual information expressed in bits for the trine and tetrad states and their antistates. The values indicate the mutual information obtained in the experiment, in an ideal experiment, in an ideal experiment with detector noise analyzed using the error model described in [8], and the maximum allowed mutual information in a von Neumann measurement.

work just described (corresponding to the POM elements $|\psi_j^3\rangle\langle\psi_j^3|$) and apply it to the antitrine states $|\bar{\psi}_j^3\rangle$. This is obviously completely equivalent theoretically and more practical experimentally. State $|\bar{\psi}_j^3\rangle$ should never trigger photodetector j, but should reach either of the other two detectors with equal probability ($\frac{1}{2}$) [11].

To carry out the same generalized measurements on the tetrad states, PBS2 is replaced with a non-polarizing beam splitter (NPBS), and WP5 rotated to an angle of 27.37° clockwise from the horizontal. The theoretical result is that state $|\psi_1^4\rangle$ should reach photodetector PD1 with probability $\frac{1}{2}$ and each of the remaining photodetectors with probability $\frac{1}{6}$. Similarly each of the remaining three states should trigger its associated photodetector with probability $\frac{1}{2}$ and the others with probability $\frac{1}{6}$. The maximum mutual information measurement is realized in the same way as for the trine ensemble, by using the antitetrad states as input. If the antitetrad states are introduced into this network then state $|\bar{\psi}_j^4\rangle$ should never trigger photodetector j. It will, however, lead to detection in any of the remaining three photodetectors with equal probability ($\frac{1}{3}$).

All the experimental results obtained [8] are in good agreement with the theoretical predictions. The overall RMS deviation was 3.8% fr the trine and 2.9% for the tetrad. For the antitrine and antitetrad states where the count rate is theoretically zero, the average count rates were 1.6% and is 0.9% respectively. The low count rates are very significant for the mutual information that can be obtained.

For both trine and tetrad experiments, the mutual information was calculated in bits ($\log_2$) using Eq. (7), using experimental values for $P(y_j|\psi_k^N)$. The results, summarized in Table 1, are in excellent agreement with the theoretical predictions using a model for detector noise described in [8]. The mutual information of both sets of antistates is significantly greater than the best possible von Neumann measurement.

4. CONCLUSIONS

It has long been recognised that von Neumann measurements do not always provide the optimum detection strategy. We have described our experimental realizations of generalized measurements of optical polarization. Our experiments have demonstrated unambiuous discrimination between two non-orthogonal states and detection of the trine and tetrad ensembles with the minimum error probability. We have achieved for the first time a mutual information that is significantly higher than the maximum attainable using von Neumann measurements. Our work suggests that practical commmunications using POMS are feasible.

Acknowledgments

This work was funded by the UK Engineering and Physical Sciences Research Council grant numbers GR/L55216, GR/M60712 and GR/N17393. SMB, AC and MS thank the British Council for financial support.

References

[1] A. S. Holevo, Problemy Peredachi Informatsii **9**, 31 (1973).

[2] C. W. Helstrom, *Quantum Detection and Estimation Theory* (Academic Press, New York, 1976).

[3] A. Peres, *Quantum Theory: Concepts and Methods* (Kluwer, Dordrecht, 1993).

[4] S. M. Barnett and E. Riis, J. Mod. Opt. **44**, 1061 (1997).

[5] I. D. Ivanovic, Phys. Lett. A **123**, 257 (1987); D. Dieks, Phys. Lett. A **126**, 303 (1988); A. Peres, Phys. Lett. A **128**, 19 (1988).

[6] B. Huttner *et al.*, Phys. Rev. A **54**, 3783 (1996).

[7] R. B. M. Clarke, A. Chefles, S. M. Barnett, and E. Riis, sub. to Phys. Rev. Lett..

[8] R. B. M. Clarke *et al.*, quant-ph/0008028, submitted to Phys. Rev. A.

[9] A. S. Holevo, J. Multivariate Analysis **3**, 337 (1973).

[10] H. P. Yuen, R. S. Kennedy, and M. Lax, IEEE Trans. Info. Th. **21**, 125 (1975).

[11] S. J. D. Phoenix, S. M. Barnett, and A. Chefles, J. Mod. Opt. **47**, 507 (2000).

[12] A. Peres and W. Wootters, Phys. Rev. Lett. **66**, 1119 (1992).

[13] E. B. Davies, IEEE Trans. Inform. Theory **24**, 596 (1978).

[14] M. Ban *et al.*, Int. J. Theor. Phys. **36**, 1269 (1997).

[15] P. Hausladen *et al.*, Phys. Rev. A **54**, 1869 (1996).

[16] M. Sasaki *et al.*, Phys. Rev. A **59**, 3325 (1999).

[17] M. Osaki, *Quantum Communication, Measurement, and Computation* (Plenum press, New York, 2000).

[18] M. A. Naimark, Izv. Akad. Nauk. SSSR Ser. Mat. **4**, 277 (1940).

SPONTANEOUS INTRINSIC DECOHERENCE IN RABI OSCILLATIONS EXPERIMENTS

Rodolfo Bonifacio, Stefano Olivares
Dip. di Fisica, Università di Milano, INFN and INFM, Sezione di Milano, via Celoria 16, 20133, Milano, Italy

Abstract We present a simple theoretical description of a recent QED experiment where damping of Rabi oscillations cannot be attributed to *environment-induced decoherence* (EID). We generalize a previous formalism to describe also non-exponential decay. The possibility of *spontaneous intrinsic decoherence* (SID), i.e., not due to interaction with environment or fluctuation of some internal parameter but of purely quantum mechanical origin, is discussed. Moreover an experimental cavity QED test of SID is suggested.

1. INTRODUCTION

Even though decoherence is a very general phenomenon [1], it is very difficult to verify it experimentally because most often the physical nature of the environmental degrees of freedom, responsible for the decoherence process, remains unknown. The only controlled experimental verification of decoherence has been given by the experiment of Ref. [2], in which the progressive transformation of a linear superposition of two coherent states of a microwave cavity mode into the corresponding statistical mixture has been monitored. In this case, the environmental decoherence has been checked with no fitting parameters because its physical origin, i.e., photon leakage out of the cavity, was easily recognizable and measurable. In this case, it is even possible to *control* decoherence, i.e., to considerably suppress its effects, for example by using appropriately designed feedback schemes [3].

In some cases however, the mechanisms responsible for decoherence are not easily individuated and examples are provided by a recent experiment which observed Rabi oscillations between two circular states of a Rydberg atom in a high-Q cavity [4]. In this case one observes damped oscillations to a steady state in which the population of each of the two levels approaches $1/2$. A number of candidates have been already considered as possible physical sources of decoherence in this case: dark counts of the atomic detectors, dephasing colli-

Quantum Communication, Computing, and Measurement 3
Edited by P. Tombesi and O. Hirota, Kluwer Academic/Plenum Publishers, 2001

sions with background gas or stray magnetic fields within the cavity [4, 5] have been proposed as possible sources of decoherence. The only established fact is that, differently from Ref. [2], in this case, decoherence has a *non-dissipative* origin. In fact, the observed decay of the Rabi oscillations is much faster than the energy relaxation rate in this experimental configuration. Moreover, the fact that the population of each of the two levels asymptotically approaches $1/2$ cannot be explained in terms of dissipative mechanisms as the photon leakage out of the cavity, which would takes all the atoms in the ground state. A *quantitative* explanation of the observed decay rate of the Rabi oscillations was just presented in Ref. [6], where we used a different approach to decoherence, proposed in Ref. [7], where a *model-independent* formalism has been derived to describe decoherence. The idea underlying the approach of Ref. [7] is the fact that the interaction time, i.e., the time interval in which the effective Hamiltonian evolution takes place, is a random variable. This randomness can have different origins depending on the studied system. In the case of the Rydberg atom experiment [4], the interaction time is determined by the transit time of the velocity-selected atom through the high-Q microwave cavity. This interaction time is random, due to fluctuations of the atomic velocities, which can have classical origin [6] or purely quantum mechanical origin as suggested in this paper. This randomness implies random phases $e^{-iE_n t/\hbar}$ in the energy eigenstates basis. The experimental results unavoidably average over these random phases and this leads to decoherence, i.e., to the decay of off-diagonal matrix elements of the density operator in the energy basis. Therefore, as we shall see, our approach will give a generalized phase-destroying master equation, able to describe many situations in which decoherence is associated with random phases, originating for example from some frequency or interaction time fluctuations.

2. THE FORMALISM

First of all we recall the main points of the new formalism (see also Ref. [6, 7]). Let us consider an initial state described by the density operator $\rho(0)$ and consider the case of a random evolution time. The experimentally observed state is not described by the usual density matrix of the whole system $\rho(t)$, but by its time averaged counterpart [7]

$$\bar{\rho}(t) = \int_0^\infty dt' P(t, t', \tau) \rho(t') , \tag{1}$$

where $\rho(t') = \exp\{-iLt'\}\rho(0)$ is the usual unitarily evolved density operator from the initial state and $L \ldots = [H, \ldots]/\hbar$. Hence one can write

$$\bar{\rho}(t) = V(t)\rho(0) , \tag{2}$$

where

$$V(t) = \int_0^\infty dt' P(t,t',\tau) e^{-iLt'} \tag{3}$$

In Ref. [7], the function $P(t,t',\tau)$ has been determined so to satisfy the following conditions: i) $\bar{\rho}(t)$ must be a density operator, i.e., it must be self-adjoint, positive-definite, and with unit-trace. This leads to the condition that $P(t,t',\tau)$ must be non-negative and normalized, i.e., a probability density in t' so that Eq. (1) is a completely positive mapping. ii) $V(t)$ satisfies the semi-group property $V(t_1+t_2) = V(t_1)V(t_2)$, with $t_1, t_2 \geq 0$. These requirements are satisfied by [7]

$$V(t) = (1+iL\tau)^{-t/\tau} \tag{4}$$

$$P(t,t',\tau) = \frac{e^{-t'/\tau}}{\tau} \frac{(t'/\tau)^{(t/\tau)-1}}{\Gamma(t/\tau)}, \tag{5}$$

The above expressions $V(t)$ and $P(t,t',\tau)$ satisfy Eq. (3) according to the Γ-function integral identity [8, 9]. $P(t,t',\tau)$ is the well known positive definite Γ-distribution function for the random variable t' and it parametrically depends on the *clock time t* and on the scaling time τ. The parameter τ characterizes the strength of the evolution time fluctuations. When $\tau \to 0$, $P(t,t',\tau) \to \delta(t-t')$ so that $\bar{\rho}(t) = \rho(t)$ and $V(t) = \exp\{-iLt\}$ is the usual unitary evolution. However, for finite τ, the evolution operator $V(t)$ of Eq. (4) describes decoherence in the energy representation (i.e., the approach to diagonal form), whereas the diagonal matrix elements remain constant, i.e., the energy is still a constant of motion (*non dissipative decoherence*). Furthermore by differentiating with respect to time Eq. (2) and using (4), one gets the following master equation for $\bar{\rho}(t)$

$$\dot{\bar{\rho}}(t) = -\frac{1}{\tau} \log(1+iL\tau)\, \bar{\rho}(t) \tag{6}$$

If one expands the logarithm at second order in τ, one obtains

$$\dot{\bar{\rho}}(t) = -\frac{i}{\hbar}\left[H, \bar{\rho}(t)\right] - \frac{\tau}{2\hbar^2}\left[H, \left[H, \bar{\rho}(t)\right]\right], \tag{7}$$

which is the well-known phase-destroying master equation [10]. Hence Eq. (6) appears as a *generalized* phase-destroying master equation taking into account higher order terms in τ. Notice, however, that the present approach is different from the usual master equation approach in the sense that no perturbative and specific statistical assumptions are made.

3. SID IN RABI OSCILLATIONS EXPERIMENTS

In a previous work [6] we applied this formalism to the two experiments of Refs. [4, 11]. Now we focus our attention on the one of Ref. [4], where the resonant interaction between a quantized mode in a high-Q microwave cavity (with annihilation operator a) and two circular Rydberg states ($|e\rangle$ and $|g\rangle$) of a Rb atom is studied. This interaction is well described by the usual Jaynes-Cummings model [12], which in the interaction picture reads

$$H = \hbar\Omega_R \left(|e\rangle\langle g|a + |g\rangle\langle e|a^\dagger \right) , \tag{8}$$

where Ω_R is the Rabi frequency. The Rabi oscillations, describing the exchange of excitations between atom and cavity mode, are studied by injecting the velocity-selected Rydberg atom, prepared in the excited state $|e\rangle$, in the high-Q cavity and measuring the population of the lower atomic level g, $P_{eg}(t)$ as a function of the interaction time t, which is varied by changing the Rydberg atom velocity. In the case of vacuum state induced Rabi oscillations, the decoherence effect is particularly evident and the Hamiltonian evolution according to Eq. (8) predicts

$$P_{eg}(t) = \frac{1}{2}\left(1 - \cos\left(2\Omega_R t\right)\right) . \tag{9}$$

Experimentally instead, damped oscillations are observed, which are fitted by

$$P_{eg}^{exp}(t) = \frac{1}{2}\left(1 - e^{-\gamma t}\cos\left(2\Omega_R t\right)\right) , \tag{10}$$

where the decay time fitting the experimental data is [5] $\gamma^{-1} \approx 40\mu$sec and the corresponding Rabi frequency is $\Omega_R/2\pi = 25$ kHz. This decay of quantum coherence cannot be associated with photon leakage out of the cavity because the cavity relaxation time is larger (220 μsec) and also because in this case one would have an asymptotic limit $P_{eg}^{exp}(\infty) = 1$. The damped behavior of Eq. (10) is instead easily obtained if one applies the approach described above. In fact, from the linearity of Eq. (1), one has that the time averaging procedure is also valid for mean values and matrix elements of each subsystem. Therefore one has

$$\bar{P}_{eg}(t) = \int_0^\infty dt' P(t,t',\tau) P_{eg}(t') . \tag{11}$$

Using Eqs. (2), (4), (5) and (9), Eq. (11) can be rewritten in the same form of Eq. (10):

$$\begin{aligned} \bar{P}_{eg}(t) &= \frac{1}{2}\left[1 - \frac{\cos(\nu t)}{\left(1 + 4\Omega_R^2\tau^2\right)^{t/2\tau}}\right] \\ &= \frac{1}{2}\left(1 - e^{-\gamma t}\cos\left(\nu t\right)\right) , \end{aligned} \tag{12}$$

where

$$\gamma = \frac{1}{2\tau}\log\left(1 + 4\Omega_R^2\tau^2\right) \tag{13}$$

$$\nu = \frac{1}{\tau} \arctan(2\Omega_R \tau) \tag{14}$$

We note that in general the time averaging procedure introduces not only a damping of the probability oscillations but also a frequency shift. However, if the characteristic time τ is sufficiently small, i.e., $\Omega_R\tau \ll 1$, there is no phase shift, $\nu \simeq 2\Omega_R$, and

$$\gamma = 2\Omega_R^2 \tau \tag{15}$$

The fact that in Ref. [4] the Rabi oscillation frequency essentially coincides with the theoretically expected one, suggests that the time τ characterizing the fluctuations of the interaction time is sufficiently small so that it is reasonable to use Eq. (15). Using the above values for γ and Ω_R, one can derive an estimate for τ, so to get $\tau \simeq 0.5$ μsec. This estimate is consistent with the assumption $\Omega_R\tau \ll 1$ we have made, but, more importantly, it turns out to be comparable to the experimental value of the *uncertainty in the interaction time*. In fact, the fluctuations of the interaction time [4] $t = \sqrt{\pi}w/v$ (w is the cavity mode waist) are mainly due to the experimental uncertainty of the atomic velocity v. Since $w = 0.6$ cm, the mean velocity is $\bar{v} \simeq 300$ m/sec and the velocity uncertainty is $\delta v/v = 1\%$ (see Ref. [4]), one has $\bar{t} = \sqrt{\pi}w/\bar{v} \simeq 50$ μsec and $\tau \simeq \delta t = \bar{t}\delta v/v = 0.5$ μsec, which is just the estimate we have derived from the experimental values. This simple and approximate argument supports the interpretation that the decoherence observed in Ref. [4] is essentially due to the randomness of the interaction time.

In reality the above argument is just a rough estimate of τ. A more precise evaluation requires a generalization of the previous theory. In fact, at the beginning of this paper we exposed our formalism with a constant τ. However we notice that complete positive mapping, described by Eqs. (1)–(5) and Eqs. (11)–(14), can be maintained also when τ is a function of t. [6] In such a case the semigroup property is dropped.

As a matter of fact in the experiment of Ref. [4] each experimental point at a given time t is a different experiment, i.e., one selects with a proper laser excitation a velocity group with mean velocity $\overline{v}$ such that

$$t = \sqrt{\pi}\frac{w}{\overline{v}} \tag{16}$$

sending one particle at a time across the cavity. In this way different values of t are obtained selecting different mean velocity group $\overline{v}$. The spread of each velocity group, Δ_v, depends on the laser excitation. Hence the experimental uncertainty in interaction time t can be written as

$$\tau \approx at \tag{17}$$

where we defined

$$a \equiv \frac{\Delta_v}{\overline{v}} \tag{18}$$

In the experimental situation of Ref. [4] $a \approx 0.01$ for all velocity group.

The above estimate of $\tau = 0.5$ μsec corresponds to replace t in Eq. (17) a sort of *mean* interaction time of 50 μsec . Furthermore one should not forget that beside this classical velocity spread, Δ_v, there is an intrinsic velocity spread, σ_v, due to Heisenberg uncertainty principle (HUP)

$$\sigma_v = \frac{\hbar}{2m\sigma_0} \tag{19}$$

where m is the atomic mass and we have assumed that each atom can be described by a minimum uncertainty wave packet with initial position uncertainty σ_0. The total position spread at time t can be written as

$$\Delta_x(t) = \sqrt{\sigma_0^2 + \sigma_v^2 t^2 + \Delta_v^2 t^2} \tag{20}$$

Here, the first two terms describes the particle wave packet Schrödinger spread and the last term is due to the classical velocity spread described above.

Eq. (20) can be justified by first principle as follows. Let us assume that atomic translational degree of freedom is represented by the following statistical operator

$$\rho(t) = \int_{-\infty}^{\infty} dv\, \mathcal{P}(v)\, |v,t\rangle\langle v,t| \tag{21}$$

where $|v,t\rangle$ is a free particle gaussian wave packet with average velocity v, which corresponds to the *classical* velocity, i.e.,

$$|\langle x|v,t\rangle|^2 = |\psi(x,t)|^2 = \frac{1}{\sqrt{2\pi}\sigma_t} \exp\left\{-\frac{(x-vt)^2}{2\sigma_t^2}\right\} \tag{22}$$

where $\sigma_t^2 = \sigma_0^2 + \sigma_v^2 t^2$ is the well known Schrödinger free particle spread. $\mathcal{P}(v)$ represents the classical velocity distribution, which we assume to be gaussian:

$$\mathcal{P}(v) = \frac{1}{\sqrt{2\pi}\Delta_v} \exp\left\{-\frac{(v-\overline{v})^2}{2\Delta_v^2}\right\} \tag{23}$$

Hence, with obvious notation,

$$P(x,t) \equiv \langle x|\rho(t)|x\rangle = \int_{-\infty}^{\infty} dv\, \mathcal{P}(v)\, |\langle x|v,t\rangle|^2 \tag{24}$$

Using Eq. (22) and (23) and performing the integral (24), we obtain

$$P(x,t) = \frac{1}{\sqrt{2\pi}\Delta_x(t)} \exp\left\{-\frac{(x-\overline{v}t)^2}{2\Delta_x(t)^2}\right\} \tag{25}$$

where $\Delta_x(t)$ is the same of Eq. (20).

We propose to take as value of τ the uncertainty of interaction time, so that

$$\tau \equiv \tau(t) = \frac{\Delta_x(t)}{\overline{v}} \tag{26}$$

where t and $\overline{v}$ are related by Eq. (16). Eq. (26) corresponds to the uncertainty of the arrival time [13] of the atoms at the end of the cavity, i.e., at the time t given by Eq. (16). For long enough times, the initial spread σ_0 can be neglected in Eq. (20), so that one can write $\tau \approx at$ (see Eq. (17)), where

$$a = \frac{\sqrt{\sigma_v^2 + \Delta_v^2}}{\overline{v}} \tag{27}$$

Let us notice that if the classical contribution is dominating, as it is in Ref. [4], Eqs. (26) and (27) reduce to Eqs. (17) and (18). If σ_v dominates Δ_v, then it is possible to see SID.

In general Eqs. (13) and (14) become

$$\gamma(t) = \frac{1}{2at}\log\left(1 + 4\Omega_R^2 a^2 t^2\right) \tag{28}$$

$$\nu(t) = \frac{1}{at}\arctan\left(2\Omega_R at\right) \tag{29}$$

where a is given by (27).

For short enough times, such that one has $2\Omega_R at \ll 1$, one obtains $\nu \approx \Omega_R$ and $\gamma \approx 2\Omega_R^2 at$ so that damping becomes gaussian (i.e., non exponential) and

$$\bar{P}_{eg}(t) = \frac{1}{2}\left(1 - e^{-2\Omega_R^2 at^2}\cos\left(2\Omega_R t\right)\right) \tag{30}$$

For long enough times (i.e., short enough velocity group), this last approximation is no more valid and one has a power law decay given by Eq. (12) with $\tau = at$.

Due to uncertainty of experimental points of Ref. [4], especially for the long time behaviour, it is difficult to decide which one is better between the exponential fit, described before and in Ref. [6], or the non-exponential one.

However, from Eq. (26) one sees that τ can never be taken arbitrarily small, since [13] $\tau \geq \sigma_0/\overline{v} \approx \hbar/2\sigma(K)$, where $\sigma(K)$ is the standard deviation of the kinetic energy $K = p^2/2m$ of the atomic center of mass. However, since the *total* Hamiltonian of the system is the sum $\mathcal{H} = K + H$, we have $\sigma(\mathcal{H}) \geq \sigma(K)$ so that $\tau \geq \hbar/\sigma(\mathcal{H})$. This result is in agreement with the suggestion of Ref. [7]. In any case, even if the classical dispersion of velocity group is arbitrarily small, the limit $\tau \to 0$ cannot be taken, in fact Eq. (26) says that there is a finite, intrinsic value of τ due to Schrödinger spread, whose minimum value is

$$\tau_{min} = \sqrt{\frac{\hbar}{m\overline{v}^2}t} \tag{31}$$

obtained for $\sigma_0 = \sqrt{\hbar t/2m}$. When the classical contribution in Eq. (20) is negligible, one should observe *spontaneous intrinsic decoherence* (SID), i.e. decoherence which is not induced by environment or by fluctuations of some experimental parameters (spontaneous) but of purely quantum mechanical origin (intrinsic). Hence SID should be observable at all times in the experimental apparatus of Ref. [4] if the classical velocity spread Δ_v is reduced and it becomes comparable or smaller than the quantum one, $\sigma(v)$, in Eq. (20).

In conclusion, referring to a specific experimental situation where environment induced decoherence is negligible, we have explicitally shown that there is another source of decoherence due to fluctuation of the interaction time. These fluctuations can be originated by classical velocity spread or by intrinsic quantum velocity spread, due to the Heisenberg uncertainty principle: when this is dominant, SID should be observed.

Acknowledgments

We like to thank Dr. D. Vitali and Prof. P. Tombesi for helpful suggestions and Prof. H. Walther and Dr. B. Varcoe for stimulating discussions.

References

[1] W. H. Zurek, Phys. Today **44**(10), 36 (1991), and references therein.

[2] M. Brune *et al.*, Phys. Rev. Lett. **77**, 4887 (1996).

[3] D. Vitali, P. Tombesi, G. J. Milburn, Phys. Rev. Lett. **79** 2442 (1997); Phys. Rev. A **57** 4930 (1998).

[4] M. Brune *et al.*, Phys. Rev. Lett. **76**, 1800 (1996).

[5] J. M. Raimond, private communication.

[6] R. Bonifacio, S. Olivares, P. Tombesi, D. Vitali, Phys. Rev. **A61**, 053802 (2000); LANL e-print archive quant-ph/9911100.

[7] R. Bonifacio, Nuovo Cimento **114B**, 473 (1999); LANL e-print archive quant-ph/9901063; in *Mysteries, Puzzles and Paradoxes in Quantum Mechanics*, edited by R. Bonifacio, AIP, Woodbury, 1999, pag. 122.

[8] I. S. Gradshteyn and I.M. Ryzhik, *Table of Integrals and Series*, Academic, Orlando, 1980, pag. 317.

[9] In Ref. [7] a more general expression for $P(t,t')$ and $V(t)$, depending on two parameters τ_1 and τ_2 is derived. We choose $\tau_1 = \tau_2 = \tau$ because in the experiments considered here, the effective interaction time $\langle t' \rangle = t\tau_1/\tau_2$ [7] has to coincide with the "laboratory time" t, implying therefore $\tau_1 = \tau_2$.

[10] D. F. Walls and G. J. Milburn, *Quantum Optics*, Springer, Berlin, 1996.

[11] D. M. Meekhof *et al.*, Phys. Rev. Lett. **76**, 1796 (1996).

[12] E. T. Jaynes and F. W. Cummings, Proc. IEEE **51**, 89 (1963).

[13] A. Messiah, *Quantum Mechanics*, Vol. 1, Ch. IV § 10, pag. 136, North–Holland Publishing Company, Amsterdam, 1961.

QUANTUM TOMOGRAPHY, TELEPORTATION, AND CLONING

Giacomo Mauro D'Ariano
INFM Unità di Pavia, via Bassi 6, I-27100 Pavia, Italy
dariano@qubit.it

Introduction

In this paper, in a simple unifying matrix framework, I will present general classification of all possible tomography methods, teleportation schemes, and optimal quantum cloning maps. We will see how every tomographic method or teleportation scheme corresponds to a choice of operator spanning sets, and how this framework also leads to methods for engineering new Bell measurements. On the other hand, the classification of all possible covariant cloning maps (that are optimal for a given criterion) includes all known types of cloning, and leads to methods for engineering new cloning machines, which can be physically realized through unitary transformation with ancilla, and/or via probabilistic quantum operations. Fidelity criteria for POVM's can be exploited to achieve joint POVM's via cloning. I will give concrete physical realizations in the paper.

The matrix formalism

Monopartite quantum systems. In the following $\mathcal{H}$ and $\mathcal{K}$ will denote two Hilbert spaces with dim$(\mathcal{K}) = n \geq m =dim(\mathcal{H})$, for which we fix orthonormal *standard basis* (SB) $\{|f(i)\rangle\} \in \mathcal{H}$ and $\{|e(j)\rangle\} \in \mathcal{K}$, respectively [when there is no ambiguity we'll also use the loose notation $\{|i\rangle\}$ for either SB]. By the same matrix symbol M we'll denote: a) the $m \times n$ matrix itself $M = [M(j)] \equiv [M(1)\, M(2)\, \ldots, M(n)\,] \equiv \{M_{ij}\}$, $M(j)$ column vectors; b) the operator $\hat{M} \in \mathcal{L}(\mathcal{K},\mathcal{H})$ from $\mathcal{K}$ to $\mathcal{H}$: $\hat{M} = \sum_{i=1}^{m}\sum_{j=1}^{n} M_{ij}|f(i)\rangle\langle e(j)|$ [when identifying the operator with its matrix—i.e. dropping the "hat"—we must remember that the one-to-one correspondence needs keeping the SB as fixed. Hence, when considering basis e', f' different from the SB, the operator must be written in outer-product form as $M' = \sum_{i=1}^{m}\sum_{j=1}^{n} M_{ij}|f'(i)\rangle\langle e'(j)|$]; c) the vector set (VS) $|M(j)\rangle \doteq \sum_{i=1}^{m} M_{ij}|f(i)\rangle \in \mathcal{H}$, in term of which the operator writes equivalently as

$$M = \sum_{j=1}^{n} \sum_{j=1}^{n} |M(j)\rangle\langle e(j)| = \sum_{i=1}^{m} \sum_{j=1}^{n} |f(i)\rangle\langle M^{\dagger}(i)| \,.$$

The VS M is also a new basis if $n = m = \text{rank}(M)$. When $n > m = \text{rank}(M)$ strictly, we call the (complete) VS M a *spanning set* (SS). The relative scalar product between two VS X, Y is given by $\langle x(l)|y(s)\rangle = (X^{\dagger}Y)_{ls}$. Orthonormal basis correspond to unitary matrices. For a generic SS C the inner product is the positive matrix $P^2 = C^{\dagger}C$ called *Gram matrix*. If C is a (non orthogonal) basis, one can find a *biorthogonal* or *dual* basis B, such that $\langle b(l)|c(s)\rangle = (B^{\dagger}C)_{ls} = \delta_{ls}$, by matrix inversion $B = (C^{\dagger})^{-1}$, shortly: $|b(l)\rangle = |c^{-1\dagger}(l)\rangle$ (orthonormal sets are trivially selfdual). The completeness relation writes $C^{\dagger}B = \sum_l = |c(l)\rangle\langle b(l)| = I_{\mathcal{H}}$, where $I_{\mathcal{H}}$ denotes the identity operator in $\mathcal{L}(\mathcal{H})$. From a biorthogonal couple (C, B) one can obtain another biorthogonal couple (C', B') with the same Gram matrix as $C' = UCV^{\dagger}$ and $B' = UBV^{\dagger}$, U and V being unitary matrices. Finally, by the QR algorithm (which is based on the Gram-Schmidt orthogonalization procedure), one achieves the factorization $C = QR$, where Q is unitary and R is upper triangular (with positive diagonal), and in this way the orthogonal basis Q is extracted from the nonorthogonal SS C.

Bipartite quantum systems. We now consider a bipartite quantum system with Hilbert space $\mathcal{H} \otimes \mathcal{K}$. The SB will be denoted by $|E(h)\rangle\rangle \equiv |E(ij)\rangle\rangle = |f(i)\rangle \otimes |e(j)\rangle$, $h \equiv (ij)$ polyindex. We introduce a matrix notation[1] which exploits the isomorphism $\mathcal{H} \otimes \mathcal{K} \simeq \mathcal{L}(\mathcal{K}, \mathcal{H})$. Every vector in $\mathcal{H} \otimes \mathcal{K}$ can be written in the matrix form $|A\rangle\rangle = \sum_{i=1}^{m} \sum_{j=1}^{n} A_{ij} |f(i)\rangle \otimes |e(j)\rangle$, where $A = [a(j)]$ is a $m \times n$ matrix. Also, we have $|A\rangle\rangle = \sum_{j=1}^{n} |a(j)\rangle \otimes |e(j)\rangle \equiv \sum_{i=1}^{m} |f(i)\rangle \otimes |a^T(i)\rangle$ (in the following, we'll use T for transposition and $*$ for complex conjugation, both with respect to the fixed SB, e.g. by O^* we denote the operator $O^* = \sum_{i=1}^{m} \sum_{j=1}^{n} O^*_{ij} |f(i)\rangle\langle e(j)|$. Notice the following simple rules: $A \otimes B|C\rangle\rangle = |ACB^T\rangle\rangle$, where $A \in \mathcal{L}(\mathcal{H})$ and $B \in \mathcal{L}(\mathcal{K})$, $\langle\langle B|A\rangle\rangle = \text{Tr}[B^{\dagger}A]$. For $\mathcal{H} \simeq \mathcal{K}$ one can also write $|A\rangle\rangle = A \otimes I|I\rangle\rangle = I \otimes A^T|I\rangle\rangle$, and $\langle\langle B|A\rangle\rangle = \langle\langle I|A \otimes B^*|I\rangle\rangle$. We also have the rules for partial traces $\text{Tr}_{\mathcal{K}}[|A\rangle\rangle\langle\langle B|] = AB^{\dagger} \in \mathcal{L}(\mathcal{H})$ and $\text{Tr}_{\mathcal{H}}[|A\rangle\rangle\langle\langle B|] = A^T B^* \in \mathcal{L}(\mathcal{K})$. To a biorthogonal SS $\sum_l |c(l)\rangle\rangle\langle\langle b(l)| = I_{\mathcal{H}\otimes\mathcal{K}}$ it will correspond a biorthogonal SS of operators $\{b(l), c(l)\} \in \mathcal{L}(\mathcal{K}, \mathcal{H})$ such that every operator $A \in \mathcal{L}(\mathcal{K}, \mathcal{H})$ can be expanded as $A = \sum_l \text{Tr}[b^{\dagger}(l)A]c(l)$. The completeness and biorthogonality of the SS is equivalent to the so-called *orthogonality relations* $\sum_l b^{\dagger}(l)_{pq} c(l)_{rs} = \delta_{ps}\delta_{qr}$, i.e. $P = \sum_l b^{\dagger}(l) \otimes c(l)$ is the operator that permutes the two Hilbert spaces in the tensor product $\mathcal{H} \otimes \mathcal{K}$. The orthogonality relations lead to identities of the form $\sum_l b^{\dagger}(l) A c^T(l) = A^T$ and its transposed/dagger-ed, and, for $\mathcal{H} \simeq \mathcal{K}$ one also has $\sum_l b^\dagger(l) A c(l) = \mathrm{Tr}[A]\, I$, and $\sum_l b^\dagger(l) \otimes c^T(l) \equiv |I\rangle\rangle\langle\langle I|$ and transposed/dagger-ed relations. Analogously to the monopartite case, changes of basis are in correspondence with matrices (tensors), e.g. $|c(l)\rangle\rangle = \sum_h c_{hl}|E(h)\rangle\rangle = \sum_{ij} c(l)_{ij}|E(ij)\rangle\rangle$, $h \equiv (ij)$ polyindex. One can see that finding the dual SS $\{b(l)\}$ of $\{c(l)\}$ resorts to the matrix-tensor inversion $B = (C^\dagger)^{-1}$, where $C = [c(l)] \equiv \{c_{hl}\}$, h polyindex [here, we consider $\mathcal{H} \simeq \mathcal{K}$ for simplicity: the generalization to different spaces is trivial, i.e. inverting the rank=m square part of the rectangular matrix and/or adding orthogonal vectors]. As in the monopartite case one can use the QR algorithm to factorize the matrix C and find an orthogonal basis Q of operators. Notice that an orthogonal basis $\{|q(l)\rangle\rangle$ will correspond to unitary matrix-tensors $Q = [q(l)]$, however the operators $q(l)$ are not unitary. >From the partial traces $\mathrm{Tr}_{\mathcal{H}}[|q(i)\rangle\rangle\langle\langle q(j)|] = q(i)q(j)^\dagger$ and $\mathrm{Tr}_{\mathcal{K}}[|q(i)\rangle\rangle\langle\langle q(j)|] = (q(j)^\dagger q(i))^T$, we see that choosing *maximally entangled* vectors (i.e. with partial trace proportional to the identity), gives $\sqrt{m}q(i)$ unitary (obviously for $\mathcal{H} \simeq \mathcal{K}$), with the additional orthogonality condition for the set $\mathrm{Tr}[q(i)q(j)^\dagger] = \delta_{ij}$. We'll call such operator set *unitary spanning sets*. Notice that in the present matrix formalism the *Schmidt form* $|A\rangle\rangle = \sum_{l=1}^k \lambda_l |v(l)\rangle|w^*(l)\rangle$, with $\lambda_l > 0$, $\sum_{l=1}^k \lambda_l^2 = 1$, $\{|v(l)\rangle\}$, $\{|w^*(l)\rangle\}$ orthonormal sets, is noting but the so-called *singular value decomposition* $A = V\Sigma W^\dagger$ of the matrix A, with V and W unitary, and $\Sigma = [diag(\lambda_l), 0]$, $k \equiv$rank(A) the *Schmidt number*. Other forms of the bipartite vector are related to other matrix decompositions, as, for example $|A\rangle\rangle = \sum_{i=1}^m |p(i)\rangle|u^T(i)\rangle$, with $P \geq 0$, $P^2 = AA^\dagger$, $UU^\dagger = I_m$ is just the *polar decomposition* $A = PU$ of the matrix A.

Group representations. A simple way to construct unitary SS is to consider a (generally projective) unitary irreducible representation (UIR) of a group $\mathbf{G}$. We'll denote by $[\mathbf{G}, U, \mathcal{H}]$ the unitary representation of $\mathbf{G}$ on $\mathcal{H}$, $U(g)$ being the unitary representative of the group element $g \in \mathbf{G}$. Then, an operator SS $\{u(g)\} \in \mathcal{L}(\mathcal{H})$ is simply $u(g) \propto U(g)$, with the proportionality constant depending on the group representation. This is just a consequence of the orthogonality relations $\int \mathrm{d}g\, U(g) \otimes U^\dagger(g) = P$, which follow from the Schur lemmas, $\mathrm{d}g$ denoting an invariant Haar measure.[1] Corresponding to the operator SS the vectors $|u(g)\rangle\rangle$ will constitute an orthonormal basis of maximally entangled vectors, e. g. a Bell POVM on $\mathcal{H} \otimes \mathcal{H}$ [in the infinite dimensional case we can keep the vectors $|u(g)\rangle\rangle$ maximally entangled with Dirac-delta normalization]. Such Bell POVM will obviously be covariant under $[\mathbf{G}, U, \mathcal{H}]$.

Completely positive maps. The matrix formalism for entangled vectors can be used to exploit the one-to-one correspondence $\mathcal{E} \leftrightarrow R_{\mathcal{E}}$ between completely-positive (CP) maps $\mathcal{E} \in \mathcal{L}(\mathcal{L}(\mathcal{H}), \mathcal{L}(\mathcal{K}))$, and operators $0 \leq R_{\mathcal{E}} \in$

$\mathcal{L}(\mathcal{H} \otimes \mathcal{K})$[2]. The correspondence is given by $R_\mathcal{E} = \mathcal{E} \otimes \mathcal{I}(|I\rangle\!\rangle\langle\!\langle I|) \geq 0$, and inversely $\mathcal{E}(\rho) = \mathrm{Tr}_2[I \otimes \rho^T R_\mathcal{E}]$, $\mathcal{I}$ denoting the trivial CP-map. In the following we'll call $R_\mathcal{E}$ the "R-matrix" or "R-operator" of $\mathcal{E}$. In this fashion the Krauss decomposition of the CP-map $\mathcal{E}(\rho) = \sum_l K_l \rho K_l^\dagger$ is just the diagonal form of its R-operator $\mathcal{E} = \sum_l |K_l\rangle\!\rangle\langle\!\langle K_l|$, and the trace-preserving condition $\sum_l K_l^\dagger K_l = I_\mathcal{K}$ becomes the partial-trace normalization $\mathrm{Tr}_\mathcal{K}[R_\mathcal{E}] = I_\mathcal{H}$. Similarly, Stinespring dilations of $\mathcal{E}$ are nothing but purifications of the R-operator in an extended Hilbert space. In general, all various matrix forms for the R-matrix will lead to different forms for the CP-map. The dual CP-map $\mathcal{E}^\vee \in \mathcal{L}(\mathcal{L}(\mathcal{K}), \mathcal{L}(\mathcal{H}))$ e.g. $X \to \mathcal{E}^\vee(X)$, $X \in \mathcal{L}(\mathcal{K})$ is obtained as $\mathcal{E}^\vee(X) = \mathrm{Tr}_1[X \otimes I R_\mathcal{E}^{T_2}]$, T_2 denoting partial transposition on the second Hilbert space, and we'll consistently call $R_\mathcal{E}^{T_2}$ the "dual R-matrix".

A main virtue of the R-matrix representation resides in the possibility of classifying CP-maps via the Cholesky decomposition of R-matrices, and, in particular, classifying group-covariant CP-maps in terms of group-invariant R-matrices [3], thus resorting to a simple application of the Schur lemma. We call the CP-map $\mathcal{E}$ $\mathbf{G}$-covariant, when $\mathcal{E}(U_g \rho U_g^\dagger) = V_g \mathcal{E}(\rho) V_g^\dagger$, U_g and V_g being unitary representatives of $g \in \mathbf{G}$ over $\mathcal{H}$ and $\mathcal{K}$, respectively. Then, $\mathcal{E}$ is $\mathbf{G}$-covariant, iff $[R_\mathcal{E}, U_g \otimes V_g^*] = 0$. An operator $Q \in \mathcal{L}(\mathcal{K}, \mathcal{H})$ is invariant under $[\mathbf{G}, U, \mathcal{H}]$ iff $Q \equiv [Q_0]_U^\mathbf{G}$, where $[Q]_U^\mathbf{G} \doteq \int dg\, U_g^\dagger Q U_g$ denotes group averaging. Then, one can prove that the general form of the R-matrix of the $\mathbf{G}$-covariant CP-map is $R_\mathcal{E} = A_\mathcal{E}^\dagger A_\mathcal{E}$, with $A_\mathcal{E}$ (complex lower triangular) matrix itself invariant, which can be written in the block form $A_\mathcal{E} = [A_0]_{UV^*}^\mathbf{G} \equiv \oplus_\nu a_\nu$, each block a_ν comprising a full set of equivalent irreducible components of the representation.

The R-representation of CP maps turns out to be very useful for engineering new quantum channels and new POVM's. For example, approximating a new POVM $n(l)$ starting from a given one $o(l)$ via a CP-map, is equivalent to maximizing the POVM-fidelity $F = \mathrm{Tr}[(n(l) \otimes o(l)) R_\mathcal{E}^{T_2}]$ (or other measure of distance between operators) over all possible R-matrices $R_\mathcal{E}$. In particular, for $\mathbf{G}$-covariant POVM's, i.e. $n(g) = U_g^\dagger n(e) U_g$ and $o(g) = V_g^\dagger o(e) V_g$ (e the identity element of $\mathbf{G}$) a CP-map covariant under $[\mathbf{G}, UV^*, \mathcal{H}]$ must be used, and the fidelity only between the POVM's "seeds" $F = \mathrm{Tr}[n(e) \otimes o(e) R_\mathcal{E}^{T_2}]$ needs to be optimized, which is perfectly analogous to the conventional fidelity for states.

Tomography, Teleportation, and Cloning

Quantum Tomography. Quantum tomography is a method to estimate the ensemble average $\langle A \rangle$ of any linear operator $A \in \mathcal{L}(\mathcal{H})$ by using only measurement outcomes of *quorum* of observables $\{c(l)\}$. One can immediately recognize that a *quorum* is nothing but a SS $\{c(l)\}$ made of observables[2]. Then, the tomographic estimation of the ensemble average $\langle A\rangle$ is obtained as the double average—over both the ensemble and the quorum—of the *unbiased estimator*[3] $\mathrm{Tr}[b^\dagger A]c(l)$, namely $\langle A\rangle = \sum_l \mathrm{Tr}[b^\dagger(l)A]\langle c(l)\rangle$. Hence, of the biorthogonal couple, the $c(l)$'s are the *quorum*, and the $b(l)$'s give the unbiased estimator.[4] Examples of applications are given in Table 1.

	l	$\mathrm{d}\mu(l)$	$\mathcal{D}$	$c(l), b(l)$
1a	α	$\mathrm{d}^2\alpha$	$\mathbb{C}$	$\pi^{-1/2}D(\alpha)$
1b	(k,ϕ)	$\mathrm{d}k\vert k\vert\mathrm{d}\phi/(4\pi)$	$\mathbb{R}\times[0,\pi]$	$\exp(ikX_\phi)$
1c	(x,ϕ)	$\mathrm{d}x\mathrm{d}\phi/\pi$	$\mathbb{R}\times[0,\pi]$	$E^\phi_x,\ -\frac{1}{2}\mathrm{P}\frac{1}{(x-X_\phi)^2}$
2	α	$\mathrm{d}^2\alpha$	$\mathbb{C}$	$\pi^{-1/2}(-)^{a^\dagger a}D(2\alpha)$
3	(x,t)	$\mathrm{d}x\,\mathrm{d}t$	$\mathbb{R}\times\mathbb{R}$	$e^{-ip^2t}\vert q\rangle\langle q\vert e^{ip^2t}$
4	$(\psi,\vec{n})$	$\frac{2J+1}{8\pi}\mathrm{d}\vec{n}\,\mathrm{d}\psi\,\sin^2\frac{\psi}{2}$	$S^2\times S^1$	$\exp(i\psi\vec{J}\cdot\vec{n})$
5	ν	1	$\{x,y,z,t\}$	σ_ν
6	(n,ϕ)	$\mathrm{d}\phi/(2\pi)$	$\mathbb{Z}\times S^1$	$e^{\vert n\vert}_{s(n)}e^{ia^\dagger a\phi}$
7	(ψ,ϕ)	$\mathrm{d}\phi\mathrm{d}\psi/(4\pi^2)$	$S^1\times S^1$	$e^{-ia^\dagger a\psi}\vert e^{i\phi}\rangle\langle e^{i\phi}\vert e^{-ia^\dagger a\psi}$

Table 1 Examples of spanning sets used in different tomographic schemes. $\mathcal{D}$ denotes the domain of the index l, $\mathrm{d}\mu(l)$ the integration measure for continuous l. The dual basis $b(l)$ is not specified when $c(l)$ is self adjoint or unitary. Examples 1 are various versions of quantum homodyne tomography[5]. E^ϕ_x denotes the projector over the quadrature eigenvector at phase ϕ, with eigenvalue x. Example 2 is the parity photon-counting tomography over displaced states or observables [6, 7]. Example 3 is the tomography of a free particle[7], p and q denoting momentum and position respectively, and t the time. Example 4 is the angular momentum tomography[4]. Example 5 is the Pauli tomography[4]. Example 6 and 7 are the nonlinear and the nonunitary phase-tomographies of Ref. [7] (both have a quorum of generalized observables, i.e. nonorthogonal POVM's). $|e^{i\phi}\rangle = \sum_{n=0}^\infty e^{in\phi}|n\rangle$ denotes the Susskind-Glogower phase vector, $e_\pm$ the raising/lowering photon operators, $s(n)$ is the sign of n. Other examples, as the multimode homodyne tomography, can be found in Ref. [4]. For more details the reader is addressed to the extensive literature on the subject.

Quantum Teleportation. In a general teleportation scheme, Alice and Bob share the entangled pure state $|A\rangle\!\rangle\langle\!\langle A|$ on the Hilbert space $\mathcal{H}_2\otimes\mathcal{H}_3$, while Alice performs a joint measurement on $\mathcal{H}_1\otimes\mathcal{H}_2$, corresponding to the POVM $|q(h)\rangle\!\rangle\langle\!\langle q(h)|$. The state that emerges at $\mathcal{H}_3$ at the Bob side after the Alice measurement is given by $\rho_l = \mathrm{Tr}_{12}[(\rho)_1\ (|A\rangle\!\rangle\langle\!\langle A|)_{23}\ (|q(h)\rangle\!\rangle\langle\!\langle q(h)|)_{12}] = A^* q^\dagger(h)\,\rho\, q(h) A^T$, for measurement outcome l. Alice transmits l to Bob, and if the quantum operation $A^* q^\dagger(l)$ is invertible[8]—as if unitary, when both $|A\rangle\!\rangle\langle\!\langle A|$ and $|q(h)\rangle\!\rangle\langle\!\langle q(h)|$ are maximally-entangled (the POVM is Bell)—then Bob can apply the inverse operation of $A^* q^\dagger(l)$ and recover the original state. *E voilà*: that's all! In Ref. [9] the teleportation schemes corresponding to SS from UIR of groups were presented, showing that all known teleportation schemes correspond to abelian groups. The original teleportation scheme [10] for one qubit corresponds to the projective UIR of the dihedral group D_2. The generalization to d dimension of the same Ref. [10] is the projective UIR of the group $\mathbb{Z}_d \times \mathbb{Z}_d$.[5] The continuous variables teleportation of Ref. [11] is an example of infinite dimensional teleportation, with the group being the abelian group of translations over the complex plane, with projective UIR given by the Weyl-Heisenberg group of displacement operators $D(z)$.[6] This last case is particularly interesting, since from it we can learn a general procedure to engineer Bell measurements from local measurements. Here the shared resource (from parametric downconversion) cannot be maximally entangled due to the infinite dimension of the Hilbert space, and the output state comes out distorted as $D_A^\dagger(z)\,\rho\,D_A(z)$, where $D_A(z) = D(z)\,A\,D^\dagger(z)$. The Bell measurement is achieved through the global unitary transformation $V = \exp\left(a^\dagger b - ab^\dagger\right)$ performed by the beam splitter over local (i.e. single-mode) homodyne joint measurements, resulting in $\frac{1}{\sqrt{\pi}}|D(x+iy)\rangle\!\rangle = V|2^{-1/2}x\rangle_0 \otimes |2^{-1/2}y\rangle_{-\pi/2}$, where $|v\rangle_\phi$ denotes an eigenvector of the quadrature operator $X_\phi = \frac{1}{2}\left(a^\dagger e^{i\phi} + ae^{-i\phi}\right)$ with eigenvalue v. Perfect Teleportation is achieved in the limit of vanishing distortion for $A_{ij} = \lim_{\xi\to 1}(1-|\xi|^2)^{1/2}\xi^j\delta_{ij}$. The present Bell measurement scheme maybe generalized to arbitrary quantum system by devising a single global unitary operator which achieves the transformation $V|i\rangle \otimes |j\rangle = |q(ij)\rangle\!\rangle$ from the local observable $|i\rangle \otimes |j\rangle$ to the Bell one $|q(ij)\rangle\!\rangle$, where $\{q(h)\}$ $h = (ij)$ is a unitary SS (i.e. maximally entangled). Modulo local unitary transformations, this resort to find the solution V of the factorization equation $V^\dagger\, q(ij) \otimes I\, V = m^{-1/2} U(i) \otimes U(j)$, where $U(ij) = \sqrt{m}q(ij)$, m =dim($\mathcal{H}$) (we choose $U(0,0) = I$), and the unitary local operators $U(j)$ connect the local POVM to a fixed reference vector $|0\rangle$ as $|j\rangle = U(j)|0\rangle$. For a single qubit a solution is given by the phase-shift operator $V = \exp\left(i\frac{\pi}{4}\sigma_y \otimes \sigma_y\right)$. For arbitrary quantum system, the general solution is unknown.

Quantum Optimal Cloning[3]. A quantum cloning map $\mathcal{C}$ from N to M copies is just a CP-map in $\mathcal{L}(\mathcal{L}(\mathcal{H}^{\otimes N}), \mathcal{L}(\mathcal{H}^{\otimes M}))$ which outputs a permutation-invariant state, i.e., invariant under S_M (which doesn't necessarily mean that the state is Bose or Fermi). A G-covariant cloning is a quantum cloning with R-matrix invariant under $U_g^{\otimes N} V_g^{*\otimes N}$. We call the quantum cloning optimal when it satisfies some optimality condition, typically it minimizes a given cost-function. The cloning which is covariant under the full unitary group $U(m)$, m =dim($\mathcal{H}$), and minimizes the fidelity $F = \mathrm{Tr}[\sigma^{\otimes M}\mathcal{C}(\sigma^{\otimes N})]$ for pure states σ is the universal cloning of Werner[12]. While proving optimality, one can also see that here the output state $\mathcal{C}(\sigma^{\otimes N})$ is bosonic. A $N = 1 \to M = 2$ cloning which achieves the optimal joint measurement of two conjugated quadratures of radiation (analogous of position and momentum of a particle) is the cloning of Cerf[13], which is covariant under the WH group. It minimizes the seed POVM fidelity $F = \mathrm{Tr}[(E_0^0 \otimes E_0^{\pi/2} \otimes v)R_C^{T_3}]$, where the vacuum state v "seeds" the coherent state POVM (optimal joint measurement), which is WH-covariant, and $E_0^0 \otimes E_0^{\pi/2}$ seeds the joint local measurements of commuting quadratures over clones under the joint displacement $D^{\otimes 2}$. Stinespring dilations of the cloning map $\mathcal{C}$ in the form $\mathcal{C}(\rho) = S^\dagger(\rho \otimes I_{\mathcal{A}})S$ for $\rho \in \mathcal{L}(\mathcal{H}^{\otimes N})$ for some ancillary Hilbert space $\mathcal{A}$, with $S \in \mathcal{L}(\mathcal{H}^{\otimes N}, \mathcal{H}^{\otimes N} \otimes \mathcal{A})$ generally not unitary, naturally provide quantum operations which physically achieve the cloning. For the Cerf cloning map[13] ($N = 1 \to M2$) one has $\mathcal{C}(\rho) = S(\rho \otimes I)S$, with $S = \frac{1}{\sqrt{2}}V(v \otimes I)V^\dagger$, V being the beam splitter transformation given above. This map can be achieved either probabilistically or unitarily (on a further extended Hilbert space). The probabilistic way is mimicked by just a couples of beam splitters in a row [14], with an ancillary chaotic input radiation (which imitates the identity in the Stinespring extension). A unitary realization is given in Ref. [15] through a network of three parametric amplifiers. Extensions of the probabilistic map to generic N and M can be obtained using multi-splitters and chaotic ancillary radiation, however the efficiency is going down exponentially vs M (for example, for $N = 1$ and large M it goes asymptotically as $M\overline{n}^{1-M}$ [14], $\overline{n}$ being the thermal photon number of the ancillary state).

Notes

1. Throughout the paper, $\int dg$ will represent a discrete sum for $\mathbf{G}$ discrete.

2. We call *observable* a generally complex operator with orthonormal spectral resolution, or, in other words, $c(l)$ is a generally complex function of a single self adjoint operator

3. Notice that the general method of noise deconvolution given in Ref. [4]—where the deconvolved estimator is achieved by evaluating the inverse CP-map of the noise over the estimator—is just equivalent to finding the biorthogonal basis of the noisy quorum.

4. The *quorum* obtained with the method of operator SS can be used also for quantum tomographic strategies different from the averaging one, e.g. the maximum-likelihood strategy. This method works only for the estimation of the density matrix itself (or for any set of unknown parameters of the density matrix), and it is restricted to finite dimensions. The likelihood function is just $L(\rho) = \sum_{j=1}^N \log \mathrm{Tr}[\rho p(l_j)(x_j)] - N\mathrm{Tr}(\rho)$, where j runs over the N measurements in the sample, $p(l)(x)$ denotes the (noisy) POVM corresponding to the l-th element of the quorum, with outcome x, and the search of the maximum is made over the parameters θ which parameterize the density matrix $\rho = \rho_\theta$. A full reconstruction of the density matrix can be achieved with the Cholesky parametrization $\rho = \tau^\dagger\tau$, searching over matrix elements (real on the diagonal) of the lower-triangular matrix τ, through a downhill simplex method.

5. $\mathbb{Z}_d \times \mathbb{Z}_d$ denotes the abelian group of discrete translations over a lattice embedded in a torus. The m-dimensional projective UIR is given by: $Z(j,l) = \sum_k e^{2\pi i k j/m}|k\rangle\langle k \oplus l|$, $Z(l,j)\,Z(l',j') = e^{2\pi i j l'/m} Z(l \oplus l', j \oplus j')$; The particular case for $m = 2$ is the dihedral group D_2 of π-rotations around three perpendicular axes, with the projective representation isomorphic to the nonabelian group of the three Pauli matrices plus identity. More explicitly, one has $U(0,0) = I$, $U(0,1) = \sigma_x$, $U(1,0) = \sigma_z$, $U(1,1) = i\sigma_y$.

6. The abelian group of translations over the complex plane, with projective UIR given by the Weyl-Heisenberg group (WH) of displacement operators ($z \in \mathbb{C}$): $D(z) = \exp(za^\dagger - z^* a)$, $D(z)D(w) = D(z+w)e^{\frac{1}{2}(zw^* - z^* w)}$.

References

[1] G. M. D'Ariano, P. Lo Presti, M. Sacchi, Phys. Lett. A **272** 32 (2000)

[2] A. Jamiolkowski, Rep. of Math. Phy. No. 4, **3** (1972)

[3] G. M. D'Ariano and P. Lo Presti, unpublished.

[4] G. M. D'Ariano, Phys. Lett. A **268** 151 (2000).

[5] G. M. D'Ariano, *Measuring Quantum States*, in *Quantum Optics and Spectroscopy of Solids*, ed. by T. Hakioğlu and A. S. Shumovsky, (Kluwer Academic Publisher, Amsterdam 1997), p. 175-202

[6] K. Banaszek and K. Wodkiewicz, Phys. Rev. Lett. **76**, 4344 (1996).

[7] G. M. D'Ariano, L. Maccone and M. G. A. Paris, quant-ph/0006006

[8] M. A. Nielsen and C. Caves, Phys. Rev. A **55**, 2547 (1997).

[9] S. L. Braunstein, G. M. D'Ariano, G. J. Milburn, and M. F. Sacchi, Phys. Rev. Lett. **84**, 3486 (2000).

[10] C. H. Bennett, G. Brassard, C. Crepeau, R. Jozsa, A. Peres, and W. K. Wotters, Phys. Rev. Lett. **70**, 1895 (1993).

[11] S. L. Braunstein and H. J. Kimble, Phys. Rev. Lett. **80**, 869 (1998).

[12] R. Werner, Phys. Rev. A **58**, 1827 (1998).

[13] N. J. Cerf, A. Ipe, and X. Rottenberg, quant-ph/9909037.

[14] G. M. D'Ariano, unpublished.

[15] G. M. D'Ariano, F. De Martini, and M. F. Sacchi, Phys. Rev. Lett. in press.

DECOHERENCE VERSUS THE IDEALIZATION OF MICROSYSTEMS AS CORRELATION CARRIERS BETWEEN MACROSYSTEMS

Ludovico Lanz, Bassano Vacchini
Dipartimento di Fisica dell'Università di Milano and INFN, Sezione di Milano
Via Celoria 16, I-20133, Milano, Italy
lanz@mi.infn.it, vacchini@mi.infn.it

Olaf Melsheimer
Fachbereich Physik, Philipps–Universität
Renthof 7, D-35032, Marburg, Germany
melsheim@mailer.uni-marburg.de

Keywords: Decoherence, macroscopic system, non-equilibrium statistical mechanics

Abstract It is argued that the appropriate framework to describe a microsystem as a correlation carrier between a source and a detector is non-equilibrium statistical mechanics for the compound source-detector system. An attempt is given to elucidate how this idealized notion of microsystem might arise inside a field theoretical description of isolated macrosystems: then decoherence appears as the natural limit of this idealization.

1. PREMINENT ROLE OF FIELD THEORY

Even if it is generally accepted that quantum field theory must be used in high energy physics, questions on foundations of quantum mechanics, description of measuring process and discussion of decoherence are usually addressed to in the context of the N particle generalization of the Schrödinger equation, while in that context quantum field theory is often only appreciated as a more refined tool to accommodate relativity and to account for particlelike aspects of electromagnetism. This is deeply rooted in mechanics and in the atomistic picture of matter. However one runs into difficulties and puzzles: objective properties for particles cannot be reconciled with quantum mechanics, quantum mechanical models of the measuring process are hardly compatible with objective description of macrosystems, decoherence must be supplied to the Schrödinger equation, either due to lack of isolation in any system, or by some additional stochasticity. We stress the point of view that the concept of a physical process running inside a suitably prepared isolated system and displayed by a certain set of relevant variables must be the starting point and that the theoretical description should be based on quantum field theory of finite systems. This point of view is much closer to thermodynamics than to mechanics: the basic ideas linked to atomistic structure of matter are however kept into account in a more subtle way by the quantization of the fields underlying the physical model and by the locality or quasi-locality of their interaction. The concept of a particle arises only in an unsharp way when one is pursuing universal features arising from locality of field theory and polishing away what comes from boundary conditions and residual interactions. In the most striking way a particle emerges when a process can be performed in which a *source part* of a macrosystem affects a *detecting part* of it through a microchannel consisting of a one- or few-particle system produced by the first part and directed to the second one. Looking at the problem this way, decoherence is obviously already inside the description: actually it is a hard theoretical job to drive it back from the microchannel, completely in tune with what experimentalists do. On the contrary the usual theoretical setting seems strange since it makes theorists work putting decoherence in, while experimentalists work hard to drive it back. This point of view about particles goes back to Ludwig's approach to quantum mechanics. By suitable axiomatization of general features of particular processes which give evidence of particles he succeeded in obtaining as a description of these processes quantum mechanics already in the modern form [1] (p.o.v. measures, operations, instruments) that is now generally recognized [2] as the formalism adequate to describe in a realistic way processes due to microsystems. Obviously when highly idealized schematizations can be applied, typically if decoherence can be neglected, and space-time symmetry for the microsystem can be assumed, the more schematic Dirac's book axiomatics emerges in all its geometrical neatness. While Ludwig pointed to a new theory encompassing microsystems and macrosystems in order to set the duality micro-macrosystems, we try to do this remaining inside quantum field theory, only improving somehow non-equilibrium theory for isolated systems. In § 2 we simply describe how a microchannel can arise, in § 3 the general structure of non-equilibrium theory is recalled and compatibility of the general dynamics of the system with the presence of the microchannel is indicated. The physical model we will use is a self-interacting spinless Schrödinger field confined inside a finite region ω. It can be trivially improved using several interacting fields with spin and should be amenable to the treatment of bound states and resonances between them. However all this is a very primitive stage since no intermediate gauge fields are introduced. To the space region ω a set of *normal modes* $\{u_r(\mathbf{x})\}$ is associated. They are an orthonormal, complete set of solutions of the stationary state equation:

$$-\frac{\hbar^2}{2m}\Delta_2 u_n(\mathbf{x}) + \mathcal{V}(\mathbf{x})u_n(\mathbf{x}) = W_n u_n(\mathbf{x}) \quad u_n(\mathbf{x}) = 0,\, \mathbf{x} \in \delta\omega,\; u_n \in L^2(\omega); \tag{1}$$

$\mathcal{V}(\mathbf{x})$ being a suitable potential for external and internal effective forces. The Schrödinger field is defined by:

$$\hat{\Phi}(\mathbf{x}) = \sum_n \hat{a}_n u_n(\mathbf{x}), \qquad [\hat{a}_n, \hat{a}^\dagger_{n'}]_\mp = \delta_{nn'} \tag{2}$$

with $\hat{a}_n$ Bose or Fermi annihilation operators on the Fock-space of the system. The Hamiltonian of the system is assumed as:

$$\hat{H} = \sum_n W_n \hat{a}^\dagger_n \hat{a}_n + \frac{1}{2}\int_\omega d^3\mathbf{x} d^3\mathbf{y}\, \hat{\Phi}^\dagger(\mathbf{x})\hat{\Phi}^\dagger(\mathbf{y})V(|\mathbf{x}-\mathbf{y}|)\hat{\Phi}(\mathbf{y})\hat{\Phi}(\mathbf{x}), \tag{3}$$

$V(r)$ being the basic microphysical input, a short-range function which gives the quasi-local form of interaction and will finally represent two-body interaction between the particles of the system.

2. THE MICROCHANNEL

Postponing a more technical sketch of the treatment of a non-equilibrium system, we now come to the main point: the microchannel. For this issue we choose a bundle of normal modes $r \in M$, M being a suitable subset of the indexes n: M are the normal modes, M^C the remaining ones. The field operator $\hat{\Phi}(\mathbf{x})$ contains them both

$$\hat{\Phi}(\mathbf{x}) = \hat{\Phi}_M(\mathbf{x}) + \hat{\Phi}_{M^C}(\mathbf{x}), \tag{1}$$

$$\hat{\Phi}_M(\mathbf{x}) = \sum_{r\in M} \hat{a}_r u_r(\mathbf{x}), \qquad \hat{\Phi}_{M^C}(\mathbf{x}) = \sum_{s\in M^C} \hat{a}_s u_s(\mathbf{x}).$$

The idea of a microchannel is formalized assuming that during a time interval $[t_0, t_1]$ the channel modes are depleted, so that the contribution to the dynamics of the system due to interaction between channel-modes is negligible. Then there is a possible dynamics of the system with unfeeded channel, described by a statistical operator $\hat{\rho}^0_t$ satisfying:

$$\hat{a}_r \hat{\rho}^0_t = 0, \qquad \forall r \in M \tag{2}$$

i.e., without excitations of M-modes and evolving according to $d\hat{\rho}_t^0/dt = -i/\hbar[\hat{H}, \hat{\rho}_t^0]$. There is however also a possible dynamics with feeded channel, described by a statistical operator

$$\hat{\rho}_t^{(1)} = \sum_{r,r' \in M} w_{rr'}(t) \hat{a}_r^\dagger \hat{\rho}_t^0 \hat{a}_{r'}, \tag{3}$$

with one excitation related to M. $\hat{\rho}_t^{(1)}$ is a positive operator if $w_{rr'}(t)$ is a positive matrix and by (2) it is normalized if $\sum_r w_{rr}(t) = 1$. For $t \in [t_0, t_1]$ the following representation of the statistical operator for a system endeavored with a microchannel should hold:

$$\hat{\rho}_t = (1 - \lambda)\, \hat{\rho}_t^0 + \lambda \sum_{r,r' \in M} w_{rr'}(t) \hat{a}_r^\dagger \hat{\rho}_t^0 \hat{a}_{r'}, \qquad 0 < \lambda < 1, \tag{4}$$

λ giving the probability that the microchannel is feeded. For the statistical operator (4) by (2) the interaction between modes in M is negligible. We assume at first that also the interaction between a mode $r \in M$ and the modes $s \in M^C$ can be neglected at least in the time interval $[t_0, t_1]$: then Liouville von Neumann equation for $\hat{\rho}_t$ implies that

$$\frac{dw_{rr'}}{dt}(t) = -\frac{i}{\hbar}(W_r - W_{r'}) w_{rr'}(t). \tag{5}$$

Eq. (5) can be considered as the evolution equation of a statistical operator $W^{(1)}(t)$ defined as

$$W^{(1)}(t) = \sum_{r,r' \in M} |r\rangle w_{rr'}(t) \langle r'|, \tag{6}$$

describing the microsystem inside the microchannel:

$$\frac{dW^{(1)}}{dt}(t) = -\frac{i}{\hbar}[H_0^{(1)}, W^{(1)}(t)], \qquad H_0^{(1)} = \sum_r |r\rangle W_r \langle r|,$$

while $|r\rangle$ is a basis in the one-particle Hilbert space $\mathcal{H}^M \subset L_2(\omega)$ spanned by $r \in M$, $\langle \mathbf{x}|r\rangle = u_r(\mathbf{x})$. Taking an observable $\hat{A}$ of the system or more in particular an element $\hat{E}^{\mathbf{A}}(S)$ of the spectral measure of some commuting set of self-adjoint operators on some σ-algebra of sets S, expectations or probability measures are given by expressions:

$$\mathrm{Tr}(\hat{A}\hat{\rho}_t) = (1 - \lambda)\mathrm{Tr}(\hat{A}\hat{\rho}_t^0) + \lambda \sum_{r,r' \in M} w_{rr'}(t) \mathrm{Tr}(\hat{a}_{r'} \hat{A} \hat{a}_r^\dagger \hat{\rho}_t^0) \tag{7}$$

$$\mathrm{Tr}(\hat{E}^{\mathbf{A}}(S)\hat{\rho}_t) = (1 - \lambda)\mathrm{Tr}(\hat{E}^{\mathbf{A}}(S)\hat{\rho}_t^0) + \lambda \sum_{r,r' \in M} w_{rr'}(t) \mathrm{Tr}(\hat{a}_{r'} \hat{E}^{\mathbf{A}}(S) \hat{a}_r^\dagger \hat{\rho}_t^0).$$

Setting $\text{Tr}(\hat{a}_{r'}\hat{A}\hat{a}_r^\dagger\hat{\rho}_t^0) = \langle r'|A_t^{(1)}|r\rangle$, $\text{Tr}(\hat{a}_{r'}\hat{E}^{\mathbf{A}}(S)\hat{a}_r^\dagger\hat{\rho}_t^0) = \langle r'|F_t^{(1)}(S)|r\rangle$, the r.h.s. of eq. (7) related to the microchannel can be written:

$$\sum_{r,r'\in M} w_{rr'}(t)\langle r'|A_t^{(1)}|r\rangle = \text{Tr}_{\mathcal{H}^M}(W^{(1)}(t)A_t^{(1)}) \tag{8}$$

$$\sum_{r,r'\in M} w_{rr'}(t)\langle r'|F_t^{(1)}(S)|r\rangle = \text{Tr}_{\mathcal{H}^M}(W^{(1)}(t)F_t^{(1)}(S)),$$

showing the typical mathematical structure of *one-particle* quantum mechanics formulated in the Hilbert space $\mathcal{H}^M$. Let us notice that even if $\hat{E}^{\mathbf{A}}(S)$ is a Fock-space projection valued measure, the related $\hat{F}^{\mathbf{A}}(S)$ is in general a p.o.v. normalized measure (positive operator valued or *effect* valued), according to modern axiomatics. Let us assume that $[\hat{E}^{\mathbf{A}}, \hat{N}_M]$, with $\hat{N}_M = \sum_{r\in M}\hat{a}_r^\dagger\hat{a}_r$ and furthermore

$$\begin{aligned}
&\sum_r \langle r_1|F_t^{(1)}(S)|r\rangle\langle r|F_t^{(1)}(S)|r_2\rangle \\
&\quad = \sum_r \text{Tr}(\hat{a}_{r_1}\hat{E}^{\mathbf{A}}(S)\hat{a}_r^\dagger\hat{\rho}_t^0)\text{Tr}(\hat{a}_r\hat{E}^{\mathbf{A}}(S)\hat{a}_{r_2}^\dagger\hat{\rho}_t^0) \\
&\quad \approx \sum_r \text{Tr}(\hat{a}_{r_1}\hat{E}^{\mathbf{A}}(S)\hat{a}_r^\dagger\hat{a}_r\hat{E}^{\mathbf{A}}(S)\hat{a}_{r_2}^\dagger\hat{\rho}_t^0) \\
&\quad = \text{Tr}(\hat{a}_{r_1}\hat{E}^{\mathbf{A}}(S)\hat{N}_M\hat{a}_{r_2}^\dagger\hat{\rho}_t^0) = \text{Tr}(\hat{a}_{r_1}\hat{E}^{\mathbf{A}}(S)\hat{a}_{r_2}^\dagger\hat{\rho}_t^0).
\end{aligned}$$

Then $\hat{F}_t^{(1)} = (\hat{F}_t^{(1)})^2$ turns out to be the spectral measure of a self-adjoint operator in $\mathcal{H}^M$, representing an observable $A_t^{(1)}$ of the microsystem. Let us stress that in this construction an explicit time dependence of $F_t^{(1)}$ ($A_t^{(1)}$ in the more particular case) arises in a quite natural way, since in addition to the microchannel a macrosystem dependent dynamics cannot be in general avoided. However right now the concept of a *good detecting part* inside the system can be easily formulated assuming that in the relevant time interval $[t_0, t_1]$ the explicit time dependence of $F_t^{(1)}$ ($A_t^{(1)}$) is either negligible or well-known on the basis of macroscopic phenomenology. We shall simply forget this time dependence setting $\hat{F}_t^{(1)}(S) \approx \hat{F}_{t_0}^{(1)}(S) \approx \hat{F}^{(1)}(S)$ ($\hat{A}_t^{(1)} \approx \hat{A}_{t_0}^{(1)} \approx \hat{A}^{(1)}$). Then r.h.s. of (8) becomes the basic formula for probability distribution of an observable given in general by a p.o.v. measure or by a self-adjoint operator $A^{(1)}$ in a more idealized situation, related to a microsystem associated with the statistical operator $W^{(1)}(t)$ and produced, living and detected inside the macrosystem: $\mathcal{H}^M$ is its Hilbert space and $H^{(1)} = \sum_{r\in M}|r\rangle W_r\langle r|$ its Hamiltonian. In this neat picture there is however a fundamental flaw: interaction with M^C modes has been neglected. Experimental particle physics shows us that this is indeed allowed when experimental physicists have been clever enough, but what we have described can never be more than an approximation. Corrections to this picture can be calculated: when they are small enough to preserve the basic picture, the concept of a microsystem undergoing an unavoidable decoherence arises in a very natural way. Let us take a statistical operator $\hat{\rho}^0_{t_0}$ of the form (4)

$$\hat{\rho}_t = (1-\lambda)\hat{\rho}^0_t + \lambda \sum_{r,r' \in M} w_{rr'}(t_0) e^{-\frac{i}{\hbar}\hat{H}(t-t_0)} \hat{a}^\dagger_r e^{+\frac{i}{\hbar}\hat{H}(t-t_0)} \hat{\rho}^0_t e^{-\frac{i}{\hbar}\hat{H}(t-t_0)} \hat{a}_{r'} e^{+\frac{i}{\hbar}\hat{H}(t-t_0)},$$

by the assumption of a *good detecting part* we can replace $\hat{\rho}^0_t$ with $\hat{\rho}^0_{t_0}$ in the second term. To take interaction with M^C modes into account a suitable time scale $\tau \approx \frac{\hbar}{\Delta W}$, where ΔW is the width of the energy band of the microchannel, must be considered and τ must be large enough to allow for a treatment of M, M^C interaction by a formalism similar to scattering theory, in which states are replaced by operators, the Hamiltonian by the Liouvillean $\mathcal{H} = -\frac{i}{\hbar}[\hat{H},\cdot]$ and the scattering operator by a scattering map $\mathcal{T}(z)$. In fact setting $\mathcal{H} = \mathcal{H}_0 + \mathcal{V}$, with $\mathcal{H}_0 = \frac{i}{\hbar}[\hat{H}_0,\cdot]$, $\hat{H}_0 = \sum_r W_r \hat{a}^\dagger_r \hat{a}_r$ one has:

$$e^{-\frac{i}{\hbar}\hat{H}(t-t_0)} \hat{a}^\dagger_r e^{+\frac{i}{\hbar}\hat{H}(t-t_0)} = \int_{-i\infty+\eta}^{+i\infty+\eta} \frac{dz}{2\pi i} e^{z\tau} (z-\mathcal{H})^{-1} \hat{a}^\dagger_r$$

$$= e^{-\frac{i}{\hbar}W_r(t-t_0)} \hat{a}^\dagger_r + \int_{-i\infty+\eta}^{+i\infty+\eta} \frac{dz}{2\pi i} e^{z\tau} \left[(z-\mathcal{H}_0)^{-1}\mathcal{T}(z)(z-\mathcal{H}_0)^{-1}\right] \hat{a}^\dagger_r$$

and similarly for the adjoint operator. The part depending on $\mathcal{T}(z)$ is responsible for decoherence. If $\hat{\rho}^0_{t_0}$ is an equilibrium state the treatment of this part gives in a perspicuous way the theory of Brownian motion [3] and in the limit of small momentum transfers the typical dynamics of a particle undergoing friction and position and momentum diffusion is found. One can expect that also in the case of a non-equilibrium $\hat{\rho}^0_{t_0}$ of the kind that will be discussed in § 3 a similar approach can be fruitful.

3. EMBEDDING OF MICROCHANNEL IN THE DYNAMICS OF A MACROSYSTEM

In our approach microsystems are derived entities and are no longer the basic elementary starting point of the physical description: then this description must stand on its own legs by a suitable reformulation of quantum mechanics of finite isolated non-equilibrium system. Let us briefly recall some main points about this general description of macrosystems [4]. The very claim that a physical system is isolated implies the choice of a subset of observables that are *under control* by a suitable preparation procedure performed on the system during a preparation time interval $[T, t_0]$. These observables are suitable slowly varying quantities, typically densities of conserved charges $\hat{A}_j(\boldsymbol{\xi})$ related to symmetry properties of the underlying local field theoretical structure. By their expectations $\{\langle\hat{A}_j(\boldsymbol{\xi})\rangle_t\}_j$ a set of classical fields $\{\zeta_j(\boldsymbol{\xi},t)\}_j$ is determined when these expectations are reproduced by means of a maximal von Neumann entropy state $\hat{w}[\zeta(t)]$. This is a generalized Gibbs state, induced at any time t by the statistical operator $\hat{\varrho}_t$ via the expectations $\langle\hat{A}_j(\boldsymbol{\xi})\rangle_t = \mathrm{Tr}\,(\hat{A}_j(\boldsymbol{\xi})\hat{\varrho}_t)$. It depends on the operators $\hat{A}_j(\boldsymbol{\xi})$ and the fields $\zeta_j(\boldsymbol{\xi},t)$ and provides an entropy for the classical state $\{\zeta_j(\boldsymbol{\xi},t)\}_j$. Such classical state, though related to statistical properties of the system that has been prepared in the time interval $[T,t]$, is taken as an objective property of the system at time t. This is already done, perhaps without complete awareness, when a velocity, a temperature or a chemical potential field is associated to a massive continuum. The statistical operator $\hat{\varrho}_t$, representing preparation until time $t \geq t_0$, shows the spontaneous dynamics of the isolated system in the time interval $[t_0,t]$, given by $\hat{\varrho}_t = e^{-\frac{i}{\hbar}\hat{H}(t-t_0)}\hat{\varrho}_{t_0}e^{+\frac{i}{\hbar}\hat{H}(t-t_0)}$, and can be written in the form

$$\hat{\varrho}_t = \exp\{-\zeta_0(t)\hat{\mathbf{1}} - \sum_j \int d\boldsymbol{\xi}\, \zeta_j(\boldsymbol{\xi},t)\hat{A}_j(\boldsymbol{\xi}) + \int_T^t dt'\, \hat{S}_t(t')\} \qquad (1)$$

where the history part $\hat{S}_t(t')$ for $t' \in [T,t_0]$ describes the preparation procedure and for $t' \in [t_0,t]$ can be simply given in terms of state variables $\zeta_j(\boldsymbol{\xi},t')$, $\dot{\zeta}_j(\boldsymbol{\xi},t')$ and the related density and current operators given at time $-(t-t')$ in Heisenberg picture. Since the first term alone in the exponent already exactly gives $\langle\hat{A}_j(\boldsymbol{\xi})\rangle_t$, an expansion with respect to the history term becomes very natural and e.g., at first order leads to an evolution equation

$$\frac{d}{dt}\langle\hat{A}_j(\boldsymbol{\xi})\rangle_t = \mathrm{Tr}\,(\dot{\hat{A}}_j(\boldsymbol{\xi})\hat{w}[\zeta(t)]) + \int_T^t dt'\, \langle\dot{\hat{A}}_j(\boldsymbol{\xi}),\hat{S}_t(t')\rangle_{\hat{w}[\zeta(t)]} + \cdots \qquad (2)$$

where $\hat{w}[\zeta(t)]$ is the generalized Gibbs state associated to the classical state at time t. To the expectation values of the operators $\dot{\hat{A}}_j(\boldsymbol{\xi})$ calculated with $\hat{w}[\zeta(t)]$ corrections responsible for irreversibility arise by the history term, which brings in foreground an integral over t' of the two point Kubo correlation functions between operators $\dot{\hat{A}}_j(\boldsymbol{\xi})$ and operators $\dot{\hat{A}}_j(\boldsymbol{\xi}',-(t-t'))$, $\hat{A}_j(\boldsymbol{\xi}',-(t-t'))$ in the macrostate $\hat{w}[\zeta(t)]$. Now the possibility of a great simplification imposes on our attention: as at equilibrium, these correlation functions, at least inside a time integral with well shaped classical state parameters, could practically vanish if $t' < t-\tau$, τ being a characteristic decay time; then $\int_T^t dt' \to \int_{t-\tau}^t dt'$, thus eliminating memory of the preparation procedure for $t > t_0+\tau$ and memory of previous classical state if it variates slowly enough during a time interval τ. We call such a situation *simple dynamics*: it dominates a large part of equilibrium thermodynamics. However we also discover a large arena where a behavior different from *simple dynamics* can arise.

One expects that when the fields $\zeta_j(\boldsymbol{\xi}, t_1)$ are inhomogeneous enough around time t_1, depletion of certain modes can arise: $\hat{a}_r \hat{w}[\zeta(t_1)] \approx 0$ if $r \in M$, then the part of previous history related to creation of these modes might present a slowly decaying contribution. Let us write:

$$\hat{\mathcal{S}}_{t_1}(t') = \hat{\mathcal{S}}_{t_1}^{(S)}(t') + \hat{\mathcal{S}}_{t_1}^{(M)}(t') \tag{3}$$

where $\hat{\mathcal{S}}_{t_1}^{(S)}(t')$ does not create particles in the M modes, thus yielding through (1) (with $\hat{\mathcal{S}}_{t_1}^{(S)}(t')$ at place of $\hat{\mathcal{S}}_t(t')$) a statistical operator $\hat{\varrho}_{t_1}^{(S)}$ with *simple dynamics*, while the full statistical operator $\hat{\varrho}_{t_1}$ can be written by an expansion with respect to $\hat{\mathcal{S}}_{t_1}^{(M)}(t')$, preserving positivity:

$$\hat{\varrho}_{t_1} = \lambda \left[\hat{\mathbf{1}} + \int_{t_1-\tau}^{t_1} dt' \, \hat{\mathrm{S}}_{t_1}^{(M)}(t') \right] \hat{\varrho}_{t_1}^{(S)} \left[\hat{\mathbf{1}} + \int_{t_1-\tau}^{t_1} dt' \, \hat{\mathrm{S}}_{t}^{(M)\dagger}(t') \right], \tag{4}$$

where $\hat{\mathrm{S}}_{t_1}^{(M)}(t')$ is essentially determined by $\hat{\mathcal{S}}_{t_1}^{(M)}(t')$. Let us write:

$$\int_{t_1-\tau}^{t_1} dt' \, \hat{\mathrm{S}}_{t_1}^{(M)}(t') = \sum_{r \in M} \hat{a}_r^\dagger \int_{t_1-\tau}^{t_1} dt' \int_{\omega_s} d\boldsymbol{\xi}' \, \hat{A}_r(-(t_1 - t'), \boldsymbol{\xi}'),$$

where the field operator $\hat{A}_r(-(t_1 - t'), \boldsymbol{\xi}')$ acts as annihilation operator typically for $\boldsymbol{\xi}'$ inside some space region ω_s. If a local observable $\hat{B}(\boldsymbol{\xi}, t)$ is considered at time t, such that correlations between the space-time point $(\boldsymbol{\xi}, t)$ and region ω_s are negligible, one can write

$$\mathrm{Tr}\,[\hat{B}(\boldsymbol{\xi}, t) \sum_{r,r' \in M} \hat{a}_r^\dagger \int_{t_1-\tau}^{t_1} dt' \int_{\omega_s} d\boldsymbol{\xi}' \, \hat{A}_r(-(t_1 - t'), \boldsymbol{\xi}') \hat{\varrho}_{t_1}^{(S)}$$
$$\cdot \int_{t_1-\tau}^{t_1} dt' \int_{\omega_s} d\boldsymbol{\xi}' \, \hat{A}_r^\dagger(-(t_1 - t'), \boldsymbol{\xi}') \hat{a}_r] \approx \mathrm{Tr}\,[\hat{B}(\boldsymbol{\xi}, t) \sum_{r,r' \in M} \hat{a}_r^\dagger \hat{\varrho}_{t_1}^{(S)} \hat{a}_r] \sigma_{r'r}$$

with

$$\sigma_{r'r} = \mathrm{Tr}\,[\int_{t_1-\tau}^{t_1} dt' \int_{\omega_s} d\boldsymbol{\xi}' \, \hat{A}_r(-(t_1 - t'), \boldsymbol{\xi}') \hat{\varrho}_{t_1}^{(S)} \int_{t_1-\tau}^{t_1} dt' \int_{\omega_s} d\boldsymbol{\xi}' \, \hat{A}_r^\dagger(-(t_1 - t'), \boldsymbol{\xi}')],$$

being a positive matrix describing the way in which the normal modes of M are feeded by destruction of particles in ω_s: in this way we are recovering the starting point of § 2.

Acknowledgments

L. L. and B. V. gratefully acknowledge financial support by MURST under Cofinanziamento. B. V. also acknowledges support by MURST under Progetto Giovani.

References

[1] Ludwig, G. (1985). *An Axiomatic Basis for Quantum Mechanics*, Springer, Berlin; (1983). *Foundations of Quantum Mechanics*, Springer, Berlin.

[2] Davies, E. B. (1976). *Quantum Theory of Open Systems*, Academic Press, London; Holevo, A. S. (1982). *Probabilistic and Statistical Aspects of Quantum Theory*, North Holland, Amsterdam; Kraus, K. (1983). *States, Effects and Operations*, in Lecture Notes in Physics, Volume 190, Springer, Berlin; Busch, P., Grabowski, M., and Lahti, P. J. (1995). Operational Quantum Physics, in *Lecture Notes in Physics*, Vol. m31, Springer, Berlin.

[3] Vacchini, B. (2000). *Phys. Rev. Lett.* **84**, 1374.

[4] Lanz, L., Melsheimer, O., Vacchini, B. (2000). *Rep. Math. Phys.* to appear.

QUANTUM MEASUREMENT, INFORMATION, AND COMPLETELY POSITIVE MAPS

Masanao Ozawa
School of Informatics and Sciences
Nagoya University
Chikusa-ku, Nagoya 464-8601, Japan

Keywords: Quantum measurement, quantum information, completely positive maps

Abstract Axiomatic approach to measurement theory is developed. All the possible statistical properties of apparatuses measuring an observable with nondegenerate spectrum allowed in standard quantum mechanics are characterized.

1. INTRODUCTION

Every measuring apparatus inputs the state ρ of the measured system and outputs the classical output $\mathbf{x}$ and the state $\rho_{\{\mathbf{x}=x\}}$ of the measured system conditional upon the outcome $\mathbf{x} = x$. In the conventional approach, the probability distribution of the classical output $\mathbf{x}$ and the output state $\rho_{\{\mathbf{x}=x\}}$ are determined by the spectral projections of the measured observable by the Born statistical formula and the projection postulate respectively. This description has been a very familiar principle in quantum mechanics, but is much more restrictive than what quantum mechanics allows. In the modern measurement theory, the problem has been investigated as to what is the most general description of measurement allowed by quantum mechanics. This paper investigates the problem of the determination of all the possible measurements of observables with nondegenerate spectrum and shows that the following conditions are equivalent for measurements of nondegenerate observables: (i) The joint probability distribution of the outcomes of successive measurements depends affinely on the initial state. (ii) The apparatus has an indirect measurement model. (iii) The state change is described by a positive superoperator valued measure. (iv) The state change is described by a completely positive superoperator valued measure. (v) The family of output states is a Borel family of density operators independent of the input state and can be arbitrarily chosen by the choice of the apparatus.

2. MEASUREMENT SCHEMES

Let $\mathcal{H}$ be a separable Hilbert space. The *state space* of $\mathcal{H}$ is the set $\mathcal{S}(\mathcal{H})$ of density operators on $\mathcal{H}$. In what follows, we shall give a general mathematical formulation for the statistical properties of measuring apparatuses. For heuristics, we shall consider a measuring apparatus which measures the quantum system **S** described by the Hilbert space $\mathcal{H}$. Every measuring apparatus has the output variable that gives the outcome on each measurement. We assume that the output variable takes values in a standard Borel space which is specified by each measuring apparatus. We shall denote by $\mathbf{A}(\mathbf{x})$ the measuring apparatus with the output variable $\mathbf{x}$ taking values in a standard Borel space Λ with the Borel σ-field $\mathcal{B}(\Lambda)$. The statistical property of the apparatus $\mathbf{A}(\mathbf{x})$ consists of the output distribution $\Pr\{\mathbf{x} \in \Delta \| \rho\}$ and the state reduction $\rho \mapsto \rho_{\{\mathbf{x}=x\}}$. The output distribution $\Pr\{\mathbf{x} \in \Delta \| \rho\}$ describes the probability distribution of the output variable $\mathbf{x}$ when the input state is $\rho \in \mathcal{S}(\mathcal{H})$, where $\Delta \in \mathcal{B}(\Lambda)$. The state reduction $\rho \mapsto \rho_{\{bx=x\}}$ describes the state change from the input state ρ to the output state $\rho_{\{\mathbf{x}=x\}}$, when the measurement leads to the output $\mathbf{x} = x$. The output state $\rho_{\{\mathbf{x}=x\}}$ is determined up to probability one with respect to the output distribution. The state reduction determines the collective state reduction $\rho \mapsto \rho_{\{\mathbf{x}\in\Delta\}}$ that describes the output state $\rho_{\{\mathbf{x}\in\Delta\}}$ given that the output of the measurement is in a Borel set Δ. The collective state reduction is naturally related to the state reduction by the integral formula

$$\rho_{\{\mathbf{x}\in\Delta\}} = \frac{1}{\Pr\{\mathbf{x} \in \Delta \| \rho\}} \int_{\Delta} \rho_{\{\mathbf{x}=x\}} \Pr\{\mathbf{x} \in dx \| \rho\}.$$

Formal description of the statistical properties of measuring apparatuses will be given as follows. Let Λ be a standard Borel space with σ-field $\mathcal{B}(\Lambda)$. Denote by $B(\Lambda, \mathcal{S}(\mathcal{H}))$ the space of Borel families of states for $(\mathcal{H}, \Lambda)$. The *state space* of Λ is the set $\mathcal{S}(\Lambda)$ of probability measures on $\mathcal{B}(\Lambda)$. A *measurement scheme* for $(\mathcal{H}, \Lambda)$ is the pair $(\mathcal{P}, \mathcal{Q})$ of a function $\mathcal{P}$ from $\mathcal{S}(\mathcal{H})$ to $\mathcal{S}(\Lambda)$ and a function $\mathcal{Q}$ from $\mathcal{S}(\mathcal{H})$ to $B(\Lambda, \mathcal{S}(\mathcal{H}))$. The function $\mathcal{P}$ is called the *output probability scheme*, and $\mathcal{Q}$ is called the *state reduction scheme*. Two measurement schemes $(\mathcal{P}, \mathcal{Q})$ and $(\mathcal{P}', \mathcal{Q}')$ are said to be *equivalent*, in symbols $(\mathcal{P}, \mathcal{Q}) \cong (\mathcal{P}', \mathcal{Q}')$, if $\mathcal{P} = \mathcal{P}'$ and $\mathcal{Q}\rho$ and $\mathcal{Q}'\rho$ differ only on a null set of the probability measure $\mathcal{P}\rho$, i.e.,

$$(\mathcal{P}\rho)\{x \in \Lambda |\ (\mathcal{Q}\rho)(x) \neq (\mathcal{Q}'\rho)(x)\} = 0,$$

for all $\rho \in \mathcal{S}(\mathcal{H})$.

The set of equivalence classes of measurement schemes for $(\mathcal{H}, \Lambda)$ is denoted by $\mathcal{M}(\mathcal{H}, \Lambda)$ and we define $\mathcal{M}(\mathcal{H}) = \bigcup_{\Lambda} \mathcal{M}(\mathcal{H}, \Lambda)$. A *measurement theory* for $\mathcal{H}$ is a pair $(\mathcal{A}, \mathbf{M})$ consisting of a nonempty set $\mathcal{A}$ and a function $\mathbf{M}$ from $\mathcal{A}$ to $\mathcal{M}(\mathcal{H})$. An element of $\mathcal{A}$ is called an *apparatus*. Every apparatus has its distinctive *output variable*. We denote the apparatuses with output variables $\mathbf{x}, \mathbf{y}, \ldots$ by $\mathbf{A}(\mathbf{x}), \mathbf{A}(\mathbf{y}), \ldots$, respectively. We assume that $\mathbf{x} = \mathbf{y}$ if and only if $\mathbf{A}(\mathbf{x}) = \mathbf{A}(\mathbf{y})$. The image of $\mathbf{A}(\mathbf{x})$ by $\mathbf{M}$ is denoted by $\mathbf{M}(\mathbf{x})$ instead of $\mathbf{M}(\mathbf{A}(\mathbf{x}))$ for simplicity, so that $\mathbf{M}(\mathbf{x})$ denotes the equivalence class of the measurement scheme corresponding to the apparatus $\mathbf{A}(\mathbf{x})$. The apparatus $\mathbf{A}(\mathbf{x})$ is called Λ*-valued* if $\mathbf{M}(\mathbf{x}) \in \mathcal{M}(\mathcal{H}, \Lambda)$. For any $\mathbf{A}(\mathbf{x}) \in \mathcal{A}$, the equivalence class $\mathbf{M}(\mathbf{x})$ of the measurement scheme is called the *statistical property of* $\mathbf{A}(\mathbf{x})$. We shall denote by $(\mathcal{P}_{\mathbf{x}}, \mathcal{Q}_{\mathbf{x}})$ a representative of $\mathbf{M}(\mathbf{x})$; in this case, we shall also write $\mathbf{M}(\mathbf{x}) = [\mathcal{P}_{\mathbf{x}}, \mathcal{Q}_{\mathbf{x}}]$. The function $\mathcal{P}_{\mathbf{x}}$ is called the *output probability* of $\mathbf{A}(\mathbf{x})$ and $\mathcal{Q}_{\mathbf{x}}$ the *state reduction* of $\mathbf{A}(\mathbf{x})$ which is determined uniquely up to the output probability one. Two apparatuses $\mathbf{A}(\mathbf{x})$ and $\mathbf{A}(\mathbf{y})$ are said to be *statistically equivalent*, in symbols $\mathbf{A}(\mathbf{x}) \cong \mathbf{A}(\mathbf{y})$, if they have the same statistical property, i.e., $\mathbf{M}(\mathbf{x}) = \mathbf{M}(\mathbf{y})$.

Suppose that a measurement theory $(\mathcal{A}, \mathbf{M})$ is given. Let $\mathbf{A}(\mathbf{x}) \in \mathcal{A}$ and $\mathbf{M}(\mathbf{x}) = [\mathcal{P}_{\mathbf{x}}, \mathcal{Q}_{\mathbf{x}}]$. The *probability distribution* of the output variable $\mathbf{x}$ in the state ρ is defined by

$$\Pr\{\mathbf{x} \in \Delta \| \rho\} = (\mathcal{P}_{\mathbf{x}}\rho)(\Delta)$$

for all $\rho \in \mathcal{S}(\mathcal{H})$ and $\Delta \in \mathcal{B}(\Lambda)$. This probability distribution is called the *output distribution* of $\mathbf{A}(\mathbf{x})$ in ρ. The *output states* $\{\rho_{\{\mathbf{x}=x\}}\}_{x\in\Lambda}$ of $\mathbf{A}(\mathbf{x})$ in ρ is defined by

$$\rho_{\{\mathbf{x}=x\}} = (\mathcal{Q}_{\mathbf{x}}\rho)(x)$$

for all $\rho \in \mathcal{S}(\mathcal{H})$ and $x \in \Lambda$.

Let $\Lambda_1, \ldots, \Lambda_n$ be standard Borel spaces. For $j = 1, \ldots, n$, let $\mathbf{A}(\mathbf{x}_j)$ be a Λ_j-valued apparatus. A *successive measurement in the input state* ρ is a sequence of measurements using $\mathbf{A}(\mathbf{x}_1), \ldots, \mathbf{A}(\mathbf{x}_n)$ such that the input state of the apparatus $\mathbf{A}(\mathbf{x}_1)$ is ρ and the input state of the apparatus $\mathbf{A}(\mathbf{x}_{j+1})$ is the output state of the apparatus $\mathbf{A}(\mathbf{x}_j)$ for $j = 1, \ldots, n-1$. The joint probability distribution of the outcomes of the successive measurements using $\mathbf{A}(\mathbf{x}_1), \ldots, \mathbf{A}(\mathbf{x}_n)$ in the input state ρ is naturally defined recursively by

$$\begin{aligned} &\Pr\{\mathbf{x}_1 \in \Delta_1, \ldots, \mathbf{x}_n \in \Delta_n \| \rho\} \\ &\quad = \int_{\Delta_1} \Pr\{\mathbf{x}_2 \in \Delta_2, \ldots, \mathbf{x}_n \in \Delta_n \| \rho_{\{\mathbf{x}_1 = x_1\}}\} \Pr\{\mathbf{x}_1 \in dx_1 \| \rho\} \quad (1) \end{aligned}$$

for $\Delta_1 \in \mathcal{B}(\Lambda_1), \ldots, \Delta_n \in \mathcal{B}(\Lambda_n)$.

Now, for a measurement theory $(\mathcal{A}, \mathbf{M})$, we consider the following requirement:

Mixing law of the n-joint probability distributions (nMLPD): *For any sequence $\mathbf{A}(\mathbf{x}_1), \ldots, \mathbf{A}(\mathbf{x}_n)$ of apparatuses with values in $\Lambda_1, \ldots, \Lambda_n$, respectively, if the input state ρ is the mixture of ρ_1 and ρ_2 such that $\rho =$*

$\alpha\rho_1 + (1-\alpha)\rho_2$ *with* $0 < \alpha < 1$ *then we have*

$$\begin{aligned}
&\Pr\{\mathbf{x}_1 \in \Delta_1, \ldots, \mathbf{x}_n \in \Delta_n \| \rho\} \\
&\quad = \alpha \Pr\{\mathbf{x}_1 \in \Delta_1, \ldots, \mathbf{x}_n \in \Delta_n \| \rho_1\} \\
&\qquad + (1-\alpha) \Pr\{\mathbf{x}_1 \in \Delta_1, \ldots, \mathbf{x}_n \in \Delta_n \| \rho_2\}
\end{aligned} \tag{2}$$

for all $\rho \in \mathcal{S}(\mathcal{H})$ *and* $\Delta_1 \in \mathcal{B}(\Lambda_1), \ldots, \Delta_n \in \mathcal{B}(\Lambda_n)$.

An *observable* of $\mathcal{H}$ is a self-adjoint operator (densely defined) on $\mathcal{H}$. We denote by E^A the spectral measure of an observable A. According to the Born statistical formula, we say that an $\mathbf{R}$-valued apparatus $\mathbf{A}(\mathbf{x})$ *measures an observable* A if

$$\Pr\{\mathbf{x} \in \Delta \| \rho\} = \mathrm{Tr}[E^A(\Delta)\rho] \tag{3}$$

for all $\rho \in \mathcal{S}(\mathcal{H})$ and $\Delta \in \mathcal{B}(\mathbf{R})$, where $\mathbf{R}$ stands for the real number field. A measurement theory $(\mathcal{A}, \mathbf{M})$ is called *nonsuperselective* if for any observable A there is at least one apparatus measuring A.

A Λ-*valued observable* of $\mathcal{H}$ is a projection valued measure E from $\mathcal{B}(\Lambda)$ to $\mathcal{L}(\mathcal{H})$ such that $E(\Lambda) = I$. The Born statistical formula is generalized as follows. We say that a Λ-valued apparatus $\mathbf{A}(\mathbf{x})$ *measures a* Λ-*valued observable* E if

$$\Pr\{\mathbf{x} \in \Delta \| \rho\} = \mathrm{Tr}[E(\Delta)\rho] \tag{4}$$

for all $\rho \in \mathcal{S}(\mathcal{H})$ and $\Delta \in \mathcal{B}(\Lambda)$. A *probability operator valued measure (POVM)* for $(\mathcal{H}, \Lambda)$ is a positive operator valued measure F from $\mathcal{B}(\Lambda)$ to $\mathcal{L}(\mathcal{H})$ such that $F(\Lambda) = I$. We say that a Λ-valued apparatus $\mathbf{A}(\mathbf{x})$ *measures a POVM* F for $(\mathcal{H}, \Lambda)$ if

$$\Pr\{\mathbf{x} \in \Delta \| \rho\} = \mathrm{Tr}[F(\Delta)\rho] \tag{5}$$

for all $\rho \in \mathcal{S}(\mathcal{H})$ and $\Delta \in \mathcal{B}(\Lambda)$. Conventional measurement theory is devoted to measurements of observables but modern theory extends the notion of measurements to measurements of POVMs [1–5]. We shall describe in the following requirement the essential feature of the modern approach.

Existence of probability operator valued measures (EPOVM): *For any apparatus* $\mathbf{A}(\mathbf{x})$, *there exists a POVM* $F_{\mathbf{x}}$ *uniquely such that* $\mathbf{A}(\mathbf{x})$ *measures* $F_{\mathbf{x}}$.

The EPOVM is justified by the following theorem proved in [3].

Theorem 1 *For any measurement theory* $(\mathcal{A}, \mathbf{M})$, *the EPOVM is equivalent to the 1MLPD.*

3. COLLECTIVE MEASUREMENT SCHEMES

In order to provide an alternative definition of measurement schemes, we call a pair $(\mathcal{P}, \mathcal{R})$ as a *collective measurement scheme* if $\mathcal{P}$ is a function from $\mathcal{S}(\mathcal{H})$ to $\mathcal{S}(\Lambda)$ and $\mathcal{R}$ is a function from $\mathcal{B}(\Lambda) \times \mathcal{S}(\mathcal{H})$ to $\mathcal{S}(\mathcal{H})$ satisfying

$$\sum_n (\mathcal{P}\rho)(\Delta_n)\mathcal{R}(\Delta_n, \rho) = \mathcal{R}(\Lambda, \rho) \tag{6}$$

for any countable Borel partition $\{\Delta_1, \Delta_2, \ldots\}$ of Λ and $\rho \in \mathcal{S}(\mathcal{H})$, where the sum is convergent in the trace norm. The function $\mathcal{R}$ is called the *collective reduction scheme*. Two collective measurement schemes $(\mathcal{P}, \mathcal{R})$ and $(\mathcal{P}', \mathcal{R}')$ are said to be *equivalent*, in symbols $(\mathcal{P}, \mathcal{R}) \cong (\mathcal{P}', \mathcal{R}')$, if $\mathcal{P} = \mathcal{P}'$ and $\mathcal{R}(\Delta, \rho) = \mathcal{R}'(\Delta, \rho)$ for all $\Delta \in \mathcal{B}(\Lambda)$ with $(\mathcal{P}\rho)(\Delta) > 0$.

Theorem 2 *The relation*

$$(\mathcal{P}\rho)(\Delta)\mathcal{R}(\Delta, \rho) = \int_\Delta (\mathcal{Q}\rho)(x)\, d(\mathcal{P}\rho)(x) \tag{7}$$

where $\Delta \in \mathcal{B}(\Lambda)$ and $\rho \in \mathcal{S}(\mathcal{H})$, sets up a one-to-one correspondence between the equivalence classes of measurement schemes $(\mathcal{P}, \mathcal{Q})$ and the equivalence classes of collective measurement schemes $(\mathcal{P}, \mathcal{R})$.

Let $(\mathcal{P}, \mathcal{Q})$ be a measurement scheme for $(\mathcal{H}, \Lambda)$. The collective measurement scheme $(\mathcal{P}, \mathcal{R})$ defined by (7) up to equivalence is called the *collective measurement scheme induced by* $(\mathcal{P}, \mathcal{Q})$ and the function $\mathcal{R}$ is called the *collective reduction scheme induced by* $(\mathcal{P}, \mathcal{Q})$.

Let $(\mathcal{A}, \mathbf{M})$ be a measurement theory satisfying the 1MLPD. For any $\mathbf{A}(\mathbf{x}) \in \mathcal{A}$, define $\mathcal{R}_\mathbf{x}$ to be the collective reduction scheme induced by $(\mathcal{P}_\mathbf{x}, \mathcal{Q}_\mathbf{x})$. The functions $\mathcal{R}_\mathbf{x}$ is called the *collective reduction* of the apparatus $\mathbf{A}(\mathbf{x})$. The *collective output states* $\{\rho_{\{\mathbf{x}\in\Delta\}}\}_{\Delta\in\mathcal{B}(\Lambda)}$ of $\mathbf{A}(\mathbf{x})$ in ρ is defined by

$$\rho_{\{\mathbf{x}\in\Delta\}} = \mathcal{R}_\mathbf{x}(\Delta, \rho)$$

for all $\rho \in \mathcal{S}(\mathcal{H})$ and $\Delta \in \mathcal{B}(\Lambda)$.

4. DAVIES-LEWIS POSTULATE

In what follows, we shall introduce some mathematical terminology independent of particular measurement theory. A *superoperator* for $\mathcal{H}$ is a bounded linear transformation on the space $\tau c(\mathcal{H})$ of trace class operators on $\mathcal{H}$. The *dual* of a superoperator L is the dual superoperator L^* defined by $\langle L^*A, \rho\rangle = \langle A, L\rho\rangle$ for all $A \in \mathcal{L}(\mathcal{H})$ and $\rho \in \tau c(\mathcal{H})$, where $\langle\cdot,\cdot\rangle$ stands for the duality pairing defined by $\langle A, \rho\rangle = \mathrm{Tr}[A\rho]$ for all $A \in \mathcal{L}(\mathcal{H})$ and $\rho \in \tau c(\mathcal{H})$. A superoperator is called *positive* if it maps positive operators to positive operators. We denote by $\mathcal{P}(\tau c(\mathcal{H}))$ the space of positive superoperators. Positive contractive superoperators are called *operations* [6].

A *positive superoperator valued (PSV) measure* is a mapping $\mathcal{I}$ from $\mathcal{B}(\Lambda)$ to $\mathcal{P}(\tau c(\mathcal{H}))$ such that if $\{\Delta_1, \Delta_2, \ldots\}$ is a countable Borel partition of Λ, then we have

$$\mathcal{I}(\Lambda)\rho = \sum_n \mathcal{I}(\Delta_n)\rho$$

for any $\rho \in \tau c(\mathcal{H})$, where the sum is convergent in the trace norm. The PSV measure $\mathcal{I}$ is said to be *normalized* if it satisfies the further condition

$$\mathrm{Tr}[\mathcal{I}(\Lambda)\rho] = \mathrm{Tr}[\rho]$$

for any $\rho \in \tau c(\mathcal{H})$. Normalized PSV measures are called *instruments* [2,8] for short.

Let $\mathcal{I}$ be an instrument for $(\Lambda, \mathcal{H})$. The relation

$$X(\Delta) = \mathcal{I}(\Delta)^* I \tag{8}$$

for all $\Delta \in \mathcal{B}(\Lambda)$, defines a POVM for $(\mathcal{H}, \Lambda)$, called the *POVM* of $\mathcal{I}$. The relation $T = \mathcal{I}(\Lambda)$ defines a trace preserving operation, called the *total operation* of $\mathcal{I}$.

A measurement theory $(\mathcal{A}, \mathbf{M})$ is said to satisfy the *Davies-Lewis postulate* if it satisfies the follows postulate.

Davies-Lewis postulate: *For any apparatus* $\mathbf{A}(\mathbf{x})$*, there is a normalized PSV measure* $\mathcal{I}_{\mathbf{x}}$ *satisfying the following relations for any* $\rho \in \mathcal{S}(\mathcal{H})$ *and Borel set* $\Delta \in \mathcal{B}(\Lambda)$*:*

(DL1) $\Pr\{\mathbf{x} \in \Delta \| \rho\} = \mathrm{Tr}[\mathcal{I}_{\mathbf{x}}(\Delta)\rho]$.

(DL2) $\rho_{\{\mathbf{x}\in\Delta\}} = \dfrac{\mathcal{I}_{\mathbf{x}}(\Delta)\rho}{\mathrm{Tr}[\mathcal{I}_{\mathbf{x}}(\Delta)\rho]}$.

From Theorem 2, the normalized PSV measure $\mathcal{I}_{\mathbf{x}}$ determines the output state $\rho_{\{\mathbf{x}=x\}}$ uniquely up to equivalence by

(DL3) $\displaystyle\int_\Delta \rho_{\{\mathbf{x}=x\}} \mathrm{Tr}[d\mathcal{I}_{\mathbf{x}}(x)\rho] = \mathcal{I}_{\mathbf{x}}(\Delta)\rho.$

From (DL1), the Davies-Lewis postulate implies 1MLPD. Although the Davies-Lewis description of measurement is quite general, it is not clear by itself whether it is general enough to allow all the possible measurements. The following theorem shows indeed it is the case.

Theorem 3 *For any nonsuperselective measurement theory, the Davies-Lewis postulate is equivalent to the 2MLPD.*

A measurement theory $(\mathcal{A}, \mathbf{M})$ is called a *statistical measurement theory* if it is nonsuperselective and satisfies 2MLPD.

Theorem 4 *Every statistical measurement theory satisfies nMLPD for all positive integer n.*

5. MEASUREMENTS OF OBSERVABLES

An instrument $\mathcal{I}$ for $(\Lambda, \mathcal{H})$ is said *decomposable* if $\mathcal{I}(\Delta)^*A = X(\Delta)T^*(A)$ for all $\Delta \in \mathcal{B}(\Lambda)$ and $A \in \mathcal{L}(\mathcal{H})$, where X is the POVM of $\mathcal{I}$ and T the total operation. For a given Λ-valued observable E for $(\Lambda, \mathcal{H})$, an instrument $\mathcal{I}$ is said to be *E-compatible* if the POVM of $\mathcal{I}$ is E. i.e., $\mathcal{I}^*(\Delta)I = E(\Delta)$ for all $\Delta \in \mathcal{B}(\Lambda)$; such an instrument is also called *observable measuring*. For an observable A, an instrument is called *A-compatible* if it is E^A-compatible. The following theorem shows in particular that every observable measuring instrument is decomposable (see [5, Proposition 4.3] for the case of completely positive instruments).

Theorem 5 *Let E be a Λ-valued observable of $\mathcal{H}$. Let $\mathcal{I}$ be an E-compatible instrument and T its total operation. Then, we have the following statements.*

(i) *For any $\Delta \in \mathcal{B}(\Lambda)$ and $\rho \in \tau c(\mathcal{H})$, we have*

$$\mathcal{I}(\Delta)\rho = T[E(\Delta)\rho] = T[\rho E(\Delta)] = T[E(\Delta)\rho E(\Delta)]. \tag{9}$$

(ii) *For any $\Delta \in \mathcal{B}(\Lambda)$ and $B \in \mathcal{L}(\mathcal{H})$, we have*

$$\mathcal{I}(\Delta)^*B = E(\Delta)T^*(B) = T^*(B)E(\Delta) = E(\Delta)T^*(B)E(\Delta). \tag{10}$$

All the E-compatible instruments are determined as follows.

Theorem 6 *Let E be an Λ-valued observable of $\mathcal{H}$. The relation*

$$\mathcal{I}(\Delta)\rho = T[E(\Delta)\rho] \tag{11}$$

for all $\Delta \in \mathcal{B}(\Lambda)$ and $\rho \in \tau c(\mathcal{H})$ sets up a one-to-one correspondence between the E-compatible instruments $\mathcal{I}$ and the E-compatible trace preserving operations T.

From the above theorem, in every statistical measurement theory we have the following: *For any apparatus $\mathbf{A}(\mathbf{x})$ measuring a Λ-valued observable E, there is an E-compatible trace preserving operation T such that the statistical property of $\mathbf{A}(\mathbf{x})$ is represented as follows.*

$$\text{output distribution:} \qquad \Pr\{\mathbf{x} \in \Delta \| \rho\} = \mathrm{Tr}[E(\Delta)\rho] \tag{12}$$

$$\text{collective output state:} \qquad \rho_{\{\mathbf{x}\in\Delta\}} = \frac{T[E(\Delta)\rho]}{\mathrm{Tr}[E(\Delta)\rho]} \tag{13}$$

6. MEASUREMENTS OF NONDEGENERATE OBSERVABLES

Let E be a Λ-valued observable for $\mathcal{H}$. We say that E is *nondegenerate* if the commutant of E is abelian. Two Borel families $\{\varrho_x\}_{x\in\Lambda}$ and $\{\varrho'_x\}_{x\in\Lambda}$ of density operators are said to be *E-equivalent* if they differ only on an E-null set, i.e.,

$$E\{x \in \Lambda |\ \varrho_x \neq \varrho'_x\} = 0.$$

Theorem 7 *Let E be a nondegenerate Λ-valued observable of $\mathcal{H}$. The Bochner integral formula*

$$T\rho = \int_\Lambda \varrho_x \operatorname{Tr}[\rho\, dE(x)] \tag{14}$$

for all $\rho \in \tau c(\mathcal{H})$ sets up a one-to-one correspondence between the E-compatible trace preserving operations T and the E-equivalence classes of the Borel families $\{\varrho_x\}_{x\in\Lambda}$ of density operators indexed by Λ.

From the above theorem, in any statistical measurement theory we conclude the following: *For any apparatus $\mathbf{A}(\mathbf{x})$ measuring an Λ-valued observable E, there is a Borel family $\{\varrho_x\}_{x\in\Lambda}$ of density operators uniquely up to E-equivalence such that the statistical property of $\mathbf{A}(\mathbf{x})$ is represented as follows.*

$$\text{output distribution:} \qquad \Pr\{\mathbf{x} \in \Delta \| \rho\} = \operatorname{Tr}[E(\Delta)\rho] \tag{15}$$

$$\text{output state:} \qquad \rho_{\{\mathbf{x}=x\}} = \varrho_x \tag{16}$$

7. INDIRECT MEASUREMENT MODELS

An *indirect measurement model* for $(\Lambda, \mathcal{H})$ is defined a 4-tuple $(\mathcal{K}, \sigma, U, E)$ consisting of a separable Hilbert space $\mathcal{K}$, a density operator σ on $\mathcal{K}$, a unitary operator U on $\mathcal{H} \otimes \mathcal{K}$, and a Λ-valued observable E of $\mathcal{K}$.

If the apparatus $\mathbf{A}(\mathbf{x})$ has the indirect measurement model $(\mathcal{K}, \sigma, U, E)$, then $\mathbf{A}(\mathbf{x})$ has the instrument $\mathcal{I}_{\mathbf{x}}$ defined by

$$\mathcal{I}_{\mathbf{x}}(\Delta)\rho = \operatorname{Tr}_{\mathcal{K}}[(I \otimes E(\Delta))U(\rho \otimes \sigma)U^\dagger], \tag{17}$$

where $\Delta \in \mathcal{B}(\Lambda)$ and $\rho \in \tau c(\mathcal{H})$, so that the statistical property of $\mathbf{A}(\mathbf{x})$ is described by $\mathcal{I}_{\mathbf{x}}$ with relations (DL1) and (DL2). The above instrument $\mathcal{I}_{\mathbf{x}}$ is called the *instrument of* $\mathbf{A}(\mathbf{x})$.

Now, we consider the following hypothesis.

Indirect measurability hypothesis: *For any indirect measurement model $(\mathcal{K}, \sigma, U, E)$, there is an apparatus $\mathbf{A}(\mathbf{x})$ with the instrument $\mathcal{I}_{\mathbf{x}}$ defined by (17).*

In general, an instrument $\mathcal{I}_{\mathbf{x}}$ is said to be *realized* by an indirect measurement model $(\mathcal{K}, \sigma, U, E)$ if (17) holds for any $\rho \in \tau c(\mathcal{H})$. In this case, the instrument $\mathcal{I}$ is called *unitarily realizable*. Under the indirect measurability hypothesis, every unitarily realizable instrument represents the statistical property of an apparatus.

In the sequel, a statistical measurement theory $(\mathcal{A}, \mathbf{M})$ is called a *standard measurement theory*, if it satisfies the indirect measurability hypothesis. It is natural to consider that any standard measurement theory is consistent with the standard formulation of quantum mechanics.

8. COMPLETE POSITIVITY

Let $\mathcal{D} = \tau c(\mathcal{H})$ or $\mathcal{D} = \mathcal{L}(\mathcal{H})$. A linear transformation L on $\mathcal{D}$ is called *completely positive (CP)* iff for any finite sequences of operators $A_1, \ldots, A_n \in \mathcal{D}$ and vectors $\xi_1, \ldots, \xi_n \in \mathcal{H}$ we have

$$\sum_{ij} \langle \xi_i | L(A_i^\dagger A_j) | \xi_j \rangle \geq 0.$$

The above condition is equivalent to that $L \otimes I$ maps positive operators in the algebraic tensor product $\mathcal{D} \otimes \mathcal{L}(\mathcal{K})$ to positive operators in $\mathcal{D} \otimes \mathcal{L}(\mathcal{K})$ for any Hilbert space $\mathcal{K}$. Obviously, every CP superoperators are positive. A superoperator is CP if and only if its dual superoperator is CP. An instrument $\mathcal{I}$ is called *completely positive (CP)* if every operation $\mathcal{I}(\Delta)$ is CP. It can be seen easily from (17) that unitarily realizable instruments are CP. Conversely, the following theorem, proved in [5], asserts that every CP instrument is unitarily realizable.

Theorem 8 *For any CP instrument $\mathcal{I}$ for $(\Lambda, \mathcal{H})$, there is a separable Hilbert space $\mathcal{K}$, a unit vector Φ in $\mathcal{K}$, a unitary operator U on $\mathcal{H} \otimes \mathcal{K}$, and a Λ-valued observable E of $\mathcal{K}$ satisfying the relation*

$$\mathcal{I}(\Delta)\rho = \mathrm{Tr}_{\mathcal{K}}[(I \otimes E(\Delta))U(\rho \otimes |\Phi\rangle\langle\Phi|)U^\dagger]$$

for all $\Delta \in \mathcal{B}(\Lambda)$ and $\rho \in \tau c(\mathcal{H})$.

The following theorem shows that the complete positivity of observable measuring instruments is determined by their total operations.

Theorem 9 *Let E be a Λ-valued observable. Then, an E-compatible instrument is CP if and only if its total operation is CP.*

From the above theorem, in any standard measurement theory we conclude the following statement [5]: *The statistical equivalence classes of apparatuses $\mathbf{A}(\mathbf{x})$ measuring a Λ-valued observable E with indirect measurement models* *are in one-to-one correspondence with the E-compatible trace preserving CP operations, where the statistical property is represented by (12) and (13).*

For the case of nondegenerate observables, we have the following simple characterizations.

Theorem 10 *Let E be a nondegenerate Λ-valued observable. Then, every E-compatible operation is completely positive. Every E-compatible instrument is completely positive.*

From the above theorem and Theorem 8, in the statistical measurement theory we conclude: *Every apparatus measuring a nondegenerate (Λ-valued) observable is statistically equivalent to the one having an indirect measurement model.*

Every Borel family $\{\varrho_x\}$ of density operators indexed by Λ defines an E-compatible trace preserving operation by Theorem 7, and it is automatically completely positive so that it is realized by an indirect measurement model. Thus, we conclude the following: *The statistical equivalence classes of apparatuses $\mathbf{A}(\mathbf{x})$ measuring a nondegenerate Λ-valued observable E are in one-to-one correspondence with the Borel family $\{\varrho_x\}$ of density operators indexed by Λ, where the statistical property is represented by (15) and (16).*

References

[1] C. W. Helstrom, *Quantum Detection and Estimation Theory*, Academic Press, New York, 1976.

[2] E. B. Davies, *Quantum Theory of Open Systems*, Academic Press, London, 1976.

[3] M. Ozawa, *Reports on Math. Phys.* **18**, 11 (1980).

[4] A. S. Holevo, *Probabilistic and Statistical Aspects of Quantum Theory*, North-Holland, Amsterdam, 1982.

[5] M. Ozawa, *J. Math. Phys.* **25**, 79 (1984).

[6] R. Haag and D. Kastler, *J. Math. Phys.* **5**, 848 (1964).

[7] E. B. Davies and J. T. Lewis, *Commun. Math. Phys.* **17**, 239 (1970).

[8] M. Ozawa, Operations, disturbance, and simultaneous measurability, [online preprint: LANL quant-ph/0005054].

[9] M. Ozawa, in *Probability Theory and Mathematical Statistics*, edited by K. Itô and J. V. Prohorov, pages 518–525, Lecture Notes in Math. **1021**, Springer, Berlin, (1983).

ON THE NUMBER OF ELEMENTS NEEDED IN A POVM ATTAINING THE ACCESSIBLE INFORMATION

Peter W. Shor
AT&T Labs—Research
Florham Park, NJ 07932, USA
shor@research.att.com

Keywords: Quantum measurement, accessible information, POVM's

Abstract We investigate an symmetric set of three quantum states in three dimensions with interesting properties, which we call the lifted trine states. We show that for the ensemble consisting of the three lifted trine states taken with equal probabilities, the POVM measurement realizing the accessible information must contain six projectors, giving a counter-example to a conjecture of Levitin.

Accessible information was one of the first information-theoretic quantities investigated with respect to quantum systems. The accessible information of an ensemble of quantum states is the maximum mutual information obtainable between the states of the ensemble and the outcomes of a POVM (positive operator valued measurement) on these states. In this paper, we investigate how complicated a measurement which achieves the accessible information must be. Davies' theorem gives a maximum on the number of elements of the POVM needed to attain the accessible information; namely, if the ensemble being considered is contained in a d-dimensional Hilbert space, then at most d^2 elements are needed in an optimal POVM. When all the states are real, this bound can be improved to $d(d-1)/2$ [1]. C. Fuchs and A. Peres [2] have done numerical studies on ensembles containing only two elements. They found no examples where more than d states were needed; that is, they found that the optimal measurement could always be a von Neumann measurement. In two dimensions, this was proved by Levitin [3], who also conjectured that in d dimensions, if the number of quantum states in the ensemble is at most d, a von Neumann measurement is sufficient to attain the accessible information. In this paper, we give an ensemble of three real quantum states in three dimensions, where a POVM attaining the accessible information must contain at least six elements, the maximum by Davies' theorem.

Quantum Communication, Computing, and Measurement 3
Edited by P. Tombesi and O. Hirota, Kluwer Academic/Plenum Publishers, 2001

We investigate the accessible information of an ensemble consisting of three quantum states we call the *lifted trine states*, with equal probabilities on these states. The lifted trine states are obtained by starting with the two-dimensional quantum trine states: $(1,0)$, $(-1/2,\sqrt{3}/2)$, $(-1/2,-\sqrt{3}/2)$, introduced by Holevo [4] and later studied by Peres and Wootters [5]. We add a third dimension to the Hilbert space of the trine states, and lift all of the trine states out the plane into this dimension by an angle of $\arcsin\sqrt{\alpha}$, so the states become $(\sqrt{1-\alpha},0,\sqrt{\alpha})$, and so forth. We will be dealing with small α (roughly, $\alpha < .1$), so that they are close to being planar. This is the most interesting regime. When the trine states are lifted further out of the plane, they start behaving in relatively uninteresting ways until they are close to being vertical; then they start being interesting again, but this second regime is beyond the scope of this paper. The lifted trine states are thus:

$$\begin{aligned} T_0(\alpha) &= (\sqrt{1-\alpha},0,\sqrt{\alpha}) \\ T_1(\alpha) &= (-\tfrac{1}{2}\sqrt{1-\alpha},\tfrac{\sqrt{3}}{2}\sqrt{1-\alpha},\sqrt{\alpha}) \\ T_2(\alpha) &= (-\tfrac{1}{2}\sqrt{1-\alpha},-\tfrac{\sqrt{3}}{2}\sqrt{1-\alpha},\sqrt{\alpha}) \end{aligned} \tag{1}$$

When it is clear what α is, we may drop it from the notation and use T_0, T_1, and T_2.

In this section, we find the accessible information for this ensemble of lifted trine states. The accessible information is defined as the maximal mutual information between the trine states (with probabilities $\frac{1}{3}$ each) and the elements of a POVM measuring these states. Because the lifted trine states are real vectors, it follows from the version of Davies' theorem for real states [1] that there is an optimal POVM with at most six elements, all the components of which are real. The lifted trine states are three-fold symmetric, so by symmetrizing we can assume that the optimal POVM is three-fold symmetric (possibly at the cost of introducing extra POVM elements). Also, the optimal POVM can be taken to have one-dimensional elements E, so the elements can be described as vectors $|v_i\rangle$ where $E_i = |v_i\rangle\langle v_i|$. This means that there is an optimal POVM whose vectors come in triples of the form: $\sqrt{p}P_0(\phi,\theta)$, $\sqrt{p}P_1(\phi,\theta)$, $\sqrt{p}P_2(\phi,\theta)$, where p is a scalar probability and

$$\begin{aligned} P_0(\phi,\theta) &= (\cos\phi\cos\theta,\cos\phi\sin\theta,\sin\phi) \\ P_1(\phi,\theta) &= (\cos\phi\cos(\theta+2\pi/3),\cos\phi\sin(\theta+2\pi/3),\sin\phi) \\ P_2(\phi,\theta) &= (\cos\phi\cos(\theta-2\pi/3),\cos\phi\sin(\theta-2\pi/3),\sin\phi). \end{aligned} \tag{2}$$

Suppose that the optimal POVM has several such triples, which we call $\sqrt{p_1}\,P_b(\phi_1,\theta_1)$, $\sqrt{p_2}\,P_b(\phi_2,\theta_2)$, ..., $\sqrt{p_m}\,P_b(\phi_m,\theta_m)$. It is easily seen that the conditions for this set of vectors to be a POVM are that

$$\sum_{i=1}^{m} p_i \sin^2(\phi_i) = 1/3 \quad \text{and} \quad \sum_{i=1}^{m} p_i = 1. \tag{3}$$

The formula for accessible information I_A can be broken into pieces so that each triple contributes a linear amount to I_A. That is, I_A is the weighted average (weighted according to p_i) of some contribution $I(\phi, \theta)$ from each (ϕ, θ). To show this, recall that I_A is the mutual information between the input and the output, and this can be expressed as the entropy of the input less the entropy of the input given the output, $H(X_{\text{in}}) - H(X_{\text{in}}|X_{\text{out}})$. The term $H(X_{\text{in}}|X_{\text{out}})$ naturally decomposes into terms corresponding to the various POVM outcomes, and there are several ways of assigning the entropy of the input $H(X_{\text{in}})$ to the various POVM elements in order to complete this decomposition. Following this analysis eventually gives the same answer as is obtained below (and is in fact how I arrived at it). I briefly sketch this analysis so as to give the intuition behind it, and then go into detail in a second analysis, which is superior in that it explains the form of the answer.

For each ϕ, and each α, there is a θ that optimizes $I(\phi, \theta)$. This θ starts out at $\pi/6$ for $\phi = 0$, decreases until it hits 0 at some value of ϕ (which depends on α), and stays at 0 until ϕ reaches its maximum value of $\pi/2$. For a fixed α, by finding (numerically) the optimal value of θ for each ϕ and using it to obtain the contribution to I_A attributable to that ϕ, we get a curve giving the optimal contribution to I_A for each ϕ. If this curve is plotted, with the x-value being $\sin^2 \phi$ and the y-value being the contribution to I_A, an optimal POVM is obtained from the set of points on this curve whose average x-value is $1/3$ (from Eq. 3), and whose average y-value is as large as possible given this constraint on the x-values. A simple convexity argument shows that we only need at most two points from the curve to obtain this optimum, and that we will need one or two points depending on whether the relevant part of the curve is concave or convex. For small α, it turns out that the relevant piece of the curve is convex, and we need two ϕ's to achieve the maximum. Each of these ϕ's corresponds to a triple of POVM elements. One of the (ϕ, θ) pairs is $(0, \pi/6)$, and the other is $(\phi_\alpha, 0)$ for some $\phi_\alpha > \arcsin(1/\sqrt{3})$. The formula for this ϕ_α will be derived later.

The analysis in the remainder of this section shows that this six-outcome optimal POVM can be described in a different way, which unifies the optimal measurements for the different α's. For small α ($\alpha < \gamma_1$ for some constant γ_1), we first take the trine $T_b(\alpha)$ and make a partial measurement which either projects it down to the x, y plane or lifts it further out of the plane so that it becomes the trine $T_b(\gamma_1)$. (Note that γ_1 is independent of α.) If the trine was projected into the x, y plane, we make a second measurement using the POVM with outcome vectors $\sqrt{2/3}(0, 1)$ and $\sqrt{2/3}(\pm\sqrt{3}/2, -1/2)$.

This is the optimal POVM for trines in the x, y-plane. If the trine was lifted up, we use the von Neumann measurement that projects onto the basis containing $(\sqrt{2/3}, 0, \sqrt{1/3})$ and $(-\sqrt{1/6}, \pm\sqrt{1/2}, \sqrt{1/3})$. If α is larger than γ_1 (but still smaller than $8/9$) we skip the first partial measurement, and just use the above von Neumann measurement. Here, γ_1 is obtained by numerically solving a fairly complicated equation; we suspect that no closed form expression for γ_1 exists. The value of γ_1 is .061367, which is $\sin^2\phi$ for $\phi = .25033$ radians (14.343°).

We now give more details on this decomposition of the POVM into a two-step process. We first apply a partial measurement which does not extract all of the quantum information, i.e., it leaves a quantum residual state that is not completely determined by the measurement outcome. Formally, we apply one of a set of matrices A_i satisfying $\sum_i A_i^\dagger A_i = I$. If we start with a pure state $|\,v\rangle$, we observe the i'th outcome with probability $\langle v\,|\,A_i^\dagger A_i\,|\,v\rangle$, and in this case the state $|\,v\rangle$ is taken to the state $A_i\,|\,v\rangle$. For our purposes, we choose as the A_i's the matrices $\sqrt{p_i}\,M(\phi_i)$ where

$$M(\phi) = \begin{pmatrix} \sqrt{\frac{3}{2}}\cos\phi & 0 & 0 \\ 0 & \sqrt{\frac{3}{2}}\cos\phi & 0 \\ 0 & 0 & \sqrt{3}\sin\phi \end{pmatrix} \tag{4}$$

The $\sqrt{p_i}\,M(\phi_i)$ will form a valid partial measurement if and only if $\sum_i p_i \sin^2(\phi_i) = 1/3$ and $\sum_i p_i = 1$, the same conditions [Eq. (3)] as for the $P_b(\phi_i, \theta_i)$. By first applying the above $\sqrt{p_i}\,M(\phi_i)$, and then applying the von Neumann measurement with the three basis vectors

$$\begin{aligned} V_0(\theta) &= \left(\sqrt{\tfrac{2}{3}}\cos(\theta), \sqrt{\tfrac{2}{3}}\sin(\theta), \tfrac{1}{\sqrt{3}}\right) \\ V_1(\theta) &= \left(\sqrt{\tfrac{2}{3}}\cos(\theta + 2\pi/3), \sqrt{\tfrac{2}{3}}\sin(\theta + 2\pi/3), \tfrac{1}{\sqrt{3}}\right) \\ V_2(\theta) &= \left(\sqrt{\tfrac{2}{3}}\cos(\theta - 2\pi/3), \sqrt{\tfrac{2}{3}}\sin(\theta - 2\pi/3), \tfrac{1}{\sqrt{3}}\right) \end{aligned} \tag{5}$$

we obtain the POVM given by the vectors $\sqrt{p_i}\,P_b(\theta_i, \phi_i)$; checking this is simply a matter of verifying that $V_b(\theta)M(\phi) = P_b(\theta, \phi)$. Now, after applying $\sqrt{p_i}\,M(\phi_i)$ to the trine $T_0(\alpha)$, we get the vector

$$(\sqrt{3/2}\sqrt{1-\alpha}\sqrt{p_i}\cos\phi_i, 0, \sqrt{3}\sqrt{\alpha}\sqrt{p_i}\sin\phi_i). \tag{6}$$

This is just the state $\sqrt{p_i'}\,T_0(\alpha_i')$ where $T_0(\alpha_i')$ is the trine state with

$$\alpha_i' = \frac{\alpha\sin^2\phi_i}{\alpha\sin^2\phi_i + \frac{1}{2}(1-\alpha)\cos^2\phi_i} \tag{7}$$

and

$$p'_i = 3p_i \left[\alpha \sin^2 \phi + \tfrac{1}{2}(1-\alpha)\cos^2 \phi\right] \tag{8}$$

is the probability that we observe this trine state, given that we started with $T_0(\alpha)$. Similar formulae hold for the trine states T_1 and T_2. We compute that

$$\sum_i p'_i \alpha'_i = \sum_i 3p_i \alpha \sin^2(\phi_i) = \alpha. \tag{9}$$

Also notice that the first stage of this process, the partial measurement which applies the matrices $\sqrt{p_i}\, M(\phi_i)$, reveals no information about which of T_0, T_1, T_2 that we started with. Thus, by the chain rule for classical Shannon information [6], the accessible information obtained by our two-stage measurement is just the weighted average (the weights being p'_i) of the maximum over θ of the Shannon mutual information $I_{\alpha'_i}(\theta)$ between the outcome of the von Neumann measurement $V(\theta)$ and the trines $T(\alpha'_i)$. By convexity, it suffices to use only two values of α'_i to obtain this maximum. In fact, the optimum is obtained using either one or two values of α'_i depending on whether the function

$$I_{\alpha'} = \max_\theta I_{\alpha'}(\theta)$$

is concave or convex over the appropriate region. In the remainder of this section, we give the results of computing (numerically) the values of this function $I_{\alpha'}$, and we show that for small enough α it is convex, so that we need two values of α'. We will then show that obtaining this maximum requires a POVM with six outcomes.

We need to calculate the Shannon capacity of the channel whose input is one of the three trine states $T(\alpha')$, and whose output is determined by the von Neumann measurement $V(\theta)$. Because of the symmetry, we can calculate this using only the first projector V_0. The Shannon mutual information between the input and the output is $H(X_{\text{in}}) - H(X_{\text{in}}|X_{\text{out}})$, which is

$$I_{\alpha'} = \log_2 3 + \sum_{b=0}^{2} \langle V_0(\theta)|T_b(\alpha')\rangle^2 \log\left(\langle V_0(\theta)|T_b(\alpha')\rangle^2\right). \tag{10}$$

We compute that the θ giving the maximum I'_α is $\pi/6$ when $\alpha' = 0$, decreases continuously to 0 at $\alpha' = .056651$ and remains 0 for larger α'. (See Fig. 1.) This value .056651 corresponds to an angle of .24032 radians (13.769°). This θ was determined by using the computer package Maple to numerically find the point at which $dI_\alpha(\theta)/d\theta = 0$.

By plugging this optimum θ into the formula for $I_{\alpha'}$, we obtain the optimum von Neumann measurement of the form V above. We believe that this is also the optimal generic von Neumann measurement, but we have not proved this. The maximum of $I_{\alpha'}(\theta)$ over θ, and curves that show the behavior of $I_{\alpha'}(\theta)$

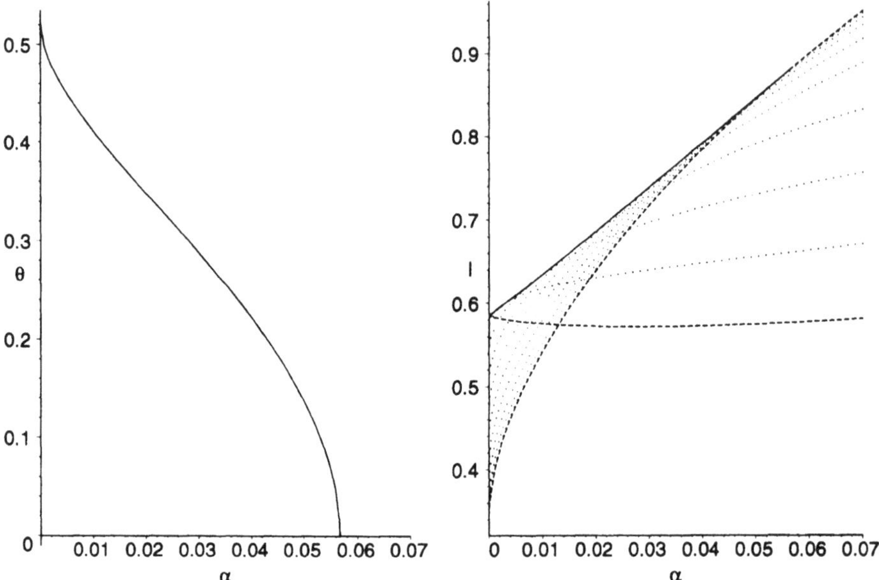

Figure 1 The value of θ maximizing I_α for α between 0 and .07. This function starts at $\pi/6$ at $\alpha = 0$, decreases until it hits 0 at $\alpha =$ .056651 and stays at 0 for larger α.

Figure 2 This plot shows $I_\alpha(\theta)$ for various θ. The dashed curves are $I_\alpha(0)$ and $I_\alpha(\pi/6)$. Note that $\theta = 0$ is optimal for $\alpha > .056651$ and $\theta = \pi/6$ is optimal for $\alpha = 0$. The dotted curves show $I_\alpha(\theta)$ for θ at intervals of 3° between 0 and $\pi/6 = 30°$. The solid curve shows $I_\alpha(\theta_{\text{opt}})$ for those α where neither 0 nor $\pi/6$ is the optimal θ. The solid curve is slightly convex; this is clearer in Fig. 3.

for constant θ, are plotted in Fig. 2. We can now observe that the first part of the curve is convex, and thus that for small α the best POVM will have six projectors, corresponding to two values of α'. We calculate that for trine states with $\alpha < .061367$, the two values of α' giving the maximum accessible information are 0 and .061367; we will let $\gamma_1 = .061367$ be this second value. The trine states $T(\gamma_1)$ make an angle of .25033 radians (14.343°) with the x-y plane. The accessible information thus obtained is plotted in Fig. 3.

We can now invert the formula for α' (Eq. 7) to obtain a formula for $\sin^2(\phi)$, and substitute the value of $\alpha' = \gamma_1$ back into the formula to obtain the optimal POVM. We find

$$\sin^2(\phi_\alpha) \quad = \quad \frac{1-\alpha}{1+\alpha\left(\frac{2-3\gamma_1}{\gamma_1}\right)}$$

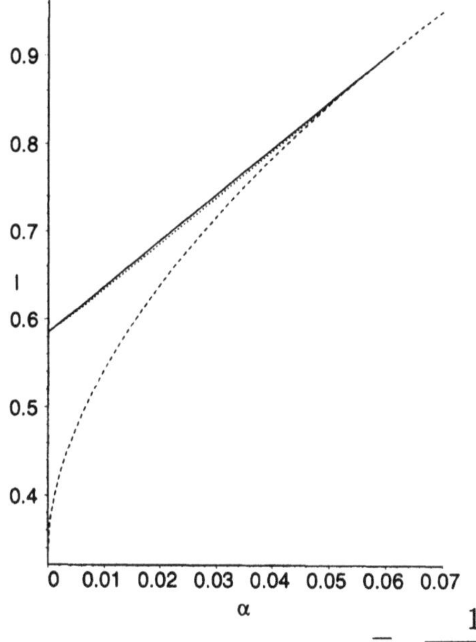

Figure 3 This graph contains three curves. The dashed curve is $I_\alpha(0)$ and the dotted curve is the maximum over θ of $I_\alpha(\theta)$ for $\alpha < .056651$. The solid curve is the convex envelope of the other two curves. This solid curve is a linear interpolation between $\alpha = 0$ and $\alpha = .061367$ and corresponds to a POVM having six elements. It gives the accessible information for the lifted trine states $T(\alpha)$ when $0 \le \alpha \le .061367$.

$$= \frac{1-\alpha}{1+29.591\alpha} \qquad (11)$$

where $\gamma_1 = .061367$ as above. Thus, the elements in the optimal POVM we have found for the trines $T(\alpha)$, when $\alpha < \gamma_1$, are the six vectors $P_b(\phi_\alpha, 0)$ and $P_b(0, \pi/6)$, where ϕ_α is given by Eq. 11 and $b = 0, 1, 2$. We must also prove there are no other POVM's which attain the same accessible information. The argument above shows that any optimal POVM must contain only projectors chosen from these six vectors: only those two values of α' can give the maximum capacity, and for each of these values of α' there are only three projectors in $V(\theta)$ which can maximize $I_{\alpha'}$ for these α'. It is easy to check that there is only one set of probabilities p_i which make the above six vectors into a POVM, and that none of these probabilities are 0 for $0 < \alpha < \gamma_1$. Thus, for the lifted trine states with $0 < \alpha < .061367$, there is only one POVM maximizing accessible information, and it contains six elements, the maximum possible for real states by a generalization of Davies' theorem [1].

Acknowledgments

I would like to thank Christopher Fuchs, Osamu Hirota, A. S. Holevo and Masao Osaki for several helpful comments on an early draft of this paper.

References

[1] M. Sasaki, S. M. Barnett, R. Jozsa, M. Osaki and O. Hirota, "Accessible information and optimal strategies for real symmetrical quantum sources,"

Phys. Rev. A, vol. 59, 3325–3335, 1999, LANL e-print quant-ph/9812062.

[2] C. Fuchs, personal communication.

[3] L. B. Levitin, "Optimal quantum measurements for two pure and mixed states," in *Quantum Communications and Measurement* (V. P. Belavkin, O. Hirota and R. L. Hudson, eds.), Plenum Press, New York, 1995, pp. 439–448.

[4] A. S. Holevo, "Information-theoretical aspects of quantum measurement," *Problemy Peredachi Informatsii* vol. 9, no. 2, pp. 31–42 1973 (in Russian); English translation: A. S. Kholevo, *Problems of Information Transmission,* vol. 9, pp. 110-118, 1973.

[5] A. Peres and W. K. Wootters, "Optimal detection of quantum information," *Phys. Rev. Lett.,* vol. 66, pp. 1119–1122 (1991).

[6] T. M. Cover and J. A. Thomas, *Elements of Information Theory,* Wiley, New York, 1991.

Peter W. Shor received his B.S. in Mathematics from Caltech in 1981 and his Ph.D. in Applied Math from M.I.T. in 1985. After a one-year postdoc at the Mathematical Sciences Research Center in Berkeley, CA, he worked at AT&T Bell Labs until AT&T split in two in 1996, after which he worked at AT&T Labs. His research has mainly been in combinatorics, algorithms, and discrete geometry. Since 1994, he has been working in the areas of quantum computing and quantum information theory. In 1998, he received the Nevanlinna Prize at the International Congress of Mathematicians in Berlin, and the Quantum Communication Award at the QCM conference in Evanston, Illinois. In 1999, he received a MacArthur Fellowship.

EINSELECTION AND DECOHERENCE FROM AN INFORMATION THEORY PERSPECTIVE

W. H. Zurek
Theory Division, T-6, MS B288, LANL, Los Alamos, NM 87545
whz@lanl.gov

Abstract We introduce and investigate a simple model of conditional quantum dynamics. It allows for a discussion of the information-theoretic aspects of quantum measurements, decoherence, and environment-induced superselection (einselection).

Introduction

Transfer of information was the focus of attention [1-3] of research on decoherence since the early days. In the intervening two decades this perspective was not forgotten [4], but the study of different mechanisms of decoherence [5-9] took precedence over considerations of information-theoretic nature. The aim of this paper is to sketch a few ideas which tie the "traditional" points of view of einselection and decoherence (especially the issue of the preferred pointer basis) to various other aspects of decoherence that have a strong connection with information-theoretic concepts.

A large part of our discussion shall be based on a simple model of conditional dynamics, which is a direct generalization of the "bit by bit" measurement introduced in [1] and studied in [3]. We shall introduce the model in Section 2 and use it to compute the "price" of information in units of action in Section 3. Section 4 defines information theoretic quantum *discord* between two classically identical definitions of mutual information. Discord can be regarded as a measure of a violation of classicality of a joint state of two quantum subsystems. Section 5 turns to the evolution of the state of the environment in course of decoherence. The *redundancy ratio* introduced there can be regarded as a measure of objectivity of quantum states. A large redundancy ratio is a sufficient condition for an effective classicality of quantum states.

Quantum Communication, Computing, and Measurement 3
Edited by P. Tombesi and O. Hirota, Kluwer Academic/Plenum Publishers, 2001

1. CONTROLLED SHIFTS FOR CONDITIONAL DYNAMICS

The simplest example of an entangling quantum evolution is known as the controlled not (c-not). It involves two bits (a "control" and a "target"). Their interaction leads to:

$$|0_C\rangle|x_T\rangle \longrightarrow |0_C\rangle|x_T\rangle \tag{1a}$$

$$|1_C\rangle|x_T\rangle \longrightarrow |1_C\rangle|\neg x_T\rangle \tag{1b}$$

where the state $|\neg x\rangle$ is defined through a basis-dependent negation:

$$\neg(\gamma|0_T\rangle + \eta|1_T\rangle) = \gamma|1_T\rangle + \eta|0_T\rangle\ . \tag{2}$$

Classical c-not "flips" the target bit whenever the control is in the state "1", but does nothing otherwise. Quantum c-not is an obvious generalization.

The distinction between the classical and quantum c-not comes from the fact that both quantum and classical bits can be in an arbitrary superposition. Thus, c-not starting from a superposition of $|0\rangle$ and $|1\rangle$ will in general lead to an entangled state. Moreover, when both the control and the target start in Hadamard-transformed states:

$$|\pm\rangle = (|0\rangle \pm |1\rangle)/\sqrt{2}\,, \tag{3}$$

c-not reverses direction:

$$|\pm\rangle|+\rangle \longrightarrow |\pm\rangle|+\rangle; \tag{4a}$$

$$|\pm\rangle|-\rangle \longrightarrow |\mp\rangle|-\rangle\ . \tag{4b}$$

Above, we have dropped labels: The original control is always to the left, as was the case in Eq. (1). We say "the original", because the Hadamard transform of Eq. (3) reverses the direction of the information flow in the quantum c-not. As can be seen in Eq. (4), the sign of the former control (left ket) flips when the former target is in the state $|-\rangle$.

Study of such simple models has led to the concept of preferred pointer states [1] and einselection [2,3]. We shall not review here these well known developments, directing the reader instead to the already available [10] or forthcoming [11] reviews of the subject.

Controlled shift (c-shift) is a straightforward generalization of c-not. The original truth table (an analogue of Eq. (1)) can be written as:

$$|s_j\rangle|A_k\rangle \longrightarrow |s_j\rangle|A_{k+j}\rangle \tag{5}$$

There is also a control and a target (which we shall more often call "the system $\mathcal{S}$" and "the apparatus $\mathcal{A}$", reflecting this nomenclature in notation). Equation

(5) implies Eq. (1) when both $\mathcal{S}$ and $\mathcal{A}$ have two-dimensional Hilbert spaces. Moreover, when $j = 0$, Eq. (1) becomes a model of a pre-measurement:

$$|\Psi_0\rangle = |\psi\rangle|A_0\rangle = \left(\sum_i a_i|s_i\rangle\right)|A_0\rangle \longrightarrow \sum_i a_i|s_i\rangle|A_i\rangle = |\Psi_t\rangle . \quad (6)$$

As in the case of c-not, in the respective bases $\{|s_i\rangle\}$ and $\{|A_k\rangle\}$, Eq. (5) seems to imply a one - way flow of information, from $\mathcal{S}$ to $\mathcal{A}$. However, a complementary basis [12,13] can be readily defined:

$$|B_k\rangle = N^{-\frac{1}{2}} \sum_{l=0}^{N-1} \exp(\frac{2\pi i}{N} kl)\, |A_l\rangle . \quad (7)$$

It is analogous to the Hadamard transform we have introduced before, but it also has an obvious affinity to the Fourier transform. We shall call it a Hadamard-Fourier Transform (HFT). It is straightforward to show that the orthonormality of $\{|A_k\rangle\}$ immediately implies:

$$\langle B_l|B_m\rangle = \delta_{lm} . \quad (8)$$

The inverse of HFT can be easily given:

$$|A_k\rangle = N^{-\frac{1}{2}} \sum_{l=0}^{N-1} \exp(-\frac{2\pi i}{N} kl)\, |B_l\rangle . \quad (9)$$

Consequently, for an arbitrary $|\psi\rangle$;

$$|\psi\rangle = \sum_n \alpha_n|A_n\rangle = \sum_k \beta_k|B_k\rangle , \quad (10)$$

where the coefficients are given by the HFT;

$$\beta_k = N^{-\frac{1}{2}} \sum_{n=0}^{N-1} \exp(-\frac{2\pi i}{N} kn)\alpha_n . \quad (11)$$

To implement the truth table of Eq. (5) we shall use observables of the apparatus:

$$\hat{A} = \sum_{k=0}^{N-1} k|A_k\rangle\langle A_k| ; \quad (12a)$$

$$\hat{B} = \sum_{l=0}^{N-1} l|B_l\rangle\langle B_l| , \quad (12b)$$

as well as the observable of the system:

$$\hat{s} = \sum_{l=0}^{N-1} l|s_l\rangle\langle s_l| \ . \tag{13}$$

The interaction Hamiltonian of the form;

$$H_{int} = g\hat{s}\hat{B} \tag{14}$$

acting over a period t will induce a transition;

$$\exp(-iH_{int}t/\hbar)|s_j\rangle|A_k\rangle = |s_j\rangle N^{-\frac{1}{2}} \sum_{l=0}^{N-1} \exp[-i(jgt/\hbar + 2\pi k/N)l]|B_l\rangle \ . \tag{15}$$

Thus, if the coupling constant g is selected so that the associated action is

$$I = gt/\hbar = G \times 2\pi/N \ . \tag{16}$$

a perfect one-to-one correlation between the states of the system and of the apparatus can be accomplished, since:

$$\exp(-iH_{int}t/\hbar)|s_j\rangle|A_k\rangle \ = \ |s_j\rangle|A_{\{k+G*j\}_N}\rangle \ . \tag{17}$$

Thus, true to its name, the interaction described here accomplishes a simple shift, Eq. (5), of the state of the apparatus, while the system acts as a control. The index $\{k+G*j\}_N$ has to be evaluated modulo N (where N is the number of the orthogonal states of the apparatus) so that when $k + G * j > N$, the apparatus states can "rotate" through $|A_0\rangle$ and stop where the interaction takes it. The integer G can be regarded as the gain factor. As Eqs. (14) - (17) imply, the adjacent states of the system (i.e., $|s_j\rangle$, $|s_{j+1}\rangle$) get mapped onto the states of the apparatus that are G apart "on the dial".

When the dimension of the Hilbert space of the system n is such that

$$nG < N \ , \tag{18}$$

the above model provides one with a simple example of amplification. It is possible to use it to study the utility of amplification in increasing signal to noise ratio in measurements [11]. It also shows why amplification can bring about decoherence and effective irreversibility (although `c-shift` is of course perfectly reversible). We shall employ `c-shift` to study the cost of information transfer, to introduce information – theoretic discord, the measure of the classicality of correlations, and to discuss objectivity of quantum states which arises from the redundancy of the records imprinted by the state of the system on its environment.

2. PLANCK'S CONSTANT AND THE PRICE OF A BIT

Transfer of information is the objective of the measurement process and an inevitable consequence of most interactions. It happens in course of decoherence. Here we shall quantify its cost in the units of action.

The consequence of the interaction between $\mathcal{S}$ and $\mathcal{A}$ is the correlated state $|\Psi_t\rangle$, Eq. (6). While the joint state of $\mathcal{AS}$ is pure, each of the subsystems is in a mixed state given by the reduced density matrix of the system

$$\rho_{\mathcal{S}} = \mathrm{Tr}_{\mathcal{A}}|\Psi_t\rangle\langle\Psi_t| = \sum_{i=0}^{N-1} |a_i|^2 |s_i\rangle\langle s_i| \; ; \tag{19a}$$

and of the apparatus

$$\rho_{\mathcal{A}} = \mathrm{Tr}_{\mathcal{S}}|\Psi_t\rangle\langle\Psi_t| = \sum_{i=0}^{N-1} |a_i|^2 |A_i\rangle\langle A_i| \, . \tag{19b}$$

The correlation brought about by the interaction, Eq. (6), leads to the loss of information about $\mathcal{S}$ and $\mathcal{A}$ individually. The entropy of each increases to

$$\mathcal{H}_{\mathcal{S}} = -\mathrm{Tr}\rho_{\mathcal{S}} \log \rho_{\mathcal{S}} = -\sum_{i=0}^{N-1} |a_i|^2 \log |a_i|^2 = -\mathrm{Tr}\rho_{\mathcal{A}} \log \rho_{\mathcal{A}} = \mathcal{H}_{\mathcal{A}} \, . \tag{20}$$

As the evolution of the whole $\mathcal{AS}$ is unitary, the entropies of the subsystems must be compensated by the decrease of their mutual entropy, i.e., by the increase of their mutual information:

$$\mathcal{I}(\mathcal{S}:\mathcal{A}) = \mathcal{H}_{\mathcal{S}} + \mathcal{H}_{\mathcal{A}} - \mathcal{H}_{\mathcal{SA}} = -2\sum_{i=0}^{N-1} |a_i|^2 \log |a_i|^2 \tag{21}$$

Above, $H_{\mathcal{SA}}$ is the joint entropy of $\mathcal{SA}$. This quantity, Eq. (21), was introduced in the quantum context as a measure of entanglement some time ago [3] and has been since rediscovered and used [14].

The cost of a bit of information in terms of some other physical quantity is an often raised question. In the context of our model we shall inquire what is the cost of a bit transfer in terms of action. Let us then consider a transition represented by Eq. (6). The associated action must be no less than

$$I = \sum_{j=0}^{N-1} |a_j|^2 \arccos |\langle A_0|A_j\rangle| \, . \tag{22}$$

When $\{|A_j\rangle\}$ are mutually orthogonal, the action is:

$$I = \pi/2 \tag{23a}$$

in Planck ($h = 2\pi\hbar$) units. This estimate can be lowered by using a judiciously chosen initial $|A_0\rangle$ which is a superposition of the outcomes $|A_j\rangle$. For a two-dimensional Hilbert space the average action can be thus brought down to $\pi\hbar/4$ [1,3]. In general, an interaction of the form

$$H_{\mathcal{SA}} = ig \sum_{k=0}^{N-1} |s_k\rangle\langle s_k| \sum_{l=0}^{N-1} (|A_k\rangle\langle A_l| - |A_l\rangle\langle A_k|) \tag{24}$$

saturates at the lower bound given by

$$I = \arcsin \sqrt{1 - 1/N}\,. \tag{23b}$$

As the dimensionality of the Hilbert spaces increases, the least action approaches $\pi/2$ per completely entangling interaction. The action per bit will be less when, for a given N, the transferred information is maximized, which happens when $|a_i|^2 = 1/N$ in Eq. (22). Then the cost of information in Planck units is

$$\iota = \frac{I}{\log N} \approx \frac{\pi}{2\log_2 N}\,. \tag{25}$$

The cost of information per bit decreases with increasing N, the dimension of the Hilbert space of the smaller of the two entangled systems.

This result is at the same time both enlightening and disappointing: It shows that the cost of information transfer is not "fixed" (as one might have hoped). Rather, the least total amount of action needed for a complete entanglement is at least asymptotically fixed as Eqs. (23) show. Consequently, the least price per bit goes down when information is transferred "wholesale", i.e., when N is large. Yet, this is enlightening, as it may indicate why in the classical continuous world (where N is effectively infinite) one may be ignorant of that price and convinced that information is free.

3. DISCORD

Mutual information can be defined either by the symmetric formula, Eq. (21), or through an asymmetric looking equation which employs conditional entropy:

$$\mathcal{J}(\mathcal{S}:\mathcal{A}) = H_{\mathcal{S}} - H_{\mathcal{S}|\mathcal{A}} \tag{26}$$

Above, $H_{\mathcal{S}|\mathcal{A}}$ expresses the average ignorance of $\mathcal{S}$ remaining after the observer has found out the state of $\mathcal{A}$. In classical physics, the two formulae, Eqs. (21) and (26), are strictly identical, so that the *discord* between them:

$$\delta\mathcal{I} = \mathcal{I}(\mathcal{S}:\mathcal{A}) - \mathcal{J}(\mathcal{S}:\mathcal{A}) = 0 \tag{27a}$$

always disappears [15]:

$$\delta\mathcal{I} = 0\,. \tag{27b}$$

In quantum physics things are never this simple: To begin with, conditional information depends on the observer finding out about one of the subsystems, which implies a measurement. So $H_{\mathcal{S}|\mathcal{A}}$ must be carefully defined before Eqs. (26) and (27) become meaningful.

Conditional entropy is non-trivial in the quantum context because, in general, in order to find out $H_{\mathcal{S}|\mathcal{A}}$ one must choose a set of projection operators Π_j and define a conditional density matrix given by the outcome corresponding to Π_j through

$$\tilde{\rho}_{\mathcal{S}|\Pi_j} = \mathrm{Tr}_{\mathcal{A}} \Pi_j \rho_{\mathcal{S}\mathcal{A}} , \tag{28}$$

where in the simplest case $\Pi_j = |C_j\rangle\langle C_j|$ is a projection operator onto a pure state of the apparatus. A normalized $\rho_{\mathcal{S}|\Pi_j}$ can be obtained by using the probability of the outcome

$$p_j = \mathrm{Tr} \tilde{\rho}_{\mathcal{S}|\Pi_j} ; \tag{29}$$

$$\rho_{\mathcal{S}|\Pi_j} = p_j^{-1} \tilde{\rho}_{\mathcal{S}|\Pi_j} . \tag{30}$$

The conditional density matrix $\rho_{\mathcal{S}|\Pi_j}$ represents the description of the system $\mathcal{S}$ available to the observer who knows that the apparatus $\mathcal{A}$ is in a subspace defined by Π_j. For a pure initial state and an exhaustive measurement the conditional density matrix will also be pure. We shall however consider a broader range of possibilities, including joint density matrices which undergo a decoherence process, so that

$$\rho^P_{\mathcal{S}\mathcal{A}} = \sum_{i,j} \alpha_i \alpha_j^* |s_i\rangle\langle s_j||A_i\rangle\langle A_j| \Longrightarrow \sum_i |\alpha_i|^2 |s_i\rangle\langle s_i||A_i\rangle\langle A_i| = \rho^D_{\mathcal{S}\mathcal{A}} . \tag{31}$$

This transition is accompanied by an increase in entropy

$$\Delta H(\rho_{\mathcal{S}\mathcal{A}}) = H(\rho^D_{\mathcal{S}\mathcal{A}}) - H(\rho^P_{\mathcal{S}\mathcal{A}})$$

and by the simultaneous disappearance of the ambiguity in what was measured [1-3]. Now, $\rho_{\mathcal{S}|\Pi_j}$ is no longer pure, unless $\Pi_j = |A_j\rangle\langle A_j|$. That is, a measurement of the apparatus in bases other than the pointer basis will leave an observer with varying degrees of ignorance about the state of the system. More general cases when the density matrix is neither a pure pre-decoherence projection operator $\rho^P_{\mathcal{S}\mathcal{A}}$ to the left of the arrow in Eq. (31) nor a completely decohered $\rho^D_{\mathcal{S}\mathcal{A}}$ state on the right are possible and typical.

To define discord $\delta\mathcal{I}$ we finalize our definition of $H_{\mathcal{S}|\mathcal{A}}$:

$$H_{\mathcal{S}|\mathcal{A}} = \sum_i p_{|C_i\rangle} H(\rho^D_{\mathcal{S}|C_i\rangle}) . \tag{32}$$

Above, we have used an obvious notation for the density matrix conditioned upon pure states $\{|C_j\rangle\}$. We emphasize again that the conditional entropy depends on $\rho_{\mathcal{S}\mathcal{A}}$, but also on the choice of the observable measured on $\mathcal{A}$. In classical physics all observables commute, so there is no such dependence. Thus, non-commutation of observables in quantum theory is the ultimate source of the information - theoretic discord.

The obvious use for the discord is to employ it as a measure of how non-classical the underlying correlation of two quantum systems is. In particular, when there exists a set of states in one of the two systems in which the discord disappears, the state represented by $\rho_{\mathcal{SA}}$ admits a classical interpretation of probabilities in that special basis. Moreover, unless the discord disappears for trivial reasons (which would happen in the absence of correlation, i.e., when $\rho_{\mathcal{SA}} = \rho_{\mathcal{S}}\rho_{\mathcal{A}}$), the basis which minimizes the discord can be regarded as "the most classical". For $\delta\mathcal{I} = 0$ the states of such preferred basis and their corresponding eigenvalues can be treated as effectively classical [11].

The vanishing discord is a stronger condition than the absence of entanglement. In effect, $\delta\mathcal{I} = 0$ implies existence of the eigenstates of $\rho_{\mathcal{SA}}$ which are products of the states of $\mathcal{S}$ and of $\mathcal{A}$. An instructive example of this situation arises as a result of decoherence: When the off-diagonal terms of $\rho_{\mathcal{SA}}$ disappear in a manner illustrated by Eq. (31), the discord disappears as well.

4. ENVIRONMENT AS A WITNESS: REDUNDANCY RATIO

The discussion of decoherence to date tends to focus on the effect of the environment on the system or on the apparatus. The destruction of quantum coherence and the emergence of preferred pointer observables whose eigenvalues are associated with the decoherence-free pointer subspaces was the focus of the investigation.

Here we shall break with this tradition. According to the theory of decoherence, the environment is monitoring the system. Therefore, its state must contain a record of the system. It is of obvious interest to analyze the nature and the role of this record. To this end, we shall use mutual information introduced before defining the *redundancy ratio*

$$\mathcal{R}_{\mathcal{I}(\otimes\mathcal{H}_{\mathcal{E}_k})} = \left(\sum_k \mathcal{I}(\mathcal{S}:\mathcal{E}_k)\right) / \mathcal{H}(\mathcal{S}) \ . \tag{33}$$

Above, we imagined a setting where the system is decohering due to the interaction with the environment which is composed of many subsystems $\mathcal{E}_k$. $\mathcal{R}_{\mathcal{I}(\{\otimes\mathcal{H}_{\mathcal{E}_k}\})}$ is a measure of how many times – how redundantly – the information about the system has been inscribed in the environment. An essentially identical formula can be introduced using the asymmetric $\mathcal{J}$, Eq. (26). It is easy to establish that the discord is always non-negative and, hence, that

$$\mathcal{R}_{\mathcal{J}_{MAX}} \leq \mathcal{R}_{\mathcal{I}_{MAX}} \ . \tag{34}$$

The subscript indicating maximization may refer to two distinct procedures: $\mathcal{R}_{\mathcal{J}}$ will obviously depend on the manner in which subsystems of the environment are measured. In fact, it is convenient to use

$$\mathcal{J}_k = \mathcal{J}(\mathcal{S} : \mathcal{E}_k) = \mathcal{H}(\mathcal{E}_k) - \mathcal{H}(\mathcal{E}_k|\mathcal{S}) \ , \tag{35}$$

to define a basis-dependent

$$\mathcal{R}_{\mathcal{J}}(\{|s\rangle\}) = \mathcal{R}_{\mathcal{J}}(\otimes\mathcal{H}_{\mathcal{E}k}) \tag{36}$$

in a manner analogous to Eq. (33). Maximizing $\mathcal{R}_{\mathcal{J}}(\{|s\rangle\})$ with respect to the choice of the choice of states $\{|s\rangle\}$ [11] is an obvious "counterpoint" to the predictability sieve [16-19], the strategy which seeks states that entangle the least with the environment.

There is one more maximization procedure which may and should be considered: The environment can be partitioned differently – for example, it may turn out that more information about the system can be extracted by measuring, say, the photon environment in some collective fashion (homodyne?) instead of directly counting the environment photons. It is clear that for such optimization to be physically significant, it should respect to some degree the natural structure of the environment.

In addition to the redundancy ratio one can define the rate at which the redundancy ratio increases. The redundancy rate is defined as

$$\dot{\mathcal{R}} = \frac{d}{dt}\mathcal{R} . \tag{37}$$

Either the basis dependent or the basis-independent versions of $\dot{\mathcal{R}}$ may be of interest. The physical significance of the redundancy ratio rate is clear: It shows how quickly the information about the system spreads throughout the environment. It is, in effect, a measure of the rate of increase of the effective number of the environment subsystems which have recorded the state of the system $\mathcal{S}$.

It is worth noting that either $\mathcal{R}_{\mathcal{I}}$ or $\mathcal{R}_{\mathcal{J}}$ can keep on increasing after the density matrix of the system has lost its off-diagonal terms in the pointer basis and after it can be therefore considered completely decohered. Indeed, direct interaction between the system and the environment is not needed for either $\mathcal{R}$ to change. For example, information about the system inscribed in the primary environment may be communicated to a secondary, tertiary, and more remote environment (which need not interact with the system at all).

It is natural to define objectivity and, therefore, classicality with the help of $\mathcal{R}$. In the limit $\mathcal{R}_{\mathcal{J}} \rightarrow \infty$ the information about the preferred states of the system is spread so widely that it can be acquired by many observers simultaneously [11]. Moreover, it already exists in multiple copies, so it can be safely cloned in spite of the no-cloning theorem [20]. Thus, a state of the system redundantly recorded in the environment has all the symptoms of "objective existence". In particular, such well-advertised states can be found out without being disturbed by approximately $\mathcal{R}$ observers acting independently [11,21] (each simply measuring the state of $\sim 1/\mathcal{R}_{\mathcal{J}}$ fraction of the environment).

5. SUMMARY AND CONCLUSIONS

Information theory offers a useful perspective on the measurement process, on decoherence and, above all, on the definitions of classicality. Discord can be used to characterize the nature of quantum correlations and to distinguish the ones that are classical. The redundancy ratio is a powerful measure of the classicality of states: While a vanishing discord is a necessary condition for classicality of correlations, the redundancy ratio is a direct measure of objective existence of quantum states. Objectivity can be defined operationally as the ability to find what the state is without disturbing it [11,21]. Objective existence of quantum states would make them effectively classical, and was the ultimate goal of the interpretation of quantum theory. We have established it here by investigating einselection from the point of view of information theory and by shifting focus from the system to the environment which is monitoring the system.

These advances clarify some of the interpretational issues which are now a century old. The relation between the epistemological and ontological significance of the quantum state vectors is now apparent: Objective existence of the quantum states is a direct consequence of the redundant records permeating the environment. Epistemology begets ontology!

Acknowledgments

This research was supported in part by NSA.

References

[1] Zurek, W. H., Phys. Rev. D **24** (1981) 1516-1524

[2] Zurek, W. H., Phys. Rev. D **26** (1982) 1862-1880

[3] Zurek, W. H., Information transfer in quantum measurements, pp. 87-116 in *Quantum Optics, Experimental Gravity, and the Measurement Theory*, P. Meystre and M. O. Scully, eds. (Plenum, New York, 1983)

[4] Schumacher, B., Westmoreland, M., and Wootters, W. K., Phys. Rev. Lett. **76** (1996) 3452-3455; Hall, M. J. H., and O'Rourke, M. J., Quant. Opt. **5** (1993) 161; Halliwell, J. J., Phys. Rev. D **60** (1999) 105031

[5] Joos, E., and Zeh, H. D., Zeits. Phys. B **59** (1985) 229

[6] Caldeira, A. O., and Leggett, A. J., Physica **121A** (1983) 587-616; Phys. Rev. A **31** (1985) 1059; Haake, F., and Reibold, R., Phys. Rev. A **32** (1985) 2462; Grabert, H., Schramm, P., and Ingold., G.-L., Phys. Rev. **168** (1988) 115-207

[7] Zurek, W. H., Reduction of the Wavepacket: How long does it take? presented at a NATO ASI *Frontiers of Nonequilibrium Statistical Mechanics* Santa Fe, June 1984, pp. 145-149 in the proceedings, G. T. Moore and M. O. Scully, eds. (Plenum, New York, 1986); Haake, F., and Walls, D. F., in *Quantum Optics IV*, J. D. Harvey and D. F. Walls, eds. (Springer, Berlin, 1986). Unruh, W. G., and Zurek, W. H., Phys. Rev. D **40** (1989) 1071-1094

[8] Hu, B. L., Paz, J. P., and Zhang, Y., Phys. Rev. D **45** (1992) 2843-2861; Zurek, W. H., Habib, S., and Paz, J. P., Phys. Rev. Lett **70** (1993) 1187-1190; Anglin, J. R., and Zurek, W. H., Phys Rev. D **53** (1996) 7327-7335; Anglin, J. R., Paz, J. P., and Zurek, W. H., Phys. Rev. A **53** (1997) 4041

[9] Brune, M. *et al.*, Phys. Rev. Lett. **77** (1996) 4887-4890; Myatt *et al.*, Nature **403** (2000) 269

[10] Giulini, D., Joos, E., Kiefer, C., Kupsch, J., Stamatescu, I.-O., and Zeh, H. D., *Decoherence and the Appearance of a Classical World in Quantum Theory*, (Springer, Berlin, 1996); Blanchard, Ph., Giulini, D., Joos, E., Kiefer, C., and Stamatescu, I.-O., eds, *Decoherence: Theoretical, Experimental, and Conceptual Problems*, (Springer, Berlin, 2000); Paz, J. P., and Zurek, W. H., in *Les Houches Lectures*, in press (2000)

[11] Zurek, W. H., Rev. Mod. Phys., submitted (2000)

[12] Ivanovic I. D., J. Phys. A **14** (1981) 3241-3245; J. Math. Phys. **24** (1983) 1199-1205

[13] Wootters, W. K., and Fields, B. D., Ann. Phys. **191** (1989) 363

[14] Barnett, S. M., and Phoenix, S. J. D., Phys. Rev. A **40** (1989) 2404-2409

[15] Cover, T. M., and Thomas, J. A., *Elements of Information Theory*, (Wiley, New York, 1991)

[16] Zurek, W. H., Progr. Theor. Phys. **89** (1993) 281-302

[17] Zurek, W. H., Habib, S., and Paz, J. P., Phys. Rev. Lett **70** (1993) 1187-1190

[18] Tegmark, M., and Shapiro, H. S., Phys. Rev. E **50** (1994) 2538-2547

[19] Gallis, M. R., Phys. Rev. A **53** (196) 655-660; Paraoanu, Gh.-S., and Scutaru, H., Phys. Lett. A **238** (1998) 219-222

[20] Wootters, W. K., and Zurek, W. H., Nature **299** (1982) 802-802; Dieks, D., Phys. Lett. **92** (1982) 271-272

[21] Zurek, W. H., Phil. Trans. Roy. Soc. Lond. **356** (1998) 1793-1821

RECONSTRUCTING THE DISCRETE WIGNER FUNCTION THROUGH COMPLEMENTARY MEASUREMENTS

Roberth Asplund, Gunnar Björk
Department of Electronics, Royal Institute of Technology (KTH), Electrum 229, SE-164 40 Kista, Sweden
gunnarb@ele.kth.se

Abstract We report a novel algorithm to reconstruct the discrete Wigner function of aquantum system from a set of non-redundant measurements on an ensemble of identically prepared systems. Not surprisingly, the necessary von Neumann measurements form a complementary set.

1. A DISCRETE WIGNER FUNCTION

Discrete valued states and operators have played an important role in quantum mechanics, and at present they are often used to form the building blocks of quantum information applications. A useful tool for characterizing these states, and to perform various computations, is the discrete Wigner function [1].

For quantum mechanical systems where the Hilbert space dimension N is a prime number, or a power of a prime, a phase space can be defined as a square lattice with N^2 phase points $\alpha = (i, j)$. At each phase point α an operator $\widehat{A}_\alpha$ can be defined having the following properties:

$$\mathrm{Tr}(\widehat{A}_\alpha \widehat{A}_\beta) = N\delta_{\alpha,\beta} \tag{1}$$

and

$$\mathrm{Tr}(\widehat{A}_\alpha) = 1. \tag{2}$$

The explicit construction of the $\widehat{A}_\alpha$:s that lay the foundation to the theory presented here can be found in [1].

An important observation is that $\mathrm{Tr}(\widehat{A}_\alpha \widehat{A}_\beta)$ is an inner product in the vector space of Hermitian matrices, so the property (1) implies that the N^2 phase-point operators form an orthogonal basis.

Hence, for every N, the phase-point operators form an orthogonal basis in the vector space of $N \times N$ Hermitian matrices. Since the density matrix is

Hermitian it can be expanded in this basis

$$\widehat{\rho} = \sum_{\alpha} W_\alpha \widehat{A}_\alpha, \qquad W_\alpha \in \mathbf{R}. \tag{3}$$

The discrete Wigner function for the state with the density matrix $\widehat{\rho}$ is then defined to be the set of coefficients $W = \{W_\alpha\}$ in this expansion. This will be a real function defined on the N^2 phase points and it will contain all information about the state. Equation (3) can be inverted, using (1), to generate the Wigner function coefficients for a given density matrix

$$W_\alpha = (1/N)\mathrm{Tr}(\widehat{\rho}\widehat{A}_\alpha). \tag{4}$$

The Wigner function will have the following properties [1]:

$$\sum_{\alpha} W_\alpha = 1 \tag{5}$$

and

$$\mathrm{Tr}(\widehat{\rho}\,\widehat{\rho}\,') = N \sum_{\alpha} W_\alpha W'_\alpha. \tag{6}$$

2. A NOVEL EXPANSION AND A NOVEL RECONSTRUCTION ALGORITHM

If N is a prime, or the power of a prime, there exist $N+1$ mutually complementary, complete sets of orthogonal states (Wootters and Fields call the sets "unbiased") [2]. These states will be labeled $|v_k^{(r)}\rangle$, where $r = 1, 2, \ldots, N+1$ denotes the set, and $k = 1, 2, \ldots, N$ labels the vector in the set. Since $\sum_{k=1}^{N} |v_k^{(r)}\rangle\langle v_k^{(r)}| = \hat{1} \;\; \forall \;\; r$, each set contains the eigenvectors of a von Neumann measurement. Furthermore, for any two vectors from different sets, that is if $r \neq s$, the construction of the vectors [2] assures that

$$|\langle v_k^{(r)}|v_l^{(s)}\rangle|^2 = 1/N, \quad \forall \;\; k, l. \tag{7}$$

That is, the all the von Neumann measurements defined by the vector sets are mutually complementary. The transition probability between an arbitrary density matrix $\widehat{\rho}$ and a measurement basis state $|v_k^{(r)}\rangle$, that is $\mathrm{Tr}(\widehat{\rho}\,\widehat{\rho}_k^{(r)})$, will be denoted $p_k^{(r)}$. This number is the probability that a measurement of a general state with density matrix $\widehat{\rho}$ in the measurement basis r will yield the outcome $|\, v_k^{(r)}\rangle$. The numbers $p_k^{(r)}$ constitute the information the measurements give about the (presumably) unknown state $\widehat{\rho}$'s properties. Since the measurement eigenvectors are known we can compute their Wigner function coefficients $W_{k(\alpha)}^{(r)}$ by

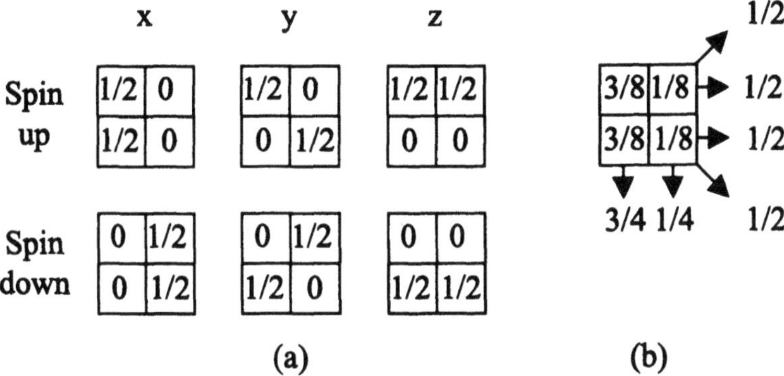

Figure 1 (a) The Wigner functions of the eigenstates of a spin 1/2 particle when measured in the x-, y-, and z-direction. (b) The Wigner function of the state $(3\,|\uparrow_x\rangle\langle\uparrow_x| + |\downarrow_x\rangle\langle\downarrow_x|)/4$. The measurement outcome probabilities for all the complementary spin measurements are found by summing the coefficients along the appropriate line. E.g. the probability of obtaining spin x up is 3/4. Using only the coefficients in (a) and the measurement probabilities in (b), all the coefficients in (b) can be calculated from (12).

using (4). The following expansion, valid for any Hermitian operator $\widehat{O}$ [3], is now introduced:

$$\widehat{O} + \widehat{1} \cdot \mathrm{Tr}(\widehat{O}) = \sum_r \sum_k \widehat{\rho}_k^{(r)} \langle v_k^{(r)} \mid \widehat{O} \mid v_k^{(r)} \rangle. \tag{8}$$

If, in particular, $\widehat{O} = \widehat{A}_\alpha$ then $\mathrm{Tr}(\widehat{O})$ will be unity and by virtue of (4) the number $\langle v_k^{(r)} | \widehat{A}_\alpha | v_k^{(r)} \rangle$ will be N times the Wigner function coefficient for the state $|v_k^{(r)}\rangle$ in the phase point α, i.e., the given number $NW_{k(\alpha)}^{(r)}$. We get

$$\widehat{A}_\alpha + \widehat{1} = N \sum_r \sum_k \widehat{\rho}_k^{(r)} W_{k(\alpha)}^{(r)}. \tag{9}$$

Assume now that $\widehat{\rho}$ is an arbitrary density matrix for which we want to determine the Wigner function coefficients W_α. Then, the following holds

$$\frac{1}{N}\mathrm{Tr}\left(\widehat{\rho}(\widehat{A}_\alpha + \widehat{1})\right) = \frac{1}{N}\mathrm{Tr}(\widehat{\rho}\widehat{A}_\alpha) + \frac{1}{N}\mathrm{Tr}(\widehat{\rho}) = W_\alpha + \frac{1}{N}. \tag{10}$$

On the other hand one can use (9) to get

$$\begin{aligned} \frac{1}{N}\mathrm{Tr}\left(\widehat{\rho}(\widehat{A}_\alpha + \widehat{1})\right) &= \frac{1}{N}\mathrm{Tr}\left(\widehat{\rho} N \sum_r \sum_k \widehat{\rho}_k^{(r)} W_{k(\alpha)}^{(r)}\right) \\ &= \sum_r \sum_k \mathrm{Tr}\left(\widehat{\rho}\,\widehat{\rho}_k^{(r)}\right) W_{k(\alpha)}^{(r)} \end{aligned}$$

$$= \sum_r \sum_k p_k^{(r)} W_{k(\alpha)}^{(r)}. \tag{11}$$

The Wigner function coefficient W_α can now be computed through (10) and (11).

$$W_\alpha = \sum_r \sum_k p_k^{(r)} W_{k(\alpha)}^{(r)} - \frac{1}{N}. \tag{12}$$

We would like to emphasize the similarity between this inverse transform and the one used for continuous quantum-state tomography. In the continuous case the transform consists of an integral over all directions in the phase space, and for each direction an integral over all possible outcomes of the measurement associated with the given direction. It is seen from (12) that in the discrete case the transform consists of a sum over a set of "directions" r, and for each r, a sum over all possible outcomes k of the measurement associated with the "direction" r. The "weights" $W_{k(\alpha)}^{(r)}$ will all be real, and in the particular case where N is a prime number, $W_{k(\alpha)}^{(r)}$ is either $1/N$ or zero. simplifies (12) dramatically.

In conclusion, we have derived a state reconstruction algorithm for any discrete state defined in a Hilbert space of dimension N, where N is a prime, or the power of a prime. The algorithm is based on a set of $N+1$ mutually complementary von Neumann measurement [2]. If N is a composite number, there also exist a reconstruction algorithm, but in this case the algorithm is less efficient in terms of the number of necessary measurements [3].

Acknowledgments

This work was supported by grants from the Swedish Research Council for Engineering Sciences (TFR) and the Swedish Natural Science Research Council (NFR).

References

[1] W. K. Wootters, Ann. Phys. (N.Y.) **176**, 1 (1987).

[2] W. K. Wootters and B. D. Fields, Ann. Phys. (N.Y.) **191**, 363 (1989).

[3] R. Asplund and G. Björk, unpublished.

STATISTICAL NOISE IN MEASURING CORRELATED PHOTON BEAMS

Stefania Castelletto, Ivo Pietro Degiovanni, Maria Luisa Rastello
Istituto Elettrotecnico Nazionale G. Ferraris
Strada delle Cacce 91- Turin- Italy
castelle@ien.it

Photon correlated beams, generated by parametric down-conversion in non-linear crystals (PDC), have been proved successfully for the measurement of quantum efficiency, η, of single photon detectors [1, 2, 3, 4] and can be used for measuring transmittance, τ, as well. The main feature relies on the realization of two correlated quantum channels yielding coincident events. To reduce uncertainty [4, 5, 6] and determine its ultimate limit, contributions to statistical noise in a real experimental set-up are here investigated, and a general model is proposed to calculate the maximum likelihood best estimators and evaluate the probability of coincidence. In typical measurement scheme a channel is set as trigger (idler, i) and attention is paid to catch at least the same amount of correlated photons in the other channel (signal, s) [4]. Unwanted uncorrelated photons due to PDC itself result in a decreased degree of correlation, noise fluctuations in coincidence measurements and asymmetry between the two channels, whose mean photon rate can be quite different. We indicate by $\alpha_x \rho_x$ with $x = s, i$, the mean rate of correlated photons in s and i channels ($0 < \alpha_x < 1$, $\alpha_s \neq \alpha_i$). We address the total mean photon rates in both channels as $\rho_x = \alpha_x \rho_x + (1 - \alpha_x)\rho_x$, where terms $1 - \alpha_x$ take into account also additive background photon rates due to straylight and detector noise. Actually statistical noise is also due to losses of correlated photons by optical elements, non ideal detectors and electronic devices in both channels. We distinguish, therefore, between optical losses (τ_x, transmittance of the correlated photon path), detection losses (η_x, detector quantum efficiency), electronic losses associated to the dead times (D_x, time during which the detection system is unable to detect photons). It is usual to identify dead times as non-extending and extending: in the first case, all photons following a revealed photon within the fixed time interval D_x are ignored, while in the second case any incoming photon produces or prolongs D_x. Eventually the finite duration of the time coincidence window, w, modifies the distribution probability of coincident events. Let us consider the distribution probability of single channel

Quantum Communication, Computing, and Measurement 3
Edited by P. Tombesi and O. Hirota, Kluwer Academic/Plenum Publishers, 2001

photons and of correlated photons as Poissonian. In the following $\mathrm{P}(m)$ will indicate general probability distribution of m events when dead time distortion effects lead to distributions different from Poissonian $\mathcal{P}(m)$.

To obtain the statistics of true coincidence we first calculate the probability distribution of m'_α correlated events in the x channel, $\mathrm{P}_x^{\mathrm{ext}}(m'_\alpha)$, for the case of extending dead time as

$$\mathrm{P}_{\mathrm{x}}^{\mathrm{ext}}(\mathrm{m}'_\alpha) = \sum_{\mathrm{m}_\alpha=\mathrm{m}'_\alpha}^{\infty} \mathcal{P}_{\mathrm{x}}(\mathrm{m}_\alpha)\,\mathrm{B}(\mathrm{m}_\alpha, \mathrm{m}'_\alpha, \pi_{\mathrm{x}}^{\mathrm{ext}}), \tag{1}$$

that results to be a Poissonian distribution and $B(m, n, p)$ is a binomial one. We identify $\pi_x^{\mathrm{ext}} = \exp(-\rho_x \eta_x \tau_x D_x)$. For the non-extending dead time the $\mathrm{P}_x^{\mathrm{next}}(m'_\alpha)$ becomes a combination of incomplete Gamma function $\gamma[m'_\alpha; \alpha_x \rho_x \eta_x \tau_x\,(T - m'_\alpha D_x)]$.

The probability distribution of true coincidence can be evaluated as $\mathrm{P}^{(c)}(m'_c) \simeq \sum_{m'_\alpha=m'_c}^{\infty} \mathrm{P}_i(m'_\alpha)\,\mathrm{B}(m'_\alpha, m'_c, \tau_s \eta_s \pi_s^*)$, where $\pi_x^* = \pi_x^{\mathrm{ext}}$ for extending dead time while $\pi_x^* = 1/(1 + \rho_x \eta_x \tau_x D_x)$ is the approximated mean rate correction for non-extending dead time. Note that so far the index s and i can be interchanged, because of the arbitrary choice of trigger channel. The calculable extending case gives $\mathcal{P}^{(c)}(m'_c)$ with mean rate $\tau_s \eta_s \alpha_i \rho_i \eta_i \tau_i \pi_s^{\mathrm{ext}} \pi_i^{\mathrm{ext}}$.

The accidental coincidence mean rate is given by two terms: (a) uncorrelated events or (b) events belonging to correlated pairs, whose twin photons have been lost by absorption or dead time. Let i–channel be the trigger one and for simplicity let us discuss first the case $D_i = 0$; $(1 - \alpha_i)\rho_i \eta_i \tau_i$ and $\alpha_i \rho_i \eta_i \tau_i (1 - \eta_s \tau_s \pi_s^*)$ are type (a) and (b) events mean rates, respectevely. The total mean rates, yielding accidental events, are $\rho_{R,i} = \rho_i \eta_i \tau_i (1 - \alpha_i \eta_s \tau_s \pi_s^*)$ and $\rho_{R,s} = \rho_s \eta_s \tau_s (1 - \alpha_s \eta_i \tau_i \pi_i^*)$. The probability to count m_R events producing accidental counts in the i channel when $D_i = 0$ is $\mathcal{P}(m_R)$ with mean rate $\rho_{R,i}$.We calculate the same probability $\mathrm{P}_i(m'_R)$ to count m'_R events in the i-channel when $D_i \neq 0$, according to Eq. (1). Let us now evaluate the probability that one or more uncorrelated events on channel s fall in w, $p_s = \sum_{m'_R=1}^{\infty} \mathrm{P}_s(m'_R)$. In the case of extending dead time $\mathcal{P}_i(m'_R)$ has a mean rate $\rho_{R,i}\pi_i^{\mathrm{ext}}$. It is easy to understand that $p_i \neq p_s$ and that the symmetry between the two channels is lost. Finally we can now calculate the probability distribution of accidental events given by $\mathrm{P}_i^{(a)}(m_a) = \sum_{m_R=m_a}^{\infty} \mathrm{P}_i(m_R)\,\mathrm{B}(m_R, m_a, p_s)$,that in the case of extending dead time is Poissonian with mean rate $\rho_{R,i}\pi_i^{\mathrm{ext}} p_s$.

The distribution probability of total coincidence counts is then

$$\mathrm{P}_{\mathrm{i}}(\mathrm{m_c}) = \sum_{\mathrm{m}_\alpha=0}^{\infty} \sum_{\mathrm{m}'_c=0}^{\infty} \delta_{\mathrm{m_c}, \mathrm{m}_\alpha+\mathrm{m}'_c} \mathrm{P}_{\mathrm{i}}^{(\mathrm{a})}(\mathrm{m_a})\mathrm{P}^{(\mathrm{c})}(\mathrm{m}'_c).$$

We calculate the maximum likelihood best estimator $\widehat{y}$ according to $\partial \mathrm{P}_i(m_c)/\partial y = 0$, and we define the relative noise fluctuations of the estimator as $\epsilon_R(\widehat{y}) = \left(\langle \widehat{y}^2 \rangle / \langle \widehat{y} \rangle^2 - 1\right)^{1/2}$, where $\langle \widehat{y}^k \rangle = \sum_{m_c=0}^{\infty} \mathrm{P}_i(m_c)\, \widehat{y}^k$. The extending dead time case can be analytically evaluated, since $\mathrm{P}_i^{\mathrm{ext}}(m_c)$ $=\mathcal{P}_i^{\mathrm{ext}}(m_c)$ with mean rate $\rho_{c,i}^{\mathrm{ext}} = \rho_i \eta_i \tau_i \pi_i^{\mathrm{ext}} - \pi_i^{\mathrm{ext}} \rho_{R,i} \exp(-\rho_{R,s} \pi_s^{\mathrm{ext}} w)$. By contrary the non-extending dead time case can be calculated exactly only numerically. Nevertheless a quite good approximation can be performed simply by supposing that the non-extending dead time distortion affects only the mean value, neglecting the presence of subpoissonian statistics, and allowing us to write $\mathrm{P}_i^{\mathrm{next}}(m_c) \approx \mathcal{P}_i^{\mathrm{next}}(m_c)$. In this case the $\rho_{c,i}^{\mathrm{next}} = \rho_i \eta_i \tau_i \pi_i^{\mathrm{next}} - \pi_i^{\mathrm{next}} \rho_{R,i} \exp(-\rho_{R,s} \pi_s^{\mathrm{next}} w)$.

The most common and easier cases are the evaluaton of the estimators $\widehat{y}$ of the detection system quantum efficiency (without dead time correction) $\widehat{\eta_s \pi_s}$, and the transmittance $\widehat{\tau}_s$. We simulated the relative noise fluctuations of the maximum likelihood best estimators $\widehat{y}$ versus pure statistical parametrs $(\alpha_i, \rho_i, \rho_s/\rho_i, \eta_i)$ and electronic ones (D_i, D_s, w), inducing different noise behaviors.We studied the dependence of relative noise fluctuations of $\widehat{\eta_s \pi_s}$ and $\widehat{\tau}_s$ on the estimators values itself, for different typical idler/trigger rates and different ratios ρ_s/ρ_i, considering also extending and non-extending dead times. The other parameters have been fixed according to typical experimental values. We observe that the extending dead time induces higher statistical noise on both estimators than the non-extending one. $\widehat{\tau}_s$ shows higher noise fluctuations with respect to $\widehat{\eta_s \pi_s}$ for the presence of dead time on both channels. The lower relative uncertainty is obviously achieved in ideal condition, i.e. $\rho_s/\rho_i = 1$, high photon rates, $\tau_s = \tau_i = 1$, $\eta_s = \eta_i = 1$, minimized D_x and narrow w. We observe that $\epsilon_R(\widehat{y})$ depends strongly on the ratio ρ_s/ρ_i when $\widehat{y}$ approaches 1 (ideal condition). The ratio $R = \epsilon_R(\widehat{y})/\epsilon_R(\,\widehat{y} = 1)$ represents the increasing noise level with respect to the ideal condition, giving in this case $R \simeq 2.5$ for $\widehat{y} = 0.2$. $\epsilon_R(\widehat{y})$ decreases strongly with higher ρ_i and increases, even if smoothly, with $\rho_s/\rho_i > 5$. The increase of idler/trigger photon rate does not reduce significantly noise in $\widehat{\tau}_s$ when $\rho_s/\rho_i > 10$.

To reduce relative noise fluctuations under 0.1% it is necessary to have trigger photons correlated better than 99.9% ($\alpha_i = 0.999$) and almost equal channels to compensate losses introduced by real detectors and electronics.

The effect of decreasing η_i on $\epsilon_R(\widehat{y}, \eta_i)$ results in an increasing noise: the trigger detector efficiency may influence dramatically the noise level of the measurement yielding a factor $R = \epsilon_R(\widehat{y}, \eta_i)/\epsilon_R(\widehat{y}, \eta_i = 0.8) \simeq 3.6$ when $\rho_s/\rho_i = 2$, $\rho_i = 200$ KHz and $\eta_i = 0.05$.

Regarding the influence of D_x in the case of a typical coincidence circuit where the idler/trigger channel present remarkable dead time, ρ_i values have to be properly adjusted to reduce noise. D_x and w generally increase uncertainty independently from ρ_s/ρ_i, because they induce a deterministic degradation on correlated photons. Aiming at reducing noise, it is recommended a dead time as low as possible in the trigger channel to increase the trigger correlated events.

Eventually we can conclude that when some experimental parameters such as $\rho_s/\rho_i, \eta_i, \alpha_i$, are much far from ideal condition, the noise fluctuations cannot be reduced under a certain level by properly adjusting the remaining parameters. The typical noise level simulated with realistic choice of experimental parameters is few parts in 10^{-3}, which corresponds to the most recent experimental improvements [2, 4].

Finally we calculated and experimentally proved the feasibility of the ultimate noise limit (0.02%) for a realistic quantum efficiency measurement, where $\eta_i = \eta_s = 0.6$, $\alpha_i = 0.999$, $D_s = D_i = 30$ ns, w=1 ns and $\rho_s/\rho_i = 1.5$ and strict accuracy requirements on w and ρ_s/ρ_i are satisfied [4]

References

[1] Kwiat P.G., Steinberg A.M., Chiao R.Y., Eberhard P.H., Petroff M.D., Appl. Opt., (1994) **33**, n.10, 1844-1853.

[2] Migdall A. L., Datla R. U., Sergienko A., Orszak J. S., Shih Y. H., Metrologia, (1996) **32**, 479-483

[3] Brida G., Castelletto S., Novero C., Rastello M.L., J. Opt. Soc. of Am B, (1999) **16**, 1623-1627

[4] Migdall A., S. Castelletto, I.P. Degiovanni, Rastello M.L., Absolute quantum efficiency measurement by correlated photons, in preparation.

[5] Hayat M.M., Joobeur A., Saleh B. E. A., J. Opt. Soc. of Am A **16**, n.2, (1999), 348-358

[6] Hayat M.M., Torres S.N., Pedrotti L.M. Opt. Comm. **169** (1999) 275-287.

RECONSTRUCTION TECHNIQUE FOR A TRAPPED ELECTRON

M. Fortunato, M. Massini, S. Mancini, D. Vitali, P. Tombesi
INFM and Dipartimento di Matematica e Fisica, Università di Camerino, I–62032 Camerino, Italy
mauro@camcat.unicam.it

Abstract We propose a measurement method for the reconstruction of entangled states (spin-cyclotron) of a trapped electron.

Entanglement is an intrinsic feature of quantum mechanics, and is also the basis of quantum information processing. A striking achievement in this rapidly expanding field has been the recent entanglement of four trapped ions [1]. However, it is also possible (and conceptually equivalent) to entangle different degrees of freedom of the same particle [2]. A single electron trapped in a Penning trap [3] is one of the most fundamental quantum systems: it allows the measurement of fundamental physical constants with striking accuracy.

Here we propose to reconstruct *entangled* states (combined cyclotron and spin states) of an electron in a Penning trap by using suitable applied fields and with the help of a tomographic reconstruction from the measured data.

In a Penning trap an electron is confined by the combination of a homogeneous magnetic field along the positive z axis and an electrostatic quadrupole potential in the xy plane [3]. The spatial part of the electronic wave function consists of three degrees of freedom, but neglecting the slow magnetron motion (whose characteristic frequency lies in the kHz region), here we only consider the axial and cyclotron motions, which are two harmonic oscillators radiating in the MHz and GHz regions, respectively. On the other hand, the spin dynamics results from the interaction between the magnetic moment of the electron and the static magnetic field, so that the free Hamiltonian reads as [3]

$$\hat{H}_{\text{free}} = \hbar\omega_z \hat{a}_z^\dagger \hat{a}_z + \hbar\omega_c \hat{a}_c^\dagger \hat{a}_c + \hbar\omega_s \hat{\sigma}_z/2 \,, \tag{1}$$

where the indices z, c, and s refer to the axial, cyclotron and spin motions, respectively.

The most general pure state of the trapped electron [5] can be cast in the form $|\Psi\rangle = c_1|\psi_1\rangle|\uparrow\rangle + c_2|\psi_2\rangle|\downarrow\rangle$, $|\psi_1\rangle$ and $|\psi_2\rangle$ being two unknown cyclotron states, and the complex coefficients c_1 and c_2 satisfying the normalization condition $|c_1|^2 + |c_2|^2 = 1$. The density operator $\hat{\rho} = |\Psi\rangle\langle\Psi|$ associated

Quantum Communication, Computing, and Measurement 3
Edited by P. Tombesi and O. Hirota, Kluwer Academic/Plenum Publishers, 2001

to the pure state $|\Psi\rangle$ can be expressed in the form

$$\hat{\rho} = \begin{pmatrix} |c_1|^2 |\psi_1\rangle\langle\psi_1| & c_1 c_2^* |\psi_1\rangle\langle\psi_2| \\ c_2 c_1^* |\psi_2\rangle\langle\psi_1| & |c_2|^2 |\psi_2\rangle\langle\psi_2| \end{pmatrix} , \tag{2}$$

whose elements $\rho^{(ij)} = c_i c_j^* |\psi_i\rangle\langle\psi_j|$ are operators in the cyclotron Hilbert space.

In order to characterize the generic state we use a simple reconstruction procedure: Adding a particular inhomogeneous magnetic field—known as the "magnetic bottle" field [3]—to that already present in the trap, it is possible to perform a simultaneous measurement of both the spin and the cyclotronic excitation numbers. The useful interaction Hamiltonian for the measurement process is then [3]

$$\hat{H}_{\text{bottle}} = \hbar\omega_b \left[\hat{a}_c^\dagger \hat{a}_c + \frac{g}{2}\frac{\hat{\sigma}_z}{2} \right] \hat{z}^2 , \tag{3}$$

where the angular frequency ω_b is directly related to the strength of the magnetic bottle field.

Eq. (3) describes the fact that the axial angular frequency is affected both by the number of cyclotron excitations $\hat{n}_c = \hat{a}_c^\dagger \hat{a}_c$ and by the eigenvalue of $\hat{\sigma}_z$. The modified (shifted) axial frequency can be experimentally measured [3] after the application of the inhomogeneous magnetic bottle field. Then, repeated measurements of this type allow us to recover the probability amplitudes associated to the two possible spin states and the cyclotron probability distribution in the Fock basis. The reconstruction of the density matrices $|\psi_i\rangle\langle\psi_i|$ $(i = 1, 2)$ in the Fock basis is then possible by employing a technique similar to the Photon Number Tomography which exploits a phase-sensitive reference field that displaces in the phase space the particular state [4].

Immediately before the measurement, we apply a pulsed standing wave tuned to ω_c in order to get a displacement γ on the cyclotron. Thus we can interpret the quantity

$$P^{(i)}(n,\gamma) = \langle n|\hat{D}^\dagger(\gamma)\hat{\rho}^{(ii)}\hat{D}(\gamma)|n\rangle = \langle n,\gamma|\hat{\rho}^{(ii)}|n,\gamma\rangle , \tag{4}$$

as the probability of finding the cyclotron state $|\psi_i\rangle$ in a displaced number state $|n,\gamma\rangle$. Expanding the density operator $\hat{\rho}^{(ii)}$ in the Fock basis, and defining N_c as an appropriate estimate of the maximum number of cyclotronic excitations (cut-off), we have

$$P^{(i)}(n,\gamma) = \sum_{k,m=0}^{N_c} \langle n,\gamma|k\rangle\langle k|\hat{\rho}^{(ii)}|m\rangle\langle m|n,\gamma\rangle . \tag{5}$$

Let us now consider, for a given value of $|\gamma|$, $P^{(i)}(n,\gamma)$ as a function of $\varphi = \arg[\gamma]$ and calculate the coefficients of the Fourier expansion

$$P^{(i)}(n,s) = \frac{1}{2\pi}\int_0^{2\pi} d\varphi\, P^{(i)}(n,\varphi)e^{is\varphi}\ , \tag{6}$$

for $s = 0, 1, 2, \ldots$. Combining Eqs. (5) and (6), we get

$$P^{(i)}(n,s) = \sum_{m=0}^{N_c-s} G^{(s)}_{n,m}(|\gamma|)\langle m+s|\hat{\rho}^{(ii)}|m\rangle\ , \tag{7}$$

where the explicit expression of the matrices G is given in Ref. [6].

We may now note that if the distribution $P^{(i)}(n,\gamma)$ is measured for $n \in [0,N]$ with $N \geq N_c$, then Eq. (7) represents for each value of s a system of $N+1$ linear equations between the $N+1$ measured quantities and the N_c+1-s unknown density matrix elements. Therefore, in order to obtain the latter, we only need to invert the system

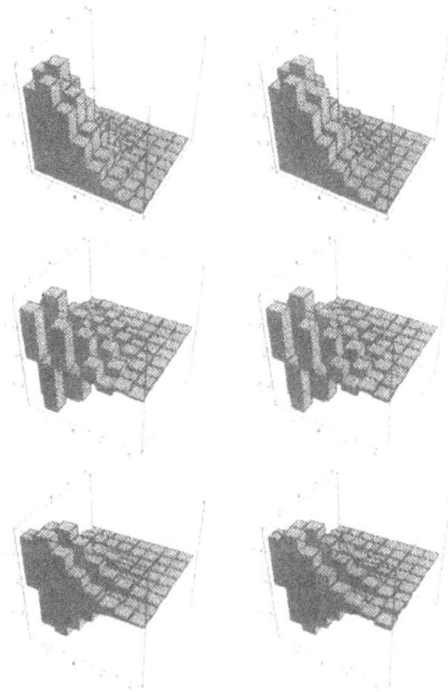

Figure 1 Simulated tomographic reconstruction of the density matrix for the entangled state given in the text with $c_1 = 1/2$, $c_2 = \sqrt{3}/2$, $\theta = \pi$, $\xi = \pi$, and $\alpha = 1$. In this simulation an amplitude $|\gamma| = 0.7$ of the applied reference field has been used. a) $(\rho_{11})_{n,m}$, b) $(\rho_{22})_{n,m}$, and c) $(\rho_{12})_{n,m}$. The true distributions are displayed on the left, whereas the reconstructed ones are displayed on the right.

$$\langle m+s|\hat{\rho}^{(ii)}|m\rangle = \sum_{n=0}^{N} M^{(s)}_{m,n}(|\gamma|)P^{(i)}(n,s) \; , \tag{8}$$

where the matrices M are given by $M = (G^T G)^{-1} G^T$.

We can now repeat the spin measurements just as we have described above in the case of the *unknown* initial state $|\Psi\rangle$. Repeating this procedure over and over again (with the same unknown initial state) for a large number of times and tracing out the cyclotron degree of freedom (no drive in this case is required), it is possible to recover the probabilities $\bar{P}^{(i)}$ associated to the two spin eigenvalues for the state $|\overline{\Psi}\rangle$.

It is important to note that the probabilities $\bar{P}^{(i)}$ can be experimentally sampled and that the modulus r and the phase β of the scalar product $\langle\psi_1|\psi_2\rangle$ can be both derived from the reconstruction of the cyclotron density matrices $\rho^{(11)}$ and $\rho^{(22)}$ [6]. Thus we are able to find the relative phase θ.

As an example of the application of the proposed reconstruction procedure, we show in Fig. 1 the results of numerical Monte-Carlo simulations of the reconstruction of the entangled state. In this simulation we have used the value $|\alpha| = 1$ which is experimentally accessible [3] and gives *mesoscopic* entangled superpositions of cyclotron coherent states with opposite phase. As it can be seen from Fig. 1, the reconstructed distributions turn out to be quite faithful.

References

[1] C. A. Sackett *et al.*, Nature (London) **404**, 256 (2000).

[2] C. Monroe *et al.*, Science **272**, 1131 (1996).

[3] L. S. Brown and G. Gabrielse, Rev. Mod. Phys. **58**, 233 (1986).

[4] S. Mancini *et al.* J. Opt. Soc. Am. A (December 2000).

[5] M. Massini *e al.* Phys. Rev. A **62**, 041401(R) (2000).

[6] M. Massini *et al.*, New J. Phys. **2**, 20.1 (2000).

QUANTUM MECHANICS WITHOUT STATISTICAL POSTULATES

H. Geiger, G. Obermair
Institut für Theoretische Physik, Universität Regensburg, D-93040 Regensburg, Germany
harald.geiger@akdb.de,gustav.obermair@physik.uni-regensburg.de

Ch. Helm
Los Alamos National Laboratory, T-11, B-262, Los Alamos NM 87545, USA
cxh@lanl.gov

Keywords: Bohmian Quantum mechanics, measurement process, statistics

Abstract The Bohmian formulation of quantum mechanics describes the measurement process in an intuitive way without a reduction postulate. Due to the chaotic motion of the hidden classical particle all statistical features of quantum mechanics during a sequence of repeated measurements can be derived in the framework of a deterministic single system theory.

1. INTRODUCTION

In the usual formulation of quantum mechanics the principal impossibility to predict the outcome of a quantum measurement and the associated probability distribution function for the possible results are usually introduced ad hoc together with the reduction of the wavefunction. In order to avoid the difficulties connected with this assumption, a deterministic formulation of quantum mechanics has been suggested by David Bohm [Bohm], where the dynamics of a nonlocal hidden variable x is determined by the wavefunction $\Psi(x,t) = R(x,t)e^{\frac{i}{\hbar}S(x,t)}$ via the potential $V_{\text{qu}}(x,t) := -\frac{\hbar^2}{2m}\frac{\frac{\partial^2}{\partial x^2}R(x,t)}{R(x,t)} + V(x,t)$. Originally an additional assumption concerning the distribution $\mathcal{P}(x,t)$ of the particles in a fictive ensemble $\{x_i\}$ had to be made, in order to recover all statistical features of the experiment [Bohm,Dürr]. In the following it will be sketched briefly, how this statistical assumption can be *derived* from the chaotic dynamics of the Bohmian particle, which shows that quantum mechanics can be understood completely on the basis of a nonstatistical formulation. Details can be found elsewhere [Geiger].

2. CHAOS IN THE MEASUREMENT

A series of von Neumann measurements with a consecutive repreparation of the initial state $|\Psi(x,t_i)|^2 := |\Psi(x,t_0)|^2$ is considered in the Bohmian formulation. This corresponds to the series of real physical experiments, which are necessary to verify quantum statistical properties in a *single* system. This sequence of position measurements and repreparation is illustrated schematically in the figure for the case of two possible outcomes $x \in [0,1[$ or $x \in [1,2[$:
(a) At the initial time $t = t_0$ an arbitrary wavefunction $\Psi(x,y,t_0)$ is to be measured by a detector with a (highdimensional) coordinate y.
(b) During the von Neumann interaction $H_{WW} = \lambda i\hat{x}\partial/\partial y$ between the system coordinate x (with accuracy $\Delta x = 1$) and detector y there is a separation of the wavefunction $\Psi(x,y,t) = \sum_n c_n\Psi_n(x)\Phi_0(y - \lambda xt)$ in two parts along the detector coordinate y. The coordinate y of the particle after the measurement is stored (in a macroscopic coordinate z) and reflects the original position of the particle x in the intervals $[0,1[, [1,2[$.
(c) After the measurement the wavepacket without the particle (here [0,1[) does not influence the particle dynamics anymore (because of the locality of $V_{qu}(x,t)$) and can therefore be neglected.
(d) For a new measurement with the same initial state (a) the remaining wavepacket has to flow from one to both intervals into the original form $|\Psi(x,y,t_0)|$.

Chaotic phenomena within the Bohmian quantum mechanics have been demonstrated repeatedly [Schwengelbeck, Dürr]. In the following it is pointed out that in this interpretation also the measurement is a deterministic process, whose chaotic properties can be studied. As the dimension of both the system x and the detector y is at least one, the Poincaré-Bendixson-Theorem, which excludes chaos in autonomous systems of dimension $n \leq 2$, is not applicable. Therefore and due to the high nonlinearity of the many particle system (including the detector) it is conceivable that the dynamics of the hidden variable (x,y) during the measurement process is in general chaotic and ergodic in particular.

One of the main consequences is the loss of predictability of the next measurement in a series of measurements. Although the position of the particle evolves deterministicaly from the initial value $x(0)$, the intrinsic inaccuracy Δx of any measurement together with the mixing property of the dynamics prevents from the complete knowledge of the system for the future.

In addition to this, the distribution $P(x)$ of measurement results in a series of measurements with repreparation of $\Psi(x,t_0)$ at times t_i is given by $P(x) = |\Psi(x,t_0)|^2$, as it is demanded by experiment.

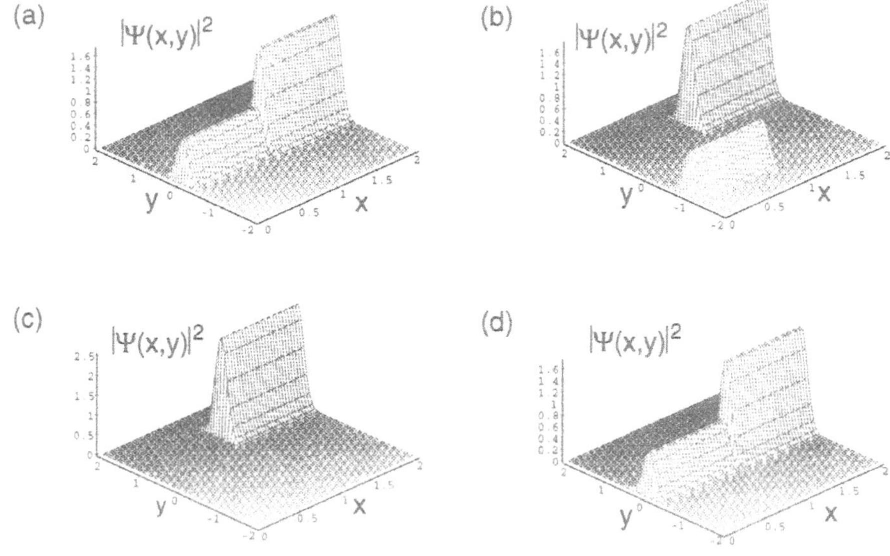

Figure 1

This argument also solves the problem of "quantum equilibrium" [Dürr] in an ensemble $\{x_i\}$ of independent systems with the same wavefunction $\Psi(x,t)$, but different positions $x_i(t)$ of the particles, which demands $\mathcal{P}(x,t) = |\Psi(x,t)|^2$ for their distribution function $\mathcal{P}(x,t)$.

Due to ergodicity of the trajectory $x(t)$ the distribution $P(x)$, which is obtained along one trajectory $x(t)$ at different times t_i, can also be expressed by the probability distribution $\mathcal{P}(x,t_0)$ of an appropriate fictive ensemble of particles $\{x_i\}$ at a fixed time t_0:

$$P(x) = \mathcal{P}(x,t_0). \tag{1}$$

Formally this identity can be concluded from the coincidence of the time average $\langle x^k \rangle_t := \frac{1}{M}\sum_i x^k(t_i)$ and the ensemble average $\langle x^k \rangle_e := \frac{1}{V}\int x^k dx$ in ergodic systems for all natural k. Note that here the ensemble $\{x_i\}$ is not introduced by an additional statistical assumption, but follows from the proven ergodicity of the dynamical system under investigation. A short calculation using the continuity equations for $\mathcal{P}(x,t)$ and $|\Psi(x,t)|^2$ shows that $\frac{d}{dt}f(x(t),t) = 0$ for any trajectory $x(t)$, $f(x,t)$ being defined by $\mathcal{P}(x,t) = f(x,t)|\Psi(x,t)|^2$. As the trajectory is ergodic, it is a dense subset of phase space and $f(x,t)$ is constant in the whole accessible area. Because of the normalisation $\int \mathcal{P}(x,t)dx = \int |\Psi(x,t)|^2 dx = 1$ it follows that $f(x,t) = 1$ and

$$\mathcal{P}(x,t) = |\Psi(x,t)|^2 \quad . \tag{2}$$

Finally we can conclude that the distribution $P(x)$ of the real experimental results during a sequence of measurements in a single system is given by

$$P(x) \overset{\text{equ.1}}{=} \mathcal{P}(x, t_0) \overset{\text{equ.2}}{=} |\Psi(x, t_0)|^2 = |\Psi(x, t_i)|^2 \quad , \tag{3}$$

where the last step follows from the consecutive repreparation of the system at times t_i in the inital state. This means that the density $P(x)$ of particle positions in a sequence of measurements is identical to $|\Psi(x, t)|^2$ of the wave-function of the single systems which is prepared before each measurement.

3. CONCLUSION

Bohmian quantum mechanics gives the usual statistical predictions of quantum mechanics without any statistical assumptions within a deterministic, single system theory. The uncertainty in the result of a quantum mechanical measurement and the probability distribution $P(x) = |\Psi(x, t)|^2$ in a sequence of measurements follows from the interplay of the chaotic motion of the hidden variable $x(t)$ and the finite accuracy Δx of any real measurement.

Acknowledgments

Work was done under the auspices of DOE (C.H.).

References

[1] D. Bohm, A suggested interpretation of the quantum theory in terms of 'hidden' variables I/II, Phys. Rev. **85,** (1952) 166 and 180; D. Bohm, B. Hiley, *The undivided universe – an ontological interpretation of quantum mechanics*, (Routledge, 1993) and references therein.

[2] D. Dürr, S. Goldstein, N. Zanghi, Quantum chaos, classical randomness and Bohmian mechanics, Journal of Statistical Physics **68** (1992) 259.

[3] H. Geiger, G. Obermair, Ch.Helm, quant-ph/9905068; H. Geiger, *Quantenmechanik ohne Paradoxa – Meßprozeß und Chaos aus der Sicht der Bohmschen Quantenmechanik* (Mainz Verlag, Aachen, 1998), Ph.D. thesis, ISBN 3-89653-452-1.

[4] U. Schwengelbeck, F. Faisal, Definition of Lyapunov exponents and KS entropy in quantum dynamics, Phys. Lett. A **199** (1995) 281.

QUANTUM RETRODICTION

J. Jeffers, S. M. Barnett
Dept. of Physics & Applied Physics, University of Strathclyde, Glasgow G4 ONG, U.K.

D. Pegg
Faculty of Science, Griffith University, Nathan, Brisbane Q 4111, Australia

O. Jedrkiewicz, R. Loudon
Dept. of Electronic Systems Engineering, University of Essex, Colchester CO4 3SQ, U.K.

The conditional probabilities of dependent events are governed by Bayes' Theorem, which can be illustrated by the following example. A physicist wishes to attend a conference, and chooses one of two airlines to get him there. The probability that he chooses airline 1 is $P(1)$ and the probability for airline 2 is $P(2) = 1 - P(1)$. Airline 1 loses luggage with high probability $P(\mathrm{L}|1) = 3/4$, whereas the corresponding probability for airline 2 is merely $P(\mathrm{L}|2) = 1/2$, where the vertical bar means "given". If the physicist arrives at the conference without his luggage, what is the probability that he travelled with airline 1? Bayes' Theorem provides an answer to this question by supplying the retrodictive conditional probability,

$$
\begin{aligned}
P(1|\mathrm{L}) &= \frac{P(\mathrm{L}|1)P(1)}{P(\mathrm{L})} = \frac{P(\mathrm{L}|1)P(1)}{P(\mathrm{L}|1)P(1) + P(\mathrm{L}|2)P(2)} \\
&= \frac{3P(1)}{3P(1) + 2P(2)},
\end{aligned} \tag{1}
$$

in terms of the known predictive probabilities. For example if the physicist chooses between the airlines at random then $P(1) = P(2) = 1/2$, so the probability that he travelled with airline 1 is $P(1|\mathrm{L}) = 3/5$.

In quantum theory state preparation and measurement are events. The preparation and measurement of particular states thus have predictive and retrodictive conditional probabilities which are related by Bayes' Theorem. Quantum theory, however, is normally only used predictively. An initially prepared state evolves and when a measurement is made the evolved state determines the likelihood of obtaining a particular result. This way of looking at things does not apply well to all experimental situations. If we know only the measured state, and have no preparation information then we have two choices. We could calculate the predictive probabilities for obtaining each measured state from all possible prepared states, and then use Bayes' Theorem to retrodict the probability that particular states were prepared. A more natural choice is to assign the system a retrodictive state based on the measurement outcome, follow the evolution of this state backwards in time, and then project it on to the preparation basis [1, 2]. Here we describe the general features of this retrodictive form of quantum theory.

Retrodiction for closed systems

Suppose that for a particular quantum system there are a set of preparation events a_i corresponding to preparation of the states $\hat{\rho}_i^{\mathrm{pred}}$. For simplicity we assume that there is no evolution before measurement. The most general description of the measurement is the probability operator measure (POM). This is an operator characterised by its elements $\hat{\Pi}_j$ which have positive or zero eigenvalues and satisfy $\sum_j \hat{\Pi}_j = \hat{1}$, where $\hat{1}$ is the unit operator in the system state space. Each POM element corresponds to a particular measurement outcome b_j. If the prepared state corresponds to the preparation event a_i the probability that a later measurement will give the outcome b_j is $P(\mathrm{b}_j|\mathrm{a}_i) = \mathrm{Tr}(\hat{\rho}_i^{\mathrm{pred}}\hat{\Pi}_j)$.

We have developed a retrodictive picture corresponding to the above[2]. For an unbiased preparation device, which satisfies $\sum_i P(\mathrm{a}_i)\hat{\rho}_i^{\mathrm{pred}} = \hat{1}/D$ (where D is the dimension of the state space), we can define a preparation POM with elements $\hat{\Xi}_i = DP(\mathrm{a}_i)\hat{\rho}_i^{\mathrm{pred}}$. Bayes' Theorem allows us to derive a retrodictive density operator such that the probability for the preparation event a_i given that the detection event was b_j, is

$$P(\mathrm{a}_i|\mathrm{b}_j) = \mathrm{Tr}(\hat{\rho}_j^{\mathrm{retr}}\hat{\Xi}_i), \tag{2}$$

and the retrodictive density operator associated with outcome b_j is the normalised measurement POM element, $\hat{\rho}_j^{\mathrm{retr}} = \hat{\Pi}_j/\mathrm{Tr}\hat{\Pi}_j$. Analogously, the predictive density operator is the normalised preparation POM.

We describe a biased preparation device using a preparation density operator, $\hat{\Lambda} = \sum_i \hat{\Lambda}_i$, with elements, $\hat{\Lambda}_i = P(\mathrm{a}_i)\hat{\rho}_i^{\mathrm{pred}}$. The same procedure as above gives the retrodictive conditional probability

$$P(\mathrm{a}_i|\mathrm{b}_j) = \frac{\mathrm{Tr}(\hat{\rho}_j^{\mathrm{retr}}\hat{\Lambda}_i)}{\sum_k \mathrm{Tr}(\hat{\rho}_j^{\mathrm{retr}}\hat{\Lambda}_k)}. \tag{3}$$

The lack of symmetry between the predictive and retrodictive conditional probabilities does not mean that there is any intrinsic time asymmetry in quantum theory. It merely reflects the fact that it is normal to use biased sources in experiments. When the source is unbiased, such as for quantum key distribution, time symmetry is restored.

If the system undergoes unitary evolution between preparation and measurement at times t_p and $t_m = t_p + \tau$ respectively, in the predictive picture the evolved density operator, $\hat{\rho}_i^{\text{pred}}(t_m) = \hat{U}(\tau)\hat{\rho}_i^{\text{pred}}(t_p)\hat{U}^\dagger(\tau)$, is projected onto the measurement POM to give the predictive conditional probability. The retrodictive density operator evolves *backwards* from the measurement to the preparation time, $\hat{\rho}_j^{\text{retr}}(t_p) = \hat{U}^\dagger(\tau)\hat{\rho}_j^{\text{retr}}(t_m)\hat{U}(\tau)$, when it is projected onto the preparation density operator.

In summary, the predictive and retrodictive quantum formalisms for closed systems can be described as follows. In the predictive formalism $\hat{\Lambda}_i$ associated with the outcome a_i of the preparation device, becomes upon normalisation the predictive density operator. This operator evolves forward in time until it is projected onto $\hat{\Pi}_j$ to determine the probability of measurement outcome b_j. In the retrodictive formalism $\hat{\Pi}_j$, associated with outcome b_j of the measuring device, becomes upon normalisation the retrodictive density operator. This operator evolves backwards in time until it is projected onto $\hat{\Lambda}_i$ to determine the probability of preparation outcome a_i.

Retrodiction for open systems

Open systems, in which an unmeasured environment E interacts with the system S, are necessarily non-unitary. For example, the evolved, reduced predictive density operator at the measurement time corresponding to the system preparation event a_i at the preparation time is

$$\hat{\rho}_{i,S}^{\text{pred}}(t_m) = \text{Tr}_E\left[\hat{U}(\tau)\hat{\rho}_i^{\text{pred}}(t_p) \otimes \hat{\rho}_E(t_p)\hat{U}^\dagger(\tau)\right]. \tag{4}$$

We know the initial environment state so this is easy to calculate.

The picture is a little more complicated for retrodiction [3, 4] as we can measure only the final system state, not the environment. Therefore we have no information about the final environment state, and the combined measurement POM is $\hat{\Pi}_j \otimes \hat{1}_E$. The reduced retrodictive density operator is defined as

$$\hat{\rho}_{j,S}^{\text{retr}}(t_p) = \text{Tr}_E\left[\hat{\rho}_E(t_p)\hat{U}^\dagger(\tau)\hat{\Pi}_j \otimes \hat{1}_E\hat{U}(\tau)\right]/N. \tag{5}$$

We can use this operator in eqs. 1.2 or 1.3 in place of $\hat{\rho}_j^{\text{retr}}$.

We have developed two methods for solving retrodictive problems in open systems, both of which are based on predictive master equantions[4]. The first approach allows us to calculate the matrix elements of the retrodictive density operator from a general initial density operator for the system. The second method allows calculation of the preparation probabilities directly by evolving forward in time the preparation POM element, and projecting it onto the measurement POM.

For closed systems the time evolution of the retrodictive density operator is the reverse of the evolution of the predictive density operator. This time reversal property does not survive for open systems because we have different information in the two cases. In the predictive formalism we know the initial state of both the system and the environment. In the retrodictive case we know the initial state of the environment and the final state of the system. The final state of the environment is an "information dump" in both formalisms.

To date we have applied retrodiction to two open systems. In quantum communication we have proved that retrodicting from a measured state which has been sent through an attenuating optical channel is equivalent to predicting the output state produced by sending the same measured state through an amplifying channel and *vice versa*. In other words, optical amplification and attenuation are predictive/retrodictive inverses [3].

The other simple system is the two-level atom interacting with the electromagnetic field. We have retrodicted prepared states from various measured atomic states. The main result is that, for preparation a long time before measurement, no matter what the state of the field, almost all measured atomic states retrodict to a 50/50 mixture of the ground and excited states, the "no information" state. We find a difference in decay rates, and different steady states, for prediction and retrodiction. The coherently excited atom also shows retrodictive Rabi oscillations [4].

Conclusion

In this paper we have described the retrodictive quantum formalism. We have defined the operator which describes preparation devices, and the retrodictive state, which is simply the normalised measurement POM element. These enable simple evaluation of preparation probabilities given the results of some later measurement event. We have shown how to apply the formalism to both closed and open quantum systems.

References

[1] D.T. Pegg and S. M. Barnett, 1999, Quantum Semiclass. Opt. **1**, 442.

[2] S.M. Barnett *et al.*, J. Mod. Opt., 2000, **47**, 1779.

[3] S.M. Barnett *et al.*, 2000, Phys. Rev. A **62**, 022313.

[4] S.M. Barnett *et al.*, 2000, J. Phys. B: At. Mol. Opt. Phys. **33**, 3047.

QUANTUM-TOMOGRAPHY METHOD IN INFORMATION PROCESSING

M. A. Man'ko
P.N. Lebedev Physical Institute,
117924 Moscow, Russia
mmanko@sci.lebedev.ru

Keywords: Noncommutative tomography, fractional Fourier transform, analytic signal, Ville–Wigner quasidistribution function, Green function

Abstract Analogy of time-dependent analytic signal to the wave function is used to apply in signal analysis the new tomographic approach developed recently in quantum mechanics and quantum optics. The tomographic probability and analytic signal are shown to be connected by the integral transform with a kernel related to the fractional Fourier transform.

Traditional methods of signal analysis are based on Fourier transform and used in the semiconductor-laser area as well [1]. The problem how to characterize as complete as possible any electromagnetic (or other) signal $f(t)$ is very important in information processing, communication lines, and quantum computing. In spite of the signal $f(t)$ is a classical c-number function, it turned out that the quantum-mechanics notions are useful to make a description of the signal's time–frequency properties. The Ville quasidistribution function [2] of two variables t and ω is a complete analog of the Wigner quasiprobability distribution function [3] and it is used in signal analysis in order to obtain characteristics of the time–frequency probability.

In view of an analogy of the complex wave function $\Psi(x)$ of quantum state in quantun mechanics to the complex analytic signal $f(t)$, we study a new method of signal analysis based on the symplectic-tomography principle [4] employed in quatum optics. The method uses noncommutative tomography of analytic signal [5]. Kernels of known integral transforms of analytic signals turn out to be identical to propagators of known simple quantum systems. The kernel of the Radon–Wigner transform of analytic signal related to the symplectic-tomography scheme is shown to coincide in the mathematical form with the Green function of the Schrödinger evolution equation for the quantum harmonic oscillator, which is the kernel of the fractional Fourier transform [6].

Let us shortly recall some useful transforms of the normalized complex analytic signal $f(t)$, i.e., $\int |f(t)|^2\,dt = 1$. The Ville–Wigner transform of analytic signal [2, 3]

$$W(t,\omega) = \int f\left(t + u/2\right) f^*\left(t - u/2\right) e^{-i\omega u}\,du \tag{1}$$

is a function of two variables and if it is known, analytic signal can be reconstructed in view of the relationship:

$$f(t)\,f^*(t') = (2\pi)^{-1} \int W\left((t+t')/2\,,\,\omega\right) e^{i\omega(t-t')}\,d\omega\,. \tag{2}$$

In quantum mechanics, there exist relationships for the density matrix of the pure state which can be expressed in terms of the following equalities for the Wigner function:

$$\int (2\pi)^{-1}\,W(t,\omega)\,dt\,d\omega = 1\,, \qquad \int (2\pi)^{-1}\,W^2(t,\omega)\,dt\,d\omega = 1\,. \tag{3}$$

The marginal distribution can be obtained in terms of analytic signal [5]

$$w\left(X,\mu,\nu\right) = \left(2\pi|\nu|\right)^{-1} \left| \int \exp\left[\left(i\mu t^2/2\nu\right) - \left(itX/\nu\right)\right] f(t)\,dt\right|^2 ; \tag{4}$$

it is normalized $\int w\left(X,\mu,\nu\right)\,dX = 1$ and contains complete information on analytic signal, since the Ville–Wigner quasidistribution $W(t,\omega)$ is connected with the marginal distribution $w\left(X,\mu,\nu\right)$,

$$w\left(X,\mu,\nu\right) = \int (2\pi)^{-2} \exp\left[-ik(X - \mu\omega - \nu t)\right] W(t,\omega)\,dk\,d\omega\,dt\,, \tag{5}$$

$$W(t,\omega) = (2\pi)^{-1} \int w\left(X,\mu,\nu\right) \exp\left[-i\left(\mu\omega + \nu t - X\right)\right]\,d\mu\,d\nu\,dX\,. \tag{6}$$

As an example, we consider a chirp signal, which is a complex Gaussian in time with linear time-dependence of its frequency,

$$f(t) = (\alpha/\pi)^{1/4} \exp\left[-\left(\alpha t^2/2\right) + \left(i\beta t/2\right) + i\,\omega_0 t\right]. \tag{7}$$

Taking the integral in (4) one obtains for the chirp signal the marginal distribution of the Gaussian form

$$w\left(X,\mu,\nu\right) = \left(2\pi\sigma_X^2\right)^{-1/2} \exp\left[-\left(X - \bar{X}\right)^2/2\sigma_X^2\right], \tag{8}$$

with parameters depending on the signal's parameters and on μ and ν

$$\sigma_X^2 = (2\alpha)^{-1}\left|\nu\left(\alpha - i\beta\right) - i\mu\right|^2, \qquad \bar{X} = \omega_0\nu\,. \tag{9}$$

One can associate the entropy $\mathcal{S}(\mu,\nu)$ and information $I(\mu,\nu)$ in a natural manner with analytic signal using the marginal distribution

$$\mathcal{S}(\mu,\nu) = -\int w\left(X,\mu,\nu\right) \ln\left[w\left(X,\mu,\nu\right)\right] dX\,, \quad I(\mu,\nu) = -\mathcal{S}(\mu,\nu). \tag{10}$$

The entropy of the Gaussian signal is determined by the dispersion

$$\mathcal{S}_{\mathrm{G}}(\mu,\nu) = \ln\left(\sigma_X\right) + \Big[1 + \ln\left(2\pi\right)\Big]/2\,, \tag{11}$$

where σ_X is given by relation (9). The smaller the dispersion of the Gaussian signal, the smaller its entropy.

In signal analysis and information processing, the fractional Fourier transform $(\mathcal{F}^a q)\,(u)$ of complex function $q\,(u)$ is used

$$(\mathcal{F}^a q)\,(u) = \int B_a\left(u,u'\right) q\left(u'\right)\,du'. \tag{12}$$

The kernel of the transform reads

$$\begin{aligned} B_a\left(u,u'\right) &= \exp\left\{-i\left[\left(\pi\hat{\Phi}/4\right) - \left(\Phi/2\right)\right]\right\} |\sin\Phi|^{-1/2} \\ &\times \exp\left[i\pi\left(u^2\cot\Phi - 2\,uu'\left(\sin\Phi\right)^{-1} + u'^2\cot\Phi\right)\right], \end{aligned} \tag{13}$$

where the angle variable is determined by the real parameter a, $\Phi = a\pi/2\,, \quad 0 < |a| < 2\,, \quad \hat{\Phi} = \mathrm{sgn}\left(\sin\Phi\right)$. On the other hand, the time evolution of the wave function of the quantum harmonic oscillator is desribed by the Green function

$$G_\omega\left(x,y,t\right) = \sqrt{\frac{m\omega}{2\pi i\hbar\sin\omega t}}\,\exp\left\{\frac{im\omega}{2\hbar}\left[\left(x^2+y^2\right)\cot\omega t - \frac{2xy}{\sin\omega t}\right]\right\}. \tag{14}$$

In view of the definition of the Green function, the oscillator's wave function at time t is expressed in terms of the initial wave function $\Psi\left(y,0\right)$ through the integral transform

$$\begin{aligned} \Psi(x,t) &= \sqrt{m\omega}\left(2\pi i\hbar\sin\omega t\right)^{-1/2}\int dy\;\Psi(y,0) \\ &\times \exp\left\{\left(im\omega/2\hbar\right)\left[\left(x^2+y^2\right)\cot\omega t - 2xy/\sin\omega t\right]\right\}. \end{aligned} \tag{15}$$

Let us now compare relations (12), (13), and (15), using the change of variables

$$u = \left(2\pi\hbar\right)^{-1/2} x\sqrt{m\omega}\,, \quad u' = \left(2\pi\hbar\right)^{-1/2} y\sqrt{m\omega}\,, \quad \omega t = \Phi\,, \tag{16}$$

and the replacements

$$(\hbar/m\omega)^{1/4}\,\Psi\left(\sqrt{2\pi\hbar}\,(m\omega)^{-1/2}\,u,\,\Phi/\omega\right) \implies (\mathcal{F}^a q)\,(u)\,,$$
$$(\hbar/m\omega)^{1/4}\,\Psi\left(\sqrt{2\pi\hbar}\,(m\omega)^{-1/2}\,u_0,\,0\right) \implies q\,(u)\,.$$

We see that relations (12), (13), and (15) coincide up to the factor $\exp{(i\,\Phi/2)} = \exp{(i\omega t/2)}$, this means that the oscillator's Green function and the kernel of the fractional Fourier transform are identical

$$B_a\left(u,u'\right) = \exp\left(i\omega t/m\omega\right) G_\omega\left(x,y,t\right), \tag{17}$$

where x, y, and t are determined by (16). The other phase factor in the kernel of the fractional Fourier transform is equal to the constant phase factor of the Green function $\exp\left(-i\,\pi\hat{\Phi}/4\right) = i^{-1/2}$.

Noncommutative tomography of analytic signal plays important role in the problem of detection and recognition of signals in the presence of noise. As analysis of noncommutative signal tomography demonstrates [5], a signal can be detected at low values of the signal-to-noise ratio since the detecting procedure chooses automatically the most favorable directions in time–frequency plane for recognizing the signal.

Acknowledgments

The author is grateful to the Organizers for kind hospitality and the Russian Foundation for Basic Research for the travel grant No. 00-02-26740. The work was partially supported by the Russian Foundation for Basic Research under Project No. 00-02-16516 and the Ministry for Science and Technology of the Russian Federation within the framework of the Program "Optics. Laser Physics."

References

[1] Bachert, H., *et al.*, (1975). Sov. J. Quantum Electron. V. 45. P. 1102; (1975). IEEE Quantum Electron. QE-11/1. Pt. 2. P. 510.

[2] Ville, J. (1948). Cables et Transmission. V. 2. P. 61.

[3] Wigner, E. (1932). Phys. Rev. V. 40. P. 749.

[4] Mancini, S., Man'ko, V.I., and Tombesi, P. (1995). Quantum Semiclass. Opt. V. 7. P. 615; (1996). Phys. Lett. A. V. 213. P. 1.

[5] Man'ko, V.I., and Mendes, R.V. (1999). Phys. Lett. A. V. 263. P. 53.

[6] Man'ko, M.A. (1999). J. Russ. Laser Res. V. 20. P. 225; (2000). J. Russ. Laser Res. V. 21. P. 411.

QUANTUM MEASUREMENT PROBLEM AND STATE DUAL REPRESENTATIONS

S. N. Mayburov
Lebedev Inst. of Physics
Leninsky Prospect 53, Moscow, Russia, 117924
maybur@sgi.lpi.msk.su

Keywords: Quantum measurements, information, representations

Abstract The information aspects of quantum measurement problem studied in the simple quantum model of information processing and memorizing device - observer O. It's argued that to describe probablistic events appearencei, i.e., 'State Collapse' consistently O complete states manifold must be tensor product of standard Hilbert space and dual to it novel set describing information acquired by O in the individual events.

The Measurement problem (or State Collapse) of Quantum Mechanics (QM) is still unresolved despite significant progress of Quantum Measurement Theory [1]. In this paper the novel approach related to quantum information described; its theoretical premises and main results are in [2]. In our model measuring system (MS) consist of measured state (particle) S and the information processing system - observer O, which in general can be human brain or some automata, but its detailed structure isn't important at this stage. We don't consider in our model detector amplification and environment E decoherence effects, but plan to account them elsewhere.

Following to some proposals dated back from Von Neuman and Wigner [3] we suppose that O can be regarded as quantum object which obeys QM despite its pôssible macroscopic scale [4]. Accordingly O state relative to some other observer O' can be described by state vector $|O\rangle$ in MS joint Hilbert space $\mathcal{H}$. O' is also quantum object, but this is unimportant for O treating. We consider in Von Neuman ansatz [1] O' description of the measurement by O of binary observable $\hat{Q}$ for S state : $\psi_s = a_1|s_1\rangle + a_2|s_2\rangle$, where $|s_{1,2}\rangle$ are Q eigenstates with eigenvalues $q_{1,2}$. Initial O state is $|O_0\rangle$ relative to O' and corresponding MS product state vector is Ψ^0_{MS}, its density matrix (DM) denoted ρ_0. S-O measuring interaction starts at $t = t_0$ and finished at t_1 and for the suitable

S-O interaction Hamiltonian H_I final state at $t > t_1$ is [1] :

$$\Psi_{MS} = a_1|s_1\rangle|O_1\rangle + a_2|s_2\rangle|O_2\rangle \quad (1)$$

and DM is ρ_{MS}. $|O_i\rangle$ are eigenstates of O observable $\hat{Q}_O$ which is the only O degree of freedom accounted in the model. Thus QM measurement theory predicts that at time $t > t_1$ for O' MS is in the pure state Ψ_{MS} which is the superposition of two states. The algebraic analysis of this MS selfmeasurement process shows that if MS states are vectors of some Hilbert space they unitarily equivalent for O and O', so O also must observe the same superposition [5]. Yet we know from the experiment that macroscopic observer O observes in each event random Q values which suits to description by mixture ρ_{mix} of Q eigenstates with probabilities $|a_{1,2}|^2$. Moreover even partial (restricted) O states $R_O(\rho) = Tr_s\rho$ differs for $\rho_{MS,mix}$ for individual events [5]. This most general formulation of measurement problem for MS system via comparison of two observers reports or 'Friend paradox' introduced by Wigner [3].

Here we'll investigate the possibility that the true MS quantum states manifold (QSM) for O differs from standard Hilbert space $\mathcal{H}$ and this results into discussed discrepancy. Really quantum systems are well studied only when O evolution isn't accounted and it can't be excluded that MS states are different from standard state vectors when they are described by O 'from inside' [1]. QSM modifications were proposed already and most famous is Namiki-Pascazio many Hilbert spaces formalism [1]. Other QSM modification applied to measurement problem are nonperturbative QFT nonequivalent representations of commutation relations [6]. Here we'll explore dual states representations, in which state vector and random Q values in individual events coexist. This states most simply defined via standard DM $\rho = |\psi\rangle\langle\psi|$. Schrodinger-Liouville (SL) equation is correct also in our formalism for arbitrary Hamiltonian $\hat{H}$:

$$\dot{\rho} = [\rho, \hat{H}] \quad (2)$$

In our model $\rho(t)$, which initial state defined alike in standard QM becomes one component of dual state $|\Phi) = \rho \bigotimes V^O$ having tensor product QSM structure where V^O describes the (random) information on measured state acquired by observer O in given event. V^O can be defined for our MS final state ρ_{MS} via O restricted (partial) state $R_O(\rho_{MS})$ after MS measurement. In O basis one finds the weights $P_j(t) = Tr_O[\hat{P}^O_j R_O(\rho_{MS})]$ where $\hat{P}^O_j$ are O_j projection operators and in this case for $t > t_1$ it gives $P_j = |a_j|^2$. Restricted state $R_O(\rho_{MS})$ and $P_j(t)$ by itself has no probabilistic meaning relative to O' or O in standard QM without additional reduction postulate [1]. In distinction we suppose that for O observation of MS 'from inside' (but not for O') $P_j(t)$ becomes probabilistic distribution which generates random restricted state $V^O_j = |O_j\rangle\langle O_j|$ in given event. It describes the subjective state perceped by O after S-O interaction finished in given event corresponding to random q_j, q_{Oj} measured values. Due to entanglement V_j^O corresponds to MS state $V_j^{MS} = |O_j\rangle\langle O_j||s_j\rangle\langle s_j|$. Before measurement starts MS dual state is $\Phi_0 = |\rho_0; V_0^O)$ where $V_0^O = |O_0\rangle\langle O_0|$ describe O initial state. When measurement finished at $t > t_1$ $|a_{1,2}|^2$ gives the probability for O to observe $q_{1,2}$ described by $|\Phi_{1,2}) = |\rho_{MS}; V_{1,2}^O)$.

The dual states manifold for O is $N_T = L_q \bigotimes L_V$, where L_V is the set of all possible V_j^O in $|O_j\rangle$ basis with $trV^O = 1$; thus dual state norm $|\Phi| = 1$. ρ manifold is $L_q = \{\rho \geq 0; Tr\rho = 1\}$, which for pure states is equivalent to $\mathcal{H}$. If MS initial state ρ_0 is mixed one should treat each mixture component separately in the described ansatz. O' has its own subjective states set L'_V of analogous structure which describes O' information acquired in the interaction with measured objects. For O' the complete manifold is $N'_T = L'_q \bigotimes L'_V$ where L'_q can be obtained from L_q by standard unitary transformation between O, O'. L_V, L'_V are independent sets and L_V is unobservable directly for O' and vice versa, because O' can measure only ρ component of Φ. But V^O, V'^O can be correlated statistically if O' measures Q_O of ρ_{MS}. For S quantum ensembles the statistics in L_V set corresponds to $|a_j|^2$ - the probabilities of particular O observation.

Our formalism in fact separates dynamical and information aspects of quantum state. In distinction from standard QM with reduction in our formalism ρ doesn't suffers the collapse stochastic jumps in the measurement. It evolves linearly and reversibly according to (2), but isn't observed directly by O. Φ subjective component V^O can change abruptly and probabilistically during measurement describing the change of O subjective information about S. We must stress that subjective component V^O is physical object which can be interpreted as ρ_{MS} random projection. It connected with O internal state described 'from inside', which lays outside of Hilbert space $\mathcal{H}$.

In general case to calculate Φ evolution for arbitrary system SL equation must be solved for given initial state ρ_{in} which derives $P_j(t)$ from $\rho(t)$. Then from $P_j(t)$ probabilities at any time one can find random V^O. So if we have several S-O rescatterings each time after it we get new V^O component. To exclude spontaneous V^O jumps without effective interactions with external world we introduce additional O identity condition : if S and O don't interact then the same V^O conserved. It means that O can observe constantly only one s_j branch of S state, despite that after measurement s_j state can evolve. This condition don't influence ρ dynamics defined by SL equation for ρ, but only subjective information V^O. Due to such MS dynamics no experiment performed by O' on MS would contradict to standard QM. In particular if in 'undoing' set-up ρ_0 restored back from ρ_{MS} then any V_j^O will be erased from O memory and initial V_0^O restored. Such 'quantum eraser' experiment on information memorization by the massive objects like molecules will be important test of collapse models including dual QM formalism (for the details see [2]).

For the conclusion dual formalism proposes possible solution of measurement problem if observer regarded as nonclassical object obeying to QM laws and in particular standard Schrodinger dynamics. It uses QSM which is larger then standard QM $\mathcal{H}$ adding to it novel set describing O information in individual events. In this approach the state collapse is 'subjective' - observer related effect depending on information transfer during the systems interaction. For regarded MS example it occurs for O and for him random Q value q_j is known. For O' it stays uncertain and MS state for him is ρ_{MS}, due to absence of O'- MS interaction. Note that Dual formalism differs principally from hidden parameters theories, and parameter j of V_j^O don't existed before S-O interaction starts. Our formalism demonstrates that probabilistic behavior is generic for QM. Wave-particle dualism always regarded as principal QM feature, but here it acquires the adequate description.

References

[1] P. Busch, P. Lahti, and P. Mittelstaedt, 'Quantum Theory of Measurements' (Springer-Verlag, Berlin, 1996)

[2] S. Mayburov quant-ph 0006104 , submitted to Int. J. Theor. Phys.

[3] E. Wigner, 'Scientist speculates' , (Heinemann, London, 1962)

[4] C. Rovelli, Int. Journ. Theor. Phys. 35, 1637 (1995); quant-ph 9609002 (1996),

[5] T. Breuer, Phyl. of Science 62, 197 (1995), Synthese 107, 1 (1996)

[6] S. Mayburov, Int. Journ. Theor. Phys. 37, 401 (1998)

HOMODYNE CHARACTERIZATION OF ACTIVE OPTICAL MEDIA

G. Mauro D'Ariano, Matteo G. A. Paris, Massimiliano F. Sacchi
Theoretical Quantum Optics Group, INFM and Dipartimento di Fisica "A. Volta"
Università di Pavia, via Bassi 6, I-27100 Pavia, ITALY

Abstract An effective maximum likelihood method is suggested to characterize the absorption/amplification properties of active optical media through homodyne detection.

The quantum characterization of optical media is an important issue in modern optical technology, since the noise in optical communications and measurements is ultimately of quantum origin. For negligible saturation effects, the propagation of an optical signal in an active media is governed by the master equation

$$\dot{\varrho} = G_1\, L[a]\, \varrho + G_2\, L[a^\dagger]\, \varrho\,, \tag{1}$$

where ϱ is the density matrix describing the quantum state of the signal mode a and $L[O]$ denotes the Lindblad superoperator $L[O]A = O\,A\,O^\dagger - \frac{1}{2}O^\dagger O\,A - \frac{1}{2}\,A\,O^\dagger O$. If we model the propagation as the interaction of a traveling wave single-mode a with a system of N identical two-level atoms, then the absorption $G_1 = \gamma N_1$ and amplification $G_2 = \gamma N_2$ parameters are related to the number N_1 and N_2 of atoms in the lower and upper level respectively. The quantity γ is a rate of the order of the atomic linewidth [1], and the propagation gain (or deamplification) is given by $\mathcal{G} = \exp[(G_2 - G_1)\,t]$.

An active medium described by the master equation (1) represents a kind of phase-insensitive optical device. In this paper, we want to evaluate the parameters G_1 and G_2 by the maximum-likelihood (ML) estimation applied to data coming from random phase homodyne detection on the signal exiting the medium. The present investigation is motivated by the fact that ML approach has been already successfully applied to estimation of the whole quantum state [2] as well as to determination of some parameters of interest in quantum optics [3].

Let us start by reviewing the ML approach. Let $p(x|\lambda)$ the probability density of a random variable x, conditioned to the value of the parameter λ. The analytical form of p is known, but the true value of the parameter λ is unknown, and should be estimated from the result of a measurement of x. In our case λ

is the couple of parameters G_1 and G_2, and p is the probability density of (random phase) homodyne data. Let $x_1, x_2, ..., x_N$ be a random sample of size N. The joint probability density of the independent random variable $x_1, x_2, ..., x_N$ (the global probability of the sample) is given by

$$\mathcal{L}(\lambda) = \Pi_{k=1}^{N} p(x_k|\lambda) \ , \tag{2}$$

and is called the likelihood function of the given data sample. The maximum-likelihood estimator of the parameter λ is defined as the quantity λ_{ml} that maximizes $\mathcal{L}(\lambda)$ for variations of λ. Since the likelihood is positive this is equivalent to maximize

$$L(\lambda) = \log \mathcal{L}(\lambda) = \sum_{k=1}^{N} \log p(x_k|\lambda) \tag{3}$$

which is the so-called log-likelihood function.

Using the Wigner representation of Eq. (1) one can easily solve the corresponding Fokker-Plank equation for the Wigner function $W(\alpha, \alpha^*; t)$. One obtains [4]

$$W(\alpha, \alpha^*; t) = \int \frac{d^2\beta}{\pi\delta^2} \exp\left(-\frac{|\alpha - g\beta|^2}{\delta^2}\right) W(\beta, \beta^*; 0) \tag{4}$$

with $g = e^{-Qt}$, $2Q = (G_1 - G_2)$, and $\delta^2 = (G_1 + G_2)(1 - g^2)/(4Q)$. The theoretical homodyne probability at phase ϕ is simply obtained as the following marginal distribution

$$p(x; \phi) = \int d\text{Im}\alpha\, W(\alpha e^{i\phi}, \alpha^* e^{-i\phi}) \qquad x = \text{Re}(\alpha) \ . \tag{5}$$

For input coherent state with amplitude α_0, one has $W(\beta, \beta^*; 0) = \frac{2}{\pi} \exp(-2|\beta - \alpha_0|^2)$ and the convolution in Eq. (4) gives

$$W(\alpha, \alpha^*; t) = \frac{1}{\pi(\delta^2 + g^2/2)} \exp\left(-\frac{|\alpha - g\alpha_0|^2}{\delta^2 + g^2/2}\right) \ . \tag{6}$$

The corresponding theoretical homodyne distribution is then given by

$$p(x; \phi) = \frac{1}{\sqrt{\pi(\delta^2 + g^2/2)}} \exp\left\{-\frac{1}{\delta^2 + g^2/2}\left[x - g\text{Re}(\alpha_0\, e^{-i\phi})\right]^2\right\} \ . \tag{7}$$

For non-unit quantum efficiency $\eta < 1$, one has the replacement

$$\delta^2 + \frac{g^2}{2} \longrightarrow \delta^2 + \frac{g^2}{2} + \frac{1-\eta}{2\eta} \ . \tag{8}$$

We applied the ML approach to determine G_1 and G_2 starting from random phase homodyne detection [ϕ in Eq. (7) randomly distributed in $(0, \pi)$]. As a input reference signal we used coherent state of fixed known amplitude. Notice that the use of coherent states is not simply a matter of computational and experimental convenience. In fact, there is no advantage in using e.g. squeezed states , because of the phase-insensitive character of the device. Compare, on the contrary, the case of phase estimation in Ref. [3]. Some results from Monte Carlo simulated experiments for both the absorption ($G_1 > G_2$) and the amplification ($G_1 < G_2$) regime are shown in Table 1. Notice also that the case $G_1 = G_2$ corresponds to the estimation of Gaussian noise, since one has the solution of Eq. (1) in the form

$$\varrho(t) = \int \frac{d^2\beta}{\pi G_1 t} e^{-\frac{|\beta|^2}{G_1 t}} D(\beta)\varrho D^\dagger(\beta) , \tag{9}$$

where $D(\beta) = e^{\beta a^\dagger - \beta^* a}$ denotes the displacement operator.

G_1	G_2	$(G_1)_{ML}$	$(\delta G_1)_{ML}$	$(G_2)_{ML}$	$(\delta G_2)_{ML}$
3.	1.	2.97250576	0.03489146	0.96966708	0.03299910
3.	2.	2.93669546	0.04629955	1.94330199	0.04412476
3.	3.	3.03023643	0.07376747	3.03199992	0.07122468
3.	4.	2.98543015	0.09926873	3.98150430	0.09763157
3.	5.	3.16888784	0.06556872	5.15783291	0.06240839

Table 1 Maximum likelihood estimation of loss (G_1) and gain (G_2) parameters of master equation (1). Input coherent state with amplitude $\alpha_0 = 4$ has been used, along with $N = 10^4$ homodyne data with quantum efficiency $\eta = 0.6$, collected after an effective time $t = 1$. δG_1 and δG_2 represent the statistical error of the estimation.

In Fig. 1 we show the behavior of the statistical errors on the maxlik determination of the parameters as a function of the number of homodyne data and the quantum efficiency of photodetectors. The robustness of the method to low quantum efficiency η is a feature of the maximum-likelihood technique [2, 3]. In the present case, however, it is not surprising [see Fig. (1)], because quantum efficiency itself can be described by master equation (1) [4]. Notice the inverse square root behaviour of the statistical errors versus the number N of data in the sample, according to the central limit theorem.

In conclusion, we applied the maximum-likelihood estimation approach to the characterization of linear active optical media through homodyne detection.

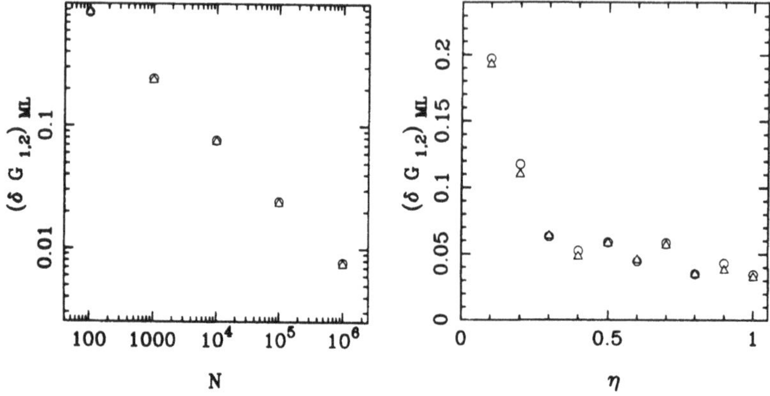

Figure 1 Behaviour of the statistical error $(\delta G_1)_{ML}$ (circle) and $(\delta G_2)_{ML}$ (triangle) versus number of homodyne data N (left, notice the scaling $\delta G \propto N^{-0.509}$) and quantum efficiency of homodyne detector (right) in the maximum likelihood estimation of G_1 and G_2 in the master equation (1). Parameters (left): $G_1 = 3, G_2 = 5, \eta = 0.6, \alpha_0 = 5, t = 1$. Parameters (right): number of data $N = 5 \times 10^3, G_1 = 2, G_2 = 1, \alpha_0 = 8, t = 1$.

The resulting method is efficient and provides a precise determination of the absorption and amplification parameters of the master equation using small homodyne data sample.

Acknowledgment

This work has been supported by the Italian Ministero dell'Università e della Ricerca Scientifica e Tecnologica (MURST) under the co-sponsored project 1999 *Quantum Information Transmission And Processing: Quantum Teleportation And Error Correction.*

References

[1] L. Mandel and E. Wolf, *Optical Coherence and Quantum Optics*, (Cambridge Univ. Press, 1995).

[2] K. Banaszek, G. M. D'Ariano, M. G. A. Paris, and M. F. Sacchi, Phys. Rev. A **61** 10304(R) (2000).

[3] G. M. D'Ariano, M. G. A. Paris, and M. F. Sacchi, Phys. Rev. A **62** 023815 (2000).

[4] G. M. D'Ariano, C. Macchiavello, and N. A. Sterpi, Quantum Semiclass. Opt. **9**, 929 (1997).

QUANTUM CLONING OPTIMAL FOR JOINT MEASUREMENTS

Giacomo Mauro D'Ariano, Massimiliano Federico Sacchi
Unità INFM and Dipartimento di Fisica "Alessandro Volta"
Università di Pavia, via A. Bassi 6, I-27100 Pavia, Italy
dariano@pv.infn.it msacchi@pv.infn.it

Abstract We show that universally covariant cloning is not optimal for achieving joint measurements of noncommuting observables with minimum added noise. For such a purpose a cloning transformation that is covariant with respect to a restricted transformation group is needed.

Introduction

Perfect cloning of unknown quantum systems is forbidden by the laws of quantum mechanics [1]. However, a universal optimal cloning has been proposed [2], which has been proved to be optimal in terms of fidelity [3, 4]. A obvious relevant application of such optimal cloning is eavesdropping in quantum cryptography [5]. However, quantum cloning can be of practical interest as a tool to engineer novel schemes of quantum measurements, in particular for joint measurements of noncommuting observables. Quite unexpectedly, as we will show in the following, the universally covariant cloning is not ideal for this purpose. Here, instead, cloning must be optimized for a reduced covariance group, depending on the desired joint measurement, in order to make the measurement over the cloned copies perfectly equivalent to an optimal joint measurement over the original.

In the following we consider: i) the case of spin 1/2, and the use of the $1 \rightarrow 3$ universally covariant cloning to achieve a joint measurement of the spin components; ii) the case of harmonic oscillator, along with a $1 \rightarrow 2$ cloning transformation that is *not* universally covariant. We show that the POVM obtained in the first case does not lead to a minimum-added-noise measurement. On the contrary, the second way allows one to achieve the ideal joint measurement of conjugated variables.

Quantum Communication, Computing, and Measurement 3
Edited by P. Tombesi and O. Hirota, Kluwer Academic/Plenum Publishers, 2001

Universally covariant cloning and joint measurement of spin components

The $1 \to 3$ cloning map is given by [4]

$$\rho_3 = \frac{1}{2} S_3 \left(|\psi\rangle\langle\psi| \otimes \mathbb{1}^{\otimes 2}\right) S_3 \,, \tag{1}$$

where $|\psi\rangle$ denotes the input state to be cloned, and S_3 is the projector on the space spanned by the vectors $\{|s_i\rangle\langle s_i|, i = 0 \div 3\}$, with $|s_0\rangle = |000\rangle$, $|s_1\rangle = 1/\sqrt{3}(|001\rangle + |010\rangle + |100\rangle)$, $|s_2\rangle = 1/\sqrt{3}(|011\rangle + |101\rangle + |110\rangle)$ and $|s_3\rangle = |111\rangle$ ($\{|0\rangle, |1\rangle\}$ being a basis for each spin 1/2 system). Independent measurements on the three copies along orthogonal axes provides the following POVM

$$\Pi(\vec{m}) = \frac{1}{2} \mathrm{Tr}_{2,3} \left[S_3 \, \Omega(\vec{m}) \, S_3\right] \,, \tag{2}$$

where

$$\Omega(\vec{m}) = \frac{1}{8} (\mathbb{1} + m_x\sigma_x) \otimes (\mathbb{1} + m_y\sigma_y) \otimes (\mathbb{1} + m_z\sigma_z) \,, \tag{3}$$

is the product of projectors on the output copies in terms of Pauli matrices σ_α, and $m_\alpha = \pm 1$ corresponds to each of the outcomes. Explicit calculation gives

$$\Pi(\vec{m}) = \frac{1}{8} \left[\mathbb{1} + \frac{5}{9} \vec{m} \cdot \vec{\sigma}\right] \,, \tag{4}$$

where the factor $5/9$ comes from the shrinking of the Bloch vector due to the cloning map (1). In order to have the measurement unbiased, the spin component outcomes $m_\alpha = \pm 1$ must be rescaled to $m_\alpha = \pm \frac{9}{5}$. In this way the sum of the variances corresponding to the three spin components $J_\alpha = \sigma_\alpha/2$ becomes

$$\langle \Delta J^2 \rangle = \frac{1}{4} \left(3 \frac{81}{25} - 1\right) = \frac{109}{50} \,. \tag{5}$$

The uncertainty for a measurement performed by projecting onto spin coherent states [6] reads [7]

$$\langle \Delta J^2 \rangle \geq 2j + 1 \,, \tag{6}$$

where for $j = 1/2$ and pure states the bound is achieved, and is equal to 2. So, the joint measurement via universally covariant cloning does not achieve the minimum added noise. Notice also that the minimum added noise would be achieved by a discrete POVM of the form $\Pi(\vec{m}) = \frac{1}{8}[\mathbb{1} + \vec{m} \cdot \vec{\sigma}]$, with $m_\alpha = \pm 1$.

Cloning for harmonic oscillator and joint measurement of conjugated variables

We consider now the $1 \to 2$ cloning transformation for a bosonic system

$$\mathcal{T}(\varrho) = \frac{1}{2} P_{c,a}(\sigma)(\varrho \otimes \mathbf{1}_a) P_{c,a}(\sigma) \,, \tag{7}$$

where ϱ denotes the initial state of the system to be cloned (in the bosonic mode c), mode a supports the second copy, and the projector $P_{c,a}(\sigma)$ is given by [7]

$$P_{c,a}(\sigma) = S_c(\ln\sigma) \otimes S_a(\ln\sigma)\, V(|0\rangle_{c\,c}\langle 0| \otimes \mathbf{1}_a) V^\dagger\, S_c^\dagger(\ln\sigma) \otimes S_a^\dagger(\ln\sigma) \tag{8}$$

with $|0\rangle$ denoting the vacuum state, $S_d(r) = \exp[r(d^{\dagger 2} - d^2)/2]$, and $V = \exp[\frac{\pi}{4}(c^\dagger a - c a^\dagger)]$. The cloning transformation in Eq. (7) can be realized by a unitary interaction of modes c and a with an ancillary system, as shown in Ref. [8]. An experimental realization of this continuous variable cloning has been proposed in Ref. [9], where the clones correspond to single-mode radiation fields and the cloning machine is a network of three parametric amplifiers under suitable gain conditions.

Upon measuring the quadrature operators $X_c = (c + c^\dagger)/2$ and $Y_a = (a - a^\dagger)/2i$ over the clones—namely projecting the output copies on the eigenstates $|x\rangle_c$ and $|y\rangle_a$—one implements the following POVM

$$\begin{aligned} F_\sigma(x,y) &= \frac{1}{2}\mathrm{Tr}_a[P_{c,a}(\sigma)|x\rangle_{c\,c}\langle x| \otimes |y\rangle_{a\,a}\langle y| P_{c,a}(\sigma)] \\ &= \frac{1}{\pi} D_c(x+iy)\, S_c(\ln\sigma)|0\rangle_{c\,c}\langle 0 S_c^\dagger(\ln\sigma)\, D_c^\dagger(x+iy) \,, \end{aligned}$$

where $D_c(\alpha) = \exp(\alpha c^\dagger - \alpha^* c)$ denotes the displacement operator for mode c. Such kind of POVM is formally a squeezed-coherent state, and it provides the optimal joint measurement of the two noncommuting quadrature operators $X_{\pm\phi} = (c^\dagger\, e^{\pm i\phi} + c\, e^{\mp i\phi})$, with $\phi = \mathrm{arctg}(\sigma^2)$. This is shown by the relations

$$\int dx \int dy\, (x\cos\phi \pm y\sin\phi)\, F_\sigma(x,y) = X_{\pm\phi} \,, \tag{9}$$

$$\begin{aligned} &\int dx \int dy\, (x\cos\phi \pm y\sin\phi)^2\, F_\sigma(x,y) \\ &= X_{\pm\phi}^2 + \frac{1}{4}\,|\sin(2\phi)\,| = X_{\pm\phi}^2 + \frac{1}{2}\,|\,[X_\phi, X_{-\phi}]\,| \,, \end{aligned} \tag{10}$$

namely the outcomes $(x\cos\phi \pm y\sin\phi)$ have the same average values as the expectations of the observables $X_{\pm\phi}$ respectively, with minimum added noise.

The cloning map in Eq. (7) is *not* universally covariant, but is covariant only under the group of unitary displacement operators, namely

$$\mathcal{T}\left(D_c(\alpha)\,\varrho\, D_c^\dagger(\alpha)\right) = D_c(\alpha) \otimes D_a(\alpha)\, \mathcal{T}(\varrho)\, D_c^\dagger(\alpha) \otimes D_a^\dagger(\alpha) \,. \tag{11}$$

Conclusions

Measures of quality other than fidelity should be used for optimality of quantum cloning, depending on the final use of the output copies. We have shown that universally covariant cloning—which maximizes the fidelity— is not optimal for engineering novel schemes of joint measurement of noncommuting observables. If one wants to use quantum cloning to achieve joint measurements, cloning must be optimized for a reduced covariance group, as shown here in the case of harmonic oscillator.

Acknowledgment

This work has been supported by the Italian Ministero dell'Università e della Ricerca Scientifica e Tecnologica (MURST) under the co-sponsored project 1999 *Quantum Information Transmission And Processing: Quantum Teleportation And Error Correction.*

References

[1] W. K. Wootters and W. H. Zurek, Nature **299**, 802 (1982); H. P. Yuen, Phys. Lett. A **113**, 405 (1986).

[2] V. Bužek and M. Hillery, Phys. Rev. A **54**, 1844 (1996); N. Gisin and S. Massar, Phys. Rev. Lett. **79**, 2153 (1997).

[3] D. Bruß, D. DiVincenzo, A. Ekert, C. Fuchs, C. Macchiavello and J. Smolin, Phys. Rev. A **57**, 2368 (1998); D. Bruß, A. Ekert and C. Macchiavello, Phys. Rev. Lett. **81**, 2598 (1998).

[4] R. Werner, Phys. Rev. A **58**, 1827 (1998).

[5] N. Gisin and S. Massar, Phys. Rev. Lett. **79**, 2153 (1997); N. Gisin and B. Huttner, Phys. Lett. **A228**, 13 (1997).

[6] A. Perelomov, *Generalized coherent states and their applications*, Springer-Verlag (1986).

[7] G. M. D'Ariano, C. Macchiavello, and M. F. Sacchi, quant-ph/0007062.

[8] N. J. Cerf, A. Ipe, and X. Rottenberg, Phys. Rev. Lett. **85**, 1754 (2000).

[9] G. M. D'Ariano, F. De Martini, and M. F. Sacchi, submitted to Phys. Rev. Lett., 2000.

[10] H. P. Yuen, Phys. Lett. **91A**, 101 (1982).

HOW MANY PROJECTIONS ARE NEEDED IN QUANTUM TOMOGRAPHY OF SPIN STATES?

Z. S. Sazonova
Physics Department, Moscow Automobile & Road Construction Institute (Technical University), 64, Leningradsky prospect, Moscow, Russia
sazonova@dataforce.net

R. Singh
Wave Research Center at General Physics Institute of Russian Academy of Sciences, 38, Vavilov street, Moscow 117942, Russia
ranjit@dataforce.net

Abstract We have studied the reconstruction of spin states, when the spin of the system takes the values $1/2$, $3/2$, and $5/2$. It is shown that as the value of the spin increases the number of projections to reconstruct the spin states increases. We have analyzed the two cases, when the spin states are not interacting with the detection field and interacting with the detection field.

1. INTRODUCTION

The Quantum Tomography methods in physics have come from X-ray computed Tomography.
Czech mathematician Johann Radon (1887–1956) discovered the mathematical solution of Tomography in 1917. Radon had introduced the integral transformation [Radon transformation (RT)], which has similar mathematical properties as in the case of Fourier transformation.
The RT came into effect in 1972 due to the presence of computers. Before 1972 mathematicians and physicists knew RT method but they could not find any technical application.
RT was used by Gel'fand in mathematics. RT is extensively used in Computer Tomography, i.e., X-ray computed Tomography.

2. DENSITY MATRICES FOR PURE/MIXED STATES

The density matrices for pure and mixed spin states:

Pure spin state:

$$\rho_{\Psi}^{pure} = |\Psi><\Psi| \tag{1}$$

where $|\Psi>$ - wave function of the spin particle.

Properties:

$$Sp(\rho_{\Psi}^{pure})^2 = Sp(\rho_{\phi}^{pure}) \tag{2}$$

$$(\rho_{\phi}^{pure})^* = \rho_{\phi}^{pure} \tag{3}$$

Mixed spin states:

$$\rho_{\Psi}^{mixed} = \sum_i \omega_i |\Psi_i><\Psi_i| \tag{4}$$

where ω_i - probability of having $|\phi>_i$ wave function in the ith spin state.

Properties:

$$Sp(\rho_{\Psi}^{mixed})^2 \neq Sp(\rho_{\phi}^{mixed}) \tag{5}$$

$$(\rho_{\phi}^{mixed})^* = \rho_{\phi}^{mixed} \tag{6}$$

3. WIGNER D-FUNCTION & ROTATION OF THE REFERENCE FRAME

We are using the Wigner D-function introduced by Wigner in 1927 for the discrete spin states. By applying D-function on any spin state we can rotate the reference frame by Euler angles. The Wigner D-function $D^j_{m,\acute{m}}$ is applied on the spin state $|\phi_{j,m}>$ with angular momentum j and its projection on z - axes.

$$|\phi_{j,\acute{m}}> = \sum_{m=-j}^{j} |\phi_{j,m}> D^j_{m,\acute{m}}(\alpha, \beta, \gamma) \tag{7}$$

Where $D^j_{m,\acute{m}}(\alpha,\beta,\gamma) = exp(-im\alpha)d^j_{m,\acute{m}}exp(-i\acute{m}\gamma)$ and $d^{1/2}_{m,m'}(\beta) = \begin{pmatrix} \cos(\frac{\beta}{2}) & -sin(\frac{\beta}{2}) \\ \sin(\frac{\beta}{2}) & \cos(\frac{\beta}{2}) \end{pmatrix}$

$$d^1_{m,m'}(\beta) = \begin{pmatrix} \frac{1+\cos(\beta/2)}{2} & -\frac{\sin(\frac{\beta}{2})}{\sqrt{2}} & \frac{1-\cos(\beta)}{2} \\ \frac{\sin(\beta)}{\sqrt{2}} & \cos(\beta) & -\frac{\sin(\beta)}{\sqrt{2}} \\ \frac{1-\cos(\beta)}{2} & \frac{\sin(\beta)}{\sqrt{2}} & \frac{1+\cos(\beta)}{2} \end{pmatrix}$$

The reference frame of $|\phi_{j,\acute{m}}>$ is rotated by Euler angles α, β, γ but the spin state has the same properties as before the rotation.

The probability of spin j having projection $\acute{m}$ is given by

$$\omega(j,\alpha,\beta,\gamma) = \rho_{\hat{j},j} = |\phi_{j,\acute{m}}> \tag{8}$$

4. HOW MANY PROJECTIONS ARE NEEDED TO RECONSTRUCT THE DENSITY MATRICES?

Pure & mixed spin states: Any complex density matrix of order $N \times N$ contains N^2 elements. Each element of density matrix contains two real numbers i.e., we have $2N^2$ independent elements.
Since, the density matrix is Hermitian and normalized i.e.,

$$Re(< +1/2|\rho| - 1/2 >) = Re(< -1/2|\rho| + 1/2 >). \tag{9}$$

$$Im(< +1/2|\rho| - 1/2 >) = Im(< -1/2|\rho| + 1/2 >). \tag{10}$$

$$Sp(\rho) = 1. \tag{11}$$

So we can reconstruct the density matrix by knowing three projections or probabilities of having spin projection on different axes by rotations of the reference frame.

Pure spin state:
For the construction of density matrix of spin 1/2, we need only 3 projections. Analogically, we need 8 projections for spin 1 to reconstruct the density matrix. e.g.

$$\rho^{pure} == \begin{pmatrix} 1 & 0 \\ 0 & 0 \end{pmatrix},$$

i.e., $\omega^{pure}(1/2,\beta) = \cos^2(\beta/2)$.

Mixed spin state: For the mixed spin state, we need 2 projections (unpolarized spin state) by rotatons of the reference frame to reconstruct the density matrix

$$\rho^{mixed} == \begin{pmatrix} 1/2 & 0 \\ 0 & 1/2 \end{pmatrix},$$

i.e., $\omega^{mixed}(1/2,\beta) = 1/2cos^2(\beta) + 1/2sin^2(\beta)$.

5. POLARIZATION TENSORS

Another way to characterize the density matrix is the polarization vector and polarization tensors (for spin/s >1/2). We can decompose the density matrix:

$$\rho = \sum_{L,M} < T(j)^+_{LM} > T(j)_{LM}. \tag{12}$$

$$< T(j)^+_{LM} >= Sp(\rho T_{L,M}(j)). \tag{13}$$

Where $T_{L,M}(j) = \sqrt{(2L+1)/(2j+1)} C^{j,\acute{m}}_{j,m,L,M} |\phi_{j,\acute{m}} >< \phi_{j,m}|^+$. and $C^{j,\acute{m}}_{j,m,L,M}$ are the Clebsch-Gordan coefficients. L amd M take value between $0 \leq L \leq 2j, -L \leq M \leq L$
As an example we have taken spin j=1/2 and j=1:

$$\rho = \sum_{L,M} < T(1/2)^+_{LM} > T(1/2)_{LM}. \tag{14}$$

$$\rho = 1/2E + \sum_{M} < T(1/2)^+_{1M} > T(1/2)_{1M}. \tag{15}$$

or

$$\rho = 1/\sqrt(2) + \sum_{M} < T(1/2)^+_{1M} > T(1/2)_{1M}. \tag{16}$$

Where $< T(1/2)_{00} >= 1/\sqrt{2}, < T(1/2)_{1,M} >= P_M/\sqrt{2}$.
or

$$\rho = 1/2[E + P_x\sigma_x + P_y\sigma_y + P_z\sigma_z]. \tag{17}$$

where $P_i = Sp(\rho\sigma_i)$ is the polarization vector. σ_i - are the Pauli matrices.
For the spin j=1, we have the decomposition of density matrix:

$$\rho = 1/3E + \sum_{M} < T(1)^+_{1M} > T(1)_{1M} + \sum_{M} < T(1)^+_{2M} > T(1)_{2M}. \tag{18}$$

Where $< T(1)_{00} >= 1/\sqrt{3}, < T(1)_{1,M} >= P_M/\sqrt{2}$ - are spherical tensors of rank 1 or orientation vector i.e. dipole moment and $< T(1)_{2,M} >$ - spherical tensors of rank 2 or alignment tensor i.e. qadrupol moment.

It is clear from the orientation vector that when spin 1 is perpendicular to the z-axis then it takes 0 projection on the z-axis. There is no concrete direction where the spin 1 is oriented. So we need the tensors of higher order i.e. 2nd rank to determine the polarization direction.

Similarly can also find the polarization tensors for the higher order spin for 3/2, 2, 5/2 and so on.

6. SPIN IN THE MAGNETIC FIELD

To detect the spin state we need magnetic field. We have applied constant magnetic filed in the z-axis. By applying magnetic field the spin starts precession around z-axis. The interaction Hamiltonian is

$$H_{int} = -\mu.H = -\sum_{j=x,y,z} (1/2)\gamma\hbar\sigma_j H_j. \tag{19}$$

Where μ is the magnetic moment of the spin, and H is the constant magnetic field.

7. CONCLUSIONS

We need 3 projections for the reconstruction of density matrix of pure spin state 1/2. With polarization vector, we need 2 projections for the case of pure spin 1/2. For mixed states we need minimum 2 projections or more. To reconstruct the density matrix for spin 1, we need 8 projections. With polarization tensor, we need for the case of pure Spin State 1 we need 5 projections. When we include constant magnetic filed to detect the spin state, we need minimum 2 projections to detect the spin state with spin 1/2 and 5 projections to detect the spin state with spin 1.

References

[1] Man'ko V. I., and Man'ko O. V., 1997, *JETP*, 85, 430.

[2] Landau L. D., Lifshitz E. M., 1989, *Quantum Mechanics*, (Moscow: Nauka) (in Russian).

[3] Varshalovich D. A., Moskalev A. N., and Khersonsky V. K., 1975, *Quantum Theory of Angular Momentum* (Leningrad: Nauka) (in Russian).

[4] Davidov A. S., 1973, *Quantum Mechanics*, (Moscow: Nauka) (in Russian).

III

QUANTUM COMPUTING

THE REPRESENTATION OF NUMBERS BY STATES IN QUANTUM MECHANICS

Paul Benioff
Physics Division, Argonne National Laboratory
Argonne, IL 60439
pbenioff@anl.gov

Keywords: Number representation, quantum states, quantum computers

Abstract The representation of numbers by tensor product states of composite quantum systems is examined. Consideration is limitedto $k - ary$ representations of length L and arithmetic $\bmod k^L$. An abstract representation on an L fold tensor product Hilbert space $\mathcal{H}^{arith}$ of number states and operators for the basic arithmetic operation is described. Unitary maps onto a physical parameter based tensor product space $\mathcal{H}^{phy}$ are defined and the relations between these two spaces and the dependence of algorithm dynamics on the unitary maps is discussed. The important condition of efficient implementation by physically realizable Hamiltonians of the basic arithmetic operations is also discussed.

INTRODUCTION

The representation of numbers by states of physical systems is basic and widespread in science. However, this representation is assumed and used implicitly, with little effort devoted to exactly what assumptions and conditions are implied. Here this question is examined for the nonnegative integers. Consideration will be limited to $k - ary$ representations of length L of numbers by tensor product states and to arithmetic modulo k^L.

Based on the universality of quantum mechanics, all physical systems of interest are quantum systems described by pure or mixed quantum states. This is the case whether the systems are microscopic, as is the case for quantum computers, or macroscopic, as is the case for all presently existing computers. Microscopic systems are those for which the ratio $t_{dec}/t_{sw} \gg 1$ where t_{dec} and t_{sw} are the decoherence and switching times [1]. In this case the system remains isolated from the environment for a time duration of many switching steps. If $t_{dec}/t_{sw} < 1$ then the system is macroscopic, and environmental interactions stabilize some system states (the pointer states) [2] for a time duration of many switching steps. The emphasis here is on microscopic systems although much of the material also holds for macroscopic systems.

One route to the exact meaning of the representation of numbers by states of physical systems begins with pure mathematics. A nonempty set is a set of numbers if and only if it satisfies (i.e. is a model of) the axioms of number theory or arithmetic [3]. Here these axioms need to be modified by inclusion of relevant axioms for a commutative ring with identity as these axioms are satisfied by modular arithmetic [4]. Details of the axioms are not important here. However, necessary conditions for a nonempty set to be a model include the existence of functions or operators with the properties of basic arithmetic operations, the successor S (or $+1$), $+$, and $\times$, given by the axioms. The ordering relation and the induction schema will not be discussed here.

If the axioms are consistent then there are many mathematical models of the axioms. Included are models containing tensor product states and unitary operators on a product Hilbert space of states. A model on an abstract product Hilbert space $\mathcal{H}^{arith}$ is described in Section 1.

The connection to physics is made in two steps. First a Hilbert space $\mathcal{H}^{phy}$ based on two sets of physical parameters is described. Models, based on $\mathcal{H}^{phy}$, of the axioms are described in Section 2. These are generated by unitary operators from $\mathcal{H}^{arith}$ to $\mathcal{H}^{phy}$.

The second step is the requirement that there exist physical models of the axioms in which the basic arithmetic operations are efficiently implementable. The widespread existence of computers shows that this existence requirement is satisfied, at least for macroscopic systems.

This requirement is discussed in Section 3 and applied to models based on $\mathcal{H}^{phy}$. Due to space limitations the discussion in this and other sections is brief. Details and proofs are provided elsewhere [5].

1. MODELS BASED ON $\mathcal{H}^{ARITH}$

Let $\mathcal{H}^{arith} = \otimes_{j=1}^{L}\mathcal{H}_j$ where $\mathcal{H}_j$ is a k dimensional Hilbert space spanned by states $|h, j\rangle$ where $0 \leq h \leq k-1$ and j is fixed. States in the corresponding basis spanning $\mathcal{H}^{arith}$ have the form $|\underline{s}\rangle = \otimes_{j=1}^{L}|\underline{s}(j), j\rangle$ where $\underline{s}$ is any function from $1, \cdots L$ to $0, \cdots, k-1$. The value of j distinguishes the component states (or qubytes) and h ranges over the k possible values of the states for each component.

The reason j is part of the state and not a subscript, as in $\otimes_{j=1}^{L}|\underline{s}(j)\rangle_j$, is that the action of operators to be defined depends on the value of j. Expression of this dependence is not possible if j appears as a subscript and not between $|$ and $\rangle$.

Definitions of S, $+$, $\times$ are required by the axioms. The efficient implementation requirement necessitates the definition of L different successor operators, V_j^{+1}, for $j = 1, \cdots, L$. These operators are defined to correspond to the addition of k^{j-1} mod k^L where V_1^{+1} corresponds to S.

To define the V_j^{+1} let u_j be a cyclic shift of period k that acts on the states $|h, j\rangle$ according to $u_j|h, j\rangle = |h + 1 \bmod k, j\rangle$. u_j is the identity on all states $|m, j'\rangle$ where $j' \neq j$. Define V_j^{+1} by

$$V_j^{+1} = \sum_{n=j}^{L} u_n P_{(\neq k-1),n} \prod_{\ell=j}^{n-1} u_\ell P_{(k-1),\ell} + \prod_{\ell=j}^{L} u_\ell P_{(k-1),\ell} \quad (1)$$

Here $P_{(k-1),j} = |k-1, j\rangle\langle k-1, j| \otimes 1_{\neq j}$ is the projection operator for finding the j component state $|k - 1, j\rangle$ and the other components in any state. $P_{m,j}$ and u_j satisfy the commutation relation $u_j P_{m,j} = P_{m+1,j} u_j \bmod k$ for $m = 0, \cdots, k - 1$. Also $P_{(\neq k-1),j} = 1 - P_{(k-1),j}$. In this equation the unordered product is used because for any p, q, $u_m P_{p,m}$ commutes with $u_n P_{q,n}$ for $m \neq n$. Also for $n = j$ the product factor with $j \leq \ell \leq n - 1$ equals 1.

The operator $+$ is defined on $\mathcal{H}^{arith} \otimes \mathcal{H}^{arith}$ by $+|\underline{s}\rangle|\underline{w}\rangle = |\underline{s}\rangle|\underline{s+w}\rangle$ where

$$|\underline{s+w}\rangle = V_L^{+\underline{s}(L)} V_{L-1}^{+\underline{s}(L-1)}, \cdots, V_1^{+\underline{s}(1)}|\underline{w}\rangle \quad (2)$$

and $V_j^{+\underline{s}(j)} = (V_j^{+1})^{\underline{s}(j)}$.

Note that for pairs of product states, which are first introduced here, the domains of the functions $\underline{s}$ and $\underline{w}$ must be different. That is $|\underline{s}\rangle|\underline{w}\rangle = |\underline{s * w}\rangle$ where $*$ denotes concatenation and $\underline{s * w}$ is a function from $1, \cdots, 2L$ to $0, \cdots, k - 1$. This follows from the requirement that all components of $|\underline{w}\rangle$ must be distinguished from those in $|\underline{s}\rangle$.

There are some basic properties the operators V_j^{+1} must have: they are cyclic shifts on $\mathcal{H}^{arith}$ and they satisfy

$$V_{j+1}^{+1} = (V_j^{+1})^k \text{ for } 1 \leq j < L\text{:} \quad (V_L^{+1})^k = 1 \quad (3)$$

This shows the exponential dependence on j and the need for separate definitions and efficient implementation of each of these operators. Also both the V_j^{+1} and $+$ are unitary. Proofs of these and other properties and a definition of $\times$ are given elsewhere [5].

The proof that the operators V_j^{+1}, $+$, $\times$ and states of the form $|\underline{s}\rangle$ in $\mathcal{H}^{arith}$ are a model of modular arithmetic consists in showing that the appropriate axioms are satisfied. Some details of this are given in [5]. Note that $\mathcal{H}^{arith}$ is also a model of the Hilbert space axioms.

2. MODELS BASED ON $\mathcal{H}^{PHY}$

The model of modular arithmetic described so far is abstract. No connection to quantum physics is provided. To remedy this one needs to describe models based on physical parameters. Let A and B be sets of L and k physical parameters for physical observables $\hat{A}$, $\hat{B}$ for an L component quantum system. The parameters in A distinguish the L component systems from one another and the parameters of B refer to k different internal physical states of the component systems. Examples of A include a set of locations in space or a set of excitation energies, as is used in NMR quantum computers [6]. Examples of B include spin projections along an axis or energy levels of a particle in a potential well.

The physical parameter based Hilbert space $\mathcal{H}^{phy} = \otimes_{a\epsilon A}\mathcal{H}_a$ where $\mathcal{H}_a$ is spanned by states of the form $|b, a\rangle$ with $b\epsilon B$ and a fixed. $\mathcal{H}^{phy}$ is spanned by states $|\underline{t}\rangle = \otimes_{a\epsilon A}|\underline{t}(a), a\rangle$ where $\underline{t}$ is a function from A to B.

The presence of the a component in the state $|\underline{t}(a), a\rangle$ property in the state is essential in that the state of the composite quantum system contains all the quantum information available to any physical process or algorithm. It is used by algorithms to distinguish the different components or qubytes. This is especially the case for any algorithm whose dynamics is described by a Hamiltonian that is selfadjoint and time independent. This is an example of Landauer's dictum "Information is Physical" [7].

The goal here is for physical parameter states such as $|\underline{t}\rangle$ to represent numbers. However, it is clear that the product states $|\underline{t}\rangle = \otimes_{a\epsilon A}|\underline{t}(a), a\rangle$ do not represent numbers. The reason is that there is no association between the labels a and powers of k; also there is no association between the range set B of $\underline{t}$ and numbers.

This can be remedied by use of unitary maps from $\mathcal{H}^{arith}$ to $\mathcal{H}^{phy}$ that preserve the tensor product structure. Let g and d be bijections (one-one onto) maps from $1, \cdots, L$ to A and from $0, \cdots, k-1$ to B. For each g, d, and j, $w_{g,d,j}$ is a unitary operator that maps states $|h, j\rangle$ in $\mathcal{H}_j$ to states in $\mathcal{H}_{g(j)}$ according to $w_{g,d,j}|h, j\rangle = |d(h), g(j)\rangle$. This induces a unitary operator $W_{g,d} = \otimes_{j=1}^{L} w_{g,d,j}$ from $\mathcal{H}^{arith}$ to $\mathcal{H}^{phy}$ where

$$\begin{aligned} W_{g,d}|\underline{s}\rangle = & \otimes_{j=1}^{L} w_{g,d,j}|\underline{s}(j), j\rangle \\ = & \otimes_{j=1}^{L}|d(\underline{s}(j)), g(j)\rangle = |\underline{s}_g^d\rangle. \end{aligned} \quad (4)$$

Here $|\underline{s}_g^d\rangle$ is the physical parameter based state in $\mathcal{H}^{phy}$ that corresponds, under $W_{g,d}$ to the number state $|\underline{s}\rangle$ in $\mathcal{H}^{arith}$.

Conversely the adjoint operator $W_{g,d}^{\dagger}$ relates physical parameter states in $\mathcal{H}^{phy}$ to number states in $\mathcal{H}^{arith}$:

$$W_{g,d}^{\dagger}|\underline{t}\rangle = \otimes_{a\epsilon A} w_{g^{-1},d^{-1},a}|\underline{t}(a), a\rangle$$

$$= \otimes_{a\epsilon A}|d^{-1}(\underline{t}(a)), g^{-1}(a)\rangle = |\underline{t}^{d^{-1}}_{g^{-1}}\rangle. \tag{5}$$

Here $|\underline{t}^{d^{-1}}_{g^{-1}}\rangle$ is the number state in $\mathcal{H}^{arith}$ that corresponds to the physical state $|\underline{t}\rangle$. Note that $W^{\dagger}_{g,d} = W_{g^{-1},d^{-1}}$ where g^{-1}, d^{-1} are the inverses of g and d, and $w_{g^{-1},d^{-1},a} = w^{\dagger}_{g,d,g^{-1}(a)}$.

The operators $W_{g,d}$ also induce representations of the V^{+1}_j, $+$, and $\times$ operators on the physical parameter states. For the V^{+1}_j one defines $V^{d,+1}_{g,j}$ by

$$V^{d,+1}_{g,j} = W_{g,d}V^{+1}_j W^{\dagger}_{g,d}. \tag{6}$$

An equivalent definition can be obtained from Eq. 1 by replacing u_ℓ by $u^d_{g(\ell)} = w_{g,d,\ell}u_\ell w^{\dagger}_{g,d,\ell}$ and $P_{k-1,\ell}$ by $P_{d(k-1),g(\ell)} = w_{g,d,\ell}P_{k-1,\ell}w^{\dagger}_{g,d,\ell}$.

In a similar fashion the $W_{g,d}$ are used to define $+_{g,d}$ acting on the physical parameter states in $\mathcal{H}^{phy} \otimes \mathcal{H}^{phy}$. The operator $+_{g,d}$ is defined in terms of the operator $+$ on $\mathcal{H}^{arith} \otimes \mathcal{H}^{arith}$ by $+_{g,d} = (W_{g,d} \otimes W_{g,d}) + (W^{\dagger}_{g,d} \otimes W^{\dagger}_{g,d})$. $\times_{g,d}$ is defined similarly.

There are a large number of tensor product preserving unitary maps from $\mathcal{H}^{arith}$ to $\mathcal{H}^{phy}$. Each of these induces a model for the axioms of modular arithmetic on $\mathcal{H}^{phy}$. There are $L!k!$ of these maps restricted to the form given by Eq. 4 for $W_{g,d}$ as there are $L!$ bijections g and $k!$ bijections d. Thus there is no unique correspondence between number states $|\underline{s}\rangle$ in $\mathcal{H}^{arith}$ and physical parameter states $|\underline{t}\rangle$ in $\mathcal{H}^{phy}$. In general the physical state $|\underline{s}^d_g\rangle$ corresponding to the number state $|\underline{s}\rangle$ depends on g and d. Conversely, by Eq. 5, the number state $|\underline{t}^{d^{-1}}_{g^{-1}}\rangle$ corresponding to the physical state $|\underline{t}\rangle$ in $\mathcal{H}^{phy}$ depends on g and d.

The question arises regarding the dependence of the dynamics of a quantum algorithm on the $W_{g,d}$. Some algorithms are independent of these maps; others are not. In general, since the the dynamics of any algorithm is physical, it must be described on $\mathcal{H}^{phy}$. If the algorithm can also be defined on $\mathcal{H}^{phy}$, then the dynamics is independent of these maps. Grover's Algorithm [8] is an example of this type of algorithm as it can be described and implemented on states in $\mathcal{H}^{phy}$ that are linear superpositions of $|\underline{t}\rangle$. No reference to states in $\mathcal{H}^{arith}$ is needed.

However, arithmetic algorithms must be defined on $\mathcal{H}^{arith}$. For these the dynamics does depend on these maps. Shor's Algorithm [9] is an example of this as it describes the computation of a numerical function $f_m(x) = m^x$ mod M where M and m are relatively prime.

So far the discussion has been limited to models of modular arithmetic on $\mathcal{H}^{arith}$ and on $\mathcal{H}^{phy}$. However the full connection to physics has not yet been established. This is given by the condition of efficient implementation of the basic arithmetic operations.

3. EFFICIENT IMPLEMENTABILITY OF THE BASIC ARITHMETIC OPERATIONS

The meaning of this important requirement is that it must be possible to physically implement the basic operations and that the implementation must be efficient. That is, for the $V_{g,j}^{d,+1}$, for each j there must exist a physically realizable Hamiltonian $H_{g,j}^{d}$ and a time t_j such that

$$e^{-iH_{g,j}^{d}t_j} = V_{g,j}^{d,+1}. \tag{7}$$

This definition is quite general in that the Hamiltonian can depend on j. However for many systems and dynamics a Hamiltonian H_g^d that implements $V_{g,j}^{d,+1}$ and is independent of j is realizable.

The requirement of efficiency means that both the space-time and the thermodynamic resources required to physically implement the operations must be at most polynomial in L, k. This condition excludes $k = 1$ or unary representations as all arithmetic operations are exponentially hard in this case. Large values of k are also excluded as distinguishing among a large set of symbols and carrying out simple arithmetic operations becomes thermodynamically expensive. Also there are physical limitations on the amount of information that can be reliably stored and distinguished per unit space time volume [10].

Thermodynamic resources are needed to protect the system from errors resulting from operation in a noisy environment. Microscopic systems also need protection from decoherence [11]. Methods include the use of quantum error correction codes [12], EPR pairs [13], and decoherence free subspaces [14]. Protection of macroscopic systems is less difficult since one takes advantage of decoherence to give stabilized "pointer" [2] states that represent numbers in a macroscopic computer.

The reason for separate definitions of the $V_{g,j}^{d,+1}$ for each j is that the requirement means that each of these operators for $j = 1, \cdots, L$ must be efficiently implemented. If the $V_{g,j}^{d,+1}$ were defined in terms of iterations of $V_{g,1}^{d,+1}$, then implementation of $V_{g,j}^{d,+1}$ would require k^{j-1} iterations of $V_{g,1}^{d,+1}$. This is not efficient even if $V_{g,1}^{d,+1}$ can be efficiently implemented.

Since the $V_{g,j}^{d,+1}$, $+_{g,d}$, and $\times_{g,d}$ operators are many system nonlocal operators, many dynamical steps would be needed to implement these operators by a realizable two particle local Hamiltonian, Eq. 7. As is well known, there are many methods of efficiently implementing these operators, at least in macroscopic computers. For the $V_{g,j}^{d,+1}$ methods include moving the procedure for efficiently implementing $V_{g,1}^{d,+1}$ along the path g in A to the site $g(j)$. For $+_{g,d}$, methods are based on iterations of the $V_{g,j}^{d,+1}$, Eq. 2.

The thermodynamic resources required to physically implement the $V_{g,j}^{d,+1}$ and other arithmetic operations also depend on the path g. In general paths are chosen that respect the topological or neighborhood properties of A as this reduces the resources required. But in principle any path is possible as the resource dependence on the path choice is not exponential.

4. DISCUSSION

The importance of the efficient implementability condition must be emphasized. Besides excluding $k = 1$ and large values of k, it excludes most unitary maps from $\mathcal{H}^{arith}$ to $\mathcal{H}^{phy}$. To see this one notes that for any unitary U, tensor product preserving or not, if the states $|\underline{s}\rangle$ and operators V_j^{+1}, $+$, $\times$ satisfy the axioms of modular arithmetic, so do the states $U|\underline{s}\rangle$ and the operators $UV_j^{+1}U^\dagger$, $(U \otimes U) + (U^\dagger \otimes U^\dagger)$ and an operator for $\times$. However, for most U, these operators are not efficiently implementable. Also the states $U|\underline{s}\rangle$ may not be efficiently preparable. This is the main reason the U were taken to have the form $W_{g,d}$.

Much remains to be done. Future work includes dropping the modulo limitation and considering other types of numbers. The use of annihilation and creation operators to represent states needs examination. Also the exact meaning of physical realizability needs to be clarified.

Acknowledgments

This work is supported by the U.S. Department of Energy, Nuclear Physics Division, under contract W-31-109-ENG-38.

References

[1] DiVincenzo, D. P. Science **270**: 255, (1995); Los Alamos Archives quant-ph/**0002077**.

[2] W. H. Zurek, Phys. Rev. D **24**: 1516, (1981); **26**: 1862 (1982); E. Joos and H. D. Zeh, Z. Phys. B **59**: 23, (1985); H. D. Zeh quant-ph/**9905004**; E Joos, quant-ph/**9808008**.

[3] J. R. Shoenfield, *Mathematical Logic* (Addison-Weseley, Reading, MA 1967); R. Smullyan, *Gödel's Incompleteness Theorems* (Oxford University Press, Oxford, 1992).

[4] I. T. Adamson, *Introduction to Field Theory*, 2nd. Edition, Cambridge University Press, London, 1982.

[5] P. Benioff, Los Alamos Archives Preprint quant-ph/**0003063**.

[6] N.A. Gershenfeld, Science **275**: 350 (1997); D.G. Cory, A.F. Fahmy, and T.F. Havel, Proc. Natl. Acad. Sci. **94**: 1634 (1997).

[7] R. Landauer, Physics Today **44**: No 5, 23, (1991);Physics Letters A **217**: 188, (1996); in *Feynman and Computation, Exploring the Limits of Computers*, A.J.G.Hey, Ed., (Perseus Books, Reading MA, 1998).

[8] L.K.Grover, in *Proceedings of 28th Annual ACM Symposium on Theory of Computing* ACM Press New York 1996, p. 212; Phys. Rev. Letters, **79**: 325 (1997); Phys. Rev. Letters, **80**: 4329 (1998).

[9] P. W. Shor, in *Proceedings, 35th Annual Symposium on the Foundations of Computer Science*, S. Goldwasser (Ed), IEEE Computer Society Press, Los Alamitos, CA, 1994, pp 124-134; SIAM J. Computing, **26**: 1481 (1997).

[10] S. Lloyd, Los Alamos Archives preprint quant-ph/9908043; Y. J. Ng, Los Alamos Archives Preprint quant-ph/0006105.

[11] W. H. Zurek, Physics Today **44**: No. 10, 36 (1991); J.R. Anglin, J. Paz, and W. H. Zurek, Phys. Rev A **55**: 4041 (1997); W. G. Unruh, Phys. Rev. A **51**: 992 (1995).

[12] R. Laflamme, C. Miquel, J. P. Paz, and W. H. Zurek, Phys. Rev. Letters **77**: 198 (1996); D. P. DiVincenzo and P. W. Shor, Phys. Rev. Letters **77**: 3260 (1996); E. M. Raines, R. H. Hardin. P. W. Shor, and N. J. A. Sloane, Phys. Rev. Letters **79**: 954 (1997); E. Knill, R. Laflamme, and W. H. Zurek, Science **279**: 342 (1998).

[13] C.H. Bennett in *Feynman and Computation, Exploring the Limits of Computers* A.J.G.Hey, Ed., (Perseus Books, Reading MA, 1998); C.H.Bennett D.P.DiVincenzo, C.A.Fuchs, T.Mor, E.Rains, P.W.Shor, J.A.Smolin, and W.K.Wooters, Rev. A **59**: 1070 (1999).

[14] D. Bacon, D. A. Lidar and K. B. Whaley, Phys.Rev. **A60**: (1999) 1944.

TOWARDS QUANTUM COMPUTATION WITH TRAPPED CALCIUM IONS

D. Leibfried
Institut für Experimentalphysik, Universität Innsbruck, A-6020 Innsbruck, Austria

C. Roos
Ecole Normale Superieure, Paris, France

P. Barton, H. Rohde, S. Gulde, A. B. Mundt,
F. Schmidt-Kaler, J. Eschner, R. Blatt
Institut für Experimentalphysik, Universität Innsbruck, A-6020 Innsbruck, Austria
Rainer.Blatt@uibk.ac.at

Keywords: Ion traps, laser cooling, quantum information

Abstract For quantum information experiments we have cooled one and two $^{40}Ca^+$ ions to the ground state of vibration with up to 99.9% probability, using resolved sideband cooling on the optical $S_{1/2}$ -$D_{5/2}$ quadrupole transition. Implementing a novel cooling scheme based on electromagnetically induced transparency on the $S_{1/2}$ -$P_{1/2}$ manifold we have achieved simultaneous ground state cooling of two motional sidebands 1.7 MHz apart. Coherent quantum state manipulation on the $S_{1/2}$ -$D_{5/2}$ quadrupole transition at 729 nm was demonstrated and up to 30 Rabi oscillations within 1.4 ms have been observed starting from the motional ground state and from the $n = 1$ Fock state. In a linear rf-trap two ions were cooled to the ground state of motion and individual addressing of the ions with laser pulses is achieved.

1. INTRODUCTION

Quantum information processing operations put severe demands on the experimental techniques. For example, a two-qubit quantum gate requires two strongly interacting quantum systems, non-interacting with the environment. A quantum logic gate based on trapped ions was proposed by Cirac and Zoller. This scheme requires the ions to be trapped in a linear radio-frequency (Paul) trap and cooled to the motional ground state of their (collective) motion [1].

The work reported here is based on $^{40}Ca^+$ ions. The light sources for all transitions involved are derived from diode and solid state lasers. With two different traps we have investigated several ways to cool one or two ions to the ground state of motion. Starting from this pure initial state, we perform coherent quantum manipulations of a single ion and of individual ions in an ion string.

2. EXPERIMENTAL SETUP

Quantum information processing requires that atomic transitions are available with two long lived states which can serve to encode the quantum bit (qubit). The relevant levels of Ca^+ are shown schematically in Fig. 1. Doppler cooling is achieved by driving the $S_{1/2} - P_{1/2}$ dipole transition at 397 nm with a frequency-doubled Ti:Sapphire laser that is red detuned from the transition line center by about 20 MHz, the natural line width of the $P_{1/2}$ level. To prevent pumping to the $D_{3/2}$ level, the ions are simultaneously irradiated with a diode laser at 866 nm driving the $D_{3/2}$-$P_{1/2}$ transition (Fig. 1 (a)). With a magnetic field of about 4 Gauss we produce a well defined quantization axis and split the Zeeman sublevels. Using an additional σ^+ polarized beam at 397 nm the pure electronic state $S_{1/2}, m = 1/2$ of the ion(s) can be prepared.

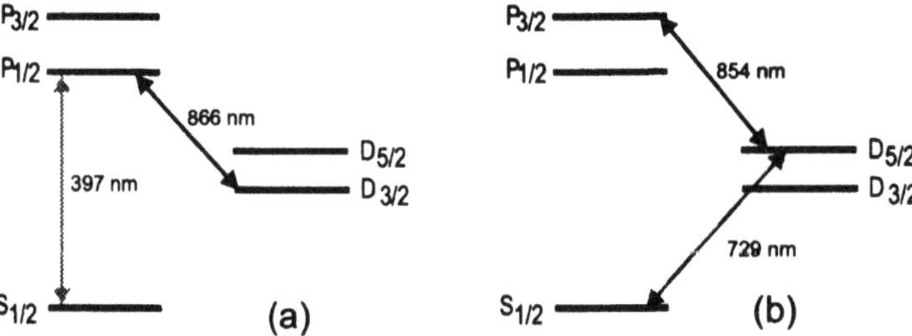

Figure 1 Relevant levels and transitions in $^{40}Ca^+$ for (a) Doppler cooling and state detection and for (b) resolved sideband cooling and coherent manipulations which are performed on the effective two level system (qubit) formed by $S_{1/2}$ and $D_{5/2}$ levels.

The $S_{1/2}$ ground state and the metastable $D_{5/2}$ state with a natural lifetime of about 1 s serve as the two qubit levels (Fig. 1(b)). Quadrupole transitions between these levels are driven with a Ti:Sapphire laser at 729 nm, stabilized with the Pound-Drever-Hall method to a high finesse cavity. To maintain the coherence necessary for qubit manipulations this laser has to be highly stable. We have determined an upper bound of 76(5) Hz (FWHM) for the effective linewidth of our laser system, by observing the fringe contrast in high resolution Ramsey spectroscopy on the $S_{1/2}, m = -1/2$ – $D_{5/2}, m = -5/2$ transition as a function of the time delay between the two excitation pulses [2]. The laser beams at 397 nm and 866 nm used for Doppler cooling also provide highly efficient state detection with the quantum jump technique [3]. Many photons are scattered and observed at 397 nm if the ion is in the ground state. On the other hand, if the ion is in $D_{5/2}$, this level is decoupled from the excitation and no fluorescence photons will be emitted. Although only a small fraction (ca. 10^{-2}) of these fluorescence photons is collected by a lens and imaged onto a photomultiplier with a quantum efficiency of about 10%, one can distinguish the two qubit states within 2 ms of detection time. This allows us to measure the state of our qubit with practically 100% efficiency.

In our experiments we use two different ion traps. The first trap is a regular spherical Paul trap with motional frequencies of up to 4.5 MHz and 2 MHz along the axis of symmetry and in the ring plane, respectively. Doppler cooling leads to average occupation numbers $\bar{n} = 2(4)$ for the axial (radial) harmonic oscillator, making this trap a good test bed for cooling and coherent control techniques with just one ion. For quantum information processing we use a linear quadrupole trap with secular frequencies of 2 MHz in the radial direction and up to 700 kHz in the axial direction. For a string of up to 4 ions this leads to inter-ion distances $\geq 5\mu$m, well above the diffraction limit of our laser beams. Under these conditions we are able to individually address ions within such a string, at the expense of a higher average occupation number ($\bar{n} \simeq 25$ in the axial direction) after Doppler cooling.

3. INDIVIDUAL ADDRESSING

For the Cirac-Zoller [1] gate the internal states of ions in a string have to be manipulated individually. One obvious limitation of this approach is that the size of the focus is limited by diffraction to roughly one micron and so the minimum distance between ions has to be larger than that number. A given minimum spacing of ions restricts the maximum center of mass (COM) frequency for a given number of ions along the axis of symmetry [4]. If four ions of Ca^+ should not be closer together than 5 μm, the maximum COM frequency is about 700 kHz The spatial resolution of our imaging system [5] is about 2 μm. For individual addressing we use the imaging lens in reverse and superimpose the addressing beam at 729 nm with the imaging channel on a dichroic mirror [6]. The beam is steered over the ions with an electrooptic deflector driven by a high voltage amplifier stage which allows us to switch from one ion to the other in a few μs. We have checked the beam diameter and pointing stability of our system by mapping the Rabi frequency on the $S_{1/2}$-$D_{5/2}$ transition versus the beam displacement and found a 1/e width of 3.7(0.3) μm for this excitation. If we apply a π-pulse to the ion addressed, the probability of exciting a neighboring ion in the ground state and 5 μm away would be about 1%.

4. GROUND STATE COOLING

For ground state cooling [7, 8] we use resolved sideband cooling on the $S_{1/2}$-$D_{5/2}$ quadrupole transition [9]. The slow spontaneous decay on a bare quadrupole transition would result in long cooling times. Therefore, the cooling rate is greatly enhanced by (i) strongly saturating the transition and (ii) shortening the lifetime of the excited state by coupling it to a dipole-allowed transition. For cooling a single ion in the circular trap, the $S_{1/2}(m = 1/2) - D_{5/2}(m = 5/2)$ transition, well resolved in frequency from all other possible transitions by the applied magnetic field, is excited with about 1 mW of light focused to a waist size of 30 μm at the position of the ion. At the same time, the decay rate back to the ground state is increased by exciting the $D_{5/2}(m = 5/2) - P_{3/2}(m = 3/2)$ transition. The intensity of this quenching laser at 854 nm is adjusted for optimum cooling during the experiment. Optical pumping to the $S_{1/2}(m = -1/2)$ level is prevented by occasional short laser pulses of σ^+ polarized light at 397 nm. The duration of those pulses is kept at a minimum to prevent unwanted heating. The ground state occupation is found by comparison of the on-resonance excitation probability for red and blue sideband transitions [9]. In the spherical Paul-trap we obtained up to 99.9% of motional ground state occupation within 6 ms (see Fig. 2).

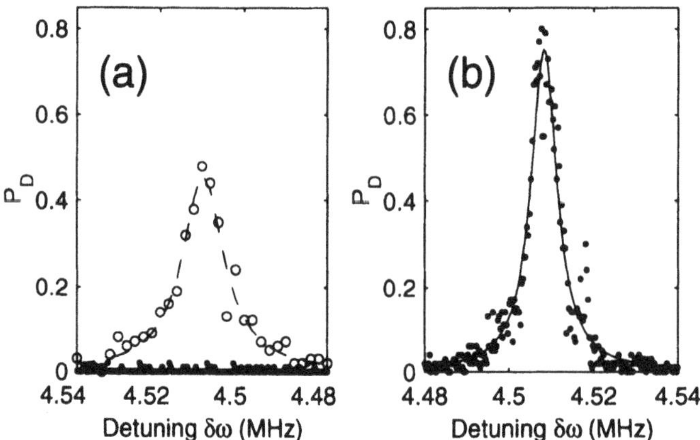

Figure 2 Red and blue sidebands of the axial mode (at 4.51 MHz) of one ion after Doppler cooling (dashed) and after resolved sideband cooling (solid). From the ratio of sideband strengths we determine 99.9% ground state occupation.

All earlier successful ground state cooling experiments were plagued by an unexpectedly high motional heating (see [10] and references therein). In our setup we find a motional heating rate of one phonon in 190 ms for a trap frequency of 4 MHz, two orders of magnitude smaller than in the trap used at NIST for the $^9Be^+$ experiment. While this is still a much higher heating rate than expected from black body radiation, it happens on a timescale much longer than the time typically needed for quantum logic gates (we estimate an upper limit of 200 μs for one CNOT gate).

In separate experiments in our linear trap, we also cooled all motional modes of two ions to the ground state [11]. Cooling only one of the 6 motional modes at a time, we achieved at least 95% ground state occupation for all modes using the addressing channel to illuminate one of the two ions with the cooling radiation. The second ion is cooled sympathetically due to the strong inter-ion Coulomb coupling [2].

5. EIT COOLING

Resolved sideband cooling only leads to very low temperatures if the red sideband is excited with a narrow excitation bandwidth. Otherwise nearby nonresonant transitions (e.g. carrier transitions) will lead to unwanted excess heating and severely increase the final temperature of the ion(s). Unless sideband frequencies are degenerate, this limits resolved sideband cooling to one motional sideband at a time, resulting in involved cooling schemes. Moreover the phonons scattered in the process of cooling one motional mode will reheat the other modes. Although for a Cirac-Zoller gate only the motional mode that is used as the 'quantum-bus' has to be cooled to a very high degree, the other motional modes must be cooled into the Lamb-Dicke regime [12]. To reach this regime by Doppler cooling would require trap frequencies of 10 MHz or higher and result in an ion spacing that is hard to optically resolve. This leads to a trade-off between addressing individual ions and sufficiently cooling all vibrational modes of a string. In our experiments with two or more ions in the linear trap we decided to maintain good conditions for individual addressing and limited our axial COM frequency to 700 kHz. Under these conditions it was desirable to find a cooling technique that is not as narrow-band as resolved sideband cooling but has a lower cooling limit than Doppler cooling. A very recent proposal to use electromagnetically induced transparency (EIT) for the cooling of trapped particles [13] promised to cool the ion deeply into the Lamb-Dicke regime for all motional degrees of freedom simultaneously. We adapted this cooling scheme for the case of the $[S_{1/2}, P_{1/2}]$ four level system in Ca^+ that we also use for Doppler cooling. The manifold is dressed with a σ^+-polarized beam at 397 nm, blue detuned by $\Delta_\sigma \simeq 70$ MHz (3.5 linewidths of the S-P transition), that connects the $S_{1/2}$, m=-1/2 with the $P_{1/2}$, m=1/2 level. Under these circumstances a second low-intensity π-polarized beam experiences an absorption (Fano-) profile as depicted in Fig. 3(a).

In addition to the usual line profile around $\Delta_\pi = 0$, a dark resonance (EIT) is created at $\Delta_\pi = \Delta_\sigma$, and a bright resonance appears at $\Delta_\pi = \Delta_\sigma + \delta$ where δ is the AC Stark shift due to the σ^+-polarized beam. For cooling the π-polarized

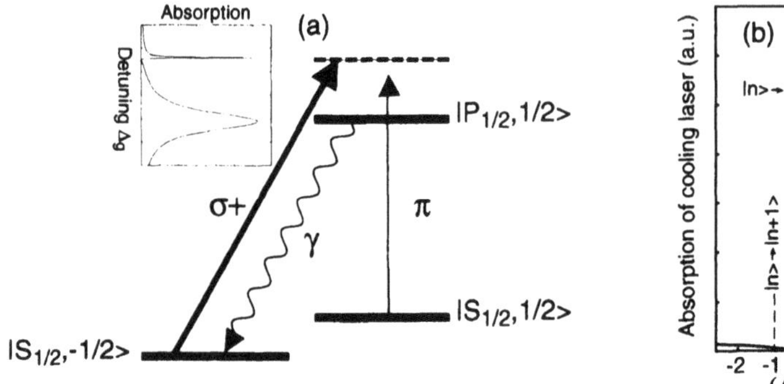

Figure 3 (a) Level scheme of the $S_{1/2}$-$P_{1/2}$ manifold and lasers used for EIT cooling. A strong σ^+ polarized beam creates a Fano-type absorption profile for a weak π-polarized beam. (b) By tuning the π-polarized beam to $\Delta_{pi} = \Delta_{sigma}$ and choosing the AC-Stark shift equal to the trap frequency, cooling red sideband transitions are much more probable than heating blue sideband transitions, and the carrier is completely suppressed.

beam is tuned to $\Delta_\pi = \Delta_\sigma$ and δ is adjusted to match the trap frequency. This creates an asymmetry in absorption for carrier and sidebands: The carrier is almost completely suppressed due to the dark resonance, the blue sideband is in the shallow wing of the profile, but the red sideband is greatly enhanced by the bright resonance, see Fig. 3(b). When we tuned the Stark-shift δ to be equal to one of the motional modes at 3.34 MHz we were able to cool this mode to 90% ground state occupation or $\bar{n} = 0.1$. Moreover, as sketched in Fig. 3(b) the bright resonance can have a substantial width and the red sideband need not necessarily coincide exactly with the maximum of the bright resonance to get a cooling effect. This opens the possibility to cool several motional modes *simultaneously*, as long as they are not too far apart in frequency.

To demonstrate simultaneous cooling of two vibrational modes with this method we chose two modes at 1.61 MHz and 3.34 MHz. The AC-Stark shift δ of the σ^+-beam was adjusted to be about 2.5 MHz, halfway between the two mode frequencies. With this settings we achieved 58% ground state occupation ($\bar{n} = 0.85$) in the mode at 1.61 MHz and 74% ($\bar{n} = 0.35$) at 3.34 MHz. From this result we estimate that we can sufficiently and simultaneously cool all axial degrees of freedom of a string of up to 5 ions with a COM mode frequency of 700 kHz.

6. COHERENT MANIPULATIONS

For quantum information processing, it is important to know for how long coherent interaction with the ion(s) is possible. For this we cooled one ion to the ground state and then irradiated the ion with light at the blue sideband

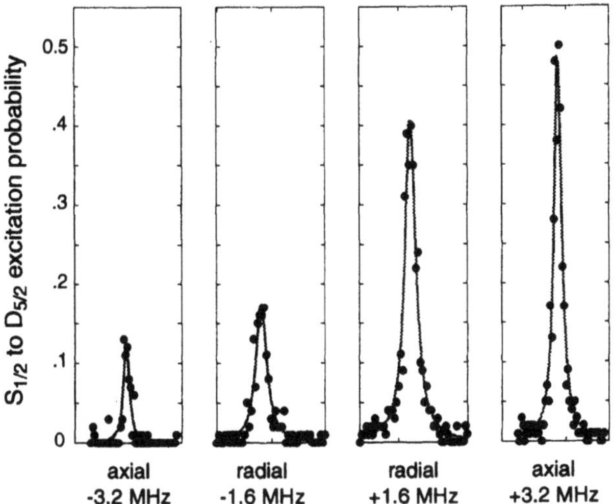

Figure 4 Simultaneous cooling of two vibrational modes with an oscillation frequency difference of 1.73 MHz. The sideband asymmetry corresponds to 58% ground state occupation ($\bar{n} = 0.85$) in the mode at 1.61 MHz and 74% ($\bar{n} = 0.35$) at 3.34 MHz.

frequency (This interaction is used in a quantum gate to transfer the internal state of a qubit into the motion) [9]. We then monitored the occupation probability of the D-state versus the pulse length on the blue sideband. The same interaction was also used after preparing the ion in the $n = 1$ motional Fock state by a π-pulse on the blue sideband followed by a repumping pulse on the $D_{5/2}$-$P_{3/2}$ transition. As Fig. 5 shows, we were able to observe Rabi-flops for

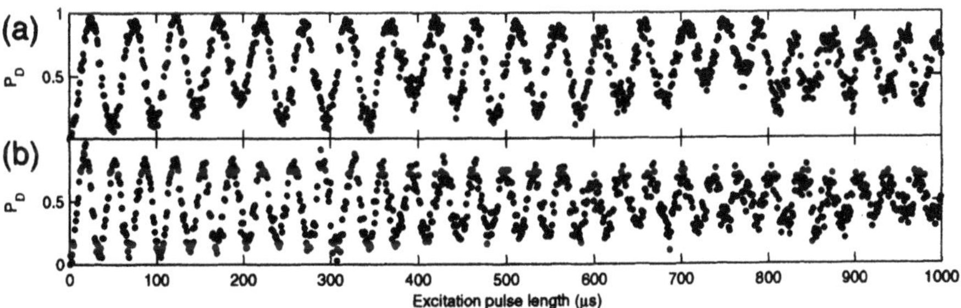

Figure 5 (a) Rabi-oscillations on the blue sideband for the initial motional state $|n = 0\rangle$. (b) Rabi-oscillations as in (a), but for $|n = 1\rangle$.

both initial motional states with a contrast of better than 50% for 1 ms. The ratio of Rabi-frequencies is $\sqrt{2}$ as expected for this kind of interaction in the Lamb-Dicke regime. These results make us confident that we should be able to apply gate pulses equivalent to at least 40 π-pulses before the fidelity of the total operation drops below 0.5. In our system motional heating is too slow to be the prime source of the observed decoherence, we rather attribute it to time dependent magnetic field fluctuations that shift the levels, to slow vibrations in our setup that introduce a fluctuating Doppler-shift of the 729 nm beam, and to laser intensity fluctuations.

7. CONCLUSIONS AND OUTLOOK

We have demonstrated that we have all necessary ingredients to perform a two-bit quantum logic gate with trapped ions. The time scales in our system are well separated. The fastest characteristic time of 1 μs is given by the harmonic motion of the ion(s). A π-pulse on a motional sideband takes about 20 μs and, as stated above, the fidelity of coherent manipulations remains above 0.5 for times smaller than 1 ms. Our laser system would allow for coherent manipulations for at least 15 ms. Motional heating begins to play a role for times around 100 ms. The ultimate source of decoherence in our experiment is the 1 s lifetime of the $D_{5/2}$ state. In the near future we plan to use our ability to individually address ions to demonstrate a CNOT quantum logic gate with two ions and to create maximally entangled states with 2 and more ions. We also envision to perform small quantum algorithms and first experiments on error corrections with up to 5 ions.

Acknowledgments

This work was supported by the Fonds zur Förderung wissenschaftlicher Forschung (FWF) within the special research grant SFB 15, by the European Commission within the TMR networks 'Quantum Information' (ERB-FMRX-CT96-0087) and 'Quantum Structures' (ERB-FMRX-CT96-0077) and by the Institut für Quanteninformation GmBH.

References

[1] J. I. Cirac, P. Zoller, Phys. Rev. Lett. **74** 4091 (1995).

[2] H. Rohde, S. T. Gulde, C. F. Roos, P. Barton, D. Leibfried, J. Eschner, F. Schmidt-Kaler, R. Blatt, quant-ph/0009031.

[3] H. Dehmelt, Bull. Am. Phys. Soc. **20** 60 (1975).

[4] A. Steane, Appl. Phys. B **64**, 623 (1998).

[5] H. C. Nägerl, W. Bechter, J. Eschner, F. Schmidt-Kaler, R. Blatt, Appl. Phys. B **66**, 603 (1998).

[6] H. C. Nägerl, D. Leibfried, H. Rohde, G. Thalhammer, J. Eschner, F. Schmidt-Kaler, R. Blatt, Phys. Rev. A **60**, 145 (1999).

[7] F. Diedrich, J. C. Bergquist, W. M. Itano, D. J. Wineland, Phys. Rev. Lett. **62**, 403 (1989).

[8] C. Monroe, D. M. Meekhof, B. E. King, S. R. Jeffers, W. M. Itano, D. J. Wineland, P. Gould, Phys. Rev. Lett. **74**, 4011 (1995).

[9] C. Roos, T. Zeiger, H. Rohde, H. C. Nägerl, J. Eschner, D. Leibfried, F. Schmidt-Kaler, R. Blatt, Phys. Rev. Lett. **83**, 4713 (1999).

[10] Q. A. Turchette, D. Kielpinski, B. E. King, D. Leibfried, D. M. Meekhof, C. J. Myatt, M. A. Rowe, C. A. Sackett, C. S. Wood, W. M. Itano, C. Monroe, D. J. Wineland, Phys. Rev. A **61**, 063418-1 (2000).

[11] F. Schmidt-Kaler, Ch. Roos, H. C. Nägerl, H. Rohde, S. Gulde, A. Mundt, M. Lederbauer, G. Thalhammer, Th. Zeiger, P. Barton, L. Hornekaer, G. Reymond, D. Leibfried, J. Eschner, R. Blatt, quant-ph/0003096 (2000).

[12] D. J. Wineland, C. Monroe, W. M. Itano, D. Leibfried, B. King, and D. M. Meekhof, Jou. Res. Nat. Inst. Stand. Tech., 103, 259 (1998).

[13] G. Morigi, J. Eschner, C. Keitel, quant-ph/0005009 (2000).

PHYSICAL LIMITS TO COMPUTATION

Seth Lloyd
Department of Mechanical Engineering
d'Arbeloff Laboratory for Information Systems and Technology
Research Laboratory of Electronics
Laboratory for Information and Decision Systems
Massachusetts Institute of Technology
Cambridge, MA 02139

Keywords: Quantum information, fundamental constants, gravitation

Abstract This paper explores the physical limits of computation as determined by the speed of light c, the quantum scale $\hbar$ and the gravitational constant G.

1. INTRODUCTION

Computers are physical systems: what they can and cannot do is dictated by the laws of physics [1]. In particular, the speed with which a physical device can process information is limited by its energy and the amount of information that it can process is limited by the number of degrees of freedom it possesses. Over the past half century, the amount of information that computers are capable of processing and the rate at which they process it has doubled every two years, a phenomenon known as Moore's law. A variety of technologies — most recently, integrated circuits — have enabled this exponential increase in information processing power. There is no particular reason why Moore's law should continue to hold: it is a law of human ingenuity, not of nature. At some point, Moore's law will break down. The question is, When? The answer to this question is to be found by applying the laws of physics to the process of computation [1, 2, 3, 4, 5, 6, 7, 8, 9]. The fundamental constants of Nature — the speed of light, $c = 2.9979 \times 10^8$ meters per second, Planck's reduced constant, $\hbar = 1.0545 \times 10^{-34}$ joule seconds, and the gravitational constant, $G = 6.673 \times 10^{-11}$ meters cubed per kilogram second squared — ultimately determine how much information a physical system can process and how rapidly it can process that information. (Other physical constants such as Boltzmann's constant, the mass and charge of the electron, the fine structure constant, etc., will also play a role in determining the computational capacity of physical systems in different regimes.) As an example, quantitative bounds are put to the computational power of an 'ultimate laptop' with a mass of one kilogram confined to a volume of one liter.

2. ENERGY LIMITS SPEED

How fast can a digital computer perform a logical operation? During such an operation, the bits in the computer on which the operation is performed go from one state to another. The problem of how much energy is required for information processing was first investigated in the context of communications theory by Levitin [4], Bremermann [5], Beckenstein [7], Margolus [9] and others, who showed that the laws of quantum mechanics determine the maximum rate at which a system with spread in energy ΔE can move from one distinguishable state to another. In particular, the correct interpretation of the time-energy Heisenberg uncertainty principle $\Delta E \Delta t \geq \hbar$ is not that it takes time Δt to measure energy to an accuracy ΔE (a fallacy that was put to rest by Aharonov and Bohm [10]) but rather that that a quantum state with spread in energy ΔE takes time at least $\Delta t = \pi\hbar/2\Delta E$ to evolve to an orthogonal (and hence distinguishable) state. More recently, Margolus and Levitin [9] extended this result to show that a quantum system such as a computer with average energy E takes time at least $\Delta t = \pi\hbar/2E$ to evolve to an orthogonal state.

Suppose that one has a certain amount of energy E to allocate to the logic gates of a computer. The more energy one allocates to a gate, the faster it can perform a logic operation. The total number of logic operations performed per second is equal to the sum over all logic gates of the operations per second per gate. (Note that this is a different bound from that of Margolus and Levitin [9], who identify the number of logic operations per second with the rate at which the entire computer goes from one state to an orthogonal state.) That is, a computer can perform no more than $\sum_\ell 1/\Delta t_\ell \leq \sum_\ell 2E_\ell/\pi\hbar = 2E/\pi\hbar$ operations per second. In other words, the rate at which a computer can compute is limited by its energy. As a consequence, a system with average energy E can perform a maximum of $2E/\pi\hbar$ logical operations per second. A one kilogram computer has average energy $E = mc^2 = 8.9874 \times 10^{16}$ joules. Accordingly, the ultimate laptop can perform a maximimum of 5.4258×10^{50} operations per second.

An interesting feature of this limit is that it is independent of computer architecture. One might have thought that a computer could be sped up by parallelization, i.e., by taking the energy and dividing it up amongst a large number of subsystems computing in parallel. This is not the case: if one spreads the energy E amongst N logic gates, each one operates at a rate $2E/\pi\hbar N$. The total number of operations per second, $N 2E/\pi\hbar N = 2E/\pi\hbar$, remains the same. If the energy is allocated to fewer logic gates (more serial operation), the rate $1/\Delta t_\ell$ at which they operate and the spread in energy per gate ΔE_ℓ go up. If the energy is allocated to more logic gates (more parallel operation) then the rate at which they operate and the spread in energy per gate go down. Note that in this parallel case, the overall spread in energy of the computer as a whole is considerably smaller than the average energy: in general $\Delta E = \sqrt{\sum_\ell \Delta E_\ell^2} \approx \sqrt{N}\Delta E_\ell$ while $E = \sum E_\ell \approx N E_\ell$. Parallelization can help perform certain computations more efficiently, but it does not alter the total number of operations per second. As will be seen below, the degree of parallelizability of the computation to be performed determines the most efficient distribution of energy among the parts of the computer. Computers in which energy is relatively evenly distributed over a larger volume are better suited for performing parallel computations. More compact computers and computers with an uneven distribution of energy are better for performing serial computations.

In sum, quantum mechanics provides a simple answer to the question of how fast information can be processed using a given amount of energy. Now it will be shown that thermodynamics and statistical mechanics provide a fundamental limit to how many bits of information can be processed using a given amount of energy confined to a given volume. Available energy necessarily limits the rate at which computer can process information. Similarly, the maximum entropy of a physical system determines the amount of information it can process. Energy limits speed. Entropy limits memory.

3. ENTROPY LIMITS MEMORY

The amount of information that a physical system can store and process is related to the number of distinct physical states accessible to the system. A collection of m two-state systems has 2^m accessible states and can register m bits of information. In general, a system with N accessible states can register $\log_2 N$ bits of information. But it has been known for more than a century that the number of accessible states of a physical system, W, is related to its thermodynamic entropy by the formula: $S = k_B \ln W$, where k_B is Boltzmann's constant.

The amount of information that can be registered by a physical system is $I = S(E)/k_B \ln 2$, where $S(E)$ is the thermodynamic entropy of a system with expectation value for the energy E. Combining this formula with the formula $2E/\pi\hbar$ for the number of logical operations that can be performed per second, we see that when it is using all its memory, the number of operations per bit per second that our ultimate laptop can perform is $k_B 2 \ln 2E/\pi\hbar S \propto k_B T/\hbar$, where $T = (\partial S/\partial E)^{-1}$ is the temperature of a kilogram of matter in a maximum entropy in a liter volume. The entropy governs the amount of information the system can register and the temperature governs the number of operations per bit per second it can perform.

Since thermodynamic entropy effectively counts the number of bits available to a physical system, the following derivation of the memory space available to the ultimate laptop is based on a thermodynamic treatment of a kilogram of matter confined to a liter volume, in a maximum entropy state. Throughout this derivation, it is important to keep in mind that although the memory space available to the computer is given by the entropy of its thermal equilibrium state, the *actual* state of the ultimate laptop as it performs a computation is completely determined, so that its entropy remains always equal to zero. As above, we assume that we have complete control over the actual state of the ultimate laptop, and are able to guide it through its logical steps while insulating it from all uncontrolled degrees of freedom. As the following discussion will make clear, such complete control will be difficult to attain.

To calculate exactly the maximum entropy for a kilogram of matter in a liter volume would require complete knowledge of the dynamics of elementary particles, quantum gravity, etc. We do not possess such knowledge. However, the maximum entropy can readily be estimated by a method reminiscent of that used to calculate thermodynamic quantities in the early universe. The idea is simple: model the volume occupied by the computer as a collection of modes of elementary particles with total average energy E. The maximum entropy is obtained by calculating the canonical ensemble over the modes. Here, we supply a simple derivation of the maximum memory space available to the ultimate laptop. A more detailed discussion of how to calculate the maximum amount of information that can be stored in a physical system can be found in the work of Bekenstein [7].

For this calculation, assume that the only conserved quantities other than the computer's energy are angular momentum and electric charge, which we take to be zero. (One might also ask that baryon number be conserved, but as will be seen below, one of the processes that could take place within the computer is black hole formation and evaporation, which does not conserve baryon number.) At a particular temperature T, the entropy is dominated by the contributions from particles with mass less than $k_B T/2c^2$. The ℓ'th such species of particle contributes energy $E = r_\ell \pi^2 V(k_B T)^4/30\hbar^3 c^3$ and entropy $S = 2r_\ell k_B \pi^2 V(k_B T)^3/45\hbar^3 c^3 = 4E/3T$ where r_ℓ is equal to the number of particles/antiparticles in the species (i.e., 1 for photons, 2 for electrons/positrons) times the number of polarizations (2 for photons, 2 for electrons/positrons) times a factor that reflects particle statistics (1 for bosons, 7/8 for fermions). As the formula for S in terms of T shows, each species contributes $(2\pi)^5 r_\ell/90 \ln 2 \approx 10^2$ bits of memory space per cubic thermal wavelength λ_T^3 where $\lambda_T = 2\pi\hbar c/k_B T$. Re-expressing the formula for entropy as a function of energy, our estimate for the maximum entropy is

$$S = (4/3)k_B(\pi^2 rV/30\hbar^3c^3)^{1/4}E^{3/4} = k_B \ln 2I \quad (1)$$

where $r = \sum_\ell r_\ell$. Note that S depends only insensitively on the total number of species with mass less than $k_BT/2c^2$.

A lower bound on the entropy can be obtained by assuming that energy and entropy are dominated by black body radiation consisting of photons. In this case, $r = 2$, and for a one kilogram computer confined to a volume of a liter we have $k_BT = 8.10 \times 10^{-15}$ joules, or $T = 5.87 \times 10^8$ K. The entropy is $S = 2.04 \times 10^8$ joule/K, which corresponds to an amount of available memory space $I = S/k_B \ln 2 = 2.13 \times 10^{31}$ bits. When the ultimate laptop is using all its memory space it can perform $2 \ln 2k_BE/\pi\hbar S = 3 \ln 2k_BT/2\pi\hbar \approx 10^{19}$ operations per bit per second. As the number of operations per second $2E/\pi\hbar$ is independent of the number of bits available, the number of operations per bit per second can be increased by using a smaller number of bits. In keeping with the prescription that the ultimate laptop operates at the absolute limits given by physics, in what follows, we assume that all available bits are used.

4. SIZE LIMITS PARALLELIZATION

Up until this point, we have assumed that our computer occupies a volume of a liter. The previous discussion, however, indicates that benefits are to be obtained by varying the volume to which the computer is confined. Generally speaking, if the computation to be performed is highly parallelizable or requires many bits of memory, the volume of the computer should be greater and the energy available to perform the computation should be spread out evenly amongst the different parts of the computer. Conversely, if the computation to be performed is highly serial and requires fewer bits of memory, the energy should be concentrated in particular parts of the computer.

A good measure of the degree of parallelization in a computer is the ratio between time it takes to communicate from one side of the computer to the other, and the average time it takes to perform a logical operation. The amount of time it takes to send a message from one side of a computer of radius R to the other is $t_{\text{com}} = 2R/c$. The average time it takes a bit to flip in the ultimate laptop is the inverse of the number of operations per bit per second calculated above: $t_{\text{flip}} = \pi\hbar S/k_B 2 \ln 2E$. Our measure of the degree of parallelization in the ultimate laptop is then

$$t_{\text{com}}/t_{\text{flip}} = k_B 4 \ln 2RE/\pi\hbar cS \propto k_BRT/\hbar c = 2\pi R/\lambda_T. \quad (3)$$

That is, the amount of time it takes to communicate from one side of the computer to the other, divided by the amount of time it takes to flip a bit, is approximately equal to the ratio between the size of the system and its thermal wavelength. For the ultimate laptop, with $2R = 10^{-1}$ meters, $2E/\pi\hbar \approx 10^{51}$ operations per second, and $S/k_B \ln 2 \approx 10^{31}$ bits, $t_{\rm com}/t_{\rm flip} \approx 10^{10}$. The ultimate laptop is highly parallel. A greater degree of serial computation can be obtained at the cost of decreasing memory space by compressing the size of the computer or making the distribution of energy more uneven. As ordinary matter obeys the Beckenstein bound [7], $k_B RE/\hbar c S > 1/2\pi$, however, as one compresses the computer $t_{\rm com}/t_{\rm flip} \approx k_B RE/\hbar c S$ will remain greater than one: i.e., the operation will still be somewhat parallel. Only at the ultimate limit of compression — a black hole — is the computation entirely serial.

5. COMPRESSING THE COMPUTER FOR MORE SERIAL COMPUTATION

Suppose that one wants to perform a highly serial computation on few bits. Then it is advantageous to compress the size of the computer so that it takes less time to send signals from one side of the computer to the other at the speed of light. As the computer gets smaller, keeping the energy fixed, the energy density inside the computer goes up. As the energy density in the computer goes up, different regimes in high energy physics are necessarily explored in the course of the computation. First the weak unification scale is reached, then the grand unification scale. Finally, as the linear size of the computer approaches its Schwarzchild radius, the Planck scale is reached. (No known technology could possibly achieve such compression.) At the Planck scale, gravitational effects and quantum effects are both important: the Compton wavelength of a particle of mass m, $\lambda_C = 2\pi\hbar/mc$ is on the order of its Schwarzschild radius, $2Gm/c^2$. In other words, to describe behavior at length scales of the size $\ell_P = \sqrt{\hbar G/c^3} = 1.616 \times 10^{-35}$ meter, time scales $t_P = \sqrt{\hbar G/c^5} = 5.391 \times 10^{-44}$ second, and mass scales of $m_P = \sqrt{\hbar c/G} = 2.177 \times 10^{-8}$ kilograms, a unified theory of quantum gravity is required. We do not currently possess such a theory. Nonetheless, although we do not know the exact number of bits that can be registered by a one kilogram computer confined to a volume of a liter, we do know the exact number of bits that can be registered by a one kilogram computer that has been compressed to the size of a black hole. This is because the entropy of a black hole has a well-defined value.

In the following discussion, we use the properties of black holes to place limits on the speed, memory space, and degree of serial computation that could be approached by compressing a computer to the smallest possible size. Whether or not these limits could be attained, even in principle, is a question whose answer will have to await a unified theory of quantum gravity.

The Schwarzschild radius of a 1 kilogram computer is $R_S = 2Gm/c^2 = 1.485 \times 10^{-27}$ meters. The entropy of a black hole is Boltzmann's constant times its area divided by 4, as measured in Planck units. Accordingly, the amount of information that can be stored in a black hole is $I = 4\pi G m^2/\ln 2\hbar c = 4\pi m^2/\ln 2 m_P^2$. The amount of information that can be stored by the 1 kilogram computer in the black-hole limit is 3.827×10^{16} bits. A computer compressed to the size of a black hole can perform 5.4258×10^{50} operations per second, the same as the 1 liter computer.

In a computer that has been compressed to its Schwarzschild radius, the energy per bit is $E/I = mc^2/I = \ln 2\hbar c^3/4\pi m G = \ln 2 k_B T/2$, where $T = (\partial S/\partial E)^{-1} = \hbar c/4\pi k_B R_S$ is the temperature of the Hawking radiation emitted by the hole. As a result, the time it takes to flip a bit on average is $t_{\text{flip}} = \pi\hbar I/2E = \pi^2 R_S/c\ln 2$. In other words, according to a distant observer, the amount of time it takes to flip a bit, t_{flip}, is on the same order as the amount of time $t_{\text{com}} = \pi R_S/c$ it takes to communicate from one side of the hole to the other by going around the horizon: $t_{\text{com}}/t_{\text{flip}} = \ln 2/\pi$. In contrast to computation at lesser densities, which is highly parallel as noted above, computation at the horizon of a black hole is highly serial: every bit is essentially connected to every other bit over the course of a single logic operation. As noted above, the serial nature of computation at the black-hole limit can be deduced from the fact that black holes attain the Beckenstein bound [7], $k_B R E/\hbar c S = 1/2\pi$.

6. ERROR CORRECTION AND POWER

Note that the limit on how fast our computer can operate is given by the computer's energy. There are also power limitations on our computer. If all operations were carried out in a completely efficient and error free fashion, then the computer could run forever on a finite charge to its ultimate 'battery' (whatever such a battery might look like): the processor borrows energy from the battery, uses that energy to perform logical operations, then returns it to the battery when the computation is finished. In principle, computation does not require dissipation. In practice, however, any computer – even our ultimate laptop – will dissipate energy.

If the computer dissipates a fraction of energy during computation, however, then it needs a continual supply of free energy to operate. Since the rate at which a logical operation can be performed is proportional to the energy devoted to performing it, if a fixed fraction of that energy is dissipated in each operation, the rate at which energy is dissipated goes as the square of the number of operations per second [11].

In addition, Landauer's principle [2] that erasing a bit of information involves an energetic cost $k_B T \ln 2$, implies that if our computer is subject to an error rate of ϵ bits per second, then error-correcting codes must be used to detect those errors and reject them to the environment at a dissipative cost of $\epsilon k_B T_E \ln 2$ joules per second, where T_E is the temperature of the environment.

As the discussion of error correction above indicates, the rate at which errors can be detected and rejected to the environment by error correction routines puts a fundamental limit on the rate at which errors can be committed. Suppose that each logical operation performed by the ultimate computer has a probability ϵ of being erroneous. The total number of errors committed by the ultimate computer per second is then $2\epsilon E/\pi\hbar$. The maximum rate at which information can be rejected to the environment is, up to a geometric factor, $\ln 2cS/R$ (all bits in the computer moving outward at the speed of light). Accordingly, the maximum error rate that the ultimate computer can tolerate is $\epsilon \leq \pi \ln 2\hbar cS/2ER = 2t_{\text{flip}}/t_{\text{com}}$. That is, the maximum error rate that can be tolerated by the ultimate computer is the inverse of its degree of parallelization.

7. CONSTRUCTING ULTIMATE COMPUTERS

Throughout this entire discussion of the physical limits to computation, no mention has been made of how to construct a computer that operates at those limits. In fact, contemporary quantum 'microcomputers' such as those constructed using nuclear magnetic resonance [12, 13] do indeed operate at the limits of speed and memory space described above. Information is stored on nuclear spins, with one spin registering one bit. The time it takes a bit to flip from a state $|\uparrow\rangle$ to an orthogonal state $|\downarrow\rangle$ is given by $\pi\hbar/2\mu B = \pi\hbar/2E$, where μ is the spin's magnetic moment, B is the magnetic field, and $E = \mu B$ is the average energy of interaction between the spin and the magnetic field. To perform a quantum logic operation between two spins takes a time $\pi\hbar/2E_\gamma$, where E_γ is the energy of interaction between the two spins.

Although NMR quantum computers already operate at the limits to computation set by physics, they are nonetheless much slower and process much less information than the ultimate laptop described above. This is because their energy is largely locked up in mass, thereby limiting both their speed and their memory. Unlocking this energy is of course possible, as a thermonuclear explosion indicates. Controlling such an 'unlocked' system is another question, however. In discussing the computational power of physical systems in which all energy is put to use, we assumed that such control is possible in principle, although it is certainly not possible in current practice. All current designs for quantum computers operate at low energy levels and temperatures, exactly so that precise control can be exerted on their parts.

At the greater extremes of a black hole computer, we assumed that whatever theory (string theory, M theory?) turns out to be the correct theory of quantum matter and gravity, it is possible to prepare initial states of such systems that causes their natural time evolution to carry out a computation. What assurance do we have that such preparations exist, even in principle?

Physical systems that can be programmed to perform arbitrary digital computations are called computationally universal. Although computational universality might at first seem to be a stringent demand on a physical system, a wide variety of physical systems — ranging from nearest neighbor Ising models to quantum electrodynamics and conformal field theories — are known to be computationally universal. Indeed, computational universality seems to be the rule rather than the exception. Essentially any quantum system that admits controllable nonlinear interactions can be shown to be computationally universal.

In addition to the theoretical evidence that most systems are computationally universal, the computer on which I am writing this article provides strong experimental evidence that whatever the correct underlying theory of physics is, it supports universal computation. Whether or not it is possible to make computation take place in the extreme regimes envisaged in this paper is an open question. The answer to this question lies in future technological development, which is difficult to predict. If, as seems highly unlikely, it is possible to extrapolate the exponential progress of Moore's law into the future, then it will only take two hundred and fifty years to make up the forty orders of magnitude in performance between current computers that perform 10^{10} operations per second on 10^{10} bits and our one kilogram ultimate laptop that performs 10^{51} operations per second on 10^{31} bits.

Acknowledgments

This research was supported in part by DARPA under the QUIC program and by the U.S. Army Research Office. The author acknowledges fruitful technical discussions with Norm Margolus, Lev Levitov, and Rolf Landauer.

References

[1] Lloyd, S., *Nature* **406**, 1047-1054 (2000).

[2] Landauer, R., 'Irreversibility and Heat Generation in the Computing Process,' *IBM J. Res. Develop.* **5**, 183-191 (1961). Keyes, R.W., Landauer, R., 'Minimal Energy Dissipation in Logic,' *IBM Journal of Research and Development*, **14**, no. 2, 152-157 (1970). Landauer, R., 'Dissipation and Noise-Immunity in Computation and Communication,' *Nature*, **335**, 779-784 (1988). Landauer, R., 'Information is Physical,' *Physics Today*, May 1991, 23-29 (1991). Landauer, R., 'The physical nature of information,' *Phys. Lett. A*, **217**, 188-193 (1996).

[3] von Neumann, J., 'Theory of Self-Reproducing Automata,' Lecture 3 (University of Illinois Press, Urbana), 1966.

[4] Lebedev, D.S., Levitin, L.B., 'Information Transmission by Electromagnetic Field,' *Information and Control* **9**, 1-22 (1966). Levitin, L.B., in 'Proceedings of the 3rd International Symposium on Radio Electronics,' part 3, 1-15 (Varna, Bulgaria), 1970. L.B. Levitin, 'Physical Limitations of Rate, Depth, and Minimum Energy in Information Processing,' *Int. J. Theor. Phys.* **21**, 299-309 (1982). Levitin, L.B. 'Energy cost of information transmission (along the path to understanding),' *Physica D* **120**, 162-167 (1998).

[5] Bremermann, H.J., in 'Proceedings of the Fifth Berkeley Symposium on Mathematical Statistics and Probability,' (University of California Press, Berkeley), 1967. Bremermann, H.J., *Int. J. Theor. Phys.* **21**, 203-217 (1982).

[6] Bennett, C.H., 'Logical Reversibility of Computation,' *IBM J. Res. Develop.* **17**, 525-532 (1973). Bennett, C.H., 'Thermodynamics of Computation—a Review,' *Int. J. Theor. Phys.* **21**, 905-940 (1982).

[7] Beckenstein, J.D., 'Universal Upper Bound on the Entropy-to-Energy Ratio for Bounded Systems,' *Phys. Rev. D.* **23**, 287-298 (1981). Beckenstein, J.D., 'Energy Cost of Information Transfer,' *Phys. Rev. Letters* **46**, 623-626 (1981). Beckenstein, J.D., 'Entropy Content and Information Flow in Systems With Limited Energy,' *Phys. Rev. D.* **30**, 1669-1679 (1984).

[8] Benioff, P., 'Quantum Mechanical Models of Turing Machines that Dissipate No Energy,' *Phys. Rev. Lett.* **48**, 1581-1585 (1982).

[9] Margolus, N., Levitin, L.B., 'The Maximum Speed of Dynamical Evolution,' in *PhysComp96*, T. Toffoli, M. Biafore, J. Leao, eds. (NECSI, Boston) 1996. Margolus, N., Levitin, L.B., 'The Maximum Speed of Dynamical Evolution,' *Physica D* **120**, 188-195 (1998).

[10] Aharonov, Y., Bohm, D., 'Time in the Quantum Theory and the Uncertainty Relation for the Time and Energy Domain,' *Phys. Rev.* **122**, 1649-1658 (1961). Aharonov, Y., Bohm, D., *Phys. Rev. B.* **134**, 1417-1418 (1964).

[11] Mundici, D., *Nuovo Cimento* **61B**, 297-305 (1981).

[12] Cory, D.G., Fahmy, A.F., Havel, T.F., 'Nuclear Magnetic Resonance Spectroscopy: an experimentally accessible paradigm for quantum computing,' in *PhysComp96, Proceedings of the Fourth Workshop on Physics and Computation*, T. Toffoli, M. Biafore, J. Leão, eds., (New England Complex Systems Institute, Boston) 1996 87-91.

[13] Gershenfeld, N.A., Chuang, I.L., *Science* **275**, 350-356 (1997).

REALISING QUANTUM COMPUTING: PHYSICAL SYSTEMS AND ROBUSTNESS

Andrew M. Steane
Centre for Quantum Computation, Clarendon Laboratory,
University of Oxford, Parks Road, Oxford OX1 3PU, England.

Keywords: quantum error correction, computer

Abstract The physical realisation of a large quantum computer, i.e. one which could perform calculations beyond the capabilities of classical computers, is discussed. It is necessary to consider both the physical mechanisms of the hardware and the noise tolerance of quantum error correction (QEC) methods. Estimates for noise tolerance which involve fewer simplifying assumptions than were previously employed are given, and the scaling of logic gate rate with logic gate precision is discussed. It is found that QEC is fast compared to methods such as adiabatic passage.

This work is motivated by the desire to realise a working quantum computer which could perform useful computations beyond the capabilities of the best classical computers. Current technology is far from achieving this, but progress over the last five years has made such a prospect become more realistic. The chief reason for cautious optimism is the much greater understanding we now have of methods to control noise in a quantum computer. The concepts of quantum error correction (QEC) in particular are now quite well understood. This is not to say that QEC is the only method we might envisage employing in order to stabilise a computer, since other methods such as adiabatic passage and ideas based on geometric phase can also be powerful. However, QEC is more efficient that adiabatic methods (see section 3), and shares a common mathematical framework with some methods based on geometric phases [Kitaev, 1997, Knill et al., 2000]. Therefore there is much interest in understanding exactly what degree of noise tolerance QEC offers.

One of the remarkable features of QEC is the *threshold result*, which which states that in order for arbitrarily long computations to succeed with probability greater than some finite value (say, one half), it is sufficient for elementary gate operations and quantum memory to have precision below some finite threshold γ_0 which does not depend on the size of the computation. The threshold is found to be of order 10^{-4} (see section 2), and this result has lead to a mistaken belief that once logic operations in some proposed computer design are at this level of precision, then robust large-scale computing can follow. This is a mistake because the value of the threshold depends strongly on a set of assumptions about the types of operations which the physical hardware of the computer allows. This leads to a variation of γ_0 over several orders of magnitude, depending on the type of computer one envisages. In particular, if only nearest neighbour qubits can be coupled, then the swap operations between neighbouring qubits need to be a lot more precise than the threshold value for a computer which does not have this limitation.

To understand what is needed for a robust quantum computer, we find therefore that it is necessary to consider both the abstract methods of fault-tolerant QEC, and also the physical details of particular proposals for the computing system. This paper will address in section 1 some of the physical issues. Section 2 considers the noise tolerance of a quantum computer, and briefly reevaluates previous estimates. Section 3 then discusses the trade-off between computer speed and precision. The elementary gates on the physical qubits have a maximum speed R, at given precision p, which typically varies as $R \sim p^\alpha$ where the exponent α can take on values from $1/2$ to 2 depending on the physical mechanism. This means that some published estimates for gate rates in quantum computer proposals may be inaccurate by six orders of magnitude, because of the slow-down required to achieve $p \simeq 10^{-4}$ or better.

1. PHYSICAL SYSTEMS FOR QUANTUM COMPUTING

It will emerge in section 2 that a large quantum computer will need a memory with a very high degree of passive stability. For this reason I will concentrate attention here on methods based on nuclear spins as the physical qubits of the computer. Nuclear spins offer the most stable and precisely controllable simple quantum systems. For example this is why the standard of time is based on the hyperfine structure of a free atom. The atomic electrons provide a very strong and very precise magnetic field on the nucleus, and they allow us to manipulate the nuclear spins, both by spatially locating them where we want, and also enabling controlled qubit-qubit coupling for quantum logic gates.

The next most obvious candidate for a qubit is the electron spin. Here the coupling strength is about $m_p/m_e \simeq 2000$ times stronger, making both noise and possible gate rates 2000 times larger. For both nuclear and electron spins, the case of a free or quasi-free atom provides a more precise magnetic field on the magnetic dipole of the spin than the case of spins embedded or moving in a solid. On the other hand, a solid material provides a convenient way to locate electrodes close to the qubits.

Several quantum computing proposals based on nuclear spins have been put forward; for a thorough discussion of them the reader is referred to the recent special issue of the *Fortschritte der Physik*, Braunstein and Lo, eds, 2000. These include bulk liquid-state nuclear magnetic resonance (NMR) experiments, trapped atoms and ions, and spins of dopant nuclei implanted in a semiconductor. Apart from NMR, these proposals use electrons to couple to the nuclear spins and hence allow quantum gates. For trapped ions or atoms the electrons are located in individual almost free atoms, which are coupled via laser-stimulated exchange of either photons or phonons. In the semiconductor proposals, the electron wavefunction is spread over much greater distance scales inside a solid, and controlled coupling is proposed through controlling the position of such an electron using voltages on electrodes deposited on the solid surface.

I have put forward elsewhere [Steane and Lucas, 2000] arguments for identifying light as a promising candidate for coupling electrons in a controlled way. The essential physics involved is the coupling of a single atom to a mode containing a single photon. The coupling strength is given by $g = Ed/\hbar = d(2\omega/\epsilon_0 \hbar V)^{1/2}$ where d is the electric dipole moment of the atom, V is the mode volume and ω is the angular frequency of the transition. The photon mode must be confined in a high finesse cavity to avoid relaxation, the cavity decay rate is $\kappa = c\pi/\mathcal{F}L$ where $\mathcal{F}$, L are the finesse and length of the cavity, respectively. An analysis of two possible methods is given in [Steane and Lucas, 2000]. The first method is to use a stimulated Raman transition in which one photon is provided by the cavity mode, and the other by a laser field. The gate failure probability p is the sum of the probabilities for the atom to emit a spontaneous photon, and for the cavity photon to decay, during a gate. Adjusting the laser parameters to minimise this, one finds for the logic gate rate (inverse of the time for one 2-qubit gate)

$$R = \frac{g^2 p}{8\pi^2 \Gamma}. \tag{1}$$

where Γ is the linewidth of the intermediate level and the minimal probability $p = 2\pi(2\kappa\Gamma/g^2)^{1/2}$. The essential parameter is therefore g^2/Γ which for a free atom is $g^2/\Gamma \simeq c\lambda^2/V$, ignoring a numerical factor of order 1. For small p we require large g^2/Γ and this is available in the microwave domain. However, it is likely that in order to control very many (i.e. thousands of) qubits, use of microwaves will not be feasible owing to the difficulty of addressing individual qubits with microwave radiation. In the optical domain, the best cavities based on dielectric coated mirrors currently offer $p \simeq 0.8$, and for whispering gallery modes of a microsphere one finds $p \simeq 0.3$. Therefore optical cavities are a long way from offering the 10^{-4} or better stability we need for quantum logic gates.

The scaling of p with cavity finesse is $p \sim \mathcal{F}^{-1/2}$, so very great improvements in finesse will be needed.

The second method is to avoid the the relaxation problems by using dark states and adiabatic passage [Pellizzari et al., 1995, Imamoḡlu et al., 1999]. The main error mechanisms are now non-adiabaticity and decay of the cavity mode. Adjusting the gate time to minimise p, one finds

$$R \simeq \frac{1}{9} p^2 \frac{g^2}{\kappa}. \tag{2}$$

In this expression p is now a free parameter, so we can obtain gates as precise as we wish (up to a limit), at a cost in processor speed. Putting in values for g and κ available from the best optical cavities we obtain a gate rate of the order of 1 kHz, for $p = 10^{-3}$.

The main conclusion of this analysis is that, although the idea of using photons for quantum gates has some advantages, in practice the gate rates one obtains are rather low, and no faster than those available by other methods which may be easier to implement.

The original proposal for quantum information processing in an ion trap [Cirac and Zoller, 1995] was based on laser-induced coupling to the shared motional degree of freedom of a string of trapped ions. The speed limit of the motional method has been carefully considered in [Steane et al., 2000, Jonathan et al., 2000, Sørensen and Mølmer, 2000]. The fastest gate method is an elaboration of the original Cirac-Zoller scheme, where the Rabi frequency of the laser beams exciting any trapped ion is chosen so that the a.c. Stark shift is equal to the separation ω_z of the vibrational energy levels of the trapped ion string [Jonathan et al., 2000]. This leads to a gate rate

$$R = \eta \frac{\omega_z}{2\pi} \simeq (2p)^{1/2} \frac{\omega_z}{2\pi} \tag{3}$$

for error probability p, where $\eta = (E_R/\hbar\omega_z)^{1/2}$ is the Lamb-Dicke parameter.

Current technology permits trap frequencies ω_z up to tens of MHz. If we wish to obtain $p = 10^{-3}$ the recoil frequency must then be of the order of tens of kHz, which is available for typical candidate ions such as Beryllium or Calcium. We then obtain a gate rate of order 100 kHz.

Although motional (phonon) coupling of trapped ions offers a good method for quantum logic gates, it is clear that in itself this does not offer a large scale quantum computer of thousands of qubits. For this reason it has been said that ion trap methods are 'not scalable.' This is incorrect, because we can readily envisage methods to connect many ion traps together, and the fabrication of multiple ion traps is of no great technical difficulty.

We have found that photon-based methods are slow for the logic gates between physically close qubits, but photons do provide a promising candidate for communication between relatively distant parts of a computer. Methods to couple atoms at separate locations, via photons and single-mode optical fibres, have been put forward [Cirac et al., 1997, van Enk et al., 1997]. Overall, the quantum computer model which I propose as most promising is a hybrid, in which one physical effect offers a good way to implement logic gates between qubits which are physically close together (such as in a single trap), and one or a few qubits in each group can be controllably coupled to a cavity mode and optical fibre, and hence quantum communication between groups is implemented.

2. NOISE TOLERANCE

The currently best understood method to stabilise quantum computing is that of fault-tolerant QEC. The estimates we have of the noise which can be tolerated are based on considering the propagation of errors around quantum networks, and the properties of the QEC codes. The single most significant assumption made in such estimates is that noise acts independently on different qubits and at different times. The degree of noise tolerance depends on exactly what capabilities the physical hardware of the quantum computer offers. Initial estimates have involved some simplifying assumptions about the hardware. In particular, it has been assumed that measurement of a single qubit can take place as fast as a logic gate, that gates can couple qubits arbitrarily far apart in the computer at no cost in speed, and optimistic assumptions have been made about the degree of parallelism available. We would like to relax all these assumptions.

I have reexamined the estimates given in [Steane, 1999] for the noise tolerance of a quantum computer based on efficient QEC codes. Let ${}^{C}X^{k}$ signify a controlled-multiple-not gate, in which a single control bit influences k target bits. Such gates are much used in fault-tolerant QEC, for preparing and verifying desired entangled states. The assumption concerning parallelism made in [Steane, 1999] was that elementary gates within different blocks could act simultaneously, which is reasonable, but also that a ${}^{C}X^{k}$ gate within a single block could take place in a time independent of k, which is not always possible.

In what follows we will assume that controlled-not (ie ${}^{C}X^{1}$) can take place in a single time-step between any pair of qubits in a block, and the time for ${}^{C}X^{k}$ is proportional to k. We will allow an order of magnitude reduction in speed for gates between blocks, compared to gates within a block, and allow measurements on qubits to be 1000 times slower than the fastest gates. The details of the noise analysis for this more realistic case is work in progress, it requires careful reconsideration of the precise structure of the computer. Overall, preliminary results suggest that the tolerated noise per elementary gate is, as before, at the level of $\gamma \simeq 10^{-5}$, in order to allow success of a large computation, but the tolerated memory noise is reduced to $\epsilon \simeq 10^{-8}$ per qubit per time step, where the time step is that required for a single controlled-not gate (or one equivalent up to single bit rotations) within a block. These figures are for the case of a large computation, $KQ \simeq 10^{16}$, where K is the number of logical qubits, Q is the number of logical Toffoli gates in the algorithm. γ and ϵ scale as $(KQ)^{1/8}$ for the 7-error-correcting code which forms the basis of the analysis.

The memory requirement is severe. It can be relaxed at the sacrifice of using less efficient coding methods, i.e. more physical qubits per logical qubit. At the extreme of a concatenated code, the scale-up N/K rises to 10^4 and 10^5 after three or four levels of concatenation respectively, compared to 22 for the efficient case just considered. Preliminary results for the noise tolerance in the case of concatenated codes are $(\gamma, \epsilon) \simeq (10^{-5}, 5 \times 10^{-7})$ and $(10^{-4}, 5 \times 10^{-6})$ for the case of 3 and 4 levels of concatenation, respectively, at $KQ \simeq 10^{16}$.

These figures imply that the choice for realising a working quantum computer is between either a relatively small machine with an extremely low-noise memory, or a much (1000 to 10,000 times) larger machine with a memory less perfect but still very good.

3. SPEED VERSUS ERROR RATE

Equations (1), (2) and (3) all show gate rates varying with gate precision as $R \sim p^{\alpha}$. It is notable that the dark state adiabatic passage method in cavity QED has an especially poor (quadratic) scaling. It would be interesting to discover whether this is a universal feature of adiabatic passage processes. The linear scaling of equation (1) is typical of a process involving off-resonant scattering, and the square-root law of equation (3) is typical of off-resonant driving of a coherent (stimulated) transition.

QEC also involves a slow-down, however this is more like a threshold effect than a scaling law. As a result QEC is comparatively fast once the error rates are below threshold. For example with elementary gates of precision 10^{-5} the QEC protocols offer a gain in precision of 10 or more orders of magnitude with a slow-down by a factor of only 200 to 5000. This is much faster than adiabatic passage or slow Rabi flopping.

It is instructive to examine the time scale for the all the processes in a large quantum computer. Figure 1 gives an illustration for the case of a quantum computer based on multiple ion traps linked by cavity QED methods. The hyperfine structure gives the starting point of 3 GHz for the frequency scale, and all the gate rates are reduced from this by the considerations of gate precision and coupling of qubits. The downward arrows in the figure give the reasons for the reductions in rate. In the ion trap there is a gap of two and a half orders of magnitude between the rates of the single-bit gates and the 2-bit gates,

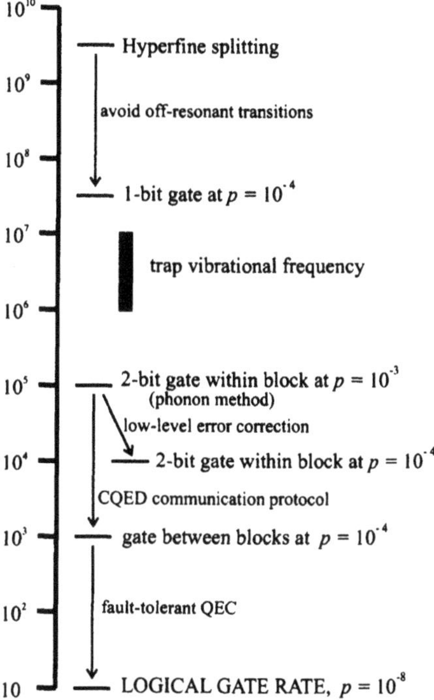

Figure 1 Rates (Hz) for various types of gate in a single quantum computer, showing the slow-down associated with gains in precision and coupling of distant qubits.

owing to the need to couple quantum information into the phonon mode. It may be possible to avoid such a large gap in other physical devices. The other speed reductions can be expected in any system, however. The case illustrated is based on the goal of 10^{-8} precision for the logical gates, rather than 10^{-16}. This is available by QEC using gates on the physical qubits of precision approximately 10^{-4}. Modest improvements in the physical gate precision will then allow the logical gates to be very much more precise.

Acknowledgment

I thank the organisers of the conference. This work was supported by EPSRC and the European Community network QUBITS.

References

[Cirac and Zoller, 1995] Cirac, J. I. and Zoller, P. (1995). Quantum computations with cold trapped ions. *Phys. Rev. Lett.*, 74(20):4091–4094.

[Cirac et al., 1997] Cirac, J. I., Zoller, P., Kimble, H. J., and Mabuchi, H. (1997). Quantum state transfer and entanglement distribution among distant nodes in a quantum network. *Phys. Rev. Lett.*, 78:3221.

[Imamoḡlu et al., 1999] Imamoḡlu, A., Awschalom, D. D., Burkard, G., DiVincenzo, D. P., Loss, D., Sherwin, M., and Small, A. (1999). Quantum information processing using quantum dot spins and cavity QED. *Phys. Rev. Lett.*, 83(20):4204–4207.

[Jonathan et al., 2000] Jonathan, D., Plenio, M. B., and Knight, P. L. (2000). Fast quantum gates for cold trapped ions. *Phys. Rev. A*, 62:042307. quant-ph/0002092.

[Kitaev, 1997] Kitaev, A. Y. (1997). Fault tolerant quantum computation with anyons. *Preprint:*, quant-ph/9707021.

[Knill et al., 2000] Knill, E., Laflamme, R., and Viola, L. (2000). Theory of quantum error correction for general noise. *Phys. Rev. Lett.*, 84:2525–2528.

[Pellizzari et al., 1995] Pellizzari, T., Gardiner, S. A., Cirac, J. I., and Zoller, P. (1995). Decoherence, continuous observation, and quantum computing: a cavity QED model. *Phys. Rev. Lett.*, 75(21):3788–3791.

[Sørensen and Mølmer, 2000] Sørensen, A. and Mølmer, K. (2000). Entanglement and quantum computation with ions in thermal motion. quant-ph/0002024.

[Steane, 1999] Steane, A. M. (1999). Efficient fault-tolerant quantum computing. *Nature*, 399:124–126. quant-ph/9809054.

[Steane and Lucas, 2000] Steane, A. M. and Lucas, D. M. (2000). Quantum computing with trapped ions, atoms and light. *Fortschritte der Physik*, 48:839–858.

[Steane et al., 2000] Steane, A. M., Roos, C. F., Stevens, D., Mundt, A., Leibfried, D., Schmidt-Kaler, F., and Blatt, R. (2000). Speed of ion trap quantum information processors. *Phys. Rev. A*, 62:042305.

[van Enk et al., 1997] van Enk, S. J., Cirac, J. I., and Zoller, P. (1997). Purifying two-bit quantum gates and joint measurements in cavity QED. *Phys. Rev. Lett.*, 79:5178–5181.

INFORMATION ANALYSIS OF QUANTUM GATES FOR SIMULATION OF QUANTUM ALGORITHMS ON CLASSICAL COMPUTERS

S. V. Ulyanov, S. A. Panfilov, I. Kurawaki, A. V. Yazenin
Yamaha Motor Europe N.V. R&D Office, Polo Didattico e di Ricerca di CREMA
Via Bramante, 65, 26013, Crema (CR), Italy
Tel.: +39 0373 204438 / Fax: +39 0373 204518
ulyanov@tin.it

Abstract Quantum algorithm dynamics of quantum computation states are analyzed from classical and quantum information theory standpoint. The information measure of intelligence for a quantum algorithm states with respect to a subset of qubits is defined. The intelligence of a state is maximal if the gap between the Shannon and the Von Neumann entropy for the chosen result qubits is minimal. We prove the Quantum Fourier Transform creates maximally intelligent states with respect to the first n qubits for Shor's problem, since it annihilates the gap between classical and quantum entropy for the first n qubits of every output state. Dynamic and information analysis of Shor's algorithm are described.

1. INTRODUCTION

Most of the applications of quantum information theory have been developed in the domain of quantum communication systems, in particular in quantum source coding, quantum data compression and quantum error-correcting codes [1-4]. In parallel, quantum algorithms have been deeply studied as computational processes [5-9]. Great analysis effort has been directed toward Deutsch-Jozsa algorithm [5], Grover quantum search algorithm [6,9], and Shor algorithm for factorizing great integer numbers, concentrating attention on its dynamics and ignoring the information aspects involved in quantum computation [7,9]. In this sense, only the information analysis of Toffoli gate [10], Shor's quantum algorithm gate [11] and communication capacity of quantum computation [12] has been developed.

In this article, the possibility to employ tools and techniques from quantum information theory [1-4,9,13] in the domain of quantum algorithm synthesis and simulation is investigated. On this purpose, as Benchmark, the analysis

Quantum Communication, Computing, and Measurement 3
Edited by P. Tombesi and O. Hirota, Kluwer Academic/Plenum Publishers, 2001

of the classical and quantum information flow in Shor algorithm is carried on. It is shown that the quantum gate G, based on superposition of states, quantum entanglement and interference, when acting on the input vector, stores information into the system state, minimizing the gap between the classical Shannon entropy and the quantum Von Neumann entropy. This principle is fairly general, suggesting both a methodology to design a quantum gate and a technique to simulate (efficiently) its behavior on a classical computer [14].

2. COMPUTATION DYNAMICS OF SHOR QUANTUM GATE

In the Shor algorithm [7] an integer number $n > 0$ and a function $f : \{0,1\}^n \rightarrow \{0,1\}^n$ are given such that f has period r, namely f is such that $\forall x \in \{0,1\}^n : f(x) \equiv f(x+r) \bmod n$ and f is injective with respect to its period. The problem is to find r. Function f is first encoded into the injective function $F : \{0,1\}^n \times \{0,1\}^n \rightarrow \{0,1\}^n \times \{0,1\}^n$ such that:

$$F(x,y) = (x, y \oplus f(x)) \tag{1}$$

where $\oplus$ is the bitwise XOR operator. F is then encoded into a unitary operator U_F. This purpose is fulfilled by mapping every binary input string of length $2n$ into a vector in a Hilbert space of dimension 2^{2n} according to the following recursive encoding scheme τ:

$$\tau(0) = |0\rangle, \quad \tau(1) = |1\rangle, \quad \tau(z_1 \dots z_{2n}) = |z_1 \dots z_{2n}\rangle \tag{2}$$

where $|0\rangle$ and $|1\rangle$ denote vectors $\begin{pmatrix} 1 \\ 0 \end{pmatrix}$ and $\begin{pmatrix} 0 \\ 1 \end{pmatrix}$ respectivelly, and $|z_1 \dots z_{2n}\rangle$ stands for the tensor product $|z_1\rangle \otimes \dots \otimes |z_{2n}\rangle$, being $z_1, ..., z_{2n} \in \{0,1\}$. We note that every bit value in a string is mapped into a vector of a Hilbert subspace of dimension 2. This subspace is called a qubit. Similary, a sequense of successive bit values of length l is mapped into a vector of dimension 2^l. We call this subspace a quantum register of length l.

Using this scheme, the operator U_F is defined as a squared matrix of order 2^{2n} such that:

$$\forall z \in \{0,1\}^{2n} : U_F|z\rangle = |F(z)\rangle \tag{3}$$

Practically, the operator U_F is as follows:

$$[U_F]_{i,j} = \delta_{i,1+[F([j-1]_{(2)})]_{(10)}} \tag{4}$$

where $[q]_{(b)}$ is the basis b representation of number q, $\delta_{i,j}$ is the Kronecker delta.

The idea of encoding a function f into a unitary operator is not a peculiarity of the Shor algorithm, but is typical [14] of all known quantum algorithms [5-9]. In general, U_F contains the whole information about function f needed to solve the problem.

In order to extract the information stored in U_F more efficiently, we must change our perspective. The operator U_F must in fact be used in order to transfer as much information as possible from the operator to the input vector each time U_F works. To this purpose, it is embedded into another unitary operator G, called the *quantum gate*, having the following general form:

$$G = (IF \otimes I_m) \cdot U_F \cdot (IF \otimes I_m) \tag{5}$$

where IF stands for unitary squared matrix of order 2^n and I_m for the identity matrix of order 2^m. In the case of Shor algorithm, U_F is embedded into the unitary quantum gate [11], [14]

$$G = (QFT_n \otimes I_n) \cdot U_F \cdot (QFT_n \otimes I_n) \tag{6}$$

The symbol QFT_n denotes the unitary quantum Fourier transform of order n:

$$[QFT_n]_{ij} = 1/2^{n/2} \exp(2\pi J[(i-1)(j-1)/2^n]) \tag{7}$$

where J is the imaginary unit. The corresponding computation is described by the following steps:

$$\text{Step 0} : |input\rangle = \underbrace{|0...0\rangle}_{n} \otimes \underbrace{|0...0\rangle}_{n} \tag{8}$$

$$\text{Step 1} : |\psi_1\rangle = (QFT_n \otimes I_n)|input\rangle = \frac{1}{2^{n/2}} \sum_{i_1,\dots,i_n} |i_1...i_n\rangle \otimes |\underbrace{0...0}_{n}\rangle \tag{9}$$

$$\text{Step 2} : |\psi_2\rangle = U_F|\psi_1\rangle = \frac{1}{2^{n/2}} \sum_{i_1,\dots,i_n} |i_1...i_n\rangle \otimes |f(i_1...i_n)\rangle \tag{10}$$

$$\begin{aligned} \text{Step 3} : \quad & |output\rangle = (QFT_n \otimes I_n)|\psi_2\rangle = \\ & \frac{1}{2^{n/2}} \sum_{\substack{j_1,\dots,j_n \\ i_1,\dots,i_n}} a_{i_1...i_n}^{j_1...j_n} |j_1...j_n\rangle \otimes |f(i_1...i_n)\rangle \end{aligned} \tag{11}$$

where

$$a_{i_1...i_n}^{j_1...j_n} = \frac{1}{2^{n/2}} \exp(2\pi J[i_1...i_n]_{(10)}[j_1...j_n]_{(10)}/2^n)$$

If $k = 2^n/r$ is an integer number, the output state can be written as

$$|output\rangle = \frac{1}{r} \sum_{p=0}^{r-1} \sum_{s=0}^{r-1} \exp(2\pi J s [i_1...i_n]_{(10)} l_p/r) |[s2^n/r]_{(2)}\rangle \otimes |[p]_{(2)}\rangle \tag{12}$$

where l_p is an integer positive number and binary representations are obtained using n bits. Therefore, the first n bits of the output state generates a periodical probability distribution with period k for every possible vector of the second quantum register. By repeating the algorithm a number of times polynomial in n and by performing a measurement each time, we can reconstruct the value of r [7,8].

In order to illustrate the Shor algorithm, we propose two different computations in Example 1.

Example 1. Let $n = 3$ and f_1, f_2 be defined as:

$$\begin{cases} x = 000, 001, 010, 011, 100, 101, 110, 111 \\ f_1(x) = 001, 111, 001, 111, 001, 111, 001, 111 \\ f_2(x) = 000, 010, 100, 110, 000, 010, 100, 110 \end{cases} \tag{13}$$

Then

$$U_{F_1} = I_2 \otimes \begin{pmatrix} I_2 & 0 \\ 0 & C_2 \end{pmatrix} \otimes C, \quad C = \begin{pmatrix} 0 & 1 \\ 1 & 0 \end{pmatrix} \tag{14}$$

and

$$U_{F_2} = I \otimes \begin{pmatrix} I_2 & 0 & 0 & 0 \\ 0 & I \otimes C & 0 & 0 \\ 0 & 0 & C \otimes I & 0 \\ 0 & 0 & 0 & C_2 \end{pmatrix} \otimes I \tag{15}$$

The computation involved by these two operators is resumed in table 1 and table 2.

Table 1 Shor quantum gate state dynamics with $f = f_1$

Step	State
Input	$\lvert 000\rangle \otimes \lvert 000\rangle$
Step 1	$\frac{\lvert 0\rangle+\lvert 1\rangle}{2^{1/2}} \otimes \frac{\lvert 0\rangle+\lvert 1\rangle}{2^{1/2}} \otimes \frac{\lvert 0\rangle+\lvert 1\rangle}{2^{1/2}} \otimes \lvert 000\rangle$
Step 2	$\frac{\lvert 0\rangle+\lvert 1\rangle}{2^{1/2}} \otimes \frac{\lvert 0\rangle+\lvert 1\rangle}{2^{1/2}} \otimes \frac{\lvert 000\rangle+\lvert 111\rangle}{2^{1/2}} \otimes \lvert 1\rangle$
Step 3	$\frac{1}{2^{1/2}}(\frac{\lvert 0\rangle+\lvert 1\rangle}{2^{1/2}} \otimes \lvert 0000\rangle + \frac{\lvert 0\rangle-\lvert 1\rangle}{2^{1/2}} \otimes \lvert 0011\rangle) \otimes \lvert 1\rangle$

Table 2 Shor quantum gate state dynamics with $f = f_2$

Step	State
Input	$\|000\rangle \otimes \|000\rangle$
Step 1	$\frac{\|0\rangle+\|1\rangle}{2^{1/2}} \otimes \frac{\|0\rangle+\|1\rangle}{2^{1/2}} \otimes \frac{\|0\rangle+\|1\rangle}{2^{1/2}} \otimes \|000\rangle$
Step 2	$\frac{\|0\rangle+\|1\rangle}{2^{1/2}} \otimes \frac{1}{2}(\|0000\rangle + \|0101\rangle + \|1010\rangle + \|1111\rangle) \otimes \|0\rangle$
Step 3	$\frac{1}{2}\left(\begin{array}{l} \|000\rangle \otimes \frac{\|0\rangle+\|1\rangle}{2^{1/2}} \otimes \frac{\|0\rangle+\|1\rangle}{2^{1/2}} + \|010\rangle \otimes \frac{\|0\rangle-\|1\rangle}{2^{1/2}} \otimes \frac{\|0\rangle+J\|1\rangle}{2^{1/2}} + \\ +\|100\rangle \otimes \frac{\|0\rangle+\|1\rangle}{2^{1/2}} \otimes \frac{\|0\rangle-\|1\rangle}{2^{1/2}} + \|110\rangle \otimes \frac{\|0\rangle-\|1\rangle}{2^{1/2}} \otimes \frac{\|0\rangle-J\|1\rangle}{2^{1/2}} \end{array} \right) \otimes \|0\rangle$

3. SHANNON AND VON NEUMANN ENTROPY

A vector in a Hilbert space of dimension 2^k acts as a classical information source if the measurement with respect to a given orthonormal basis is performed. The possible outputs are the 2^k basis vectors, each one with probability given by the squared modulus of its probability amplitude. More in general, given a vector

$$|\psi\rangle = \sum_{i_1 \dots, i_n \in \{0,1\}} c_{i_1 \dots i_n} |i_1\rangle \otimes |i_2\rangle \otimes \dots \otimes |i_n\rangle \tag{16}$$

in a Hilbert space of dimension 2^n, let $T = \{ j_1, \dots, j_k \} \subseteq \{ 1, \dots, n \}$ and $\{ 1, \dots, n \} - T = \{ l_1, \dots, l_{n-k} \}$. We define

$$|\psi\rangle\langle\psi|_T = \sum_{\substack{t_{j_1}, \dots, t_{j_k} \\ i_{j_1}, \dots, i_{j_k}}} b_{i_{j_1} \dots i_{j_k}}^{t_{j_1} \dots t_{j_k}} |i_{j_1} \dots i_{j_k}\rangle\langle t_{j_1} \dots t_{j_k}| \tag{17}$$

where

$$b_{i_{j_1} \dots i_{j_k}}^{t_{j_1} \dots t_{j_k}} = \sum_{i_{l_1} = t_{l_1} \dots \, i_{l_{n-k}} = t_{l_{n-k}}} c_{i_1 \dots i_n} c^*_{t_1 \dots t_n} \tag{18}$$

Choosing T means to select a subspace of the Hilbert space of $|\psi\rangle$. If $T = \{ j \}$, this subspace has dimension 2 and we shall call it the subspace of the qubit j. Similarly, if $T = \{ j_1, \dots, j_k \}$, we say that we are dealing with the

subspace of qubits $j_1, ..., j_k$. $|\psi\rangle\langle\psi|_T$ describes the projection of the *density matrix* corresponding to $|\psi\rangle$ on this subspace.
We define the *Shannon entropy* of T in $|\psi\rangle$ with respect to the basis $B = \{ |i_1\rangle \otimes ... \otimes |i_n\rangle \}$ as

$$E_T^{Sh}(|\psi\rangle) = -\sum_{i=1}^{2^k} [|\psi\rangle\langle\psi|_T]_{ii} \log_2 [|\psi\rangle\langle\psi|_T]_{ii} \quad (19)$$

The Shannon entropy can be interpreted as the degree of disorder involved by vector $|\psi\rangle$ when the qubits in T are measured.

Vector $|\psi\rangle$ does not act only as a classical information source. On the contrary, it stores also some other kind of informaion in non-local quantum correlation, that is through entanglement. In order to measure the quantity of entanglement of a set $T = \{ j_1, ..., j_k \}$ of qubits in $|\psi\rangle$ we employ the *Von Neumann entropy* of $|\psi\rangle$ with respect to T, which is defined as follows

$$E_T^{VN}(|\psi\rangle) = -\mathrm{tr}(|\psi\rangle\langle\psi|_T \log_2 |\psi\rangle\langle\psi|_T) \quad (20)$$

where tr denotes the *trace operator*. The Von Neumann entropy of the qubits in T is intepreted as the measure of the degree of entanglement of these qubits (in pure states [11]) with the rest of the system [13].

The *intelligence* $\Im_T(|\psi\rangle)$ of a state $|\psi\rangle$ with respect to the qubits in T and to the basis $B = \{ |i_1\rangle \otimes ... \otimes |i_n\rangle \}$ is

$$\Im_T(|\psi\rangle) = 1 - \frac{E_T^{Sh}(|\psi\rangle) - E_T^{VN}(|\psi\rangle)}{|T|} \quad (21)$$

The intelligence of a state with respect to T and B is minimal (i.e. 0) when $E_T^{VN}(|\psi\rangle) = 0$ and $E_T^{Sh}(|\psi\rangle) = |T|$, it is maximal (i.e. 1) when $E_T^{VN}(|\psi\rangle) = E_T^{Sh}(|\psi\rangle)$. The information analysis of two computations (14) and (15) of the Shor algorithm is reported in Example 2.

Example 2. The two operators represented in (13) produce the information flow reported in table 3. It is worth observing how the intelligence of the state increases and decreases while the algorithm evolves. The intelligence of a quantum algorithm state is maximal if the gap between the Shannon and the Von Neumann entropy for the chosen result qubits is minimal. The quantum Fourier transform creates maximally intelligent states with respect to the first n qubits for Shor's problem.

4. CONCLUSIONS

The methods in quantum algorithm gate design based on qualitative measures: 1) analysis of quantum algorithm dynamics and structure gate design;

Table 3 Shor quantum gate information flows with $f = f_1$ (left columns) and with $f = f_2$ (right columns)

Step	$E_T^{Sh}(\|\psi\rangle)$		$E_T^{VN}(\|\psi\rangle)$		$\Im_T(\|\psi\rangle)$	
Input	0	0	0	0	1	1
Step 1	3	3	0	0	1	0
Step 2	3	3	1	2	1/3	2/3
Step 3	1	2	1	2	1	1

2) analysis of information flow; and 3) simulation of intelligent quantum algorithms on classical computers are presented. The developed analysis of quantum algorithm dynamics is the background for the solvable guarantee of necessary conditions that the possibility of successful solution of the investigated problem is existent and the solution is unique with the desired probability. Analysis of information flow in quantum algorithm gates is the background of the guarantee for sufficient conditions that the unique solution is existent with the desired accuracy and the reliability of solutions can be achieved with higher probability. Intelligence of quantum algorithms is achieved with the principle of minimum information distance between Shannon and von Neumann entropy. From the physical point of view the output states of quantum algorithms as the solution of expected problems are the intelligent states with minimum entropic relations of uncertainty (coherent superposition states). The successful results of quantum algorithm computing are robust to noise excitations in quantum gates, and intelligent quantum operations are fault-tolerant in quantum computing. Three quantum operators as superposition, entanglement, and interference together with operators of genetic algorithm are the background of quantum computations of qualitative and quantitative measures of fitness in quantum soft computing.

References

[1] Schumacher B., "Quantum coding", Physical Review, Vol. A51, 1995, 2738

[2] Bennett C.H. and Shor P., "Quantum information theory", IEEE Trans. Information Theory, Vol. 44, No 6, 1998, 2724

[3] Bouwmeester D., Ekert A. and Zeilinger A., "The physics of quantum information", Springer Verlag, N.Y., 2000

[4] Nielsen M.A. and Chuang I.L., "Quantum computation and quantum information", Cambridge University Press, Cambridge, 2000

[5] Deutsh D. and Jozsa R., "Rapid solution of problems by quantum computing", Proc. R. Soc. Lond., Vol. 439, No 1907, 1992, 553

[6] Grover L.K., "Quantum mechanics helps in searching for a needle in a haystack", Phys. Rev. Lett., Vol. 78, 1997, 325; "How fast can a quantum computer search?", arXiv: quant-ph/ 9809029, 13 Apr, 1999

[7] Shor P., "Introduction to quantum algorithms", quant-ph/0005003, 29 Apr 2000

[8] SIAM J. Comp. (Special Issue on Quantum Computing), Vol. 26, No 5, 1997; Jozsa R., "Quantum algorithms and Fourier transform", Proc. Royal Soc. Lond., Vol. 454A, 1998, 323

[9] Preskill J., "Lectures Notes for Physics 229: Quantum information and computation", California Institute of Technology, 1998, available online at http: // www. theory.caltech.edu/people/preskill/ph229/

[10] Ohya M. and Watanabe N., "On the mathematical treatment of the Fredkin - Toffoli - Milburn gate", Physica, Vol. 120 D, 1998, 206

[11] Ghisi F. and Ulyanov S.V., "The information role of entanglement and interference operators in Shor's a quantum algorithm gate dynamics", J. Mod. Opt. , 2000 (in print)

[12] Bose S., Rallan L. and Vedral V., "Communication capacity of quantum computation", arXiv: quant - ph/ 0003072, 17 Mar 2000

[13] Cerf N.J. and Adami C., "Information theory of quantum entanglement and measurement", Physica, Vol. 120D, 1998, 62

[14] Ulyanov S.V., Ghisi F., Panfilov S.A. et al., "Simulation of quantum algorithms on classical computers", Note del Polo - Ricerca, Università degli Studi di Milano Publ., Crema, Vol. 32, 2000

QUANTUM PROBABILISTIC SUBROUTINES AND PROBLEMS IN NUMBER THEORY

A. Carlini, A. Hosoya
Department of Physics, Tokyo Institute of Technology
Oh-Okayama, Meguro-ku, Tokyo 152, Japan

Keywords: Quantum algorithms, number theory

Abstract We describe a quantum version of the classical probabilistic algorithms *à* la Rabin.The quantum probabilistic algorithm is fully unitary and reversible, and can be used as a subroutine in larger quantum computations. As an example, we describe a polynomial time algorithm for counting the number of primes smaller than a given integer.

1. INTRODUCTION

Two of the most important quantum algorithms known at present are Shor's algorithm for factoring integers [1] and Grover's algorithmfor the unstructured database search [2] which achieve, respectively, an exponential and square root speed up compared to their classical analogues. These two basic unitary blocks are exploited in conjunction in the algorithm COUNT by Brassard et al. [3], which can count the cardinality t of a set of states present in a flat superposition of N items in a time polynomial in N/t with an exponential accuracy. An extended use of this algorithm can be exploited to construct unitary and fully reversible quantum operators [4] which are able to emulate classical probabilistic algorithms. One of the prototypes of these classical algorithms is that of Rabin [5] for testing the primality of a given number k. The algorithm tests a certain condition $W_k(a)$, for $1 \leq a < k$. When $W_k(a) \equiv 0$ a is said to be a witness to the compositeness of k. For a composite number $k \equiv k_C$ the number t_{k_C} of witnesses is $t_{k_C} \geq 3(k_C - 1)/4$. For a prime number $k \equiv k_P$, instead, none of the a is a witness (i.e. $W_{k_P}(a) = 1$ for all $1 \leq a < k_P$). The main idea underlying our quantum algorithm to the test of primality for a given number k is based on the repeated use of the counting algorithm COUNT for estimating the number of witnesses to the compositness of k, which is summarized by the sequence of operations [3, 4]: 1) $[W|0>][W|0>] = \sum_{m=0}^{P-1}|m> \sum_a |a>$; 2) $\rightarrow [F \otimes I][\sum_{m=0}^{P-1}|m> G^m(\sum_a |a>)]$, where Grover's operator is $G = -WS_0WS_1$, with $W|a> \equiv \sum_{b=0}^{k-1}(-1)^{a \cdot b}|b> /\sqrt{k}$, $S_0 \equiv I - 2|0><0|$ and $S_1 \equiv I - 2\sum_{W_k(a)=0}|a><a|$, and Shor's operator is $F|a> \equiv \sum_{b=0}^{k-1} e^{2i\pi ab/k}|b> /\sqrt{k}$. [1]

2. THE PRIME NUMBER THEOREM

Then, the test of the so called 'Prime Number Theorem', according to which the total number t_N of primes k_P smaller than a given number N is given by the formula

$$t_N \equiv \pi(N) \simeq \frac{N}{\log N}, \tag{1}$$

can be done by a quantum algorithm consisting of a sub-loop which checks for the primality of a given $k < N$ by counting its witnesses, a main loop for the counting of primes less than N, and a final measurement of an ancilla qubit. Schematically:

MAIN-LOOP:
Count $\sharp\{k|k = k_P < N\}$ *using COUNT with* $G \rightarrow \tilde{G}$ *and* $S_1 \rightarrow \tilde{S}_1 \equiv 1 - 2\sum_{k_P}|k_P><k_P|$ *(parameter Q)*
SUB-LOOP:
Parallel primality tests $\forall\ k < N$ *(parameter P) and (approximate) construction of* $\tilde{S}_1$

In order to construct the unitary transform $\tilde{S}_1$ we start from the state

$$|\bar{\psi}_1> \equiv \frac{1}{\sqrt{N}}\sum_{k=0}^{N-1}|k> \frac{\sum_{m=0}^{P-1}|m>}{\sqrt{P}} \frac{\sum_{a=0}^{k-1}|a>}{\sqrt{k}} \tag{2}$$

(with $P \simeq O[poly(\log N)]$) and operate with an $|m>$-'controlled' Grover transform G^m on the last ancilla qubit $|a>$ followed by a Fourier transform F and a phase change operator S_0 on the first ancilla qubit $|m>$, and finally undo again all the previous operations (F, G^m, and the two initial Fs necessary to build $|\bar{\psi}_1>$ from $|\bar{\psi}_0> \equiv \sum_{k=0}^{N-1}|k>|0>|0>/\sqrt{N}$), obtaining the state $|\bar{\psi}_2> \equiv \sum_{k=0}^{N-1}|k>[|0>|0> -2|C_k>]/\sqrt{N}$. Using the property that [4], for a prime k_P, $|C_{k_P}> = |0>|0>$ and, for a composite k_C, $<C_{k_C}|C_{k_C}> \simeq O[P^{-2}]$, we then have

$$\tilde{S}_1|\bar{\psi}_0> = |\bar{\psi}_2> \equiv |\Psi> + |E>, \tag{3}$$

where $|\Psi> \equiv \sum_{k=0}^{N-1}(-1)^{F_k}|k> |0> |0> /\sqrt{N}$ (with $F_k \equiv 1$ for a prime $k = k_P$ and $F_k \equiv 0$ for a composite $k = k_C$), and $|E>$ is a correction term whose norm is upper bounded by $< E|E > \leq O[P^{-2}]$.

Let us now consider the MAIN-LOOP of the algorithm, i.e. that counting the total number of $k_P < N$. Grover's transform $\tilde{G}$ entering this part of the algorithm is written as $\tilde{G} \equiv U_2\ \tilde{S}_1$, with $U_2 \equiv -W^{(k)}S_0^{(k)}W^{(k)}$, and $W^{(k)}$ and $S_0^{(k)}$ acting on the qubit $|k>$. The quantum algorithm starts from the state

$$|\bar{\psi}_3> \equiv \frac{1}{\sqrt{Q}}\sum_{m=0}^{Q-1}|m>|\bar{\psi}_0> \tag{4}$$

(with $Q \simeq O[poly(\log N)]$), then acts on $|\bar{\psi}_0>$ with the $|m>$-'controlled' $\tilde{G}^m$ and with F on $|m>$, producing [4]

$$\begin{aligned} |\bar{\psi}_4> &\simeq \frac{1}{2}\sum_{n=0}^{Q-1}|n>[e^{-i\pi f_Q}s_{n-}^{(Q)}|D_+> + e^{i\pi f_Q}s_{n+}^{(Q)}|D_->] \\ &+ \frac{1}{Q}\sum_{m,n=0}^{Q-1}|n>|E_m>, \end{aligned} \tag{5}$$

where $f_Q \equiv Q\theta_N/\pi$, $\sin\theta_N \equiv \sqrt{t_N/N}$, $s_{n\pm}^{(Q)} \equiv [\sin\pi(n \pm f_Q)]/[Q\sin\pi(n \pm f_Q)/Q]$, $|D_\pm> \equiv [\pm i\sum_{k_P}|k> /\sqrt{t_N} + \sum_{k_C}|k> /\sqrt{N-t_N}]$, and $|E_n>$ is a correction term whose norm is upper bounded by $< E_n|E_n > \leq O[n^2] < E|E >$. The last step of the algorithm consists in measuring the first qubit in the state $|\bar{\psi}_4>$. Using the expected estimate that $\theta_N \simeq O[1/\sqrt{\log N}]$ and by choosing $Q \simeq O[(\log N)^\beta]$; $\beta > 1/2$ and $P \simeq O[(\log N)^\gamma]$; $\gamma > \beta$, the probability to obtain any of the states $|f_->$, $|f_+>$, $|Q-f_->$ or $|Q-f_+>$ (where $f_- \equiv [f_Q]+\delta f$, $f_+ \equiv f_-+1$ and $0 < \delta f < 1$) in the final measurement of the ancilla qubit is given by $W \geq [8/\pi^2]\{1 - O[(\log N)^{-(\gamma-\beta)}]\}$. This means that with a high probability and precision [2] one will always be able to evaluate the number t_N from f_Q and θ_N. The computational complexity of the quantum algorithm can be shown to be polynomial in $\log N$ [4].

3. DISCUSSION

We have shown a method to build a quantum version of the classical probabilistic algorithms *à* la Rabin. Our quantum algorithms make essential use of some of the basic blocks of quantum networks known so far, i.e. Grover's operator for the quantum search of a database, Shor's Fourier transform for extracting the periodicity of a function and their combination in the counting algorithm of Ref. [3]. The most important feature of our quantum probabilistic algorithms is that the coin tossing used in the correspondent classical probabilistic ones is replaced here by a unitary and reversible operation, so that the quantum algorithm can even be used as a subroutine in larger and more complicated networks.

Acknowledgments

A.H.'s research was partially supported by the Ministry of Education, Science, Sports and Culture of Japan, under grant n. 09640341. A.C.'s research was supported by the EU under the S.T.F. Programme in Japan, grant n. ERBIC17CT970007.

Notes

1. The parameter P determines the precision of the estimate t_k and the computational complexity of the COUNT algorithm.

2. In general, f_Q is not an integer and the estimate has an error $|\Delta t_N|_{exp} \leq \pi N \left[\pi/Q + 2\sqrt{t_N/N}\right]/Q \simeq O[N(\log N)^{-\beta-1/2}]$, see Ref. [3].

References

[1] P.W. Shor, in *Proceedings of the 35th Annual Symposium on Foundations of Computer Science* (ed. S. Goldwater, IEEE Computer Society Press, New York, 1994), p. 124.

[2] L.K. Grover, in *Proceedings of the 28th Annual Symposium on the Theory of Computing* (ACM Press, New York, 1996), p. 212.

[3] G. Brassard, P. Hoyer and A. Tapp, Los Alamos e-print quant-ph/9805082; M. Boyer, G. Brassard, P. Hoyer and A. Tapp, in *Proceedings of the 4th Workshop on Physics and Computation* (ed. T. Toffoli et al., New England Complex Systems Institute, Boston, 1996), p. 36.

[4] A. Carlini and A. Hosoya, Phys. Rev. A **62** (2000), 032312.

[5] M.O. Rabin, Journ. Num. Th. **12** (1980), 128.

THEORY OF THE QUANTUM SPEED UP

Giuseppe Castagnoli
Elsag, 16154 Genova, Italy

David Ritz Finkelstein
Georgia Institute of Technology, Atlanta, USA

Abstract Insofar as quantum computation is faster than classical, it appears to be irreversible. In all quantum algorithms found so far the speed-up depends onthe extra-dynamical irreversible projection representing quantum measurement. Quantum measurement performs a computation that dynamical computation cannot accomplish as efficiently.

The quantum algorithms are sometimes faster than their classical counterparts. We show that this quantum speed-up results from a succession of entanglement and disentanglement, the former due to dynamical quantum-parallel computation, the latter to the extra-dynamical projection of quantum measurement. Thus the quantum speed-up implies irreversibility.

Some standard notions concerning problem solving must be modified to understand the speed-up. Standard problem solving has three stages: (i) *State the problem.* This defines the problem solution, usually implicitly. E.g. consider the problem of finding two primes x, y (unknown) such that $x \cdot y = c$ (known). This equation implicitly defines the values of x and y which satisfy it. An implicit definition does not represent the process required to compute the solution. (ii) *Program the computation.* Change the implicit definition into an explicit, finite step-by-step logical procedure for constructing the solution. This procedure is specified by the solution algorithm. (iii) *Run the program.* The execution is dynamical in character. By dynamics we mean, here and in the following, deterministic dynamics[1].

The standard assumption is that the solution of a problem must be computed by a dynamical development. Step (ii) changes a definition which does not represent a dynamical process, into one which represents it.

Quantum computation does not fit this scheme. Consider Shor's algorithm (Shor, 1994). The problem is to efficiently find the period r of a hard-to-reverse function $f(x)$ from $\{0,1\}^n$ to $\{0,1\}^n$. Fig. 1 gives the algorithm block diagram.

Quantum Communication, Computing, and Measurement 3
Edited by P. Tombesi and O. Hirota, Kluwer Academic/Plenum Publishers, 2001

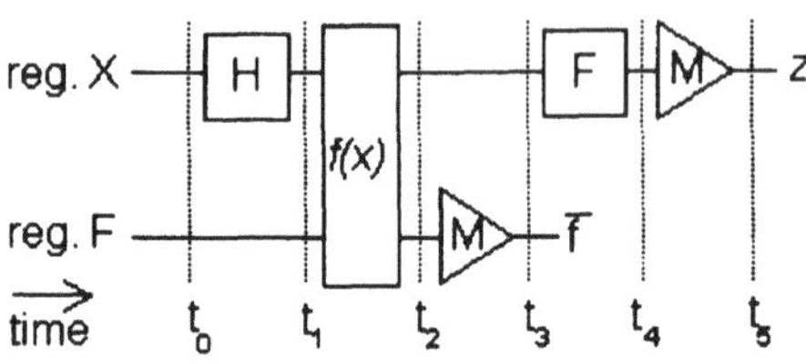

Figure 1

H is the Hadamard and F the digital Fourier transform, M denotes measurement of a register content. We need to consider only two steps of the algorithm (see Castagnoli et al., 2000).

(I) The process of computing $f(x)$ for all possible x, in quantum superposition, puts two n-qubit registers X and F into the state [2]

$|\psi, t_2\rangle_{XF} = \frac{1}{\sqrt{N}} \sum_x |x\rangle_X |f(x)\rangle_F$, x runs over $0, 1, ..., N-1$, $N = 2^n$.

(II) Let $[F]$ be the content of register F, an observable. Measuring $[F]$ in $|\psi, t_2\rangle_{XF}$ [3] and finding the result $\overline{f}$ yields the state

$|\psi, t_3\rangle_{XF} = k\left(|\overline{x}\rangle_X + |\overline{x}+r\rangle_X + |\overline{x}+2r\rangle_X + ...\right)\left|\overline{f}\right\rangle_F$; k is for normalization, $f(\overline{x}) = f(\overline{x}+r) = ... = \overline{f}$.

The transition from $|\psi, t_2\rangle_{XF}$ to $|\psi, t_3\rangle_{XF}$ obeys the quantum principle: (A) measuring an observable is extra-dynamically represented by projection on the eigenspace of one eigenvalue; (B) this eigenvalue is selected at random according to the square of the probability amplitudes. If Q and Q' are the projection operators on the state just before and after the measurement M, and P is the projection operator for the observed value of the function, then classically $Q' = Q$, while quantally $Q' = PQP$ (up to normalization), depending on both P and Q. Thus, selecting $\overline{f}$ projects into the post-measurement state all and only those tensor products of $|\psi, t_2\rangle_{XF}$ ending with $\left|\overline{f}\right\rangle_F$. Point (A) of the quantum principle, by selecting one eigenvalue, imposes a logical constraint on the output of computation. Because of this constraint, quantum measurement filters, out of an exponentially larger superposition, all and only those values of x whose function is that eigenvalue.

Therefore quantum measurement performs extra-dynamically a computation crucial for finding r, which is "readily" extracted out of $|\psi, t_3\rangle_{XF}$. Measurement time is linear in the number of qubits of register F, and is independent of the entanglement between X and F, which holds problem complexity. Disentanglement comes for free, as a by-product of quantum measurement. Filtration, together with function evaluation, is essential to speed-up. This can be better seen by comparing step by step quantum and classical computation times[4]: (I) function evaluation: poly(n) vs exp(n) ; (II) filtration: linear(n) vs exp(n); (III) extracting r out of $|\psi, t_3\rangle_{XF}$: linear(n) vs linear(n) . Speed-up is due to steps (I) and (II).

The extra-dynamical character of quantum computation is clarified by showing that Shor's algorithm does not fit standard, dynamical, problem solving. Quantum dynamics is deterministic: any state in time dynamically determines a unique successor. While function evaluation is dynamical in character, the filtration performed by quantum measurement is not. Classically the state after measurement M is the same as the state before M. In quantum fact the state after measurement is influenced by both the state before measurement and the measurement itself, by the quantum principle: $|\psi, t_3\rangle_{XF} = k\,|\overline{f}\rangle_F \langle\overline{f}|_F\,|\psi, t_2\rangle_{XF}$. $|\psi, t_2\rangle_{XF}$ is the *prior state*; the left-multiplication by $|\overline{f}\rangle_F \langle\overline{f}|_F$ represents the final constraint selecting all tensor products ending with $|\overline{f}\rangle_F$. The determination of $|\psi, t_3\rangle_{XF}$ is jointly influenced by an initial condition and a final condition. It is richer than dynamical determination, insofar as it yields the speed-up. We needed a computational context to realize this.

Extra-dynamical computation means much more than nondeterministic computation. For example, point (A) of the quantum principle does not involve randomness and yields Shor's quantum speed-up in $\sim 70\%$ of the cases, when a single run of the algorithm is sufficient to identify r [5].

Determination with joint influence is extra-dynamical, it cannot be represented by a dynamical propagation of an input into an output. Of course we could go through step (ii), and replace joint influence with a dynamical process that leads to a "filtered" state like $|\psi, t_3\rangle_{XF}$. But this would introduce programming and computation, increasing computation time exponentially in problem size. Joint influence bypasses step (ii) as well as speeding-up the computation. It yields a direct physical determination of the object of an implicit definition[6].

This can also be seen as follows. $|\psi, t_3\rangle_{XF}$ (selected with joint influence) contains the solutions x of the implicit algebraic equation $f(x) = \overline{f}$, although the reverse of f has not been computed. Thus, $f(x) = \overline{f}$ implicitly defines the solutions x while quantum measurement selects them without going through programming and dynamical computation. The extra-dynamical character of this selection is clear. It takes essentially no time.

It can be seen that the same theory of the speed-up holds for Simon's algorithm (Simon, 1994), as modified in (Cleve et al., 1996).

Until now we have assumed the intermediate measurement of $[F]$. However, as is well known, this measurement can be skipped without affecting the result of measuring $[X]$ at time t_4 (fig. 1). It was introduced by Ekert and

Jozsa (1996) to clarify the way Shor's algorithm operates; it can also clarify the speed-up. In fact skipping it is mathematically equivalent to performing it: the filtration performed by the extra-dynamical projection of quantum measurement is induced by measuring only $[X]$ at the end.

If $[F]$ measurement is skipped, the state of registers X and F at time t_4 is entangled. This establishes an equivalence between measuring $[X]$ or $[F]$. From a mathematical standpoint, the outcome of measuring $[F]$ at time t_4 can be backdated in time along the reversible process, provided that the overall state undergoes the inverse of the usual forward-time evolution. This is equivalent to having measured $[F]$ at time t_2.

We should counter the objection that register F can be annihilated immediately after function evaluation. This would leave register X in a mixture that is the partial trace over F of the density matrix of the two registers: $|\psi, t_2\rangle_X = \frac{1}{\sqrt{N}} \sum_h e^{i\delta_h} \left(|x_h\rangle_X + |x_h + r\rangle_X + |x_h + 2r\rangle_X + ...\right) |f_h\rangle_F$, where the range of h is such that f_h ranges over all the values assumed by $f(x)$, $f(x_h) = f(x_h + r) = ... = f_h$, and δ_h are random phases independent of each other[7]. For the current purposes, annihilating F is like having performed the intermediate $[F]$ measurement.

We shall now move to quantum Oracle computing. This can be seen as a competition between two players. One produces the problem, the other is challenged to produce the solution. We shall call the former player Sphinx, the latter Oedipus.

Let us consider Grover's algorithm (1994). The game is as follows. The Sphinx hides an object in drawer number k, among n drawers. Oedipus must find where it is, in the most efficient way. The chest of drawers is actually a quantum computer that, set in the mode k and given a drawer number x as the input, yields the output $f_k(x) = \delta_{k,x}$ ($\delta_{k,x} = 1$ if $k = x$ and $\delta_{k,x} = 0$ if $k \neq x$). Fig. 2a gives the usual Grover's algorithm for $n = 4$.

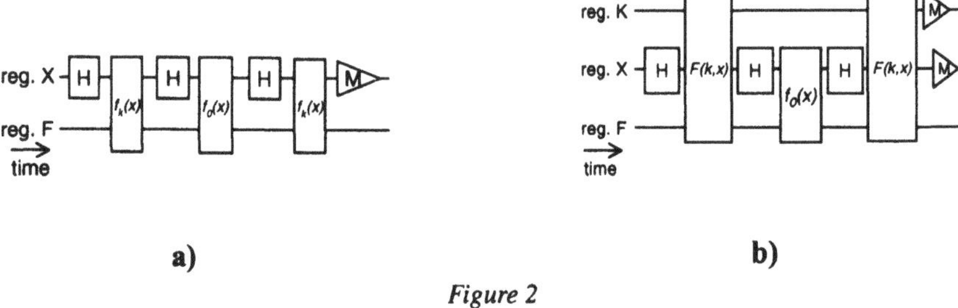

Figure 2

The Sphinx sets the mode k at random and passes the computer on to Oedipus. Oedipus must find k in the most efficient way by testing the computer input-output behaviour. Without entering into detail, we note that the computer has two registers X and F. Oedipus prepares them in the initial state $\frac{1}{\sqrt{2}}|00\rangle_X(|0\rangle_F-|1\rangle_F)$, the same for all k, and invokes the algorithm presented in fig. 2a. The state ψ before final measurement depends on the Sphinx' choice k:

$$\begin{array}{lcl} k=00 & \leftrightarrow & \psi=\frac{1}{\sqrt{2}}|00\rangle_X(|0\rangle_F-|1\rangle_F), \\ k=01 & \leftrightarrow & \psi=\frac{1}{\sqrt{2}}|01\rangle_X(|0\rangle_F-|1\rangle_F), \\ k=10 & \leftrightarrow & \psi=\frac{1}{\sqrt{2}}|10\rangle_X(|0\rangle_F-|1\rangle_F), \\ k=11 & \leftrightarrow & \psi=\frac{1}{\sqrt{2}}|11\rangle_X(|0\rangle_F-|1\rangle_F). \end{array} \tag{1}$$

Measuring $[X]$ yields Oedipus answer. This is reached in $O(\sqrt{n})$ time, versus $O(n)$ with classical computation. But it is reached in a dynamical way, without any interplay between quantum parallel computation and the extra-dynamical projection of quantum measurement.

As we have seen, this interplay is associated with an isomorphism between the problem that implicitly defines its solution and the solution determination. This obviously requires that the problem is physically represented in a complete way. Here it is not: the above possible choices of the Sphinx and the related implications are not physically represented.

This is easily altered by introducing an ancillary two-qubit register K which contains the computer mode k. Given the input k and x, the output of computation is now $F(k,x)=f_k(x)$ (fig. 2b). The preparation becomes $\frac{1}{2\sqrt{2}}\left(|00\rangle_K+e^{i\delta_1}|01\rangle_K+e^{i\delta_2}|10\rangle_K+e^{i\delta_3}|11\rangle_K\right)|00\rangle_X(|0\rangle_F-|1\rangle_F)$, where δ_1, δ_2 and δ_3 are independent random phases. To Oedipus, the Sphinx' random choice is indistinguishable from a mixture where k is a random variable with uniform distribution over 00, 01, 10, 11. Fig. 2b includes the physical representation of the problem; we can go directly to the final state before measurement:
$\frac{1}{2\sqrt{2}}\left(|00\rangle_K|00\rangle_X+e^{i\delta_1}|01\rangle_K|01\rangle_X+e^{i\delta_2}|10\rangle_K|10\rangle_X+e^{i\delta_3}|11\rangle_K|11\rangle_X\right)$ $(|0\rangle_F-|1\rangle_F)$, δ_1, δ_2 and δ_3 are independent random phases. Measuring $[K]$ gives the Sphinx' choice, measuring $[X]$ gives Oedipus answer, or vice-versa. Now that the game has been physically represented, we find again that there is the above "interplay".

In this game context, joint influence becomes the joint determination of the drawer number on the part of the two players, imposed by the quantum principle.

Why does extra-dynamical joint influence produce a speed-up? We suggest the following argument. The quantum game – yielding joint determination of the drawer number as a special quantum feature – should be as efficient as a classical game where there were joint determination of the drawer number on the part of the two players.

This cannot mean that Oedipus dictates the Sphinx' choice, or the Sphinx suggests to Oedipus the right answer: this would be unilateral determination. Joint determination is symmetrical. The classical game must be defined as follows, with reference to the square-shaped chest of drawers herebelow[8]

	0	1
0	00	01
1	10	11

The Sphinx chooses the row number, say 1. Oedipus chooses the column number, say 0. Clearly, the drawer number 10 has been jointly determined by the Sphinx and Oedipus. Now the cost of Oedipus search is $O(\sqrt{n})$ rather than $O(n)$, since he must search only the row. This is in agreement with theory. A similar analysis applies to Deutsch's algorithm (1985), as modified in (Cleve et al., 1996).

In conclusion, we have shown that quantum computation speed-up depends essentially on the extra-dynamical, irreversible projection of quantum measurement. To be sure, the entropy increase associated with speed-up is proportional only to the size of the output register, not the computation.

This extra-dynamical computation is more efficient than dynamical computation, as it yields the speed-up. It is a high level quantum feature, as it comes from a special interplay between a plurality of lower level ones (entanglement, disentanglement ...).

Earlier, attention was paid only to reversible quantum computation. The seminal well known works of Bennett, Fredkin and Toffoli, Benioff, and Feynmann demonstrated that computation can be reversible both in the classical and quantum framework. With Deutsch and others, quantum computation becomes quantum problem-solving, yields a speed-up and, we point out, ceases to be reversible. The current quantum algorithms ingeniously exploit extra-dynamical computation.

It is natural to ask whether other extra-dynamical projections than the one inherent in quantum measurement can be useful. Two come to mind at once: a statistics symmetry can be seen as an extra-dynamical projection on the "symmetric" subspace; and annealing is a projection on the ground state resulting from gradual cooling by a succession of extra-dynamical interventions. Exploiting these forms of projection might result in further speed-ups, and further reductions in the programming process.

This work developed through many discussions with Artur Ekert.

Notes

1. Classical nondeterministic computation, at the current fundamental level, will be seen as pseudorandom deterministic computation.

2. Quantum theory can be formulated praxically or ontically. The former (e.g. Finkelstein, 1996) is closest to Heisenberg's. It deals with operators dispensing with states. The ontic formulation makes quantum theory seem less time-symmetrical than it really is. In the present theory, it makes the speed-up seem to happen all at once at the end of computation. We use the ontic formulation here, misleading as it is, because it is more familiar to most physicists.

3. This intermediate measurement can be skipped, but we will show that this makes no difference.

4. Classical time is that required to derive the *symbolic description* of a quantum state (e.g. $|\psi, t_2\rangle_{XF}$, or $|\psi, t_3\rangle_{XF}$, with all x, $f(x)$, $\overline{x}$, r, etc. replaced by the proper numerical values) from the previous one by classical computation. Note that the resulting classical algorithm is reasonably efficient in itself, with times on the order of problem size.

5. In $\sim$ 30% of the cases, more than one run is needed. Randomness assures that we do not always obtain the same result.

6. Since the implicit definition is the problem, we see that, in Shor's algorithm, computation can be identified with problem solving: points (i), (ii), and (iii) are both altered and unified.

7. We are using the random phase representation. Let us exemplify it for a two-state system. The mixture $\rho = \sin^2\varphi\,|0\rangle\,\langle 0| + \cos^2\varphi\,|1\rangle\,\langle 1|$ becomes $|\psi\rangle = \sin\varphi\,|0\rangle + e^{i\delta}\cos\varphi\,|1\rangle$, where δ is a random phase with uniform distribution in $[0, 2\pi]$; ρ is the average over δ of $|\psi\rangle\,\langle\psi|$.

8. If the number of drawers is $O(n)$, the number of rows or colums is $O\left(\sqrt{n}\right)$.

References

[1] Castagnoli, Giuseppe, and Dalida Monti (2000). *Int. J. Theor. Phys.* **39**, 525. Castagnoli, G., Monti, D., and Sergienko, A., "Performing Quantum Measurement in Suitably Entangled States Originates the Quantum Computation Speed Up", arXiv: quant-ph/0005069 – 17 May 2000.

[2] Cleve, Richard, Ekert, Artur, Macchiavello, Chiara, and Mosca, Michele (1996). arXiv: quant-ph/9708016; submitted to *Proc. Roy. Soc. Lond.* A.

[3] Deutsch, David (1985). *Proc. Roy. Soc. of London* A, **400**, 97.

[4] Ekert, Artur, and Jozsa, Robert (1998). arXiv: quant-ph/9803072; to appear in *Phil. Trans. Roy. Soc. of London.*

[5] Finkelstein, David R. (1996), Quantum Relativity, Springer-Verlag Heidelberg, Berlin.

[6] Grover, L. (1996). *Proc. of 28th Annual ACM Symposium on Theory of Computing.*

[7] Shor, Peter (1994). *Proc. of the 35th Annual Symposium of the Foundation of Computer Science*, Los Alamitos, 124.

[8] Simon, D.R. (1994). *Proc. of the 35th Annual Symposium of the Foundation of Computer Science*, Santa Fe, IVM.

QUANTUM GATES USING MOTIONAL STATES IN AN OPTICAL LATTICE

Eric Charron, Eite Tiesinga, Frederick Mies, Carl Williams
Atomic Physics Division
National Institute of Standard and Technology
Gaithersburg MD 20899, USA

Abstract We study an implementation of a two-qubit universal quantum gate with neutral ^{87}Rb atoms trapped in a far-detuned two-color optical lattice. The qubit states $|0\rangle$ and $|1\rangle$ represent the ground and first excited motional states of an atom in a laser induced potential well. By varying the ratio of the laser intensities of the two colors, the barrier between two neighboring atoms is lowered, creating two-particle entanglement. We predict the duration of operation of a conditional π-phase gate with a time-dependent configuration interaction model.

Keywords: Universal quantum gate, cold atoms, optical lattice, motional states.

Introduction

The rapidly growing interest in quantum information processing [1] has led to the proposal of various physical systems as possible prototypes for qubits and the required multiple-qubit gate operations. The gates require the control of the quantum entanglement of at least two interacting two-state physical systems (the qubits), and must fulfill some very stringent conditions [2]. Trapped ions, cavity QED, nuclear spins (NMR) and cold neutral atoms have long coherence times, and are thus well known candidates. We concentrate here on a scheme where the qubits are the first two motional states of cold ^{87}Rb atoms trapped in a far-detuned optical lattice. Alternative neutral atom schemes use moving potentials [3] or dipole-dipole interaction as the source of entanglement [4, 5].

1. THE PHYSICAL SYSTEM

A 1D optical lattice is constructed from two phase locked far detuned counter propagating laser beams with linear polarization $\hat{e}$, wavelengths $\lambda_f = 10.6\,\mu$m and $\lambda_h = 5.3\,\mu$m, and intensities I_f and I_h (see Figure 1a). A neutral atom moving in this field is subjected to a periodic potential presenting a succession of double minima separated by high barriers, as shown in Figure 1b. We will assume that there are a pair of neutral ^{87}Rb atoms trapped in

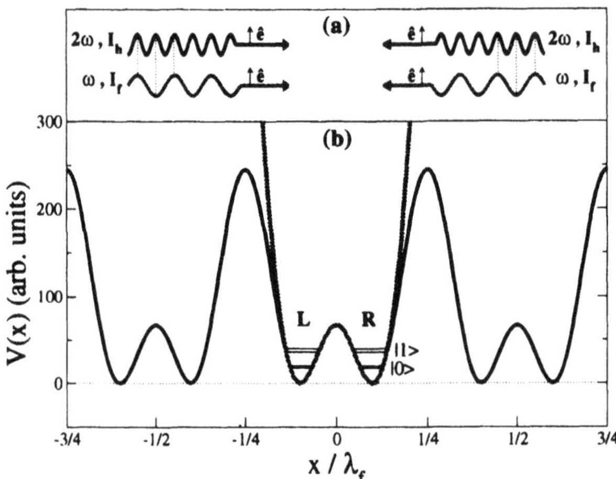

Figure 1 (a) Two-color phase locked linearly polarized counter propagating laser beams of frequency $\omega_f = \omega$ and $\omega_h = 2\omega$ (fundamental and second harmonic) with intensities I_f and I_h, respectively. (b) Black line : periodic double minima optical lattice potential. Grey line : Lower adiabat of a system of two coupled harmonic oscillators showing the same minima, trapping frequency and barrier height as the real optical potential. The first four atomic motional states are shown as thin horizontal lines with a larger tunneling splitting for the higher energy states. For an infinite barrier height, the states in the left (L) and right (R) wells can be labeled $|0\rangle$ and $|1\rangle$.

one of the double minima 1D structures, and that they have negligible interaction with atoms in adjacent double minima. We thus restrict our model to two atoms moving in one dimension (named x), and the trapping potential is defined as the lower adiabat of a system of two coupled harmonic oscillators (see the grey line of Figure 1b). One atom is assumed to be in the left well while the other is located in the right well. For an infinite barrier, the atomic motional states in the left (L) and right (R) wells are well defined. Our qubits will be the ground and first excited states in each well, which we denote as $|0\rangle$ and $|1\rangle$.

Increasing the ratio $\alpha = I_h/I_f$ of the harmonic to fundamental laser intensities increases the barrier separating the two atoms linearly, thereby decreasing their interaction exponentially. This ratio provides a control parameter which allows to switch on and off the interaction between the two qubits and thus make a gate.

2. THEORETICAL DESCRIPTION

The evolution of the time-dependent nuclear wavefunction $\Psi(x_1, x_2, t)$ describing the two atoms is ruled by the Hamiltonian

$$\hat{\mathcal{H}}(x_1, x_2, t) = \hat{h}(x_1, t) + \hat{h}(x_2, t) + V_{int}(|x_2 - x_1|) , \quad (1)$$

where x_1 and x_2 are the positions of the atoms, and V_{int} approximates the Rb - Rb interaction potential. The atomic Hamiltonian $\hat{h}(x, t)$ is

$$\hat{h}(x, t) = -\frac{\hbar^2}{2m}\frac{d^2}{dx^2} + V_{trap}(x, t) , \quad (2)$$

where m is the Rb mass, and $V_{trap}(x, t)$ is the time-dependent trapping potential. We first solve the time-independent Schrödinger equation for the monoatomic Hamiltonian $\hat{h}(x, t = 0)$ and then use the basis set of the non-interacting two-particle states to obtain the eigenstates of the total system $\Psi(x_1, x_2, t = 0)$ from a configuration interaction calculation. The atomic dynamics is then followed in a time-dependent configuration interaction calculation using the latter basis set.

3. CONSTRUCTION OF THE PHASE GATE

In the present far-off-resonance two-color optical lattice, two neighboring atoms are separated by $\lambda_f/4 \approx 2.7\,\mu$m. Single qubit operations can be performed using a near-resonant laser beam focused either on the left or on right well.

Two-qubit gates can be implemented if one lets the motional wavefunctions of adjacent atoms overlap. The atomic interaction then entangles naturally the two particles. This can be realized if the barrier separating the two atoms is lowered, for example with a slow variation of the ratio of the harmonic to fundamental laser intensities $\alpha = I_h/I_f$.

We first used a simple $\sin^2$ variation (see Figure 2a), with $\alpha(t) = \alpha_{max} - (\alpha_{max} - \alpha_{min})\sin^2(\pi t/\tau_{gate})$, and verified that for $\tau_{gate} > 50$ ms the process is adiabatic (> 0.9999) with respect to the initial state. In this case the wavefunction of the two atoms merely ends up with a different phase depending on the initial motional state. In the qubit basis set, this transformation can be written as

$$(|00\rangle, |01\rangle, |10\rangle, |11\rangle) \rightarrow \left(e^{i\varphi_1}|00\rangle\, , e^{i\varphi_2}|01\rangle\, , e^{i\varphi_3}|10\rangle\, , e^{i\varphi_4}|11\rangle\right) . \quad (3)$$

Single qubit operations on the left and right wells can then transform this last vector to $\left(|00\rangle, |01\rangle, |10\rangle, e^{i\varphi}|11\rangle\right)$, with $\varphi = \varphi_4 - \varphi_3 - \varphi_2 + \varphi_1$. If this gate is implemented for a duration $\tau_{gate} \approx 240$ ms, the phase $\varphi = \pi$ is obtained (see Figure 2b), and a universal two-qubit conditional phase gate is realized.

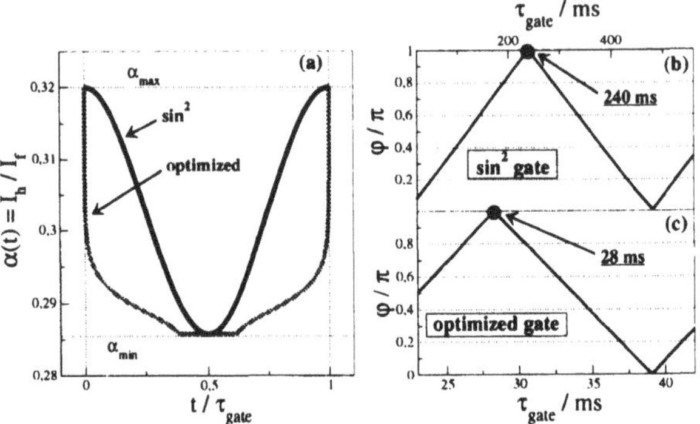

Figure 2 (a) Variation of I_h/I_f as a function of time for a $\sin^2$ gate (black line) and for the optimized gate (grey line). The operation is finished at time $t = \tau_{gate}$. (b) Phase gained in the operation of the $\sin^2$ gate as a function of its duration τ_{gate}. (c) same as (b) for the optimized gate.

During this operation, the phase φ is mainly accumulated when the barrier separating the atoms is at its lowest position, *i.e.* when the interaction is maximum. The $\sin^2$ gate remains in this situation ($\alpha = \alpha_{min}$) a very short amount of time. It can be expected that the duration of the gate can be decreased if the interaction time is maximized. Additionally, because the interaction is negligible at the beginning and at the end of the gate, it is possible to decrease and increase the barrier height at a larger rate at early and final times, without affecting the adiabaticity of the process.

Figure 2a shows an optimized variation of $\alpha(t) = I_h/I_f$, which has been obtained from the analysis of the variation of the eigenstates of the system with the barrier height. This optimized gate allows an order of magnitude improvement in the gate duration, providing a $\varphi = \pi$ phase gate with similar adiabaticity in $\tau_{gate} \approx 28\,\mathrm{ms}$.

Conclusion

In this proposed implementation of a two-qubit phase gate, the duration of operation of the gate is much shorter than the coherence time of the system : $\tau_{gate} \ll \tau_{coh} \approx 1 - 10\,\mathrm{s}$. Additionally, the interaction between the two atoms is easily controllable with the intensities of the trap laser. A drawback of this system is that it is much slower than more complex implementations using dipole-dipole interaction as the source of entanglement [5].

References

[1] A. M. Steane, Rep. Prog. Phys. **61**, 117 (1998).

[2] D. P. DiVincenzo, quant-ph/0002077.

[3] D. Jaksch *et al.*, Phys. Rev. Lett. **82**, 1975 (1999) C. Cabrillo *et al.*, Phys. Rev. A **59**, 1025 (1999). T. Calarco *et al.*, Phys. Rev. A **61**, 022304 (2000). H. J. Briegel *et al.*, J. Mod. Opt. **47**, 415 (2000).

[4] G. K. Brennen *et al.*, Phys. Rev. Lett. **82**, 1060 (1999). G. K. Brennen and I. H. Deutsch, Phys. Rev. A **61**, 062309 (2000).

[5] D. Jaksch *et al.*, Phys. Rev. Lett. **85**, 2208 (2000).

QUANTUM ERROR-CORRECTING CODE FOR BURST ERROR

Shiro Kawabata

Physical Science Division, Electrotechnical Laboratory, 1-1-4 Umezono, Tsukuba, Ibaraki 305-8568, Japan

shiro@etl.go.jp

Keywords: quantum computer, quantum communication, quantum error correction

Abstract We propose a quantum error correcting code for burst error.% By using the quantum interleaver, any quantum burst-errors that have occurred are spread over the interleaved code word, so that we can construct good quantum burst-error correcting codes without increasing the redundancy of the code.

Introduction

The potential use of quantum computer for solving certain classes of problems has recently received a considerable amount of attention. The major obstacle in the building of quantum computers, however, is the coupling of the computer with its environment or the decoherence, which rapidly destroys the quantum superposition at the heart of quantum algorithm. Therefore, an essential element in the realization of such quantum computers is the use of quantum error-correcting codes [1]. The key concept of the quantum error correction is to embed quantum information represented by k qubits into a larger Hilbert space of $n(> k)$ qubits. Previously, many quantum error-correcting codes have been discovered and various theories on the quantum error-correcting codes have also been developed, e.g., Calderbank-Shor-Steane (CSS) coding [2, 3] and stabilizer coding [4].

One important assumption in the theory of quantum error-correction is that qubits in the code word are randomly disturbed when transmitting or storing the states of an n qubit register. As in a classical computer [5], however, errors in quantum computers will not necessarily occur independently [6]. Although we may use the well-known quantum $random$-error-correcting codes to protect qubits from such correlated or $burst$ errors, we must increase the redundancy of the code word, i.e., $(n - k)/n$, with increasing the length of the burst. On the other hand, if details of the decoherence mechanism of qubits are known, then it might be possible to design a more efficient error-correcting method than the conventional scheme. In this paper, we propose a method for constructing quantum burst-error-correcting codes from known quantum error-correcting codes. This method is based on the classical *interleaving* technique [5]. We shall demonstrate that we can construct good burst-error-correcting codes *without increasing the redundancy* of the code word by using the quantum version of the interleaver, i.e., a *quantum interleaver* [7].

1. CLASSICAL INTERLEAVER

In the field of *classical* coding theory, much attention has been given to the correction of burst errors. A burst of length l is defined as a vector whose only nonzero components are among l successive components, the first and last of which are nonzero. A classical binary $\{n, k\}$ code C is said to have burst-error-correcting ability b, if all bursts of length $l \leq b$ are correctable. The simplest method of constructing the burst-error-correcting code is the *interleaving* method [5]. By using this method, we can easily obtain good burst-error-correcting codes without increasing the redundancy. The interleaving of m code words $(C_1, C_2, \cdots, C_m)$ is achieved by constructing the code as a two-dimensional $n \times m$ array of the form:

$$\begin{bmatrix} C_{1,1} & C_{1,2} & \cdots & C_{1,n} \\ C_{2,1} & C_{2,2} & \cdots & C_{2,n} \\ \vdots & \vdots & & \vdots \\ C_{m,1} & C_{m,2} & \cdots & C_{m,n} \end{bmatrix} . \tag{1}$$

Every row of this array is a code word in the original $\{n, k\}$ code. A burst of length bm or less can have, at most, b symbols in any row of the array in eq. (1). Since each row can correct a burst of length b or less, the code can correct all bursts of length bm or less. The parameter m is referred to as the interleaving degree.

2. QUANTUM INTERLEAVER

The *quantum* burst-error-correcting codes are defined naturally. Consider a set of quantum errors such that both bit and phase errors are bursts of length b or less. Then any quantum code that can correct such bursts is called a quantum burst-error-correcting code with burst-error collecting ability b. In order to interleave the quantum codes, we must exchange the qubits one by one, following eq. (1). Therefore, the basic step of the quantum interleaving is a swapping operation for two qubits. The swap operation U_{swap} between any two quantum states can be done by using an array of three controlled not (CNOT) gates, as shown in Fig. 1.

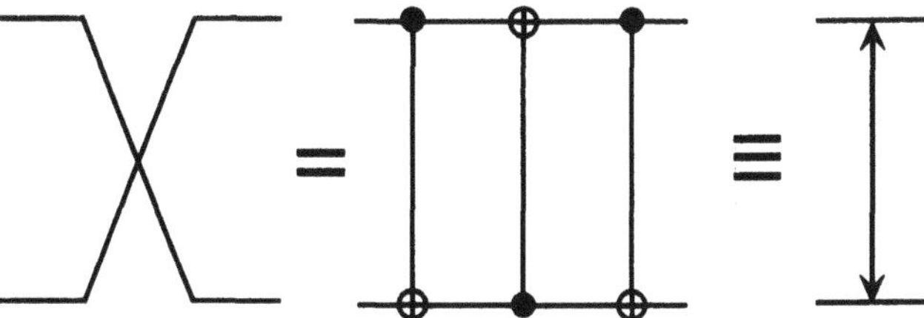

Figure 1 Quantum circuit of the swapping operation which exchanges any two quantum states.

Thus an interleaving operation U_{int} can be written as tensor products of U_{swap}. In the case of $n = m$, the interleaving operation is achieved simply by swapping a pair of qubits i and j for all i and $j(\neq i)$ as $(i, j) \leftrightarrow (j, i)$. Therefore, in this case, the resulting gate array consists of $O(n^2)$ operations. In Fig. 2 the circuit of the quantum interleaver is presented for the case of $n = m = 5$.

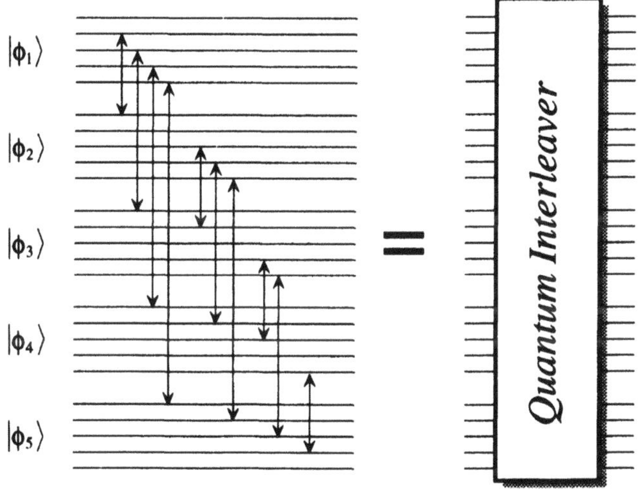

Figure 2 Quantum circuit for the quantum interleaver for the case of $n = 5$ and $m = 5$.

In the case of $n \neq m$, the number of quantum gates is given by $O(nm)$. For example, if we interleave the $\{\{5, 1\}\}$ quantum code [8] with burst-error-correcting ability $b = 1$ to the $m = 5$ degree, then we can obtain the $\{\{25, 5\}\}$ quantum code with burst-error-correcting ability $bm = 5$. Therefore, if there is an $\{\{n, k\}\}$ quantum code capable of correcting all bursts of length b or less, then interleaving this code to m degree produces an $\{\{nm, km\}\}$ quantum code with burst-error-correcting ability bm [7].

3. SUMMARY

In summary, we have proposed a simple method for constructing quantum burst-error-correcting codes without increasing redundant qubits. By using the quantum interleaving method, the quantum code words can be distributed amongst the qubit stream so that consecutive words are never next to each other. On deinterleaving they are returned to their original positions so that any errors that have occurred become widespread. This ensures that any burst (long) errors now appear as random (short) errors.

References

[1] J. Preskill, Phys. Today **52** (1999) No.6, 24.

[2] A.R. Calderbank and P.W. Shor, Phys. Rev. A **54** (1996) 1098.

[3] A.M. Steane, Proc. R. Soc. Lond. A **452** (1996) 2551.

[4] D. Gottesman, Phys. Rev. A **54** (1996) 1862.

[5] W.W. Peterson and E.J. Weldon Jr., *Error-Correcting Codes* (MIT Press, Cambridge, 1972).

[6] G.M. Palma, K.A. Suominen, and A. Ekert: Proc. R. Soc. Lond. A **452** (1996) 567.

[7] S. Kawabata, J. Phys. Soc. Jpn. **69** (2000) No.11.

[8] R. Laflamme, C. Miquel, J.P. Paz, and W.H. Zurek, Phys. Rev. Lett. **77** (1996) 198.

NON-DISSIPATIVE DECOHERENCE IN ION-TRAP QUANTUM COMPUTERS

Stefano Mancini, Rodolfo Bonifacio
INFM, Dipartimento di Fisica, Università di Milano
Via Celoria 16, I-20133 Milano
Italy

Keywords: Foundations, Quantum Information, Decoherence

Abstract We investigate the capabilities of a quantum computer based on cold trapped ions in presence of non-dissipative decoherence. The latter is accounted by using the evolution time as a random variable and then averaging on a properly defined probability distribution. Severe bounds on computational performances are found.

1. INTRODUCTION AND FORMALISM

The realization of a quantum computer in a linear ion-trap [1] was regarded as very promising. Nevertheless, its implementation on a large-scale is recognized to be a technological challenge of unprecedent proportions, mainly due to decoherence problems. Practically, there are two different types of decoherence during the computation: the dissipative one, due to the spontaneous decay of the metastable states, and the non-dissipative one due to random phase fluctuations of various nature. While the first has been considered carefully [2], the second does not received much attention. On the other hand, non-dissipative decoherence seems actually important in trapped ion based experiments [3], whose results have been recently well explained [4] by simply considering the time as a statistical variable and then averaging over a properly defined probability distribution [5]. Hence, here, we present a study [6] of the performances of an ion-trap quantum information processor by using this approach. The latter allows, e.g., to account for the fluctuations of the Raman laser intensity.

Following Ref. [5], the evolution of the system is averaged on a suitable probability distribution, and this leads to the decay of the off-diagonal elements of the density operator

$$\overline{\rho}(t) = \int_0^\infty dt'\, P(t,t')\, \rho(t') ,$$

where $\rho(t') = \exp\{-iLt'\}\rho(0)$ with $L \ldots = [H, \ldots]/\hbar$.

The function $P(t,t')$ has been determined to simply satisfy the semigroup condition for the evolution operator [5]

$$P(t,t') = \frac{1}{\tau}\frac{e^{-t'/\tau}}{\Gamma(t/\tau)}\left(\frac{t'}{\tau}\right)^{(t/\tau)-1} ,$$

where the parameter τ appears as a scaling time. Notice that $\lim_{\tau\to 0} P(t,t') = \delta(t-t')$ and that $P(t,t')$ becomes a Gaussian for $t \gg \tau$.

2. CHARACTERIZATION OF A QUANTUM PROCESS

Quantum computation ideally corresponds to a physical process $|\Psi_{\rm out}\rangle = U|\Psi_{\rm in}\rangle$, where a given input state is mapped to an output state by a unitary transformation U. To characterize such a process we assume the system initially prepared in a pure state

$$|\Psi_{\rm in}\rangle = \sum_{i=0}^{N} c_i\, |i\rangle\,, \qquad \Rightarrow \qquad \rho_{\rm in} = |\Psi_{\rm in}\rangle\langle\Psi_{\rm in}|\,,$$

where $|0\rangle$, $|1\rangle \ldots, |N\rangle$ are orthogonal states of the input Hilbert space with dimension $N+1$. Then, in the ideal case, the output state is

$$|\Psi_{\rm out}\rangle\langle\Psi_{\rm out}| \equiv \rho_{\rm out} = \sum_{i,i'=0}^{N} c_i c_{i'}^*\, R_{i',i}\,,$$

where $R_{i',i} = U|i\rangle\langle i'|U^\dagger$ are system operators not depending on the input state. Whenever the unitary transformation is given by the time evolution operator $U(t)$, the initial state represents the input while the evolved state the output. However, the formalism of Sec.1. implies the replacement

$$R_{i',i}(t) \to \overline{R}_{i',i}(t) = \int_0^\infty dt'\, P(t,t')\, U(t')|i\rangle\langle i'|U^\dagger(t') .$$

Finally, in order to see to what extend the real physical process approaches the ideal one, we use the fidelity

$$\mathcal{F} = {\rm Tr}\left\{\rho(t)\overline{\rho}(t)\right\}_{\rm ave} ,$$

where the subscript "ave" indicates the average overall possible input states [7]. Typically, $\mathcal{F}$ can be expressed in terms of the tensor $F^{i'\,i}_{j'\,j} \equiv \langle j'|U^\dagger \overline{R}_{i'\,i} U|j\rangle$ [6].

Since any computational process can be decomposed into one-bit gates and universal two-bit gates [8], we are now going to consider these two.

One-bit gate

In this case $N = 1$, and $\{|i = 0\rangle \equiv |g\rangle, |i = 1\rangle \equiv |e\rangle\}$, are the electronic states of a single ion. Then, the rotation of such qubit is obtained by a resonant laser pulse, yielding

$$U(t - t_0) = \exp\left[-i\frac{\Omega}{2}\left(|e\rangle\langle g| + |g\rangle\langle e|\right)(t - t_0)\right],$$

where Ω is the Rabi frequency. Then, it is possible to calculate the tensor elements $F_{j'\,j}^{i'\,i}$ to get the fidelity $\mathcal{F}$. In this case it depends on time, which practically determines the amount of rotation, however, we consider a π rotation, i.e. $(t - t_0) = \pi/\Omega$.

Two-bit gate

Here, $N = 3$, and $\{|i = 0\rangle \equiv |g\rangle_1|g\rangle_2, |i = 1\rangle \equiv |g\rangle_1|e\rangle_2, |i = 2\rangle \equiv |e\rangle_1|g\rangle_2, |i = 3\rangle \equiv |e\rangle_1|e\rangle_2\}$, where 1, 2 indicate the two bits and the vibrational ground state $|0\rangle$ also is employed. The universal two-bit gate $|\epsilon_1\rangle_1\,|\epsilon_2\rangle_2 \to (-1)^{\epsilon_1\epsilon_2}|\epsilon_1\rangle_1\,|\epsilon_2\rangle_2$ $(\epsilon_{1,2} = 0, 1)$, can be realized in three steps with laser pulses tuned to the first blue sideband:

I) A π laser pulse excites e.g. the first ion

$$U_I(t_1 - t_0) = \exp\left[-i\frac{\Xi}{2}\left(|e\rangle_1\langle g|a + |g\rangle_1\langle e|a^\dagger\right)(t_1 - t_0)\right], \quad (t_1 - t_0) = \frac{\pi}{\Xi}.$$

II) A 2π laser pulse (with different polarization) excite the second ion

$$U_{II}(t_2 - t_1) = \exp\left[-i\frac{\Xi}{2}\left(|e'\rangle_2\langle g|a + |g\rangle_2\langle e'|a^\dagger\right)(t_2 - t_1)\right], \quad (t_2 - t_1) = \frac{2\pi}{\Xi}.$$

III) A π laser pulse excites again the first ion

$$U_{III}(t_3 - t_2) = \exp\left[-i\frac{\Xi}{2}\left(|e\rangle_1\langle g|a + |g\rangle_1\langle e|a^\dagger\right)(t_3 - t_2)\right], \quad (t_3 - t_2) = \frac{\pi}{\Xi}.$$

Here, $\Xi = \frac{\eta\Omega}{\sqrt{N_a}}$, $a^\dagger$ and a are ladder operators of the CM phonons, η is the Lamb-Dicke parameter, N_a indicates the number of trapped ions and $|e'\rangle$ is an auxiliary electronic level [1]. Then, we can calculate

$$\overline{R}_{i'\,i} = \int_0^\infty dt_3' \int_0^\infty dt_2' \int_0^\infty dt_1' U_{III}(t_3' - t_2)U_{II}(t_2' - t_1)U_I(t_1' - t_0)|i\rangle$$
$$\langle i'|U_I^\dagger(t_1' - t_0)U_{II}^\dagger(t_2' - t_1)U_{III}^\dagger(t_3' - t_2)P(t_3, t_3')P(t_2, t_2')P(t_1, t_1'),$$

used to get the tensor elements $F_{j'\,j}^{i'\,i}$, which give $\mathcal{F}$ (in this case not depending on time).

3. RESULTS AND CONCLUSIONS

The expected probability of error in one quantum gate, for the fault tolerant quantum computation, should be of the order of 10^{-6} [9]. On the other hand, by using the value of $\tau \approx 10^{-8}$ s giving the best fit for the experimental data of Ref. [3] ($\Omega \approx 10^5$ s^{-1}, $\eta \approx 10^{-1}$), we found, for a single bit gate, an accuracy $1 - \mathcal{F} \approx 10^{-3}$, which is quite far from the desired one. In case of universal two-bit gate the accuracy is slightly better due to the presence of the factor $\eta/\sqrt{N_a}$. It turns out that the main limitations rely on the single bit rotation instead on universal two-bit gate. Anyway, based on these results, we can state that quantum information processing on a large scale is unrealistic with the present technology. As a matter of fact, the fault tolerant quantum computation requieres a value $\Omega\tau \approx 10^{-6}$. Since the finite value of τ is essentially related to the Rabi frequency fluctuations [4], this means to improve the laser stability by a factor 10^3 at least!

Furthermore, by using the described approach, it is also possible to demonstrate [6] the impossibility to factorize even a four bit number with the Shor's algorithm.

In conclusion, we have shown that non-dissipative decoherence actually constitutes a serious impediment to realize quantum computer beside dissipative decoherence.

Finally, it is worth noting that the generality of the presented approach suggests in some way the possibility that the parameter τ might have a lower *nonzero* limit, which would pose fundamental limit to quantum computation.

References

[1] J. I. Cirac and P. Zoller, Phys. Rev. Lett. **74**, 4091 (1995).

[2] M. B. Plenio and P. L. Knight, Phys. Rev. A **53**, 2986 (1996).

[3] D. M. Meekhof, *et al.*, Phys. Rev. Lett. **76**, 1796 (1996).

[4] R. Bonifacio, *et al.*, Phys. Rev. A **61**, 053803 (2000).

[5] R. Bonifacio, Il Nuovo Cimento **114 B**, 473 (1999).

[6] S. Mancini and R. Bonifacio, submitted to Phys. Rev. A.

[7] J. F. Poyatos *et al.*, Phys. Rev. Lett. **78**, 390 (1997).

[8] A. Ekert and R. Jozsa, Rev. Mod. Phys. **68**, 733 (1995).

[9] J. Preskill, Proc. R. Soc. Soc. Lond. A **454**, 385 (1988).

OPTICAL QUBIT USING LINEAR ELEMENTS

Matteo G. A. Paris
Quantum Optics Group, Unità INFM and Dipartimento di Fisica "A. Volta"
Università di Pavia, via Bassi 6, I-27100 Pavia, ITALY
paris@unipv.it

Abstract A conditional scheme to prepare optical superposition of the vacuum and one-photon states using linear elements (beam splitters and phase-shifters) and avalanche photodetectors is suggested.

In recent years, the quantum engineering of light have received much attention, which is mainly motivated by the potential improvement offered by quantum mechanics to the manipulation and the transmission of information. In particular, some conditional schemes have been suggested to prepare superpositions. Among these we mention the Fock filtering by an active Fabry-Perot cavity [1], the displacing/photon-adding scheme of Ref. [2], and the so-called optical state truncation [3]. In this paper we describe a partially interferometric conditional scheme to prepare any desired superposition $a_0|0\rangle + a_1|1\rangle$ of the vacuum and one-photon states using only linear optical components and avalanche photodetectors. For a fully interferometric setup and for more details we refer the readers to Ref. [4].

The present scheme (see Fig. 1) is built by a balanced beam splitter, fed by one-photon state in the mode a, followed by a Mach-Zehnder interferometer, with inputs consisting of one of the output from the beam splitter (mode b) and of an additional mode c excited in a weak coherent state. The two modes b and c exiting the interferometer are detected, and the situation in which a click is observed in one mode, and no clicks are seen in the other one, corresponds to the (conditional) preparation of a superposition of the vacuum and one-photon states in the mode a. The amplitudes in the superposition may be tuned by varying the internal phase-shift of the interferometer and the amplitude of the coherent state.

After the first beam splitter, the joint state of the modes a and b is given by the superposition

$$|\psi\rangle_{ab} = \frac{1}{\sqrt{2}} \left[|0\rangle_a |1\rangle_b + |1\rangle_a |0\rangle_b \right] .$$

Quantum Communication, Computing, and Measurement 3
Edited by P. Tombesi and O. Hirota, Kluwer Academic/Plenum Publishers, 2001

Mode b then enters the interferometer where it is mixed with mode c excited in a weak coherent state $|\gamma\rangle$. The evolution operator of the interferometer is given by [5] $\hat{U}(\phi) = \exp\left\{i\phi\left(b^\dagger c + c^\dagger b\right)\right\}$, where $\phi = \theta/2$. The overall output state is thus given by

$$\begin{aligned} |\psi\rangle_{out} = \hat{U}(\phi)\,|\psi\rangle_{ab}|\gamma\rangle_c \;&=\; \frac{1}{\sqrt{2}}\Big[|1\rangle_a|\gamma\cos\phi\rangle_b|\gamma\sin\phi\rangle_c \\ &+\; \sin\phi\,|0\rangle_a\, b^\dagger|\gamma\cos\phi\rangle_b|\gamma\sin\phi\rangle_c \\ &-\; \cos\phi\,|0\rangle_a|\gamma\cos\phi\rangle_b\, c^\dagger|\gamma\sin\phi\rangle_c\Big] \end{aligned} \quad (1)$$

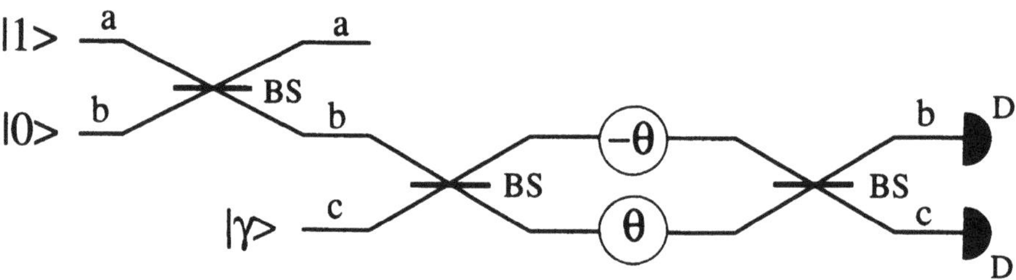

Figure 1 Schematic diagram of the setup for the generation of optical superpositions. The BS's are balanced beam splitters whereas the D's are avalanche photodetectors.

Let us now analyze the effects of the photodetection of modes b and c. The outcomes from an avalanche detector may be either YES, which means a "click", corresponding to any number of photons, or NO, which means that no photons have been recorded. This kind of measurement is described by a two-value POM

$$\hat{\Pi}_N = \sum_{p=0}^{\infty}(1-\eta)^p\,|p\rangle\langle p| \qquad \hat{\Pi}_Y = \hat{\mathbf{I}} - \hat{\Pi}_N\,, \quad (2)$$

where η is the quantum efficiency, and $\hat{\mathbf{I}}$ denotes the identity operator. For high quantum efficiency $\hat{\Pi}_N$ and $\hat{\Pi}_Y$ approach the projection operator onto the vacuum state and the orthogonal subspace respectively. In this case the event of "no clicks" corresponds exactly to the absence of photons. In general, the event of observing a click at the detector surveying the output mode b, and no photons at c, is characterized by the probability

$$P_{YN}[\eta,\gamma,\phi] = \mathrm{Tr}_{abc}\left[|\Psi_{out}\rangle\langle\Psi_{out}|\;\hat{\Pi}_Y\otimes\hat{\Pi}_N\right] = e^{-\eta|\gamma|^2\sin^2\phi}\times$$
$$\left\{1 - e^{-\eta|\gamma|^2\cos^2\phi} + \frac{\eta}{2}\left[e^{-\eta|\gamma|^2\cos^2\phi} + \cos^2\phi(\eta|\gamma|^2\sin^2\phi - 1)\right]\right\}. \quad (3)$$

The corresponding conditional output state is

$$\begin{aligned}\hat{\varrho}_{YN} &= \frac{1}{P_{YN}}\mathrm{Tr}_{bc}\left[|\Psi_{out}\rangle\langle\Psi_{out}|\ \hat{\Pi}_Y \otimes \hat{\Pi}_N\right] = \\ &= \frac{1}{P_{YN}}\left[d_{00}\ |0\rangle\langle 0| + d_{11}\ |1\rangle\langle 1| + d_{01}\ |0\rangle\langle 1| + d_{01}^*\ |1\rangle\langle 0|\right] \end{aligned} \quad (4)$$

where the coefficients are given by

$$d_{11} = \frac{1}{2}e^{-\eta|\gamma|^2\sin^2\phi}\left[1 - e^{-\eta|\gamma|^2\cos^2\phi}\right] \qquad d_{01} = e^{-\eta|\gamma|^2\sin^2\phi}\,\frac{\eta\gamma}{2}\,\sin\phi\cos\phi$$

$$d_{00} = \frac{1}{2}e^{-\eta|\gamma|^2\sin^2\phi}\left[1 - (1-\eta)e^{-\eta|\gamma|^2\cos^2\phi} + \eta\cos^2\phi(\eta|\gamma|^2\sin^2\phi - 1)\right] .$$

The symmetric case of a click observed in the mode c and no clicks in the mode b leads to an equivalent result, up to the replacement $\phi \rightarrow \phi + \pi/2$.

Due to non unit quantum efficiency of photodetectors, the conditional output state $\hat{\varrho}_{YN}$ is not a pure state. However, as we will see, there are regimes in which $\hat{\varrho}_{YN}$ approaches the desired superposition. In order to compare $\hat{\varrho}_{YN}$ with the ideal conditional output $|\psi_{10}\rangle$ we consider the fidelity $F = \langle\psi_{10}|\hat{\varrho}_{yn}|\psi_{10}\rangle$. Notice that ideal output corresponds to a conditional photodetection performed by fully efficient detectors, which are also able to discriminate among the number of photons. From previous equations we have

$$\begin{aligned}F[\eta,\gamma,\phi] &= \frac{1}{2P_{YN}}\,\frac{e^{-\eta|\gamma|^2\sin^2\phi}}{\sin^2\phi + |\gamma|^2\cos^2\phi}\left\{|\gamma|^2\sin^2\phi\left(1 - e^{-\eta|\gamma|^2\cos\phi^2}\right)\right. \\ &\quad +2\eta|\gamma|^2\sin^2\phi\cos^2\phi + \sin^2\phi\left[1 - (1-\eta)e^{-\eta|\gamma|^2\cos^2\phi}\right. \\ &\quad \left.\left. +\eta\cos^2\phi(\eta|\gamma|^2\sin^2\phi - 1)\right]\right\} . \end{aligned} \quad (5)$$

Our aim is to find regimes in which the fidelity of the conditional output state is close to unit and, at the same time, the corresponding detection probability P_{YN} does not vanish.

In Fig. 2 we show $P_{YN}[\eta,\gamma,\phi]$ and $F[\eta,\gamma,\phi]$ as a function of γ and ϕ for quantum efficiency equal to $\eta = 80\%$. As it is apparent from the plot (darker regions correspond to lower values of detection probability and fidelity) for a weak coherent signal $\gamma \leq 1$ there is, even for non unit quantum efficiency of the conditional photodetectors, a large range of values of ϕ corresponding to high fidelity and detection probability ($P_{YN} \simeq 20\%$ for the plotted case). We conclude that the present method is a reliable source of optical superpositions (qubit) employing only linear components and avalanche photodetectors.

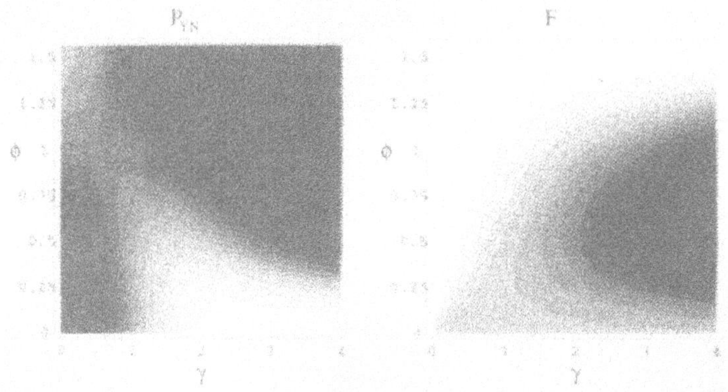

Figure 2 Detection probability and fidelity as a function of the coherent amplitude γ and the interferometric shift ϕ for quantum efficiency equal to $\eta = 80\%$.

Acknowledgments

This work has been cosponsored by C.N.R. and N.A.T.O under the 1999 Advanced Fellowship Program 215.31. The author thanks R. F. Antoni for valuable hints.

References

[1] G. M. D'Ariano, L. Maccone, M. G. A. Paris, M. F. Sacchi, Phys. Rev. A **61** 053817 (2000); Acta Phys. Slov. **49**, 659 (1999).

[2] M. Dakna, J. Clausen, L. Knöll, and D. -G. Welsch, Phys. Rev. A **59**, 1658 (1999); here the effects of non unit quantum efficiency of photodetectors are not taken into account.

[3] D. T. Pegg, L. S. Philips, S. M. Barnett, Phys. Rev. Lett. **81**, 1604 (1998); the authors employ a linear model (not avalanche) for photodetectors, thus overestimating the achievable fidelity.

[4] M. G. A. Paris, Phys. Rev. A **62**, 033815 (2000).

[5] M. G. A. Paris, Phys. Rev. A **59**, 1615 (1999).

ATOM CHIPS

Jörg Schmiedmayer, Ron Folman
Institut für Experimentalphysik, Universität Innsbruck
Physikalisches Institut, Universität Heidelberg
joerg.schmiedmayer@physi.uni-heidelberg.de

Abstract Atoms can be trapped and guided using nano-fabricated wires on surfaces, achieving the scales required by quantum information proposals. Such *Atom Chips* will form the basis for robust and widespread applications of cold atoms ranging from atom optics to fundamental questions in mesoscopic physics, and possibly quantum information systems.

In electronics and light optics, miniaturization of components and integration into networks has lead to new very powerful tools and devices, for example in quantum electronics [1] or integrated optics [2]. It is essential for the success of such designs, that the size of the structures is, at least in one dimension, comparable to the wavelength of the guided wave. Similarly we anticipate that atom optics [3], if brought to the microscopic scale, will give us a powerful tool to combine many atom optical elements into integrated quantum matter wave circuits.

Such microscopic scale atom optics can be realized by bringing cold atoms (de Broglie wavelength $\lambda_{dB} \sim 100$ nm) close to nanostructures [4, 5, 6], which can easily be designed and built using standard nanofabrication techniques as used in the semiconductor industry. Such microscopic potentials for neutral atoms can be created using:

(i) The electric interaction between a neutral, polarizable atom and a charged nanostructure, where the potential is $V_{el} = -\frac{1}{2}\alpha_{el}E^2$.

(ii) The magnetic interaction between the atomic magnetic moment μ and a magnetic field B, described by the potential $V_{mag} = -\vec{\mu} \cdot \vec{B}$.

These potentials can be combined with traditional atom optical elements like atom mirrors and evanescent light fields. Having the atoms trapped or guided, well localized near the surface, will allow integration of atom optics and light optics by way of devices such as light cavities and wave guides fabricated on the surface. Using and combining these techniques a variety of novel atom optical elements can then be constructed at the microscopic scale on such an *Atom Chip*.

Quantum Communication, Computing, and Measurement 3
Edited by P. Tombesi and O. Hirota, Kluwer Academic/Plenum Publishers, 2001

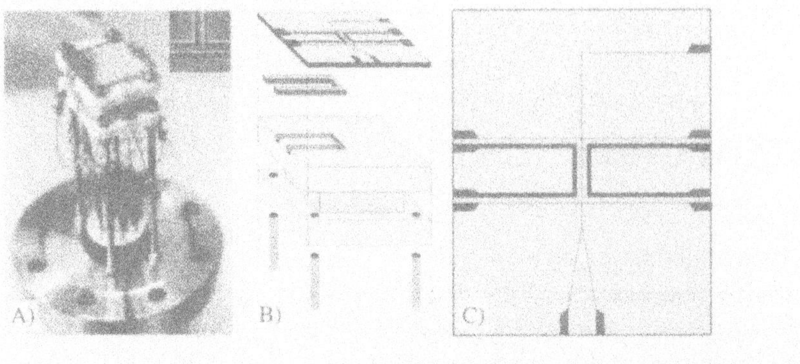

Figure 1 a) The mounted chip before it is introduced into the vacuum chamber. b) A schematic of the chip design. On top there is a GaAs-substrate with a gold layer deposited on it. In this gold layer we etch away gold to define the wires used in the experiment. Under the wafer a U-shaped silver wire is put inside a marcor block. c) Schematics of a typical *Atom Chip* layout.

To study the basic design principles we first performed experiments studying micro manipulation of atoms with small free standing current carrying structures, using thermal Na atoms [7], and later with laser cooled Li atoms from a MOT exploring various guiding and trapping geometries and their scaling laws [8, 9].

Based on the results of these experiments we demonstrated that standard nanofabrication technology can be used to build atom optical elements on such an *Atom Chip* and to miniaturize our wire experiments to allow trapping and guiding of atoms at the mesoscopic scale interesting for quantum information proposals [10]. The *Atom Chip* used in these experiments consists of a $2.5\mu m$ thick gold layer deposited onto a GaAs substrate. This gold layer is then patterned using standard nanofabrication techniques. This leaves the chip as a gold mirror (with $10\mu m$ etchings) which can be used to reflect the laser beams for the MOT during cooling and collecting of atoms. A schematic of such an *Atom Chip* assembly is shown in Fig. 1 and includes a series of magnetic traps to transfer atoms into smaller and smaller potentials: the large U-shaped wires are $200\mu m$ wide and provide a quadrupole potential if combined with a homogeneous bias field, while the thin wires are $10\mu m$ wide.

The atoms are loaded onto the *Atom Chip* using a standard procedure [11, 12]: Typically 10^8 cold ^7Li atoms are accumulated in a "reflection MOT" [13]. Transferring atoms from the MOT to the mesoscopic potentials is carried out in the following steps: Atoms are first transferred into a MOT generated by the quadrupole field of a U-wire located underneath the chip. Then, the laser light is switched off, leaving the atoms confined only by the magnetic quadrupole field of the U-wire. Atoms are then further compressed by increasing the bias field, and transferred into a magnetic trap generated by the two

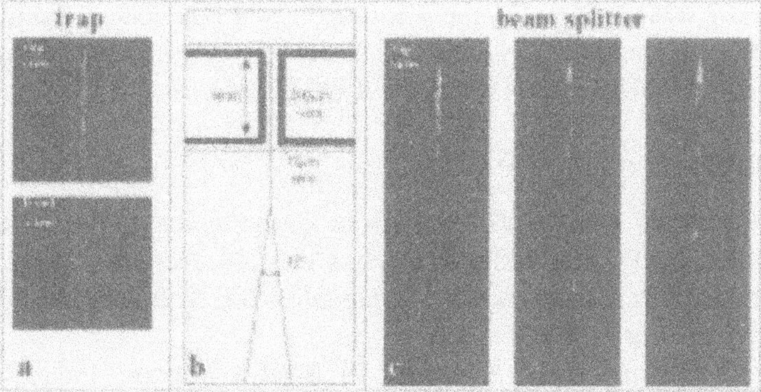

Figure 2 Experiments with an Atom Chip: (a) Atoms trapped in a Joffe-Pritchard style magnetic trap a few μm above the surface. The upper picture is taken from above, the lower one is from the front, and shows besides the atoms also their mirror image, reflected from the *Atom Chip* surface. (b) Schematic of the *Atom Chip* layout used in the experiments. (c) Images of guided atoms in the Y beam splitter. In the first two pictures we drive current only through one side of the Y, therefore guiding atoms either to the left or to the right; in the third the current is divided in equal parts and the guided atoms split into both sides. The pictures of the atoms are obtained by fluorescence imaging using a short laser pulse (typically $0.5\,ms$).

$200\mu m$ wires on the chip, compressed again and transferred into the $10\mu m$ guide. Each compression is easily achieved by first ramping up the current generating the smaller trap and then decreasing the current generating the larger trap to 0 over a time of 10 ms. Typically we transfer 10^6 into microscopic guides or traps with transverse trap frequency of up to 100 kHz and a typical ground state size down to $100nm$, surpassing the parameters required by recent quantum information proposals [10]. We have also designed and experimentally studied a simple beam splitter for guided atoms realized with a current carrying Y-shaped wire (see Fig.2). Such a beam splitter has one accessible input for the atoms, that is the central wire of the Y, and two accessible outputs corresponding to the right and left wires. Depending on how the current in the input wire is sent through the Y, atoms can be directed to the output arms with any desired ratio [14].

Integrating many elements to control atoms onto a single device, an *Atom Chip* will make atomic physics experiments more robust and simple. This will allow complicated tasks in atom manipulation to be performed in a way similar to what integration of electronic elements allowed in the development of new powerful devices. This is of special interest since it has been suggested that the high degree of control achieved over neutral atoms and their weak coupling to the environment (long decoherence time) will allow the realization of Quantum Information Processing (QIP).

This work was supported by the Austrian Science Foundation, the Jubiläums Fonds der Österreichischen Nationalbank, and by of the European Union (IST-1999-11055 ACQUIRE and HPMF-CT-1999-00235).

References

[1] For example: *Quantum Coherence in Mesoscopic Systems*, edited by B. Kramer, NATO ASI Series B: Physics Vol. 254, Plenum (1991).

[2] For example: *Fundamentals of Photonics*, B.E.A. Saleh, M.C. Teich, J. Wiley & Sons (1991).

[3] For an overview see: C.S. Adams, M. Sigel, J. Mlynek, Phys. Rep. **240**, 143 (1994); *Atom Interferometry* Ed.: P.Berman, Academic Press (1997) and references therein;

[4] J. Schmiedmayer, Eur. Phys. J. D **4**, 57 (1998).

[5] E A Hinds, I G Hughes, J. Phys. D: Appl. Phys. **32** 119 (1999)

[6] J.D. Weinstein, K. Libbrecht, Phys. Rev. A. **52**, 4004 (1995); M. Drndic *et al.*, Appl. Phys. Lett. **72**, 2906 (1998); J.H. Thywissen *et al.*, Eur. Phys. J. D **7**, 361 (1999).

[7] J. Schmiedmayer in *IQEC XVIII* Technical Digest, Series 1992, Vol. 9, 284 (1992); Phys. Rev. A **52**, R13 (1995).

[8] J. Denschlag, D. Cassettari, J. Schmiedmayer, Phys. Rev. Lett. **82**, 2014 (1999); J. Denschlag *et al.*, Appl. Phys. B 69 (1999) p. 291.

[9] A. Haase, D. Cassettari, B. Hessmo, J. Schmiedmayer, submitted to Phys. Rev. A (1999).

[10] T. Calarco, D. Jaksch, E.A. Hinds, J.Schmiedmayer, J.I. Cirac, P. Zoller, Phys. Rev. A **61**, 022304 (2000).

[11] R. Folman, P. Krüger, D. Cassettari, B. Hessmo, T. Maier, J. Schmiedmayer, Phys. Rev. Lett. **84**, 4749 (2000); D. Cassettari, *et al.* Appl. Phys. B **70**, 721-730 (2000); For experiments at larger scale see: [12, 15]

[12] J. Reichel, W. Haensel, T.W. Haensch, Phys. Rev. Lett. **83**, 3398 (1999).

[13] K.I. Lee, J.A. Kim, H.R. Noh, W. Jhe, Opt. Lett. **21**, 1177 (1996).

[14] D. Cassettari, B. Hessmo, R. Folman, T. Maier, J. Schmiedmayer, quant-ph/0003135, Phys. Rev. Lett. in print (2000).

[15] J. Fortagh *et al.*, Phys. Rev. Lett. **81**, 5310 (1998); D. Müller *et al.*, Phys. Rev. Lett **83**, 5194 (1999); N. H. Dekker *et al.*, Phys. Rev. Lett **84**, 1124 (2000).

DECOHERENCE AND FIDELITY OF SINGLE QUBIT OPERATIONS IN A SOLID STATE QUANTUM COMPUTER

C. J. Wellard, L. C. L. Hollenberg

School of Physics, University of Melbourne, 3010 Australia

cjw@physics.unimelb.edu.au

Abstract

We examine a stochastic noise process that causes phase damping of qubits in the quantum register of the solid state quantum computer proposed by Kane [1], and consider its effects on the fidelity of the single qubit operations of this QC.

1. INTRODUCTION

For a quantum computer to operate successfully it is vital that the evolution of the qubits is not only coherent, but precisely known to the operator. In this paper we investigate a situation in which the signals driving the qubit operations contain stochastic white noise. To investigate the effect this has on the operation of the computer, we determine how an ensemble of qubits evolve under the influence of this noise, and calculate its effect on the fidelity of the single qubit operations necessary for quantum computation. We find that the noise has a decohering effect on the average evolution, more precisely the process is dephasing for qubits simply in the quantum register but becomes depolarizing for qubits undergoing single qubit operations.

2. THE KANE QC

Kane has proposed a solid state quantum computer in which the qubits are spin $\frac{1}{2}$ ^{31}P nuclei in a Si substrate [1, 2]. In this proposal the Zeeman energy of the qubits can be tuned by application of a voltage biased A gate. The Hamiltonian of a qubit in the quantum register is then $H_{reg} = \gamma B_z \sigma^z$. B_z is a background magnetic field, $\gamma = \gamma_0 - \frac{\eta \hbar V}{B_z}$, where V is the applied A gate bias and $\eta = 5\pi \times 10^7 Hz$. Single qubit operations can be implemented selectively on a particular qubit by tuning γ so that it is in resonance with an oscillating transverse magnetic field B_{ac}. The Hamiltonian that produces these rotations, in the interaction picture, is then $H_{op} = B_{ac}\gamma_0\sigma^y$, where are using rotations about the y axis for illustrative purposes.

3. STOCHASTIC NOISE

We are going to consider the effect of a stochastic white noise in the applied A-gate voltage, that is we write the signal $V(t) = V_0(1 + \Delta(t))$, where $\Delta(t)$ describes a white noise process [3]. We write $\Delta(t)dt = \sqrt{\lambda}dW(t)$, where $dW(t)$ is the Weiner increment, and $\sqrt{\lambda}$ scales the noise. We can calculate λ by integrating over the duration of the voltage pulse to give the pulse area $A(t) = V_0(t) + \Delta A(t)$. Here $\Delta A(t)$ is a Gaussian random variable with a mean of zero and a variance $V_0^2\lambda t$.

4. FIDELITY OF THE QUANTUM REGISTER

To calculate the effect of the noise on the fidelity of the quantum register we rewrite the register Hamiltonian as $H = B_z\gamma(1+\xi(t))\sigma^z$, where $\xi(t)$ gives the stochastic fluctuations in the Hamiltonian and we define $\xi(t)dt = \sqrt{\epsilon}dW(t)$, where $\epsilon = (\frac{\eta\hbar V_0}{\gamma B_z})^2\lambda$. We transform to a frame rotating at the nuclear precession frequency and form the Ito stochastic Schrödinger equation for the density operator

$$\begin{aligned} d\tilde{\rho}(t) &= \frac{B_z\gamma}{i\hbar}\sqrt{\epsilon}dW(t)[\sigma^z,\tilde{\rho}(t)] \\ &- \frac{\epsilon B_z^2\gamma^2}{2\hbar^2}[\sigma^z,[\sigma^z,\tilde{\rho}(t)]]dt. \end{aligned} \tag{1}$$

This can be solved for the components of the Polarisation vector defined by $\tilde{\rho}(t) = \frac{1}{2}(1 + \vec{P}(t).\vec{\sigma})$ to give

$$\begin{aligned} P_z(t) &= P_z(0), \\ P_{x,y}(t) &= \exp[\frac{-2B_z^2\gamma^2\epsilon t}{\hbar^2}]P_{x,y}(0). \end{aligned} \tag{2}$$

We can see from this that the x and y components, which denote the coherence of the state decay exponentially while the z component remains unaltered. Thus we lose phase coherence while the population probabilities are unaltered, this is termed a dephasing process.

We can now calculate the fidelity of the operation, that is the average overlap of the state obtained via the noisy evolution, with the state we would obtain via noiseless evolution,

$$F = \frac{1}{2}(1 + P_z(0)^2 + (P_x(0)^2 + P_y(0)^2)\exp[\frac{-2B_z^2\gamma^2\epsilon t}{\hbar^2}]). \tag{3}$$

We see that the fidelity depends on the input state, because the process is only dephasing if the initial state contains no coherence then it cannot decohere. The worst case for the register then is if the input state is a maximum superposition in which case the fidelity is given by

$$F = \frac{1}{2}(1 + \exp[\frac{-2B_z^2\gamma^2\epsilon t}{\hbar^2}]). \quad (4)$$

5. FIDELITY OF SINGLE QUBIT OPERATIONS

We proceed the same way to calculate the effect of the noise on the single qubit operations, in this case the Hamiltonian becomes

$$\tilde{H_{op}} = \xi(t)B_z\gamma\sigma^z + B_{ac}\gamma_0\sigma^y, \quad (5)$$

and we are left to solve the Ito stochastic equation

$$\begin{aligned} d\tilde{\rho}(t) &= \frac{B_{ac}\gamma_0}{i\hbar}[\sigma^y, \tilde{\rho}(t)]dt + \frac{B_z\gamma\sqrt{\epsilon}dW(t)}{i\hbar}[\sigma^z, \tilde{\rho}(t)] \\ &- \frac{\epsilon B_z^2\gamma^2 dt}{2\hbar^2}[\sigma^z, [\sigma^z, \tilde{\rho}(t)]]. \end{aligned} \quad (6)$$

We solve this for the average Polarisation vector components, in the limit that the decoherence time is much greater than the time it takes to perform the operation, and find

$$\begin{aligned} P_x(t) &= \exp[\frac{-B_z^2\gamma^2\epsilon t}{\hbar^2}]\{\cos(\frac{2B_{ac}\gamma_0 t}{\hbar})P_x(0) + \sin(\frac{2B_{ac}\gamma_0 t}{\hbar})P_z(0)\}, \\ P_y(t) &= \exp[\frac{-2B_z^2\gamma^2\epsilon t}{\hbar^2}]P_y(0), \\ P_z(t) &= \exp[\frac{-B_z^2\gamma^2\epsilon t}{\hbar^2}]\{\cos(\frac{2B_{ac}\gamma_0 t}{\hbar})P_z(0) - \sin(\frac{2B_{ac}\gamma_0 t}{\hbar})P_x(0)\}. \end{aligned} \quad (7)$$

This shows that the process is depolarizing, the polarisation vector decays exponentially to zero regardless of the input state. We calculate the fidelity of this operation and find

$$F = \frac{1}{2}(1 + \exp[\frac{-B_z^2\gamma^2\epsilon t}{\hbar^2}]). \quad (8)$$

This fidelity function holds regardless of the initial state. Thus the fidelity of the single particle operations is better than that of the quantum register for highly coherent states.

6. TOLERANCE TO NOISE

To calculate the level of noise in the voltage signal that the quantum computer can tolerate in performing a typical single particle operation, the Hadamard gate we refer to Preskills [4] limit that for a quantum computer to factorize large numbers faster than a conventional computer requires an error probability of $\delta < 10^{-6}$ per operation. Using the operating parameters prescribed by Kane [1]; $B_z = 2T$ and $B_{ac} = 0.001T$ and an A-gate bias of $1V$, a value at the top of the bias range. We use the fidelity function for single particle operations, given by eq(8) to get the condition $\frac{\tau_{op}}{\tau_{dec}} < 4 \times 10^{-6}$, where $\tau_{op} = (\pi\hbar)/(4B_{ac}|\gamma_0|)$ is the time it takes to perform a Hadamard operation, and $\tau_{dec} = \hbar^2/(2B_z^2\gamma^2\epsilon)$ is the characteristic decoherence time. This imposes a bound on the ratio of the rms fluctuations in the pulse area to the mean pulse area for the A gate bias of $\Delta A(t)_{rms}/\bar{A}(t) < 6.2 \times 10^{-7}$.

Acknowledgment

CJW would like to acknowledge the support of an Australian Postgraduate Award, and a Melbourne University Postgraduate Abroad Scholarship as well as that of the Max-Planck-Institut für Kernphysik. He would also like to acknowledge valuable discussions with B.H.J.McKellar, and the theory group at the University of Queensland. LCLH wishes to thank the Alexander Von Humboldt foundation and the Max-Planck-Institut für Kernphysik.

References

[1] B.E.Kane, Nature **393**, 133 (1998)

[2] H.Goan and G.Milburn, unpublished manuscript (2000)

[3] C.W.Gardiner, *Handbook of Stochastic Methods*, (Springer, Berlin, 1985)

[4] J.Preskill, Proc. R. Soc. Lond. A **454**, 793 (1996)

IV

CRYPTOGRAPHY

LONG DISTANCE ENTANGLED STATE QUANTUM KEY DISTRIBUTION

Grégoire Ribordy, Nicolas Gisin, Hugo Zbinden
Gap-Optique, Université de Genève, 20 rue de l'Ecole-de-Médecine, 1211 Genève 4, Switzerland
gregoire.ribordy@physics.unige.ch

Abstract A quantum key distribution system based on photon pairs entangled in energy-time and optimized for long distance transmission is presented. It is based on a Franson arrangement for monitoring quantum correlations, and uses a protocol analogous to BB84. Passive state preparation is implemented by polarization multiplexing in the interferometers. We distributed a sifted key of 0.4 Mbits at a raw rate of 134 Hz and with an error rate of 8.6% over a distance of 8.5 kilometers.

1. INTRODUCTION

Quantum key distribution (QKD), the most advanced application of the new field of quantum information theory, offers the possibility for two remote parties - Alice and Bob - to securely exchange a cryptographic key. It was invented by Bennett and Brassard [1]. Although systems relying on faint laser pulses have already been thoroughly tested (see [2, 3, 4, 5] for recent experiments), the first experiments with entangled photon pairs are recent [6, 7, 8].

QKD systems relying on photon pairs offer two advantages over those using faint laser pulses (typically 0.1 photon per pulse). With these systems, most pulses are empty. Since Bob does not know which contain a photon, he must always "open" his detectors, increasing thus the overall error probability. For a given attenuation, photon pairs systems feature thus a lower QBER, since the empty pulses can be suppressed. Second, photon pairs systems allow passive state preparation by Alice. In this case, even when two (or more) photon pairs are created within a gate time, one photon of a pair is not at all correlated with a photon of another pair. The so-called photon number splitting attack [9, 10, 11], which has an important impact on the security of faint pulses systems, is thus not possible with systems using entangled photons and passive state preparation. This constitutes an important advantage.

In this paper, we present a system for QKD with entangled photon pairs optimized for long distance operation. Because of the large spectral width of entangled photons, decoherence can become critical. Energy-time entanglement was selected, because it is more robust than polarization entanglement. The system is based on an asymmetric Franson arrangement [12]. Alice has both the photon pair source and an unbalanced interferometer, while Bob has only an interferometer. The photon that Alice keeps has a wavelength of 810 nm. This allows her to use commercial silicon avalanche photodiode (APD) modules and to benefit from their high efficiency and low noise. On the other hand, the photon sent to Bob has a wavelength of 1550 nm, to minimize the attenuation in the optical fiber. Bob must thus use InGaAs APD's, whose performance are not as good and which require gated operation. When monitoring coincidences between Alice's and Bob's detections, one observes three peaks in the time domain. The central one corresponds to cases where both photons chose the same arm in the interferometers. They are indistinguishable, provided that the coherence length of the pump photon is larger than the path difference, and yield thus interference. The correlation probability depends non-locally on the sum of the phases applied by Alice and Bob in their interferometers. Ekert *et al.* proposed in [13] to use these quantum correlations to implement a protocol analogous to BB84 and distribute a key. Alice and Bob must apply randomly one of two possible phase shifts on each photon, and record in which port the photon was detected.

2. IMPLEMENTATION OF THE SYSTEM

The photon pairs source is made up of a pump laser, a beam shaping and delivery optical system, a non-linear crystal and two optical collection systems (see Fig. 1). It is realized with bulk optics. The pump laser emits 100 mW

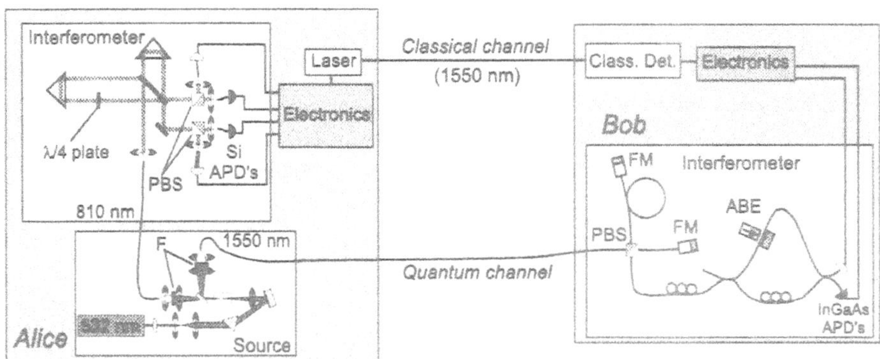

Figure 1 Asymmetric system for quantum key distribution utilizing photon pairs (PBS: polarizing beamsplitter, FM: Faraday mirror, ABE: adjustable birefringent element, F: spectral filters).

of single mode light at 532 nm, with a spectral width smaller than 10 kHz and a high frequency stability (drift smaller than 50 MHz per 10 minutes), yielding high visibility two-photons interference. The pump beam is focused on a $KNbO_3$ non-linear crystal cut with a θ-angle of 22.95°, which allows collinear down-conversion at 810 nm and 1550 nm at room temperature. The down-converted photons are then separated by a dichroic mirror aligned at 45° incidence and focused onto the cores of singlemode fibers. To characterize this source, the short wavelength output port is connected to a silicon APD and the long wavelength one to a gated InGaAs APD. We obtained a value of approximately 1.1 MHz for the single counting rate on the silicon detector. The probability to have a correlated photon at 1550 nm coupled into the fiber, knowing that one at 810 nm had been detected, was 64%. The probability to detect a photon belonging to another pair was 1%.

Alice's interferometer is realized with bulk optics, to avoid interference visibility reduction induced by chromatic dispersion. It basically consists of a symmetric beamsplitter and two trombone prisms. Instead of using a rapid phase modulator for applying the random phase, a passive scheme was devised. A birefringent element is inserted in the long arm of the interferometer, inducing a phase shift of $\pi/2$ between the horizontal and vertical polarization modes. This amounts to multiplexing two interferometers in polarization. Polarizing beamsplitters (PBS) in the output ports, are used to determine the phase experienced by a given photon. Four detectors are then necessary. When a circularly or 45°-linearly polarized photon enters such a device, it decides upon incidence on the PBS whether it experienced a phase difference of ϕ_A or $\phi_A + \frac{\pi}{2}$. The path difference in the interferometer is larger than the coherence length of the down-converted photons, to avoid single photon interference. In addition, it must be possible to observe well separated coincidence peaks in the time domain, to be able to distinguish interfering and non interfering events. Since the jitter of the detectors is as high as 800 ps, a path difference of $2 \times 0.5 = 1$ m in air was chosen.This distance must be kept stable within a fraction of a wavelength during a QKD session. The interferometer was thus placed in an insulated box, whose temperature was kept stable within 0.01°C. The length of Alice's long arm can be varied with a piezoelectric element to align the path difference of both interferometers within a fraction of a wavelength.

Bob's interferometer is similar to Alice's, except that it is implemented with optical fibers. It is realized with two 3 dB couplers connected to each other. The long arm consists of dispersion shifted fiber with minimal dispersion close to 1550 nm, to avoid visibility reduction. The optical path difference is also 1 m. A fiber loop polarization controller ensures identical polarization state transformation for both paths. Birefringence is produced by applying a constant strain on a 5 mm long uncoated section of the long arm. The two polarizations corresponding to both measurement bases are separated before the injection of the photons in the interferometer. This information is then transformed into a detection time window. A fiber optic PBS and two Faraday mirrors introduce a delay of about 200 ns between the two polarization states. As the photons are depolarized after travelling through the optical fiber line connecting Alice and Bob, the measurement basis is chosen randomly. A polarizer aligned at 45°or a polarization scrambler must however be placed in front of the PBS, to prevent Eve from forcing detection in a particular basis. The overall attenuation of Bob's interferometer is -5.2 dB. It is also placed in an insulated box. The two detectors used by Bob are InGaAs APD's. They are operated at a temperature of -60°C in gated mode [14]. Their quantum detection efficiency is approximately 9% with a thermal noise probability per gate (2 ns) of $2 \cdot 10^{-5}$. Afterpulses can be neglected.

A classical channel, providing a timing pulse for gating Bob's detectors is implemented using a second optical fiber. In order to test the system without implementing key distillation, Alice sends to Bob along this reference pulse information about the event she registered. Bob uses then an electronic circuit to compare it with his result and evaluate the quantum bit error rate (QBER).

3. EXPERIMENTAL RESULTS

Before performing QKD, the visibility of the two-photon interference fringes was measured and maximized by scanning the phase of Alice's interferometer with the piezoelectric element. A value of 91.8% $\pm$ 0.8% was obtained, after subtraction of the noise counts. Note that this measurement basically amounts at performing a Bell inequality test.

The phase was then adjusted to minimize the QBER, and a short fiber of 20 m with essentially no attenuation was connected between Alice and Bob. Nevertheless, they were located in two different rooms in order to simulate remote operation. We obtained a raw key distribution rate (after sifting, but before distillation) of 450 Hz, and a minimum QBER of 4.7% $\pm$ 0.3%. The key distribution session lasted 63 minutes and allowed the transmission of 1.7 Mbits. The average error rate was 5.9%, because of slight variations in the relative phase difference in the interferometers induced by temperature drifts. It is also possible to estimate the net rate (after distillation) using the formula presented in [5] to 178 bits per second, readily usable for encryption. In the case of the system presented here, the QBER can be separated in three contributions. The first one, $QBER_{opt}$, comes from non maximal interference visibility. It can be estimated from the fringe measurement to 4.1%. The second one, $QBER_{acc}$, is caused by accidental coincidences between uncorrelated pairs, and is estimated to 0.8%. Finally, the last one, $QBER_{det}$, corresponds to the error caused by detector noise and is evaluated to 1%. This contribution is the only one which increases with the attenuation. It limits thus the transmission distance.

An 8.45 km-long optical fiber spool was then connected between Alice and Bob. In order to avoid a reduction of the interference visibility caused by chromatic dispersion spreading, dispersion shifted fiber was selected. It featured an overall attenuation of 4.7 dB. Mode field diameter mismatch between standard and dispersion shifted fiber caused high connection losses of 1.3 dB. The attenuation was 0.25 dB / km at 1550 nm. Key was exchanged during 51 minutes at a raw rate of 134 Hz, producing 0.41 Mbits. The average QBER was 8.6% and the minimum QBER 6.6% ± 0.6%. The net rate is estimated to 32 Hz. The values of $QBER_{opt}$ (4.1%) and $QBER_{acc}$ (1.0%) remained unchanged, as expected, while $QBER_{det}$ increased to 3%. These contributions sum up to 8.1%, again slightly above the measured minimum value.

Fig. 2 depicts the QBER as a function of the attenuation of the fiber link. It shows the experimental minimum (crosses) and average (circles) values ob-

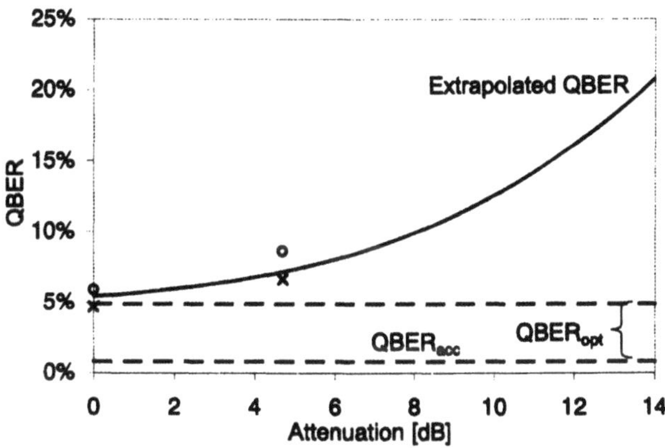

Figure 2 Experimental values of $QBER_{min}$ (circles) and $QBER_{average}$ (crosses), and extrapolation of the QBER (continuous line). The two contributions that do not depend on the distance ($QBER_{opt}$ and $QBER_{acc}$) are also shown (broken lines).

tained with and without the spool connected. The contributions $QBER_{acc}$ and $QBER_{opt}$, independent on the attenuation , are represented by the dashed lines. These results can be extrapolated to take into account the effect of additional attenuation (solid line). Assuming an attenuation coefficient of 0.25 dB / km, a QBER of 10% would be obtained with an attenuation of approximately 8.5 dB, corresponding to a fiber length of 24 km (0.25 dB / km and two connections with 1.3 dB).

4. CONCLUSION

In this article, we presented a QKD system exploiting photon pairs optimized for long distance operation. Contrary to faint laser pulses systems, it is immune to the photon number splitting attack thanks to passive state preparation. We must however acknowledge the fact that its operation is unquestionably more complex than that of the self-aligned system using faint pulses [5]. The main difficulty is the need to stabilize both interferometers. In practice, an increase in the security of a system is always accompanied by an increase in cost and complexity. Recent works suggest to distinguish three levels of security. First, a system – like the one presented in this contribution – could be immune to all attacks performed with existing, as well as future technology, including photon number splitting. Its cost and complexity may however be too high for real applications. Second, one can consider systems based on faint pulses. They are immune to existing technology, but not in all cases to future technological developments. On the other hand, they are easy to operate. Finally, one can use public key cryptography, which is considered to offer sufficient security, when implemented with suitable key length. It is also convenient to apply, as it does not require a dedicated channel, and has been in use for many years. Nevertheless, it is vulnerable to future developments in computer power, while all QKD systems can only be attacked with technology existing at the time of the key exchange. In addition, the security of public key cryptography can be jeopardized overnight by theoretical advances. In this event, QKD with faint pulses would constitute the only realistic replacement technology.

Acknowledgments

The Swiss FNRS and OFES as well as the European QuCom project (IST-1999-10033) have supported this work.

References

[1] C. Bennett and G. Brassard, in Proc. of IEEE Inter. Conf. on Computers, Systems and Signal Processing, Bangalore, (IEEE, New York, 1984), p. 175 .

[2] P. Townsend, Opt. Fiber Tech. **4**, 345 (1998).

[3] R. Hughes, G. Morgan, and C. Peterson, J. of Mod. Opt. **47**, 533 (2000).

[4] J.-M. Mérolla, Y. Mazurenko, J.-P. Goedgebuer, and W. Rhodes, Phys. Rev. Lett. **82**, 1656 (1999).

[5] G. Ribordy, J.-D. Gautier, N. Gisin, O. Guinnard, and H. Zbinden, J. of Mod. Opt. **47**, 517 (2000).

[6] W. Tittel, J. Brendel, H. Zbinden, and N. Gisin, Phys. Rev. Lett **84**, 4737 (2000).

[7] D. Naik, C. Peterson, A. White, A. Berglund, and P. Kwiat, Phys. Rev. Lett. **84**, 4733 (2000).

[8] T. Jennewein, C. Simon, G. Weihs, H. Weinfurter, and A. Zeilinger, Phys. Rev. Lett. **84**, 4729 (2000).

[9] B. Huttner, N. Imoto, N. Gisin, and T. Mor, Phys. Rev. A **51**, 1863 (1995).

[10] H. Yuen, Quantum Semiclassic. Opt. **8**, 939 (1996).

[11] G. Brassard, N. Lütkenhaus, T. Mor, and B. Sanders, Phys. Rev. Lett **85**, 1330 (2000).

[12] J. D. Franson, Phys. Rev. Lett. **62**, 2205 (1989).

[13] A. Ekert, J. Rarity, P. Tapster, and M. Palma, Phys. Rev.Lett. **69**, 1293 (1992).

[14] G. Ribordy, J.-D. Gautier, H. Zbinden, and N. Gisin, Appl. Opt. **37**, 2272 (1998).

BUNCHING AND ANTIBUNCHING FROM SINGLE NV COLOR CENTERS IN DIAMOND

A. Beveratos, R. Brouri, J.-P. Poizat, P. Grangier
Laboratoire Charles Fabry de l'Institut d'Optique, UMR 8501 du CNRS,
B.P. 147, F91403 Orsay Cedex - France
jean-philippe.poizat@iota.u-psud.fr

Abstract

We investigate correlations between fluorescence photons emitted by single N-V centers in diamond with respect to the optical excitation power. The autocorrelation function shows clear photon antibunching at short times, proving the uniqueness of the emitting center. We also report on a photon bunching effect, which involves a trapping level. An analysis using rate-equations for the populations of the N-V center levels shows the intensity dependence of the rate equation coefficients.

Introduction

Quantum cryptography relies on the fact that single quantum states can not be cloned. In this way coding information on single photons would ensure a secure transmission of encryption keys. First attempts for quantum cryptography systems [1, 2] were based on attenuated laser pulses. Such sources can actually produce isolated single photons, but the poissonian distribution of the photon number does not guarantee both the uniqueness of the emitted photon and a high bit-rate. Therefore a key milestone for efficient and secure quantum cryptography systems is the development of single photon sources.

Several pioneering experiments have already been realized in order to obtain single photon sources. Among these attempts one can cite twin-photons experiments [3, 4], coulomb blockade of electrons in quantum confined heterojunctions [5], or fluorescence emission from single molecules [6, 7]. Up to now interesting results were obtained, but considerable work is still needed to design a reliable system, working at room temperature with a good stability and well-controlled emission properties. More recently several systems have

Quantum Communication, Computing, and Measurement 3
Edited by P. Tombesi and O. Hirota, Kluwer Academic/Plenum Publishers, 2001

been proven to be possible candidates for single photon sources. Antibunching was indeed observed in CdSe nanospheres [8], thereby proving the purely quantum nature of the light emitted by these sources. In the same way photon antibunching in Nitrogen-Vacancy (N-V) colored centers in diamond was reported by our group [9] and by Kurtsiefer et al [10]. A remarkable property of these centers is that they do not photobleach at room temperature: the fluorescence level remains unchanged after several hours of continuous laser irradiation of a single center in the saturation regime. These centers are therefore very promising candidates for single photon sources.

In this paper we present further investigation of the fluorescence light emitted by NV colored centers. We analyze the system dynamic through the measurement of the autocorrelation function, showing the existence of a shelving effect which reduces the counting rate of the fluorescence emission, and gives rise to photon bunching [11]. We study the dependence of this behavior on the pumping power, and we compare the experimental results with the predictions of a three-level model using rate equations.

1. EXPERIMENTAL SETUP

Our experimental set-up is based on a home-made scanning confocal microscope.

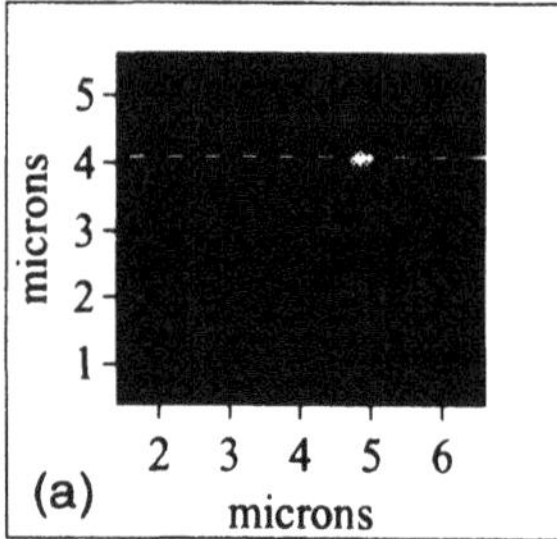

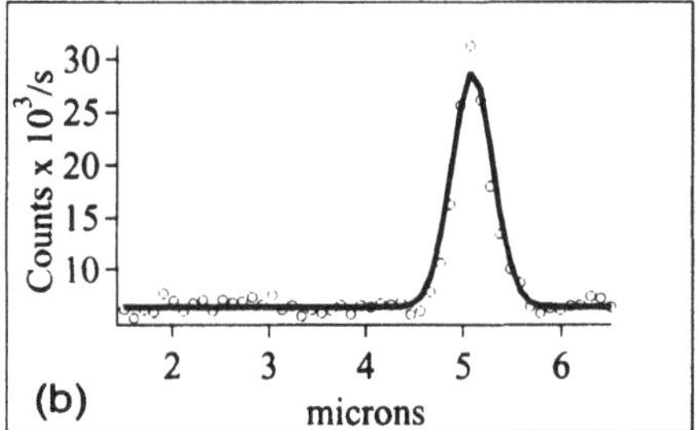

Figure 1 (a) Confocal microscopy raster scan ($5 \times 5\ \mu m^2$) of the sample performed about 10 μm below the diamond surface. The size of a pixel is 100 nm. The integration time per pixel is 32 ms. The laser intensity impinging on the sample is 15 mW . (b) Line scan along the dotted line. The data is shown together with a gaussian fit, which is used to evaluate the signal and background levels (right). Here we obtain $\rho = S/(S+B) = 0.81$

A CW frequency doubled Nd:YAG laser ($\lambda = 532nm$) is focused on the sample by a high numerical aperture (1.3) immersion objective. A PZT-mounted mirror allows a x-y scan of the sample, and a fine z-scan is obtained using another PZT.

The sample is a $0.1 \times 1.5 \times 1.5mm^3$[110] crystal of synthetic Ib diamond from Drukker International. The Nitrogen-Vacancy centers consist in a substitutional nitrogen with an adjacent vacancy, and are found with a density of about 1 μm^{-3} in Ib diamond. The centers can be seen with a signal to background ratio of about 5 by scanning the sample as shown in figure 1, and a computer-controlled servo-loop allows to focus on one center for hours.

The fluorescence is collected by the same objective, and separated from excitation light by a dichroic mirror. High rejection (10^{10}) high-pass filters remove any leftover pump light. Spatial filtering is achieved by focusing on a $50\mu m$ pinhole. The fluorescence is then analyzed by an ordinary Hanbury-Brown and Twiss set-up. The time delay between the two photons is converted by a time-to-amplitude converter (TAC) into a voltage which is digitalized by a computer board.

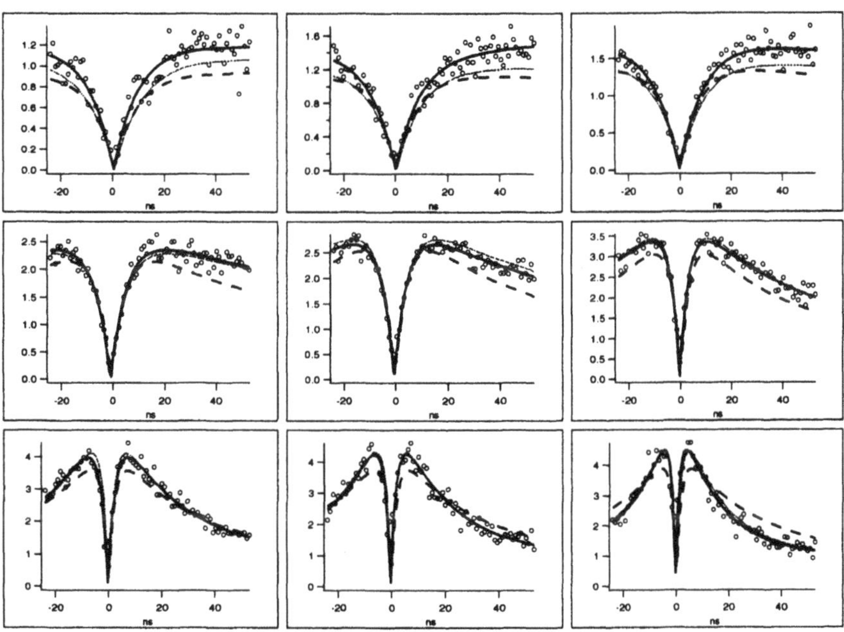

Figure 2 Experimental data of $g^{(2)}(\tau)$ for different values from 0.3 to 31 mW of pumping power. The solid line represents the fit. The dashed (resp. thin) lines are calculations of $g^{(2)}(\tau)$ for k_{23} and k_{32} (see figure 3 for definition) not depending (resp. depending) on pump power. Details are given in the text of section 2.

In a low counting regime, this system allows us to record the autocorrelation function: $g^{(2)}(\tau) = \frac{<I(t)I(t+\tau)>}{<I(t)>^2}$.

Experimental data were corrected from background noise, and normalized to the coincidence number corresponding to a poissonian source of equivalent power [9]. Different correlation functions obtained for several pumping power are shown in figure 2. These functions are centered around $\tau = 0$, but one can notice slight drifts attributed to thermal drift of our acquisition electronics. For $\tau = 0$, $g^{(2)}(\tau)$ goes clearly down to zero, which is the signature of a single emitting dipole. On the other hand, $g^{(2)}(\tau)$ goes beyond 1 at longer times, and then decays to 1. This behavior is an evidence of the presence of a trapping level. To correctly describe the system we therefore considered a 3-level system.

2. THEORETICAL BACKGROUND AND DISCUSSION

Owing to the fast damping of coherences, we use rate equations. Let us consider the 3-level scheme described in figure 3. The evolution of the populations

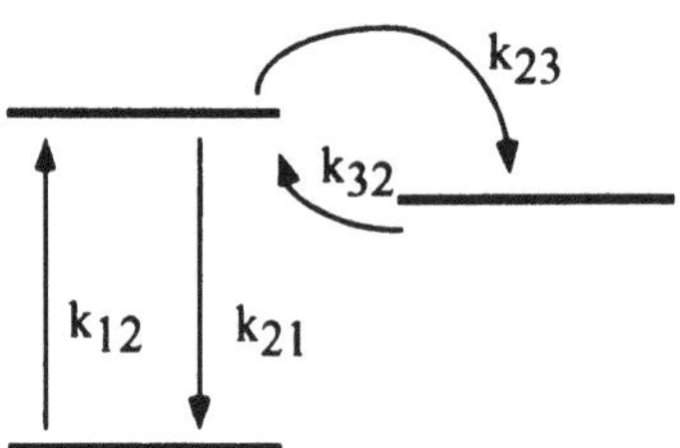

Figure 3 Three level system as used for modelisation.

are therefore given by:

$$\frac{d}{dt}\begin{pmatrix}\sigma_1\\ \sigma_2\\ \sigma_3\end{pmatrix} = \begin{pmatrix}k_{12} & k_{21} & 0\\ k_{12} & -k_{21}-k_{23} & k_{32}\\ 0 & k_{23} & -k_{32}\end{pmatrix}\begin{pmatrix}\sigma_1\\ \sigma_2\\ \sigma_3\end{pmatrix} \tag{1}$$

with the initial condition $\sigma_1 = 1$, $\sigma_2 = \sigma_3 = 0$ at $t = 0$, which means that a photon has just been emitted and the system is therefore prepared in its ground state. The decay rate from level 3 to 1 is neglected [12].

By analytically solving equations 1, one can derive

$$g^{(2)}(\tau) = \frac{\sigma_2(\tau)}{\sigma_2(\infty)} = 1 - \frac{1+g_e}{2}\exp\left(\frac{k_{tm}+k_{1m}}{2}\tau\right) - \frac{1-g_e}{2}\exp\left(-\frac{k_{tm}-k_{1m}}{2}\tau\right) \tag{2}$$

with the stationary population:

$$\sigma_2(\infty) = \sigma_{2\infty} = \frac{k_{12}k_{32}}{k_{12}k_{23} + k_{12}k_{32} + k_{21}k_{32}} \tag{3}$$

and with

$$k_{tm} = k_{12} + k_{21} + k_{23} + k_{32} \tag{4}$$

$$k_{1m}^2 = \left((k_{12} + k_{21} - k_{23} - k_{32})^2 + 4k_{21}k_{23}\right) \tag{5}$$

$$g_e = \frac{2k_{12}k_{23} + k_{32}\left(k_{12} + k_{21} - k_{23} - k_{32}\right)}{k_{1m}k_{32}} \tag{6}$$

We thus have four equations which enables us to express the four rates k_{ij} with respect to the experimental variables $g_e, k_{tm}, k_{1m}, \sigma_{2\infty}$. By fitting the experimental values of $g^{(2)}$ with expression 2 we obtain the values of g_e, k_{tm} and k_{1m} for each value of the intensity. The last value needed in order to solve the system concerns the parameter $\sigma_{2\infty}$ which is directly linked to the count rate

$$N = \eta \times k_{21} \times \sigma_{2\infty} \tag{7}$$

We assume that k_{21} doesn't depend on the pump power, and we set η to the value for which this condition is satisfied. We found $\eta = 3 \times 10^{-3}$ which is in perfect agreement with our estimated value [9]. The parameters k_{ij} are plotted as functions of pumping power in figure 4.

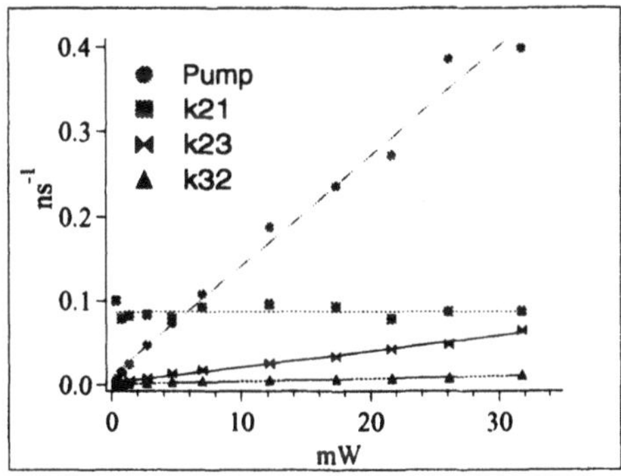

Figure 4 Evolution of k_{ij} with respect to the pump power.

Note that k_{12} is, as expected, a linear function of the pumping power, and that k_{21}^{-1} has a constant value of 11.6ns, which corresponds to results reported in literature [13].

What is noteworthy here is that k_{23} and k_{32} linearly depend on pump power, with k_{23} greater than k_{32}, which means that at high pumping power the system tends to be shelved into the third level. This intensity-dependent effect has not been reported yet. In Figure 2, three plots were superimposed to the experimental results. The solid line represents the best fit. The dashed line represents the result of equation 2 with values of k_{23} and k_{32} independent on intensity taken from reference [10]. One can see in this last case that the agreement between calculated and experimental value is not fully satisfactory. Finally, the thin line is obtained by using the values of k_{23} and k_{32} given by the linear fit of figure 4. Agreement with the experimental results is good.

This intensity dependence of k_{23} and k_{32} can also be observed on the total photon counts. In fact the number of photons emitted per second should decrease as the trapping in the metastable state increases. Figure 5 shows the

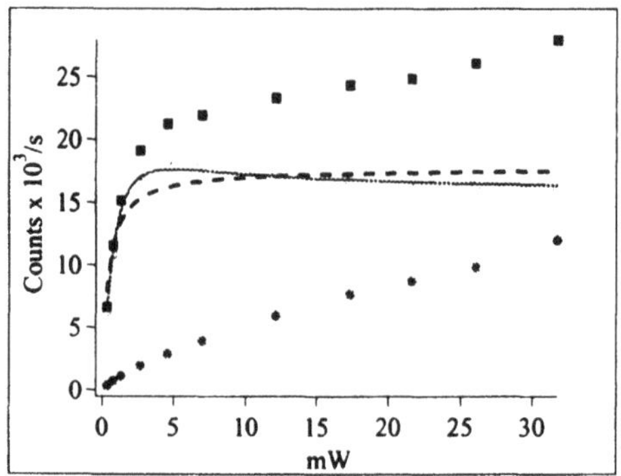

Figure 5 Saturation behaviour of the N-V center. The solid line gives the fit by using our model, and the dashed line the results for normal saturation behaviour. Values are plotted in respect to pumping power. The intensity may be deduced from the focused beam size, which in our case is not accurately known due to spherical aberrations.

count rate of the single N-V center as a function of the pump power. We clearly see a decrease in the photon counts. We can fit with a good agreement our experimental data using equations 7 and 3.

We have repeated this experiment with the $532nm$ Nd:YAG line and the $514nm$ Argon line and found similar linear dependency of the k_{23} and k_{32} rates with the pump power.

Conclusion

We have measured the autocorrelation function of the fluorescence light emitted from a single NV colored center in diamond using a confocal microscope. Our results are in very good agreement with what can be expected from rate equations in a 3-level scheme, provided that a linear dependence of some rate coefficients with pumping power is assumed. Such an effect has not been reported yet, and must be taken into account when designing a pulsed single-photon source.

References

[1] A. Muller, J. Breguet, and N. Gisin, Europhysics Letters, **23** (**6**), 383 (1993),

[2] P. D. Townsend, J. G. Rarity, and P. R. Tapster, Electronics Letters, **29** (**7**), 634 - 635 (1993)

[3] P. Grangier, G. Roger and A. Aspect, Europhysics Lett. **1**, 173 (1986).

[4] C. K. Hong and L. Mandel, Phys. Rev. Lett. **56**, 58 (1986).

[5] J. Kim, O. Benson, H. Kan, and Y. Yamamoto, Nature **397**, 500 (1999)

[6] F. De Martini, G. Di Giuseppe, and M. Marrocco, Phys . Rev. Lett. **76**, 900 (1996).

[7] C. Brunel, B. Lounis, P. Tamarat, and M. Orrit, Phys. Rev. Lett. **83**, 2722 (1999).

[8] P. Michler, A. Imamoglu; MD. Mason, P. J. Carson, G.F. Strouse, S.K. Buratto, Nature **406**, 968 (2000)

[9] R. Brouri, A. Beveratos, J.-Ph. Poizat, Ph. Grangier, Opt. Lett. **25(17)**, 1294 (2000).

[10] C. Kurtsiefer, S. Mayer, P. Zarda, and Harald Weinfurter, Phys. Rev. Lett. **85(2)**, 290 (2000).

[11] J. Bernard, L. Fleury, H. Talon, M. Orrit, Journal of Chemical Physics **98** (**2**), 850 (1993).

[12] A. Dräbenstedt, L. Fleury, C Tietz, F. Jelezko, S. Kilin, A. Nizovtev, and J. Wrachtrup, Phys. Rev. B **60**, 11503 (1999).

[13] A. T. Collins, M. F. Thomaz, and M. I. B. Jorge, J. Phys. C: Solid State Phys. **16**, 2177 (1983).

VIOLATION OF LOCALITY AND SELF-CHECKING SOURCE: A BRIEF ACCOUNT

Dominic Mayers
NEC Research Institute, Princeton, NJ, USA
mayers@research.nj.nec.com

Christian Tourenne
Maharishi University of Management, Fairfield, IA, USA

Keywords: Bell Inequalities, Quantum Cryptography.

Abstract In 1991 Ekert proposed to use Bell inequalities in the so called E91 quantum key distribution protocol. This was the first alternative to the well known BB84 protocol of Bennett and Brassard. In 1992, Bennett, Brassard and Mermin explained that the E91 protocol is no more secure than the original BB84 protocol which do not use Bell inequalities at all. So, apparently, violation of locality was not useful in quantum cryptography. In 1998 Mayers and Yao restored back violation of locality in quantum cryptography with the concept of a *self-checking source*, a source of Bell states which is provided together with testing devices. The test is designed such that, if passed, the source and the testing devices are guaranteed to be identical modulo some isomorphism to the original specification. We discuss the self-checking source of Mayers and Yao, how it is related to the E91 protocol and the fair sampling assumption which was first used to address the detection loophole in Bell inequalities experiments.

Introduction

A self-checking source allows the end-user to test at the same time the source of quantum states and the testing devices which can be provided by the manufactuer together with the source. The approach is interesting because, making use of violation of locality, the testing devices are self-tested. In particular, the test should work even if we have no guarantee that the testing devices are built to the original specification.

We denote $|\theta\rangle$ the state of a photon polarised at angle θ. The source has specification to emit two photons, which have respective state spaces H_1, H_2, in the Bell state $\Phi^+ = (|0\rangle^{H_1} \otimes |0\rangle^{H_2} + |\pi/2\rangle^{H_1} \otimes |\pi/2\rangle^{H_2})/\sqrt{2}$. Each H_i-photon is sent to a different measuring apparatus located inside a Faraday cage which then receive a random classical inputs $a_i \in \{0, \pi/8, \pi/4\}$ that is used to represent a measurement basis. We use Faraday cage metaphorically to mean a blocker of all signals. Each Faraday cage closes just after that the photon entered into the cage. We may also use relativity to guarantee this separation assumption. The H_i-photons are respectively measured in the selected bases $\{\, |a_i\rangle,\ |a_i + \pi/2\rangle \,\}$ and the classical outcomes x_i are noted.

The projections associated with these measurements on H_1 and H_2 are respectively denoted $L_{a_1}^{x_1}$ and $R_{a_2}^{x_2}$. The separation assumption means that we can use the fact that the measurements operator $L_{a_1}^{x_1}$ and $R_{a_2}^{x_2}$ associated with these two distinct measuring apparatus always commute and are only functions of (a_1, x_1) and (a_2, x_2) respectively. The separation assumption is also needed after the measurement to guarantee the privacy of the generated key, but this is another issue. If we only worry about testing the source, the separation assumption is only needed until after the test is executed. This separation assumption is only an assumption on the device or mechanism that is used to isolate the measuring apparatus; it is not a direct assumption on the measuring apparatus.

Let $p_a(x)$ be the probability of the pair of outcomes $x = (x_1, x_2)$ given the pair of measurements $a = (a_1, a_2)$ when the system respects perfectly the original specification. Let $\tilde{p}_a(x)$ be the corresponding probabilities for the actual system which might not be built to the orginal specification. The Mayers-Yao self-checking Bell state theorem states that any setting such that $\tilde{p}_a(x) = p_a(x)$ is neccessarily identical modulo some isomorphism (see journal version of [7]) to the original specification. The basic idea behind this result is that we only consider the classical data a_1, a_2, x_1, x_2 outside the cages to prove that the source and the apparatus inside the cages are secure.

If the actual system is not built to the original specification, we might have $\tilde{p}_a(x) \neq p_a(x)$. A test could eventually be executed to check how close the probabilities $\tilde{p}_a(x)$ are to the ideal case $p_a(x)$. The theorem of Mayers and Yao [7] is not a direct statement about such a test. However, using an obvious continuity argument, the theorem suggests that such a test could be useful to obtain a provably secure protocol with no or little assumptions on the measuring apparatus. Perhaps a few other assumptions in addition to the separation assumption will be required to deal with the test. This question is discussed in section 3.

There are different ways in which this self-checking source could be connected to a security proof. In particular, it can be used in the BB84 protocol in two different ways. In one way, the two measuring apparatus in our self-checking apparatus are on Alice's side. This corresponds to the original idea of a self-checking source entirely located on Alice's side to play the role of a (self-checked) BB84 source. The other approach is that the two measuring apparatus are respectively located on Alice's side and on Bob's side. The connection with the E91 protocol is discussed in sections 1. The connection with violation of locality is discussed in section 2.

1. CONNECTION WITH THE E91 PROTOCOL

The issue of imperfect apparatus in security proofs was considered partially in some papers such as [8]. In these proofs, it was implicit that the honest participants should test the quantum apparatus to verify that these apparatus respect the original specification, or else they must trust the manufacturer. The principle that was used in these tests was just the same principle that experimentalists use all the time to convince themselves that their apparatus works properly. Certainly, using violation of locality to test the quantum apparatus was not the idea at the time.

In 1991, without questionning the context that is described above, Artur Ekert proposed a protocol that was based on violation of a Bell inequality [4]. The E91 protocol corresponds to our self-checking source with one measuring apparatus on Alice's side and the other on Bob's side, except that $a_1 \in \{0, \pi/8, \pi/4\}$, $a_2 \in \{\pi/8, \pi/4, 3\pi/8\}$ and there is a test on the correlation coefficients. The test only uses the EPR pairs with $a_1 \neq a_2$, and Alice and Bob share a bit when $a_1 = a_2$. We would like to compare Ekert's idea behind his E91 protocol and the idea behind a self-checking source. In his analysis Ekert considered a strong attack in which Eve controls the EPR source. This is only a technique used in the analysis. In reality, a protocol cannot count on a cheater to operate the EPR source. The assumption that the EPR source is controled by Eve is a natural technique to address the untrusted source issue. However, the connection between this technique and the untrusted EPR source issue was missing in Ekert's paper. Techniques which give more power to the cheater are often used to simplify the theoretical analysis whithout any practical issue in mind such as an untrusted EPR source. Ekert remained silent on the untrusted EPR source issue even after the advantage of his protocol was under discussion [6]. In any case, giving control over some part of the entire apparatus (i.e. the EPR source) to Eve would only have been a partial solution.

In the self-checking approach, except for the location of each measuring apparatus and the time at which they receive their input, we assume that Eve controls both the EPR source and the measuring apparatus. The main tool used is violation of locality in the proposed setting. It is pure coincidence that Ekert also considered violation of locality. As a matter of fact, in the context of perfect apparatus which was the context in Ekert's paper, it was mentioned [6] that actually Ekert's protocol does not need violation of locality and is no more secure than the BB84 protocol. The BBBM92 analysis [6] uses the well known variation on the BB84 protocol in which the BB84 source is implemented by Alice via an EPR source and a BB84 measurement on one photon, the other photon being sent to Bob in one of the BB84 states as it should be. They explained (as it is well known now) that this variation is equivalent to the original BB84 protocol. We shall call this variation the BBM92 protocol[1]. The equivalence between the BB84 and the BBM92 protocols requires that the apparatus are perfect. Again, we see that imperfect apparatus was not the issue in these papers because otherwise the equivalence between the BB84 and BBM92 protocols used in [6] would have been easily rejected by any interested party, including the authors of [6] themselves.

The approach that we explain here brings back violation of locality in quantum cryptography. Recently, two papers reported experimental work that was based upon Ekert's protocol (or a similar protocol) [9, 10]. The purpose of [9] was to address the imperfect apparatus issue. They cite Ekert for his protocol. However, nobody knows if really Ekert's protocol fulfulls the objective (concerning imperfect apparatus). As we explained above, the basic idea used in [9] was not suggested by Ekert. A proof that this protocol addresses the imperfect apparatus issue is still missing, but [7] not only provides the basic idea but also the essential ingredient for a proof. Similarly, the work of [10] fails to realise that the work of [6] strongly suggests that violation of a Bell inequality does not help in quantum cryptography unless the purpose is to verify the quantum apparatus. In the context of trusted quantum apparatus, violation of Bell like inequalities does not provide better security. Moreover, it didn't help as a theoretical tool to obtain a better bound on Eve's information. It is not clear whether or not a violation of a Bell inequality, especially if it is only sligthly violated, implies that a private key can be generated. Therefore, it is not sufficient to detect the violation of a Bell inequality to claim that a protocol is secure.

The self-checking source approach was first proposed in the context of the unconditional security of quantum key distribution (QKD) obtained by Mayers [1] using a variation on the BB84 protocol [2]. However, the source had to be perfect [1], that is, the proof uses the fact that the source emits a single photon per signal with the exact polarisation angle specifed in the protocol. The self-checking source approach takes care of this last problem. We believe that this approach can be used together with reasonable assumptions (see section 3) in order to become practical and adapted to the current technology as in [9, 10].

2. VIOLATION OF LOCALITY

Note that the use of entanglement as a way to approximate a single photon source is well known, but no one has proposed to use Bell inequalities at this level before [7]. Recently, the old belief that parametric down conversion should be useful with regard to the imperfect apparatus issue was analysed [14]. The approach used in [14] does not take advantage of violation of locality in quantum mechanics, and thus is not as powerful as our approach.

In contrast, the self-checking approach uses violation of locality on top of entanglement as in [9, 10] to address globally the imperfect apparatus issue, not only the multiple photons per signal issue. It addresses both the imperfect source and the imperfect measuring apparatus issues in one single approach.

We do not claim that we will elucidate the concept of violation of locality in only a few paragraphs. However, we will explain in which way our approach is related to an essential aspect of this concept. Suppose that we want to reproduce the distributions of probability $p_a(x)$ with the help of a classical source. We can think that the classical source is a computer that writes the same random string on two tapes which are sent to two computers (classical or quantum) located far apart one to the other at different locations. Each classical tape replaces a photon emitted by the quantum source. The programs used by the classical source and the two computers can be anything. We say that the probabilities $p_a(x)$ do not violate locality if x_1 can be a function of a_1 and the random tape only and x_2 can be a function of a_2 and the random tape only. In such a case, only the content of the random tape would need to be sent to pass the test. Clearly, if the source can be classical and pass the test, no security is guarantee by such a test. Therefore we need violation of locality to prove that a source is secure.

An example. Consider a Φ^+ source again, but now the test uses only the bases 0 and $\pi/4$. It is not hard to see that the corresponding probabilities $p_a(x)$ can be generated by a classical source. The classical source generates two independent uniformly distributed random bits x_0 and $x_{\pi/4}$ associated with the two bases 0 and $\pi/4$ respectively. A copy of these two bits is sent to the measuring apparatus which pick the bit associated with their respective selected basis. Therefore, we know that we cannot use only the bases 0 and $\pi/4$ to design a self-checking source.

3. ON THE NECESSITY OF ASSUMPTIONS

Here we discuss two assumptions: the separation assumption and the fair sampling assumption [11]. We begin by the separation assumption, the possibility to isolate an area from the outside world, a fundamental assumption required in cryptography, quantum or classical. Typically, one tries to make this assumption looks reasonable via an appropriate implementation of the protocol, whereas the theoretical protocol remains the same. This draws a line between theory and practice and creates a clean situation for the theory side. However, as illustrated by the recent method of Kocher to break into the newest version of smart cards [13], the separation assumption (at least in the case of smart cards) is problematic; so we don't want to add other assumptions of this kind.

Our theorem suggests that, at least in principle, no extra physical assumption might be needed for quantum cryptography. This is not obvious apriori. For example, a bound on the number of photons per signal emited by a source would seem a neccessary assumption. Even if a detector is used in the protocol to evaluate how many photons per pulse are transmitted, the detector can be defectuous and detects correctly one photon and miss the others. Other possible defects might also occur. For example, information can also leak in some unused mode of the photons such as the frequency or in the time delay between two signals. Things get worst if the cheater, in order to induce such a leak, transmits strong signals inside the source using the normal path but in the opposite direction. The protocol might be secure if the original design is respected, but it might not be secure if a defective system is used or if the manufacturer used a variation on the original design.

There is also the possibility of a malicious design. In classical cryptography we could in principle use a simple (easily checked) device that will make sure that a requested computation is done by many different and independent computers. This is not possible for a computation on a quantum input. Therefore, quantum information processing require a fundamentally different level of trust which must be addressed by a new approach such as the one we propose. The interest of our approach is exactly to minimise the trust needed on the systems used.

Now, we consider the fair sampling assumption. The fair sampling assumption was found useful in the context of violation of locality [11]. A variation on this assumption was also found useful in quantum cryptography in the context of a lossy channel or bad detection rate [1]. Which variation on this assumption is required depends upon the application considered. The fair sampling assumption must at the least include that the probability of detection, i.e. of getting an outcome $x = (x_1, x_2)$, is independent of the bases used. In particular, no information is coded in the signal to determine in view of the bases used whether or not the signal will be detected. It is not too hard to check that without the above condition, in the context of a very lossy channel or very bad detection rate, the cheater can essentially picks his strategy in view of the bases chosen by Alice and Bob. There is no way one can design a secure protocol in such a context. A fair sampling assumption is needed in the BB84 protocol in the context of a lossy channel or efficient detectors, even if we have a perfect source [1].

Here, we conjecture that the self-checking theorem of Mayers-Yao [7] can be extended to the context that is mentioned above using a variation on the fair sampling assumption [11]. On the other hand, we think that a fair sampling assumption will not be required if the detection rate is sufficiently close to one, not necessarily asymptotically close to one (as a security parameter n increases). In particular, we should not need to use quantum error-correction.

Conclusion

We have explained in which way the imperfect source or detectors issue was related to Ekert's work, though the connection was disregarded by Ekert himself at the time. We also proposed a connection between 1) the fair sampling assumption [11] and 2) a possible extension of the self-checking theorem in the same context. It would also be interesting to consider the alternative "no enhancement assumption" [11, 12]. The self-checking theorem was proven in [7]. However, it only applies to the context in which we assume $p_a(x) = \tilde{p}_a(x)$. Recently, the result was extended to GHZ states [15] which is a significant improvement because we believe that violation of locality in GHZ states is more powerful for self-checking purpose in the non exact context $p_a(x) \approx \tilde{p}_a(x)$.

Acknowledgment

The author acknowledge fruitful discussions with Charles Bennett, Artur Ekert (even if it was brief) and Hitoshi Inamori.

Notes

1. Unfortunately, the BBM92 protocol is often confused with the E91 protocol.

References

[1] D. Mayers, "Quantum key distribution and string oblivious transfer in noisy channel", *Advances in Cryptology: Proceedings of Crypto'96*, Lecture Notes in Comp. Sci., vol 1109, (Springer-Verlag, 1996), pp. 343 – 357; D. Mayers, "Unconditional security in quantum cryptography", Los Alamos preprint archive quant-ph/9802025 (1998).

[2] C. H. Bennett and G. Brassard, "Quantum cryptography: Public key distribution and coin tossing", *Proceedings of IEEE International Conference on Computers, Systems and Signal Processing*, Bangalore, India, December 1984, pp. 175 – 179.

[3] Bennett, C. H., Brassard, G. and Robert, J.-M., "Privacy amplification by public discussion", SIAM Journal on Computing, vol. 17, no. 2, April 1988, pp. 210 - 229.

[4] A.K. Ekert, Quantum cryptography based on Bell's theorem, *Physical Review Letters*, vol. 67, no. 6, 5 August 1991, pp. 661 – 663.

[5] G. Brassard, N. Lütkenhaus, T. Mor and B.C. Sanders, "Security Aspects of Practical Quantum Cryptography", Los Alamos preprint archive

quant-ph/9911054 (1999).

[6] C.H. Bennett, G. Brassard and N.D. Mermin, "Quantum Cryptography without Bell's Theorem", *Physical Review Letters*, vol. 68, no. 5, 3 February 1992, pp. 557 – 559.

[7] D. Mayers and A. Yao, "Quantum Cryptography with Imperfect Apparatus", Proceedings of the 39th IEEE Conference on Foundations of Computer Science, 1998; Journal version in progress.

[8] C.H. Bennett, F. Bessette, G. Brassard, L. Salvail and J. Smolin, "Experimental quantum cryptography", *Journal of Cryptology*, vol. 5, no 1, 1992, pp. 3 – 28.

[9] D.S. Naik, C.G. Peterson, A.G. White, A.J. Berglund, P.G. Kwiat, "Entangled state quantum cryptography: Eavesdropping on the Ekert protocol", PRL (tentatively accepted); also in Los Alamos preprint archive quant-ph/9912105 (December 1999).

[10] T. Jennewein, C. Simon, G. Weihs, H. Weinfurter, A. Zeilinger, "Quantum Cryptography with Entangled Photons", PRL (tentatively accepted); also in Los Alamos preprint archive quant-ph/9912117 (December 1999).

[11] M. Freyberger, P. K. Aravind, M. A. Horne, and A. Shimony Phys. Rev. A 53, 1232-1244 (1996);

[12] J.F. Clauser, M.A. Horne, Phys. Rev. D 10, 526 (1974).

[13] Paul Kocher, Joshua Jaffe and Benjamin Jun, "Differential Power Analysis: Leaking Secrets", *Advances in Cryptology: Proceedings of Crypto '99*, Lecture Notes in Comp. Sci., vol 1666, Springer-Verlag, 1996.

[14] G. Brassard, T. Mor and B.C. Sanders, "Quantum cryptography via parametric downconversion", Los Alamos preprint archive quant-ph/9906074 (1999).

[15] Dominic Mayers, Yoshie Kohno, Yoshihiro Nambu and Akihisa Tomita, "A self-checking GHZ source" (manuscript).

THE UNCONDITIONAL SECURITY OF QUANTUM KEY DISTRIBUTION

Alice, Bob, and Eve in Quantumland

Tal Mor
Electrical Engineering
College of Judea and Samaria
Ariel, Israel
talmo@cs.technion.ac.il

Vwani Roychowdhury
Electrical Engineering
University of California at Los Angeles
Los Angeles, Cal. USA
vwani@ee.ucla.edu

Keywords: Quantum Key Distribution, Cryptography, Security

Abstract Recently Boykin, Biham, Boyer, Mor, and Roychowdhury (BBBMR) proved the security of standard quantum key distribution against the most general attacks which can be performed on the channel [1]. The attack is performed by an eavesdropper who has unlimited computation abilities, and who uses the full power allowed by the rules of classical and quantum physics. The final key can then be used to transmit secure messages in a way that their security is also unaffected in the future. We explain here some of the subtleties related to the term "proving security". We also present a simplified protocol, and we explain the role it plays in proving the security of the standard Bennett-Brassard protocol for quantum key distribution.

Quantum key distribution [2] uses the power of quantum mechanics to suggest the distribution of a key that is secure against an adversary with unlimited computation power. Such a task is beyond the ability of classical information processing; thus, it is the main success of quantum cryptography. The extra power gained by the use of quantum bits (quantum two-level systems) is due to the fact that the state of such a system cannot be cloned. In contrast, the security of conventional key distribution is based on the (unproven) existence of various one-way functions, and mainly on the difficulty of factoring large numbers, a problem which is assumed to be difficult for a classical computer, but has been proven to be easy for a hypothetical quantum computer.

Various proofs of security have previously been obtained against collective attacks [3], leading finally to a full proof of security against all collective attacks [4]. We continued this line of research in BBBMR [1] to prove the ultimate security (that is, the unconditional security) of quantum key distribution (QKD), against any attack.

In our proof, the eavesdropper (Eve) is assumed to have unlimited technology, e.g., a quantum memory and a quantum computer, while the legitimate users (Alice and Bob) use practical tools, or more precisely, simplifications of practical tools. The main goal of quantum cryptography, and one of the most important goals of quantum information processing, is to obtain a full proof of the security of a practical (or soon to be practical) QKD scheme against an unlimited adversary. Such a definition of QKD is vital, since the aim of the invention of quantum key distribution, and of quantum cryptography in general, is to obtain a *practical* scheme, which is proven secure against *any* attack, even an attack which is far from being practical with current technology.

We believe that a proof of security hence must include: a.— A precise description of the protocol which is proven secure b.— A precise description of the most general attack c.— A precise numerical bound on the eavesdropper information (and/or the probability of success in her attack) d.— A precise calculation of the complexity of the protocol.

Asymptotic proofs can also be given, i.e., proofs which assume that the number of bits goes to infinity, but are of limited importance. They are obviously simpler since they do not deal with the exact calculations of security as a function of the number of bits used in the protocol. These proofs are still important since they can potentially be extended to provide a regular proof of security.

As an alternative to the standard QKD, one could suggest an Abstract QKD (AQKD) where practicality is not a required property. In such an AQKD the legitimate users can have the same power as the eavesdropper/adversary. Relaxing the definition of QKD this way has some advantages since it enables simpler proofs of security. Such an AQKD scheme, once proven secure, still demonstrates the advantage of quantum information theory over classical information theory. However, such a scheme allows no practical applications or implementations.

To prove security of the standard QKD against a fully equipped super-strong eavesdropper, BBBMR [1] developed some important technical tools. Mainly, our proof is based on a new information versus disturbance (IvD) theorem, and on adapting a random sampling theorem. In addition to these two main ingredients there are also three reductions used to simplify the proof. One reduction allows us to deal only with the bits actually used in the protocol: the "used-bits" protocol is proven secure, and the reduction tells us that therefore the original Bennett-Brassard protocol (BB84) is also secure. A second reduction allows us to deal only with attacks symmetrical to 0 and 1: the protocol is proven secure against symmetrical attacks, and the reduction tells us that therefore the protocol is secure also against nonsymmetrical attacks. A third reduction allows us to deal with pure states in the eavesdropper's hands (when Alice sends a particular state and Bob's subsystem is traced out) due to simple bounds on accessible information. To understand the exact role of IvD, random sampling, and the three reductions, see the full paper [1] for details.

Information versus disturbance theorems usually provide the extra power of quantum cryptography over classical (modern) cryptography, hence must be used in any proof of security. The underlying reason for the IvD relations is the no-cloning theorem and indeed usually people tend to say, in more popular literature, that security is due to the no-cloning theorem. The IvD idea is that any attempt to clone, and even an attempt to perform an imperfect cloning, creates a disturbance in the quantum state, and this disturbance can be observed by the legitimate users. Thus, the more information the eavesdropper tries to obtain, the larger is the induced disturbance. In all IvD results the power of quantum information theory is manifested in an intuitive and clear way, since there is no analogous IvD in classical physics. One subtlety which endangers the correctness of any proof of security of a QKD protocol is the unintentional use of counterfactual argumentation; All IvD theorems are subtle due to this issue since they always require analysis of outcomes of performed measurements and their connections to other measurements which could have been performed instead, but had not been performed.

We believe that random sampling must also be incorporated in any proof of security, since otherwise, there is no way to take into consideration the fact that some information is not available to the eavesdropper by the time she probes the qubits sent from Alice to Bob. Thus, any proof of security which does not incorporate IvD and random sampling is suspected to be wrong.

Neither the work of [1] nor any other proof (e.g., [5, 6, 7]) has yet been verified by the quantum cryptography community in a thorough manner. Thus, there is no consensus as yet regarding the correctness, completeness, and generality of any of the proposed proofs of security. In the past, cryptography (and quantum cryptography in particular) included many cases where protocols proven secure were later found to be insecure[1] and many other cases where proofs of security were later found to be incomplete or incorrect, or limited[2] in their generality/applicability. It takes time for a consensus to be built, and it can only be built once enough people understand the proof and its subtleties. The reported proofs of security of QKD are all very recent, hence it is too early to expect a consensus.

The security result of Lo and Chau [6] uses novel techniques and is very important, but is rather limited since it is the security of an AQKD scheme and not of a standard QKD scheme. Their scheme (based on quantum privacy amplification) requires that the legitimate users have quantum memories and fault tolerant gate array quantum computers. These technologies are not yet available to the legitimate users, and are not expected within the next twenty years. In contrast, the QKD which is analyzed here (the BB84 protocol) has now been demonstrated with some partial[3] success in many labs. Even if fault-tolerant quantum computers will be built 20 years from now, these will probably be based on a nearest neighbor (nn) model and not on a gate array model, and an analysis of a fault tolerant nn quantum computer is still not completed. Some of the ideas used by Lo and Chau appeared earlier, e.g., the quantum repeaters [9], and the use of fault tolerance quantum error correction for performing quantum privacy amplification [9]. However, Lo and Chau succeeded in using them to yield a novel proof of security by incorporating classical *random sampling* techniques. As far as we know, the complexity of the protocol has not been calculated as yet. Lo and Chau's proof is claimed to be simpler than other proofs of security but this simplicity might be an artifact of the missing complexity analysis.

The security result of Mayers [5] is similar to ours in the sense that it proves the security of a much more realistic protocol against an unrestricted eavesdropper, and provides explicit bounds on the eavesdropper's information as is done in [1]. There is no doubt that Mayers was the first to understand many of the difficulties and subtle points related to the security issues. The main problems with the proof of Mayers are its complexity and strict formality, which prevent the reader from gaining much intuition regarding the proof. In addition, the basic ideas on which Mayers' proof is based remain inaccessible[4] to the majority of those attempting to understand them. As a result, there is still no consensus regarding the correctness and completeness of Mayers' proof. Recently, Mayers' proof was understood by a few researchers [10]. We hope that a consensus will be reached in the near future or once a comprehensible explanation becomes available.

In contrast, our proof is very intuitive. We analyze the density matrices which are available to the eavesdropper. We prove that the following event is extremely rare. The event consists of these density matrices carrying non-negligible information about the secret key, and at the same time, Alice and Bob agreeing that the formed secret key can be used.

Two other proofs were recently announced [7]. Shor and Preskill's proof proposes a way to extend Lo and Chau's proof so that it becomes applicable to a more practical protocol, hence bypassing the main limitation of Lo and Chau's proof.

The BBBMR proof follows the standard assumptions of QKD: 1) Alice and Bob share an unjammable classical channel. This assumption is usually replaced by the demand that Alice and Bob share a short secret key to be used for authenticating a standard classical channel. 2) Eve cannot attack Alice's and Bob's labs. She can only attack the quantum channel and listen to all transmissions on the classical channel. 3) Alice sends quantum bits (two level systems).

In the BB84 protocol Alice and Bob use four possible quantum states in two bases (using "spin" notations, and connecting them to "computation basis" notations): (i) $|0_z\rangle \equiv |0\rangle$; (ii) $|1_z\rangle \equiv |1\rangle$; (iii) $|0_x\rangle = \frac{1}{\sqrt{2}}(|0\rangle + |1\rangle)$; and (iv) $|1_x\rangle = \frac{1}{\sqrt{2}}(|0\rangle - |1\rangle)$. By comparing bases after Alice transmits such a state and Bob receives it, a common key can be created in instances when Alice and Bob used the same basis. In the conference version, BBBMR [1] proved the security of a simplified protocol in which only the relevant bits are discussed (the "used-bits-BB84"). The proof of the security of the original BB84 protocol follows logically, as we will now explain.

Eve attacks the qubits in two steps. First she lets all qubits pass through a device that tries to probe their state. Then, after receiving all the classical data from Alice and Bob, including the bases of all bits, the choice of test bits, the test bits values, the Error Correction Code (ECC), the ECC parities, and the Privacy Amplification (PA), she tries to guess the final key using her best strategy of measurement.

Let us describe the used-bits protocol, splitting it into *creating the sifted key* and *creating the final key from the sifted key*. This simplified protocol assumes that Bob has a quantum memory. I.— *Creating the sifted key*: 1.— Alice and Bob choose a large integer $n \gg 1$. The protocol uses $2n$ bits. 2.— Alice randomly selects two $2n$-bit strings, b and i which are then used to create qubits: The string b determines the basis $0 \equiv z$, and $1 \equiv x$ of the qubits. The string i determines the value (0 or 1) of each of the $2n$ qubits (in the appropriate bases). Alice generates $2n$ qubits according to her selection, and sends them to Bob via a quantum communication channel. 3.— Bob tells Alice when he receives the qubits. 4.— Alice publishes the bases she used, b; this step should be performed only after Bob received all the qubits. Bob measures the qubits in Alice's bases to obtain a $2n$-bit string j. We shall refer to the resulting $2n$-bit string as the sifted key, and it would be the same for Alice and Bob, i.e. $j = i$, if natural errors and eavesdropping did not exist. II.— *Creating the final key from the sifted key*: 1.— From the $2n$ bits, Alice selects a subset of n bits to be the test bits. Alice and Bob compare the values (i_T and j_T) of the test bits. They verify that the error rate $p_{test} = |i_T \oplus j_T|/n$ in the test bits is lower than some agreed error-rate $p_{allowed}$, and abort the protocol if the error rate is larger. The other n bits are the information bits (given by a string i_I). 2.—

The information bits are used for deriving an m-bit final key via ECC and PA techniques [1].

The differences between the protocols are only in the first part. The first part of the BB84 protocol is as follows: I.— *Creating the sifted key*: 1.— Alice and Bob choose a large integer $n \gg 1$, and a number δ_{num}, such that $1 \gg \delta_{\text{num}} \gg 1/\sqrt{(2n)}$. The protocol uses $n'' = (4 + \delta_{\text{num}})n$ bits. 2.— Alice randomly selects two n''-bit strings, b and i, which are then used to create qubits: The string b determines the basis $0 \equiv z$, and $1 \equiv x$ of the qubits. The string i determines the value (0 or 1) of each of the n'' qubits (in the appropriate bases). 3.— Bob randomly selects an n''-bit string, b', which determines Bob's later choice of bases for measuring each of the n'' qubits. 4.— Alice generates n'' qubits according to her selection of b and i, and sends them to Bob via a quantum communication channel. 5.— After receiving the qubits, Bob measures in the basis determined by b'. 6.— Alice and Bob publish the bases they used; this step should be performed only after Bob received all the qubits. 7.— All qubits with different bases are discarded by Alice and Bob. Thus, Alice and Bob finally have $n' \approx n''/2$ bits for which they used the same bases. The n'-bit string would be identical for Alice and Bob if Eve and natural noise do not interfere. 8.— Alice selects the first $2n$ bits from the n'-bit string, and the rest of the n' bits are discarded. If $n' < 2n$ the protocol is aborted.

To prove that BB84 is secure let us modify BB84 by a few steps in a way that each step can only be helpful to Eve, and the final protocol is the used-bits-BB84. Each item below describes a different protocol, obtained by modifying the previous protocol. a.— Let Bob have a quantum memory. Let Alice choose b' instead of Bob at step 3. When Bob receives the qubits at step 5, let him keep the qubits in a memory, and tell Alice he received them. In step 6, let Alice announce b' to Bob, and Bob measures in bases b'. From the announcements of b and b' Bob knows which are the used and the un-used bits, as determined in steps 7 and 8. Now Alice and Bob know all the un-used bits, so they ignore them, to be left with $2n$ bits. Note that Alice knows which are the un-used bits already at step 3 (if she wishes to know this). b.— Let Alice announce which are the un-used bits already at the end of step 3. Let her also announce their bases $b_{un-used}$ and bits-values $i_{un-used}$). Obviously, such announcements can only help Eve to gain more information (and maybe even to chose a better attack). Thus this step only reduces the security, so if the protocol defined here is secure, so is the original BB84 protocol. c.— Let Alice generate and send to Bob only the used bits in step 4, and let her ask Eve to send the un-used bits (by telling her which these are, and also the preparation data for the relevant subsets, that is—$b_{un-used}$ and $i_{un-used}$). Knowing which are the used bits, and knowing their bases $b_{un-used}$ and values $i_{un-used}$ can only help Eve in designing her attack, thus security can only be reduced by this step. Since Bob never uses the values of the unused bits in the protocol (he only ignores them), he doesn't care if Eve doesn't provide him these bits or provides them to him without following Alice's preparation request. After Bob receives the used and unused bits, let him give Eve the unused qubits (without measuring them), and ask her to measure them in bases $b'_{un-used}$. Having these qubits can only help Eve in designing her optimal final measurement, thus security can only be reduced by this step. Since Bob never uses the values of the unused bits in the rest of the protocol, he doesn't care if Eve doesn't provide him these values correctly or at all. d.— Since Alice and Bob never made any use of the unused bits, Eve could have them as part of her ancilla to start with, and Alice could just create $2n$ bits, send them to Bob, and then tell him the bases. The protocol obtained after this reduction, is a protocol in which Eve has full control on her qubits and on the unused qubits. Alice and Bob have control on the preparation and measurement of the used bits only. This is the used-bits BB84, for which we proved security in [1].

The exponentially small probability that $n' < 2n$ in Step 8 (so that the protocol is aborted due to insufficient number of bits in the sifted key) now becomes a probability that the reduction fails.

Summary. BBBMR (in the conference version [1]) proved the security of the used-bits-BB84 protocol for quantum key distribution. We showed here that the security of the standard BB84 protocol follows logically. We also explained here a few of the subtleties related to proving security of quantum key distribution.

Notes

1. E.g., quantum bit commitment was proven secure and then was proven insecure.
2. E.g., factoring is now known to be easy for a quantum computer, in contrast to the Church Thesis, hence security of RSA might be limited to classical adversaries.
3. See [8] for an explanation of the *insecurity* of the implemented schemes.
4. While the ideas are probably expressed in [5], they have not yet been well explicated.

References

[1] E. Biham, M. Boyer, P. O. Boykin, T. Mor and V. Roychowdhury, A Proof of the Security of Quantum Key Distribution. A summarized version is published in *STOC'2000*. A full version (containing the results presented here) is available; See Quant-Ph/9912053.

[2] C. H. Bennett and G. Brassard, *Proc. of IEEE Int. Conf. on Computers, Systems and Signal Processing*, pages 175–179, Bangalore, India, Dec. 1984. IEEE.

[3] E. Biham and T. Mor, *Phys. Rev. Lett.*, 78:2256–2259, 1997. E. Biham and T. Mor, *Phys. Rev. Lett.*, 79:4034–4037, 1997.

[4] E. Biham, M. Boyer, G. Brassard, J. van de Graaf, and T. Mor. Security of quantum key distribution against all collective attacks. Quant-ph/9801022, 1998.

[5] D. Mayers. Unconditional security in quantum cryptography. Quant-ph/9802025. A very preliminary draft appeared in *Proceedings of Crypto'96*, number 1109 in LNCS, pages 343–357, 1996, expanding a related result of A. Yao (1995).

[6] H.-K. Lo and H. F. Chau, *Science*, 283:2050–2056, 1999. Quantum Privacy Amplification was invented by D. Deutsch et al., *Phys. Rev. Lett.*, 77:2818–2821, 1996.

[7] P. W. Shor and J. Preskill, *Phys. Rev. Lett.* 85:441–444 (2000). Another proof was proposed by M. Ben-Or, Dec. 1999. Talk given in NEC workshop on quantum cryptography.

[8] G. Brassard, N. Lütkenhaus, T. Mor, and B. C. Sanders. *Phys. Rev. Lett.*, Accepted (2000).

[9] T. Mor. Reducing quantum errors and improving large scale quantum cryptography. Quant-ph/9608025. Quantum memory in quantum cryptography. D.Sc. (Ph.D.) Thesis, Technion, Haifa (1997). Quant/ph 9906073 (chapter 6).

[10] P. Shor, Jan. 1999. Private communication. More recently N. Lütkenhaus and L. Salvail also stated that they checked Mayers' proof and that they believe that it is correct.

ANONYMOUS-KEY QUANTUM CRYPTOGRAPHY AND UNCONDITIONALLY SECURE QUANTUM BIT COMMITMENT

Horace P. Yuen
Department of Electrical and Computer Engineering
Department of Physics and Astronomy, Northwestern University
yuen@ece.northwestern.edu

Abstract A new cryptographic tool, anonymous quantum key technique, is introduced that leads to unconditionally secure key distribution and encryption schemes that can be readily implemented experimentally in a realistic environment. If quantum memory is available, the technique would have many features of public-key cryptography; an identification protocol that does not require a shared secret key is provided as an illustration. The possibility is also indicated for obtaining unconditionally secure quantum bit commitment protocols with this technique.

This paper has the same title as my Capri talk but the contents are not identical. The portion on anonymous key is greatly expanded here, while only brief mention is made on quantum bit commitment, a detailed treatment of which is available in Ref. [1].

A classic goal of cryptography is privacy: two parties wish to communicate privately so that an adversary can learn nothing about its content. This was usually achieved through the use of a shared private key, typically a string of binary digits, for encrypting and decrypting the message data. A revolution in cryptography occurred around 1976 with the emergence of *public-key* cryptography [2], in which knowledge of a public key for encryption would not lead to knowledge of a secret private key for decryption. The concept of digital signature, the binding of a signer to an electronic digital message, was introduced via public-key technique. The idea of using quantum physics for cryptographic purpose was first proposed by Wiesner in the early 1970's [3]. It came to fruition in the work of Bennett and Brassard [4] on key distribution, culminating in an experimental prototype demonstration [5]. Despite earlier papers on the use of quantum cryptography to achieve other cryptographic goals, it turns out that key distribution is the only viable one so far [6]. Also, the lack of a quantum authentication scheme implies that some standard classical technique has to be employed which takes away some of the novelty of the quantum techniques, which at first sight seem to be public-key type protocols that do not require the prior sharing of secret information.

Consider two users, Adam and Babe, with a powerful adversary Eve who can manipulate all the communications between them. In an intruder-in-the-middle or *impersonation* attack, Eve can pretend to be Adam to Babe, and Babe to Adam, in all the known quantum protocols. If Adam and Babe do not have a prior shared secret key for message authentication, it is often assumed that a non-jammable classical public channel would prevent impersonation. This, however, is not the case [7] as there is still a user authentication (*identification*) problem — without some shared prior framework there is nothing that distinguishes Babe from an impersonator. Specifically, other than "eavesdropping" Eve may pretend to be Babe and trick Adam to tell her something that he would only tell Babe. The use of a shared secret key for authentication reduces the quantum cryptosystem to a key expansion scheme as noted in Ref. [5], without many advantages of a public-key system. In particular, a separate key is needed for each pair of users which causes major problems in a network environment. In standard cryptography there are a variety of approaches [2] to dispense with the use of shared secret keys, notably the use of digital signature for identification that is capable of preventing the identifier or verifier to pretend to be the identifiee.

In this paper a new cryptographic tool, *anonymous key encryption* (AKE), is introduced in the quantum context that has no known parallel in standard cryptography. In AKE, the encrypter does not know the value of his encrypted message. If quantum states can be stored, i.e., if quantum memory is available which is a subject of active current effort, the AKE technique can be extended to a general anonymous key technique that leads to various forms of digital signature and to a public-key type identification protocol, to be called *anonymous key identification* (AKI), which does not require any shared secret key. For key distribution, an unconditional security proof on the use of AKE would be described for qubits, and it would be indicated how a similarly secure protocol may be obtained in the presence of noise and loss by using classical error correcting codes. The possible use of large-energy coherent states would also be indicated.

Let $|\psi_A\rangle \in \mathcal{H}$, where $\mathcal{H}$ is an arbitrary quantum state space, be a state known only to Adam and transmitted by him to Babe. Depending on the message $j \in \{1, \cdots, m\} \equiv \mathcal{M}$ that Babe wants to send to Adam, she modulates $|\psi_A\rangle$ with a unitary transformation U_j^B and send $U_j^B|\psi_A\rangle$ back to Adam. From knowledge of $|\psi_A\rangle$ and the openly known U_j^B, Adam can decrypt j. The idea is that without knowing $|\psi_A\rangle$, Eve cannot tell j without significant error. The name anonymous key encryption is chosen because $|\psi_A\rangle$ acts like an encryption key for Babe to generate the encrypted signal $U_j^B|\psi_A\rangle$ with data j. Often one has n qubits $\mathcal{H} = \bigotimes^n \mathcal{H}_2$ with $\mathcal{M} = \{0,1\}^n$.

Consider the following concrete AKE system for a single qubit $m = 2, n = 1$, so that an arbitrary pure state $\rho_A = |\psi_A\rangle\langle\psi_A|$ is represented in terms of a real vector $\bar{r} \in \mathbf{R}^3$ via the Pauli matrices $\bar{\sigma}$, in component form

$$\rho_A = \frac{1}{2}(I + r_1\sigma_1 + r_2\sigma_2 + r_3\sigma_3), \qquad |\bar{r}|^2 = 1 \tag{1}$$

If ρ_A is one of M possible uniformly distributed states on the (σ_1, σ_3) great circle of the Bloch sphere (or Poincare sphere in the context of photon polarization), we have

$$r_1 = \cos\frac{2\pi\ell}{M}, r_2 = 0, r_3 = \sin\frac{2\pi\ell}{M} \qquad \ell \in \{1, \ldots, M\} \tag{2}$$

If $j = 1$, Babe rotates ρ_A by an angle $\frac{\pi}{2}$ clockwise on this great circle, and if $j = 0$, she rotates it by an angle $\frac{\pi}{2}$ counterclockwise, i.e., $U_j^B = U(\phi_j)$, the rotation matrix with $\phi_j = \pm\frac{\pi}{2}$. These two states are orthogonal in a basis known only to Adam, which he can measure to determine j. Equivalently, the rotation angles may be $\{0, \pi\}$ or some other pairs. In order that $U(\phi_j)|\psi_A\rangle$ is one of the M possible states $\rho_A(\ell)$ of (1) – (2), M is taken to be a multiple of 4. If Adam picks $\rho_A(\ell)$ randomly, the resulting density operator $\rho_B = \sum_\ell \frac{1}{M} U(\phi_j)\rho_A(\ell)U^\dagger(\phi_j)$ from B to A is the same for either j. Thus, *even if* Eve has an identical copy [8] of the state sent back to A, she can gain no information on j. This generalizes to a sequence of independent $\rho_A^i(\ell)$ with independent i, for which Eve's optimal joint attack on ρ_B just factorizes into a product of individual attacks.

The security analysis is carried out via the theory of optimal M-ary quantum detector [9, 10] in which 1 out of M possibilities, each described by a state ρ_j and a priori probability p_j, is selected to optimize a given performance criterion. The selection is based on the result of a general quantum measurement described by a positive operator-valued measure (POM), which is specified to yield the optimal performance. If Eve attempts to identify $\rho_A(\ell)$ by intercepting the transmission to Babe, the best she can do is given by the optimum M-ary quantum detector for discriminating the states (1) – (2), which has been worked out before. Lemma 3 of Ref. [10] gives the optimum quantum measurement in the form of a POM, with corresponding probability of correct identification given by $P_c' = 2/M$. However, even if Eve makes an error, her estimated state is still useful for eavesdropping purpose and a different criterion needs to be used. Generally, it is the probability P_a that Eve's estimated state is accepted to be correct by Adam as a result of his measurement.

$$P_a = \sum_{\ell,\ell'} \frac{1}{M} p(\ell'|\ell) \mathrm{tr} \rho_A(\ell) \rho_A^E(\ell') \tag{3}$$

where $p(\ell'|\ell) = \mathrm{tr}\Pi(\ell')\rho_A(\ell)$ is the probability that given $\rho_A(\ell)$ was transmitted, Eve takes it to be $\rho_A^E(\ell')$ from measuring the POM $\Pi(\ell')$. Such a criterion falls under the general optimum quantum detector formulation, and the optimum $\Pi(\ell')$ for (3) turns out [11] to be the same as that of determining $\rho_A(\ell)$ according to the error probability criterion, which is intuitively reasonable. The resulting P_a is given by 3/4 independently of M (but recall that M is a multiple of 4). Thus, if Eve measures $\Pi(\ell')$ to determine $\rho_A(\ell)$, perhaps because she cannot store the actual $\rho_A(\ell)$, and transmits $\rho_A^E(\ell')$ to Babe, determines j by measurement on $U(\phi_j)\rho_A^E(\ell')U^\dagger(\phi_j)$ from Babe, and sends the resulting state back to Adam, the probability that Adam decrypts correctly is 3/4. Pure guessing without measurement yields $P_a = 1/2$. This $P_a = 3/4$ would be reduced to 2/3 if the whole Bloch sphere is utilized, with ρ_A given by (1) with $r_1 = \sin\theta\cos\phi, r_2 = \cos\theta, r_3 = \sin\theta\sin\phi$, and e.g., $U_j^B = U(\theta_j), \theta_j = \pm\frac{\pi}{2}$ with ϕ unchanged. In the case (2), $M = 4$ is enough to yield $P_a = 3/4$, and in this case a total of $M = 6$ states [12] on the poles of any rectangular coordinate system intercepting the Bloch surface would yield $P_a = 2/3$. In both cases Eve can get these values of P_a without knowing M by measuring an orthogonal basis chosen randomly from the M possible states. Evidently P_a can be further reduced if a higher dimensional $\mathcal{H}$ is used.

If Eve could intercept and store $\rho_A(\ell)$, she could eavesdrop perfectly by sending her own $\rho_E(\ell')$ to Babe in an impersonation attack. Such manipulation can be detected with test qubits mixed into the information qubits. However, a different approach is employed here in which Babe sends her modulated qubits back to Adam in a random order. This has the advantage that all possible eavesdroppings can be thwarted without checking for disturbance, thus allowing a simple proof of protocol security for key establishment. In this scheme, Adam and Babe use AKE with $8k$ qubits to establish a key of length $4k$ while expending a shared secret key of length $2k$, resulting in a net key expansion of $2k$ as follows. For each 8 qubit block, Babe sends back the qubits in one of the following four orders equiprobably using 2 secret bits: 12345678, 87654321, 38462715, 41236587. These four sequences are chosen so that there is no qubit overlap in any position among the eight. Eve can alter the qubits from A to B in an impersonation attack, or to conduct opaque eavesdropping, or to conduct translucent eavesdropping by tapping into the communications between A and B to learn about j. The probability that Eve guesses q of the k qubit groups in the right order in an impersonation attack is given by the binomial distribution with success probability 1/4, and thus is exponentially small in q. The rest she induces an error probability $P_e = 1/2$ per qubit for Adam, and the key establishment would fail in a trial encryption.

Eve may employ an opaque eavesdropping strategy by intercepting and retransmitting the states from A to B and B to A. Instead of using disturbance detection, we merely use classical privacy amplification (CPA) [13] to eliminate

Eve's partial information. Eve's success probability P_c per qubit is bounded as follows. We grant her one copy of $\rho_A(\ell)$ and one copy of the corresponding correct $U(\phi_j)\rho_A(\ell)U^\dagger(\phi_j)$, i.e., we allow Eve to intercept both copies exactly as if there is no disturbance and the order is correct. From these two copies she can try to learn j by optimally processing both states. This is a binary detection problem with two states

$$\rho_{0,1} = \sum_{\ell=1}^{M} \frac{1}{M} \rho_A(\ell) \rho_A \left(\ell \mp \frac{M}{4} \right) \tag{4}$$

for which the optimum probability of discrimination can be obtained by diagonalizing $\rho_0 - \rho_1$ [9]. The resulting optimum probability P_c she would determine j correctly turns out to be the same as that obtained by measuring the optimum state detector on the copy $\rho_A(\ell)$ from A to B and then measuring whether the state $\rho_B(\ell)$ from B to A is clockwise or counterclockwise with respect to $\rho_A(\ell)$, which is intuitively reasonable, and is given by $P_c = P_a$. If Eve launches a joint attack by making measurements on blocks of qubits, she cannot obtain a better accuracy than that of measuring one by one — the optimum quantum detector for the bit error sum factorizes when both the states and the data probabilities of the blocks factorize into a product from the corresponding bits.

In translucent eavesdropping, Eve would try to determine the data j by correlating her tappings from A to B and B to A. She can do this in the correct order only with probability 1/4. Thus, to (loosely) bound all the possible information Eve can obtain, we let her succeed in learning the bits exactly with probability 1/4, and with probability 3/4 we let her learn the bits with probability $P_c = P_a$ as in (4) above. For $P_a = 3/4$, this yields a total of $2k$ deterministic bits and $< k$ Shannon bits, which can be eliminated by expending $4k$ bits or just $3k$ bits asymptotically [5, 13]. This completes the security proof in the ideal limit.

Note that no quantum memory is required in this scheme. We have used very loose bounds to avoid complex arguments and bounding techniques, but the resulting efficiency is still appreciable. The present AKE has no apparent classical analog because listening to both the transmissions from A to B and B to A would reveal too much about the bits in a classical system even when the bit order is random. The intrinsic statistical feature of quantum ontology, that it is impossible to determine the state of a single quantum system exactly, is directly expoited in AKE. The basic ingredients of our security guarantee are: use of qubit order randomization to thwart manipulation and correlation, use of optimum quantum detector and copies to Eve to bound her partial information which is eliminated by classical privacy amplification , and use of classical error correcting code to overcome loss and noise to be presently discussed. In particular, the explicit use of a shared secret key for key expansion, in this case in obtaining secret qubit orders, is a *new* technique that I expect to be widely applicable in many scenarios.

The major problem for quantum security proof lies in the presence of loss and noise in realistic systems. It should be clear that the above security proof does not depend on detecting small disturbance by Eve, and can thus be expected to work in a similar way in the presence of small noise and loss with some simple error correction capability. In particular, one may employ classical error correcting codes (CECC) on qubits in lieu of quantum codes. Thus, each codeword in a CECC (x_i), $x_i \in \{0, 1\}$, becomes a codeword of quantum states $(|x_i\rangle)$, where $|x_i\rangle$ is the state corresponding to 0 and 1 in the quantum modulation scheme adopted. No reconciliation[5] is needed with the use of CECC.

The protocol for AKE key distribution is in general:

(i) Adam sends enough randomly chosen $|\psi_A\rangle$'s to Babe to cover the loss and noise in transmission to Babe as well as the CECC Babe needs to use for transmission back to Adam.

(ii) Babe modulates the information qubits in a known CECC, sends the resulting qubits to Adam in a random order according to a short shared secret key.

(iii) Some form of CPA is employed by Adam to eliminate any possible leakage of information which is strictly bounded.

(iv) The resulting key is checked for correctness by a trial encryption.

There are many variations of this protocol including the use of test qubits or quantum memory in lieu of shared secret key. There are also many ways to use AKE for direct encryption. These topics and the security proof of the above protocol will be developed elsewhere.

If quantum states can be stored, some features of public-key cryptography can be obtained as follows. In classical public-key cryptography, a one-way function f: X $\rightarrow$ Y is roughly a map for which one can obtain $y = fx \in$ Y from $x \in$ X readily but it is "infeasible" to obtain x from fx. A one-way trapdoor function results if x can be readily obtained from fx with additional "trapdoor information" relating to f [14]. For a physically given $|\psi_A\rangle$, the function $\mathcal{M} \rightarrow \mathcal{H}$ with j mapped into $U_j^B|\psi_A\rangle$ can be regarded as a *quantum one-way function* with trapdoor information given by the knowledge of the actual state $|\psi_A\rangle$, to be denoted $K\psi_A$. Thus, $|\psi_A\rangle$ functions like a *quantum public key* while $K\psi_A$ is the private key. Similar to the usual one-way trapdoor function, one can obtain the physical state $U_j^B|\psi_A\rangle$ with a given public key $|\psi_A\rangle$, but cannot obtain from $U_j^B|\psi_A\rangle$ the value j without the knowledge $K\psi_A$. This is the general formulation of the anonymous quantum key technique. It is clear that AKE can be described in this way, with Adam sending Babe his public key $|\psi_A\rangle$ and Babe using $|\psi_A\rangle$ to encrypt a message j which only Adam can decrypt with $K\psi_A$. With $|\psi_A\rangle$ representing a sequence of qubits, a number of standard public key protocols can be recasted in the quantum domain. For example, one-time digital signatures and blind signatures [2] can be implemented this way. Here, we would use the anonymous key technique to obtain a quantum identification protocol AKI of the challenge-response type in which the identifier cannot pretend to be the identifiee and which is an exact analog of a protocol [15] based on classical digital signature. In AKI Adam uses his stored $|\phi_B\rangle$, ϕ_B unknown to him, to identify Babe in the following way. He modulates $|\phi_B\rangle \in \mathcal{H}_2$ with a randomly chosen ϕ_A and transmits $|\phi_B + \phi_A\rangle$ to Babe, say for states of the form (1) - (2), and asks her to return the state $|\phi_A\rangle$with ϕ_B removed which Babe is capable of doing by just adding $-\phi_B$ to the angle in $|\phi_B + \phi_A\rangle$. Adam checks by measuring the projection to $|\phi_A\rangle$. The random ϕ_A is necessary, or else Eve can just return the state $|\phi = 0\rangle$, where $\phi = 0$ is the reference angle, without using $|\phi_B\rangle$ sent by Adam. The protocol can be simply summarized:

$$\begin{aligned} &\text{(i)} \quad A \to B: \quad |\phi_B + \phi_A\rangle \\ &\text{(ii)} \quad B \to A: \quad |\phi_A\rangle \end{aligned} \tag{5}$$

The probability that Adam or Eve could successfully impersonate Babe is P_a for one qubit, which can be brought to any desired security level $P_s = (P_a)^m$ with m qubits exponentially efficiently. Apart from using quantum laws instead of number theoretic complexity assumptions, the security of this protocol is evidently the same as the conventional public-key challenge-response indentification protocol [15]. Note that the success of AKI is independent of that of AKE, with both being examples of the anonymous quantum key technique.

This technique can also be used to obtain unconditionally secure quantum bit commitment schemes, outside the framework of the impossibility proof [6], which is not sufficiently general to rule out all such schemes. In one of these, Babe sends anonymous states to Adam for bit modulation and the anonymous nature of the states prevents Adam from determining the cheating unitary transformation on his committed state. In another, the anonymous states prevent both Adam and Babe from cheating. A detailed treatment of quantum bit commitment is given in ref [1].

Some comments on possible experimental realization are in order. If (1)-(2) are realized via photon number polarization, a small M is sufficient as indicated after Eq. (3). Although our protocol is much simpler, the experimental setup would be quite similar to BB84, and the efficiency would suffer greatly in the presence of loss. In the present M-ary approach, however, it can be improved via large-energy coherent states by the use of a further new technique to be elaborated elsewhere. One underlying reason for such possibility can be explained. Consider the coherent states

$$|\alpha_0(\cos\theta_\ell + i\sin\theta_\ell\rangle, \qquad \theta_\ell = \frac{2\pi\ell}{M} \tag{6}$$

for a real positive α_0 in place of (1) – (2). Any two basis states of (6) have inner product $exp(-2\alpha_0^2) \sim 0$ for large α_0. When $M \to \infty$ or when M is unknown, one obtains $P_a = 1/2$ with heterodyne detection and $P_a < 2/3$ for the canonical phase measurement which is the maximum likelihood phase estimator [16]. This important behavior of having P_a independent of α_0 would also be obtained for a known finite $M \gg \alpha_0$, as a lower bound to the mean-square fluctuation $(\delta\theta)^2$ was obtained [17] that goes as $1/|\alpha|^2$ for coherent states $|\alpha\rangle$. In the P_a expression, this fluctuation would cancel out the α_0^2 in the form $2\alpha_0^2 \sin^2 \frac{\delta\theta}{2}$ when $M \geq 2\pi/\delta\theta$. A two-mode coherent state realization similar to (6), with $|\alpha_0 \cos\theta_\ell\rangle|\alpha_0 \sin\theta_\ell\rangle$, can also be used. In either case $M \sim 10^3$ is easily achievable in the laboratory, with much higher $M \geq 10^6$ possible, so that large α_0 can be used for overcoming loss and noise. This is not possible in previous quantum cryptosystems such as modified BB84 because large α_0 would lead to unambiguous determination of the states involved, which is not the case if there are many states $M \gg \alpha_0$. The use of (6) also allows the possibility of amplification and regeneration along the transmission path using quantum amplifiers [18], as well as routing and switching in a network. Analysis of such coherent-state systems will be given in a future publication detailing how key distribution and encryption can be carried out. It appears that they hold great promise in making secure quantum cryptography truly practical.

References

[1] H. P. Yuen, "Unconditionally Secure Quantum Bit Commitment is Possible," LANL quant-ph/0006109.

[2] For a broad and thorough discussion of standard cryptography, see A. J. Menezes, P. C. van Oorschot, and S. A. Vanstone, *Handbook of Applied Cryptography*, CRC Press, New York, 1997.

[3] But it was first published in S. Wiesner, SIGACT News *15* (1), 78 (1983).

[4] C. H. Bennett and G. Brassard, in *Proceedings of the IEEE International Conference on Computers, Systems, and Signal Processing*, IEEE Press, New York, 1984; p. 175.

[5] C. H. Bennett, F. Bessette, G. Brassard, L. Salvail, and J. Smolin, J. Cryptol. *5*, 3 (1992).

[6] H.-K. Lo and H. F. Chau, Phys. Rev. Lett. *78*, 3410 (1997); D. Mayers, ibid, p. 3414; H.-K. Lo, Phys. Rev *A56*, 1154 (1997).

[7] For the BB84 protocol, a public non-jammable channel would prevent Eve from eavesdropping via impersonation if Adam and Babe can identify themselves.

[8] Of course Eve cannot have such an identical copy from the no-clone theorem, W. K. Wooters and W. Zurek, *Nature 299*, 802 (1982), and H. P. Yuen, Phys. Lett. A *113*, 405 (1986).

[9] C.W. Helstrom, *Quantum Detection and Estimation Theory*, Academic Press, 1976, Ch. IV.

[10] H. P. Yuen, R.S. Kennedy, and M. Lax, IEEE Trans. Inform. Theory *21*, 125 (1975).

[11] The cost P_a to be optimized can be put into the same form as P_c', but with a new $\rho_A'(\ell)$ that is related to $\rho_A(\ell)$ of (1) – (2) by a constant factor of 2 on $\bar{r}$, which does not affect the optimal detector solution.

[12] This result is consistent with that of D. Brub, Phys. Rev. Lett. *81*, 3018 (1998), although both the criterion and the use of the states are different in her case.

[13] C. H. Bennett, G. Brassard, C. Crépeau, and U. M. Maurer, IEEE Trans. Inform Theory *41*, 1915 (1995).

[14] The famous RSA encryption function involves $fx = x^b$ for a positive integer b, X = Y = set of integers modulo n, $n = pq$ the product of two large primes. An f properly chosen this way is believed to be one-way, due to the computational complexity of factoring n with p, q being the trapdoor information.

[15] See p. 404 of Ref. [2].

[16] A. S. Holevo, *Probabilistic and Statistical Aspects of Quantum Theory*, North Holland, 1982, Ch. III and IV.

[17] H. P. Yuen, in *Proceedings of the Workshop on Squeezed States and Uncertainty Relations*, NASA Conference Publication 3135, pp. 13-21, 1991. See also H. P. Yuen, "Communication and Measurements with Squeezed States," in *Quantum Squeezing*, P. D. Drummond and Z. Ficek, Springer, to be published.

[18] H. P. Yuen, in *Quantum Communications and Measurements II*, ed. by P. Kumar, etc., Plenum, 2000, pp. 399-404.

QUANTUM KEY DISTRIBUTION USING MULTILEVEL ENCODING

Mohamed Bourennane, Anders Karlsson, Gunnar Björk

Department of Electronics, Laboratory of Quantum Optics and Quantum Electronics, Royal Institute of Technology (KTH), Electrum 229, 164 40 Kista, Sweden

Abstract We extend the original Bennett and Brassard for quantum key distribution protocol by using M mutually complementary bases and N orthogonalvectors in each base.

Quantum cryptography ideally provides *unconditionally* secure key distribution between two parties, Alice and Bob. In the original protocol proposed by Bennett and Brassard (BB84) [1], Alice and Bob choose randomly between two complementary (conjugate) bases and the information for each basis is encoded using two orthogonal quantum states (qubits). An extension made by Bruß[2], and Bechmann-Pasquinucci and Gisin [3] into a six-state protocol shows that an eavesdropper's information gain for a given impaired error rate is lower than in the BB84 protocol. In the present work, we generalise these results to encoding in $N-$dimensional Hilbert space using M bases and show, for each N, what is the optimal choice of M. In a $N-$dimensional Hilbert space $\mathcal{H}_N$, Alice first choses in which of M complementary bases she wants to prepare her state in and secondly decides which of the N states to send. The information in the state conveyed will from hereon denoted by quNits. Each symbol sent in the M bases and N quNits is chosen randomly with equal probability, i.e. each of the possible NM states appear with probability $1/(NM)$. We define the bases $\{\psi^a\}$ and $\{\psi^b\}$ over an $N-$dimensional space to be mutually complementary if the inner products between all possible pairs of vectors, with one state from each basis, have the same magnitude:

$$|\langle \psi_i^a \,|\, \psi_j^b \rangle| = 1/\sqrt{N} \qquad \forall \quad i,j. \tag{1}$$

If a quantum state is prepared in the $\{\psi^a\}$ basis, but measured in the complementary $\{\psi^b\}$ basis with $b \neq a$, the outcome is completely random. Wootters and Fields have shown [4] that when $N = p^k$, where p is a prime and k an integer, (which we restrict ourselves to here), then there exist a set of $M = N + 1$ mutually complementary bases [4].

First, we start with the simplest possible eavesdropping strategy, the intercept/resend. Suppose Alice prepares and sends the quantum state $|\,\psi_k^a\,\rangle$ which belongs to the basis $\{\psi^a\}$. If Eve performs her measurement in the $\{\psi^a\}$ basis she will detect the correct state $|\psi_k^a\rangle$ and she will subsequently prepare and send the correct state to Bob. Hence, Bob detects the quantum state sent by Alice, provided that he chooses the correct measurement basis. If, instead, Eve measures in the $\{\psi^b\}$ basis with $b \neq a$, her result will be completely random. In the case when Eve eavesdrops only a fraction η of the string sent by Alice, Eve's information is

$$I_{Eve}^{N,M}(\eta) = \eta \frac{\log_2(N)}{M}. \tag{2}$$

The error rate for Bob is given by

$$e_{Bob}^{N,M}(\eta) = \eta(1 - \frac{1}{M})(1 - \frac{1}{N}). \tag{3}$$

The mutual information between Bob and Alice as function of $e_{Bob}^{N,M}$ is given by

$$\begin{aligned} I_{AB}^{N,M}(\eta) &= \log_2(N) + (1 - e_{Bob}^{N,M}(\eta)) \log_2(1 - e_{Bob}^{N,M}(\eta)) \\ &+ e_{Bob}^{N,M}(\eta) \log_2(\frac{e_{Bob}^{N,M}(\eta)}{N-1}). \end{aligned} \tag{4}$$

If we want to continue on using the channel by actively removing Eve's information using privacy amplification [5], a good benchmark could be the transmitted information per symbol. Taking the sifting, error correction, and privacy amplification into account, we can define an effective transmission rate as

$$R_{AB}^{N,M}(\eta) = \frac{1}{M}(I_{AB}^{N,M}(\eta) - I_{Eve}^{N,M}(\eta)). \tag{5}$$

Giving the maximum rate at which Alice can reliably send information to Bob, such that the rate at which Eve obtain this information is arbitrarily small. From the perspective of optimising the impaired error rate for a given $R_{AB}^{N,M}(\eta)$, which tells how easy it is for Bob to detect Eve in the presence of technical noise, what is the optimum choice for M? First we note from the above, that the maximum number of mutually complementary bases are $M = N + 1$. In Fig.1 (a), we plot $R_{AB}^{N,M}(\eta)$ as function of N and M, and for a fixed Bob's error rate ($e_{Bob}^{N,M}(\eta) = 0.1$) , we observe that the optimal case correspond to a higher N and $M = 2$. We also note that for the case of no eavesdropping, that the effective transmission rate in bits per symbol after sifting cannot be more than $\log_2(N)/M$, so choosing M too large will lower the rate.

Secondly, let us discuss the eavesdropping from the perspective of quantum cloning. The basic idea is that Eve uses an Universal Quantum Cloning Machine (UQCM), introduced by Bužek and Hillery [6], to obtain two copies of

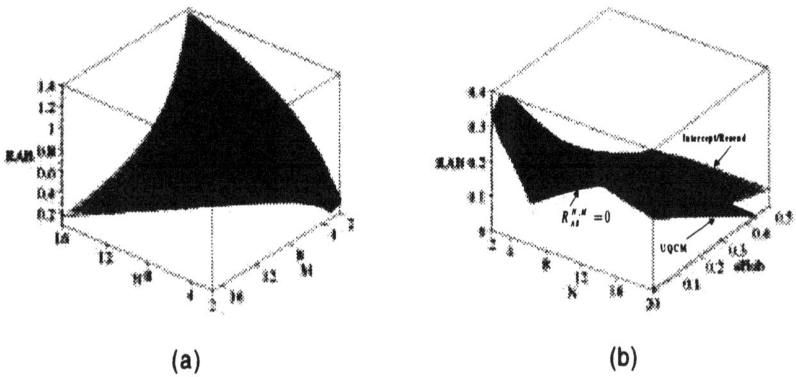

Figure 1 (a) The effective transmission rate $R_{AB}^{N,M}$ as function of the dimension of Hilbert space N, $M = N + 1$ complementary bases and for fixed Bob.'s error rate $e_{Bob}^{N,M}(\eta) = 0.1$ using the intercept/resend strategy. (b) The effective transmission rate $R_{AB}^{N,M}$ as function of the dimension of Hilbert space N, Bob.'s error rate $e_{Bob}^{N,M}(\eta)$ and when $M = N + 1$ complementary bases using either the intercept/resend or the Universal Cloning Machine eavesdropping strategies.

Alice's quantum state, keep one of the copied states for herself, and pass the other copy on to Bob. Then, after Bob and Alice has made their measurements and announced their basis, Eve does the same measurement as Bob. As shown in [6] the maximal fidelity of copying a quNit is obtained using the UQCM. This maximal value of the fidelity correspond precisely to the optimal incoherent eavesdropping strategy, as Gisin and Bechmann-Pasquinucci have shown explicitly for the ($N = 2, M = 3$) case [3]. In the N-dimensional case, from the symmetry of the problem, we conjencture that for $M = N + 1$, the optimal incoherent eavesdropping is again done using the UQCM.

The maximal value of the fidelity F_{UCM}^{N} for the optimal cloning of a quNit to two copies is given from [6] as $F_{UCM}^{N} = (N + 3)/(2(N + 1))$ and the corresponding disturbance is $D_{UCM}^{N} \equiv 1 - F_{UCM}^{N} = (N - 1)/2(N + 1)$.

Assuming that Eve listens in on a fraction η of the symbols sent to Bob, we may then define an equivalent error rate $e_{Bob}^{N,M} \equiv \eta D_{UCM}^{N}$, and the effective transmission rate becomes

$$R_{AB}^{N,M}(\eta) = \frac{1 - \eta}{N + 1} \log_2(N). \tag{6}$$

In Fig.1 (b), we have plotted the effective transmission rate $R_{AB}^{N,M}$ as a function of dimension of Hilbert space N and Bob's error rate in the case when Eve use the intercept/resend or UQCM strategy. As can be seen, the maximum rate with $M = N+1$ is for $N \approx 4$. For low error rates, as Eve gets little information, the rate is upper bounded by $\log_2(N)/(N + 1)$. In all cases, the UQCM gives the best performance (from the viewpoint of the eavesdropper) and Alice and Bob should use the UQCM estimate of information for Eve when applying privacy amplification. As for an experimental realisation of multilevel quantum key distribution, the pratical security of the protocol is bounded by the fact that the total dark count probability increases with the number of detectors. We can also generalize Ekert's quantum cryptographic protocol [7] to N-dimensional Hilbert space.

Acknowledgments

This work was supported by the Swedish Natural Science Research Council (NFR), The Swedish Technical Science Research Council (TFR) and the European Commission through the IST FET QIPC QuComm project.

References

[1] C.H. Bennett and G. Brassard, in Proceedings of IEEE International Conference on Computers, Systems and Signal Processing, Bangalore, India (IEEE, New York, 1984) 175.

[2] D. Bruss, Phys. Rev. Lett. **81**, 3018 (1998).

[3] H. Bechmann-Pasquinucci and N. Gisin, Phys. Rev. A **59**, 4238 (1999).

[4] W. K. Wootters and B. D. Fields, Ann. Phys. **191**, 363, (1989).

[5] C. Bennett, G. Brassard, C. Crepeau, and U. Maurer, IEEE Trans. Inf. Theory **41** 1915 (1995).

[6] V. Bužek and M. Hillery, Phys. Rev. Lett. **81**, 5003 (1998).

[7] A. K. Ekert, Phys. Rev. Lett. **67**, 661 (1991).

AUTHORITY-BASED USER AUTHENTICATION AND QUANTUM KEY DISTRIBUTION

Daniel Ljunggren, Mohamed Bourennane, Anders Karlsson
Laboratory of Quantum Electronics and Quantum Optics, Department of Electronics
Royal Institute of Technology, Electrum 229, SE-164 40 Kista, Sweden
andkar@ele.kth.se

Abstract We propose secure protocols for user authenticated quantum key distribution. We argue that the use of a trusted authority becomes inevitable in a realistic environment as that of multi-user key distribution. Here, we specifically look at one protocol that use nonorthogonal bases with two-qubit entanglement. As the trusted authority being the only one knowing the correlations of this entanglement he/she can protect an unauthorized user from getting the right key.

Motivated by the important task of having the key-results of quantum key distribution [1, 2, 3] moved into a wider context of (quantum) cryptography, we here present a protocol for user authentication; one of many basic cryptographic schemes [4]. By *user authentication* we mean the way of which a user's identity is proved (i.e. the origin of data). Classically, we have private key cryptography where two users need a shared secret key being distributed. QKD solves this problem of *key distribution*, where public key cryptography has been used classically.

There are two channels present in QKD; (*1*) the quantum channel that is private in the sense that it may be eavesdropped on or tampered with by no more than what is permissible by physics laws, and (2) the public channel, used to exchange classical information such as for error correction etc. It can be divided into being unjammable or jammable. The unjammable channel can be authenticated classically by hash-functions, although their security is based on computational assumptions, and the jammable channel can be actively tampered with. Now, for proof of principle of QKD the public channel needs to be unjammable. Indeed, if Mallory can control the classical public channel as well as monitor the quantum channel, QKD will inevitably fail, because Mallory can do the man-in-the-middle attack and impersonate Alice and/or Bob.

We need user authentication to protect from this. For mutual authentication between two persons, it is of course inevitable to have some initial secret. When that's not the case, classical methods uses a trusted third party to verify

Quantum Communication, Computing, and Measurement 3
Edited by P. Tombesi and O. Hirota, Kluwer Academic/Plenum Publishers, 2001

that a certain key belong to whom it is supposed – like the public key in public key cryptography. The secret you need to share now become to be with the trusted authority instead.

Crépeau and Salvail proposed that a protocol could be built on quantum-oblivious transfer [5], however it has later been shown not to be unconditionally secure. Also other protocols have been proposed that are so-called self-enforcing [4], either based on authenticated public communication and refueling QKD or entanglement catalysis [6, 7]. They all need either a classical or quantum shared secret. Therefore, we believe, in a realistic QKD environment where we have a jammable public channel, no shared initial secret between Alice and Bob, and where we need protection from man-in-the-middle attacks, also here we need to introduce an trusted authority, Trent.

We have shown in [9] that the same objectives as in [8], using an arbitrator, can be achieved in a less complex fashion using either non-entanglement or entanglement based protocols. Obviously, self-enforcing quantum protocols would be nice. However, even for some kind of "public quantum key" cryptography, we'd still need a trusted authority to authenticate the public key. Trent is indispensable without a shared secret. Yet, what is different with our protocol is that Trent will not know the final key between Alice and Bob without actively having to eavesdrop on Alice and Bob's channel, which could make an substantial difference in practice.

Suppose the protocol followed by Trent has the following property: *Trent can publicly broadcast (jammably) to Alice and Bob the results of his actions. There also exists a jammable public channel between Alice and Bob. Trent is fully trusted.*

A protocol based on entanglement

We now show the basic principle of the entanglement-based protocol with two users and Trent, using only *one* initial two-qubit entangled state per shared final bit, as compared to in [8], where they need two.

From the four well-known Bell-states we can easily obtain the following states:

$$|\phi^-\rangle_{A,B} = \frac{1}{\sqrt{2}}(|z+\rangle_A|z+\rangle_B - |z-\rangle_A|z-\rangle_B),$$

$$|\psi^+\rangle_{A,B} = \frac{1}{\sqrt{2}}(|z+\rangle_A|z-\rangle_B + |z-\rangle_A|z+\rangle_B),$$

and

$$|\Psi^+\rangle_{A,B} \equiv \frac{1}{\sqrt{2}}(|\phi^-\rangle_{A,B}+|\psi^+\rangle_{A,B}), \quad |\Phi^-\rangle_{A,B} \equiv \frac{1}{\sqrt{2}}(|\phi^-\rangle_{A,B}-|\psi^+\rangle_{A,B}).$$

where $|z+\rangle$ and $|z-\rangle$ denotes the horizontal and vertical polarization eigenstates, and A and B denotes Alice and Bob's particles respectively. Now, the set of states $|\varphi\rangle \in \{|\psi^+\rangle, |\phi^-\rangle, |\Psi^+\rangle, |\Phi^-\rangle\}$ have the feature that $\langle\psi^+|\phi^-\rangle = \langle\Psi^+|\Phi^-\rangle = 0$. Furthermore, all states are not orthogonal, as $|\langle\psi^+|\Psi^+\rangle|^2 = |\langle\psi^+|\Phi^-\rangle|^2 = 1/2$, and $|\langle\phi^-|\Psi^+\rangle|^2 = |\langle\phi^-|\Phi^-\rangle|^2 = 1/2$. This is the crucial feature that enables the detection of a eavesdropper in the protocol below. Let us as starting states pick $|\phi^-\rangle$ and $|\psi^+\rangle$ as one base, and $|\Phi^-\rangle$ and $|\Psi^+\rangle$ as the other. The user authentication and key distribution scheme illustrated with Fig. 1 is as follows: **1.** Trent sends one of the entangled states of $|\varphi\rangle \in \{|\psi^+\rangle, |\phi^-\rangle, |\Psi^+\rangle, |\Phi^-\rangle\}$, each with probability $1/4$. One photon from the entangled state is sent to Alice, and the other to Bob. Alice and Bob *locally* measures the polarization of the incoming photon by switching in random between the z-base and the x-base, where the x-base is defined as $|x\pm\rangle = (|z+\rangle \pm |z-\rangle)/\sqrt{2}$.

2. In order to test for an eavesdropper, or a dishonest party, Alice and Bob alternates in declaring a set of outcomes for some of the bits sent. They then declare the measurement basis used for all the bits.

3. Trent broadcast to Alice and Bob which of the entangled states he sent for all the bits. This information X is hidden by making XOR (in bits) of this information and the initially shared information between Trent and each user respectively.

4. Alice and Bob sort their released data in to four bins N_1 to N_4. In bin N_1, they place the states if Trent sent a $|\psi^+\rangle$ state. In this case they know that their results should be anti-correlated in the z-base and correlated in the x-base. They do the same for the other states and bins and get correlations according to Table 47. This departure from correlation to anti-correlation, gives Alice and Bob the unique signature from Trent which allow them user authentication.

5. Alice and Bob check their bits according to bins N_1 -- N_4.

6. The final step is to distribute the session-key K, which is done using the remaining secret bits from the bins $N_1 - N_4$ as before. They exchange information Y to keep the bits only where the setting of the polarizers were the same.

If the data between Alice and Bob and the settings given from Trent agree, Bob and Alice have again authenticated each other via Trent, and they have succeeded in transferring K. Let us stress the two essential ingredients for authentication; first the control of the sign of the correlation between the bits done by Trent, and second that Alice and Bob declares their bases and outcome for the test bits *before* Trent tells how the outcomes should be correlated. On the average, an eavesdropper will in a simple case impair a 25% BER, as well as induce the same BER in the channel. To check the agreement with the data one may simply check that the BER is not above a critical value. Let us furthermore stress that, as Trent does not know the outcome of Alice's and Bob's measurements, he does not know the secret key, K, established by Alice and Bob.

Acknowledgments

This work was supported by the European IST QIPC QuComm project, the Swedish Natural Science Research Council (NFR), and the Swedish Technical Science Research Council (TFR).

Alice\Bob	$z+$	$z-$	$x+$	$x-$
$z+$	ϕ^-	ψ^+	Ψ^+	Φ^-
$z-$	ψ^+	ϕ^-	Φ^-	Ψ^+
$x+$	Ψ^+	Φ^-	ψ^+	ϕ^-
$x-$	Φ^-	Ψ^+	ϕ^-	ψ^+

Table 1 Correlations of measurement outcomes given that Trent has sent a certain two-particle entangled state.

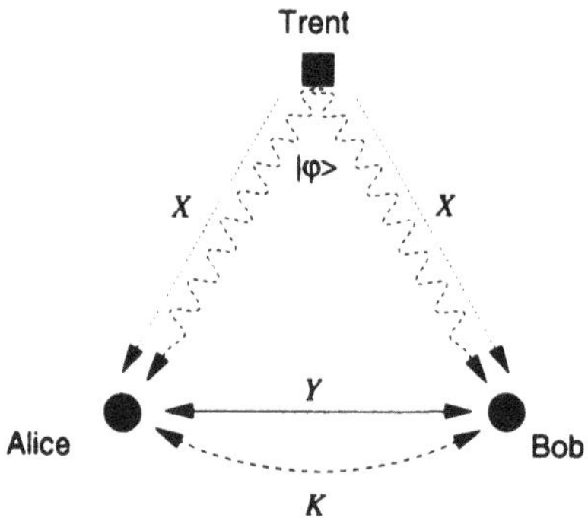

Figure 1 Channel diagram. The wavy dashed line show the entangled-state quantum channel, the dotted line show the jammable public channel, the solid the jammable public channel, and the dashed the encrypted channel. See text for details.

References

[1] C.H. Bennett, and G. Brassard, in *Proceedings of IEEE International Conference on Computers, Systems and Signal Processing*, IEEE, New York pp. 175 (1984)

[2] A.K. Ekert, Phys. Rev. Lett. **67**, pp. 661-663 (1991).

[3] M. Bourennane *et al.*, J. Mod. Opt., **47**, pp. 563-579 (2000).

[4] D.R. Stinson, CRC press, New York (1995).

[5] C. Crépeau and L. Salvail in *Advances in Cryptology: Proceedings of Eurocrypt '95*, pp. 133-14 (Springer-Verlag 1995).

[6] M. Dusek *et al.*, Phys. Rev. A **60**, pp. 149-159 (1999).

[7] H. Barnum, Los Alamos e-print archive, quant-ph/9910072.

[8] G. Zeng and W. Zhang, Phys. Rev. A **61**, 022303, (2000).

[9] D. Ljunggren *et al.*, Phys. Rev. A **62**, 022305, (2000)

IMPROVEMENT OF KEY RATE FOR YUEN-KIM CRYPTOSCHEME

Kouichi Yamazaki
Dept. of Informatics and Communication Engineering
Faculty of Engineering, Tamagawa University, Tokyo, Japan

1. INTRODUCTION

Yuen-Kim (YK) cryptoscheme is one of physical cryptoschemes, which uses some physical phenomena to guarantee their securities[1]. It provides a legitimate sender, Adam, and a receiver, Babe, a secure key distribution scheme making use of independent property of noises between Babe's receiver and that of an eavesdropper, Eve. It has great advantage over quantum key distribution (QKD) schemes for practical use that amplifiers are available to compensate transmission loss. This scheme is classical analog of the two-state quantum key distribution[3], and has a low key rate property, in nature, since much received signal would be abandoned as inconclusive results.In general, Eve can take advantage of more signal power, or more signal to noise ratio (SNR), than Babe because she may stand by Adam. Maurer proposed a scheme called "conceptual channel" to turn over the predominance of Eve's channel over that of legitimate users[2].

The purpose of this article is to improve key rate for the YK Cryptoscheme. To accomplish that, we propose to apply Maurer's conceptual channel to the scheme The effectiveness is numerically shown by comparing practical key rates of YK cryptoscheme using the conceptual channel to that using an ordinary channel.

2. CONCEPTUAL CHANNEL

Adam and Babe want to have common binary strings about which any third party has arbitrary small amount of information. They use a private channel and a noiseless public channel. Let us consider that they use classical communication system to construct their private channel. Adam uses two signals $s_0(t)(=\sqrt{S}\phi(t))$ and $s_1(t)(=-\sqrt{S}\phi(t))$ to send binary symbol 0 and 1, respectively, of duration T, where S is the signal energy over duration T and $\phi(t)$ is a carrier satisfying $\int_T \phi^2(t)dt = 1$. Babe receives

Quantum Communication, Computing, and Measurement 3
Edited by P. Tombesi and O. Hirota, Kluwer Academic/Plenum Publishers, 2001

$r(t) = s_i(t) + n(t)$. Here, $n(t)$ is assumed to be a white Gaussian noise with power spectral density σ^2. She should receive $r(t)$ with an optimum detector measuring $l = \int_T r(t)\phi(t)dt$. After that she decides the received signal using threshold $\pm l_0$ (where $l_0 \equiv m\sqrt{S}$, $m \geq 0$) in the following manner: If $l > l_0$, then $s_0(t)$ was sent, if $l < -l_0$, then $s_1(t)$ was, and if $-l_0 \leq l \leq l_0$, then she refrains to use the signal. Babe decides the received signal correctly with the probability of $P_c' = Q((m-1)(S/\sigma^2)^{1/2})$, while Babe's probability of error is $P_e' = Q((m+1)(S/\sigma^2)^{1/2})$. Here, $Q(\cdot)$ is a scaled complementary error function. The frequency F_+ of the conclusive results is the sum $P_c' + P_e'$. She informs Adam which signals are conclusive via a noiseless public channel. Then, Adam and Babe abandon all inconclusive results to have shifted strings. The error probability P_e of the shifted strings is P_e'/F_+. We assume the transmitted signal to be equiprobable for simplicity in the following.

On the other hand, Eve can also user an optimum receiver to receive $r_E(t) = \sqrt{\lambda}s_i(t) + n_E(t)$, where $\sqrt{\lambda}$ represents Eve's power advantage over Babe, and $n_E(t)$ is a white Gaussian noise with power spectral density σ_E^2. We assume that Eve's noise $n_E(t)$ and Babe's noise $n(t)$ is mutually independent. She decides the received signal by comparing one threshold (i.e. hard decision)[1], so that an error probability is $P_e^E = Q((\sqrt{\lambda}S/\sigma_E^2)^{1/2})$.

Owing to independence between Babe's noise $n(t)$ and Eve's $n_E(t)$, the legitimate users can establish more reliable channel than Eve at the sacrifice of much signals received as inconclusive results by using very weak signal and with the proper thresholds .

Let us consider that Adam transmits random binary strings to Babe with the above communication channel. Let $X \in \{0,1\}$ be a random variable chosen by Adam after discarding the inconclusive results, and Y and Z be the corresponding random variables received by Babe and Eve, respectively. Y can be represented as $Y = X + E$, with E having the property prob$(E = 1) = P_e$ representing an error of the legitimate users' channel. Similarly, Z can be represented as $Z = X + D$, where prob$(D = 1) = P_e^E$. Conventionally, the legitimate users would use X and Y as they are to make a secret key. We call this channel an ordinary channel. Babe has to settle the level of the threshold l_0 very high in order to decrease his error probability much less than that of Eve. This makes the frequency of the conclusive result very little.

Next, we propose to apply, what is called, conceptual channel[2] to YK cryptoschme to improve its key rate. The legitimate users would not use X and Y as they are to make a secret key. Instead, Babe chose another random binary variable V, and sends it to Adam by sending $W = V+Y$. This communication is carried out through the noiseless public channel. By adding X to W, he obtains $V + E$. Eve can listen, in principle, in all the information exchanged by the legitimate users through the public channel, so that she also receives W. By computing $W + Z$ she obtains $V + E + D$. Since E and D is mutually independent, the channel between Eve and Babe (Eve's conceptual channel) is more noisy than that of legitimate users (the main conceptual channel). The error probability $(P_e^E)^{con}$ of Eve's conceptual channel is given by $(P_e^E)^{con} = P_e + P_e^E - 2P_eP_e^E$. Assuming both P_e and P_e^E to be less than $\frac{1}{2}$without loss of generality, $P_e^E)^{con}$ is larger than P_e and P_e^E. On the other hand, that of the conceptual main channel is unchanged.

3. KEY RATES

The legitimate users can not directly use X and Y in the ordinary channel case, or V and $V + E$ in the conceptual channel as a secret key, since error would be contained in these strings, and Eve might know some amount of information about them. Then, they have to, first, reconcile their shifted strings, and then amplify their privacy by public discussion before using strings as a secret key. The key rate is limited by the frequency of conclusive result F_+[bit/signal], the amount of exchanged information during the reconciliation process and the amount of information abandoned for privacy amplification. The theoretical limit of the amount of information r_{rec} for the reconciliation is given by $\mathcal{H}(P_e)$ [bit/shifted bit], where $\mathcal{H}(x)$ is a binary entropy function defined by $\mathcal{H}(x) = -x\log_2 x - (1-x)\log_2(1-x)$. On the hand, that for the privacy amplification r_{PA} is given by $1+\log_2(1-2\delta+2\delta^2)$ [bit/shifted bit][7, 4], where Eve's error probability δ is P_e^E for the ordinary channel and $(P_e^E)^{con}$ for the conceptual channel. Then, key rate R is given by $F_+(1 - r_{rec} - r_{PA})$. Since, there are good algorithm performing near theoretical limit for the reconciliation[5, 6], and those for the privacy amplification[7], the key rate can be regarded to be practical.

The legitimate users would like to maximize their key rates by using appropriate transmitted signal energy S and Babe's threshold l_0. These parameters depend on signal energy S, Babe's and Eve's noise power spectral densities, and Eve's power advantage λ. To simplify the argument, we use Babe's and Eve's SNR's[2] instead of their received signal energy and their noise power spectral density. In the following, we calculate key rates of the YK cryptoscheme for given ratio of Eve's SNR (SNR_{Eve}) to Babe's one (SNR_{Babe}). Maximum key rates are obtained for the conceptual and the ordinary channels with respective optimum parameters. The maximum key rates and relevant parameters are shown for SNR_{Eve}/SNR_{Babe} of 1,2,5 and 10 in Table 1. For $SNR_{Eve}/SNR_{Babe} = 5$, for instance, YK cryptoscheme provides the maximum key rate of 0.0273[bit/signal] by using the conceptual channel, which is 17.2 times larger than that using the ordinary channel. To obtain the maximum key rate, the legitimate users have to set the signal energy S so that $SNR_{Babe}(= S/\sigma^2)$ becomes 0.11, and Babe must receive the signal using threshold l_0 of $1.98\sqrt{S}$. Parameters required for these calculations are also

Table 1 Maximum key rates R^{max} of the ordinary and the conceptual channel. Several relevant parameters are also show. The key rates are maximized w.r.t. Bob's SNR and decidion. threshold.

$\frac{SNR_{Eve}}{SNR_{Babe}}$	R_{con}^{max}	$\frac{R_{con}^{max}}{R_{ord}^{max}}$	SNR_{Babe}^{opt}	m^{opt}	P_e	P_e^{Eve}	$\left(P_e^{Eve}\right)^{con}$	F^+	r_{rec}	r_{PA}
1	0.145	1.34	0.81	0.866	0.0783	0.184	0.234	0.595	0.396	0.361
2	0.0705	2.30	0.32	1.28	0.184	0.212	0.318	0.535	0.689	0.180
5	0.0273	17.2	0.11	1.98	0.302	0.229	0.393	0.534	0.884	0.0647
10	0.0135	521.8	0.05	2.82	0.365	0.240	0.430	0.538	0.947	0.0283

shown in Table 1. We can conclude with Table 1 that the conceptual channel has a great effect for YK cryptoscheme to improve its key rate, especially, when Babe's SNR is much worse then that of Eve.

A part of this research is supported by Grant-in-Aid for Joint Work of Tamagawa University (No.1998-1).

Notes

1. She may use soft decision technique. In this article, however, we assume she decides the received signals by hard decision. Eavesdropping by soft decision will be discussed elsewhere.

2. An abbreviation "SNR" stands for the ratio of the signal energy to the noise power spectral density in this article.

References

[1] H. P. Yuen and A. M. Kim, Phys. Lett. **A241**, 135 (1998).

[2] U. M. Maurer, IEEE Trans. Inform. Theory, **39**, 733 (1993).

[3] C. H. Bennett, Phys. Rev. Lett. **68**, 3121 (1992).

[4] N. Lütkenhaus, Phys. Rev. A **54**, 97 (1996).

[5] G. Brassard and L. Salvail, Advances in Cryptology-Proc. Eurocrypt '93, 410, (1994).

[6] T. Sugimoto and K. Yamazaki, Trans. of the IEICE (*in print*).

[7] C. H. Bennett *et al.*, IEEE Trans. Inform. Theory **41**, 1915 (1995).

STABLE SOLID-STATE SOURCE OF SINGLE PHOTONS

Patrick Zarda
Max-Planck-Institut für Quantenoptik, D-85748 Garching, Germany
patrick.zarda@physik.uni-muenchen.de

Christian Kurtsiefer, Sonja Mayer
Sektion Physik, Ludwig-Maximilians-Universität, D-80797 München, Germany

Harald Weinfurter
Max-Planck Institut für Quantenoptik, D-85748 Garching, Germany
Sektion Physik, Ludwig-Maximilians-Universität, D-80797 München, Germany

Keywords: single photon source, quantum cryptography, antibunching

Abstract The controlled generation of single photons is mandatory for applications in quantum communication, in particular for secure quantum cryptography, and also for a number of fundamental problems in quantum optics. Here, we present a stable all solid-state source for single photons utilizing the fluorescence light from a single nitrogen–vacancy center (N–V center) in diamond.

1. INTRODUCTION

The generation of non-classical light and particularly of single photon states is one of the key experimental challenges in quantum communication and computation. The ideal single photon source emits light such that only one of two detectors behind a semitransparent beam splitter registers an event. A specially suited process for that is fluorescence of a single two-level quantum system.

Such two-level systems can be found with good approximation in atoms or ions and organic molecules. Experiments with single trapped atoms and ions [8] or with single quantum dots [2] require significant technical effort. Single organic dye molecules as the fluorescent emitter in solvents or polymers [3] degrade rapidly at room temperature – typically after about 10^9 emissions.

Here, we present an alternative candidate for generating single photons at room temperature. Single nitrogen–vacancy (N–V) centers in diamond combine the robustness against photo bleaching of single atoms with the simplicity of experiments with dye molecules.

2. NITROGEN-VACANCY (N–V) CENTERS

N–V centers are one of many well studied luminescent defects in diamond [4]. They are formed by a substitutional nitrogen atom with a vacancy trapped at an adjacent lattice position. We addressed single N–V centers in an untreated synthetic type Ib diamond sample ($500 \times 500 \times 200$ μm) in a confocal microscope setup shown in figure 1a.

The light of a diode pumped solid state (DPSS) laser with a wavelength of $\lambda = 532$ nm was focused into the diamond with relay optics and a standard microscope objective (M=60, NA=0.85). The fluorescence light coming from the illuminated N–V center was collected by the same microscope objective and focused into a single mode fiber. This defines the spatial mode for confocal detection and leads to a longitudinal resolution of 1.6 μm and a vertical resolution of 430 nm. The combination of a dichroic mirror and a color glass filter suppressed residual pump light for the fluorescence detection setup.

To position the focal spot within the diamond, a piezoelectric scanning unit (transverse position) and a motorized linear translation stage (longitudinal position) were used. The fluorescence light from the single N–V center was analyzed either by a Hanbury-Brown–Twiss configuration or a grating spectrometer. We finally detected the fluorescence light with passively quenched silicon avalanche photodiodes (APD) with a dark count rate of $\approx$ 250 counts per second [5].

N–V centers are identified by spectral analysis (figure 2a) of their zero phonon line at 637 nm. Additional phonon contributions result in the charac-

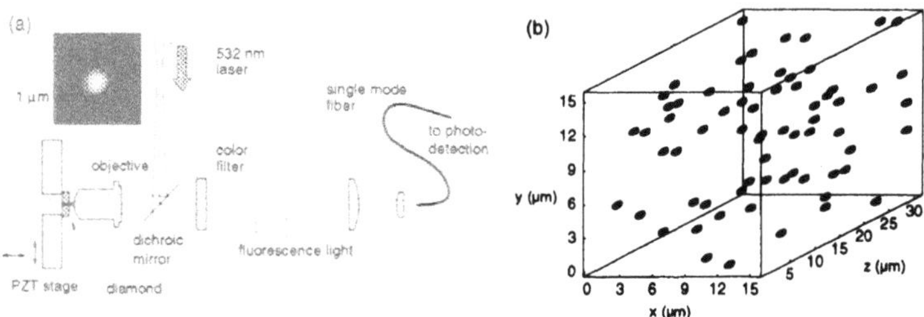

Figure 1 (a) Experimental setup: A frequency doubled DPSS laser (532 nm) is focused into a type Ib diamond crystal. Fluorescence light is collected with a confocal microscope into a single mode optical fiber, and detected with silicon APDs. The inset shows the fluorescence image of a single N–V center. (b) Distribution of N–V centers resulting in a density of 5×10^9 cm^{-3}.

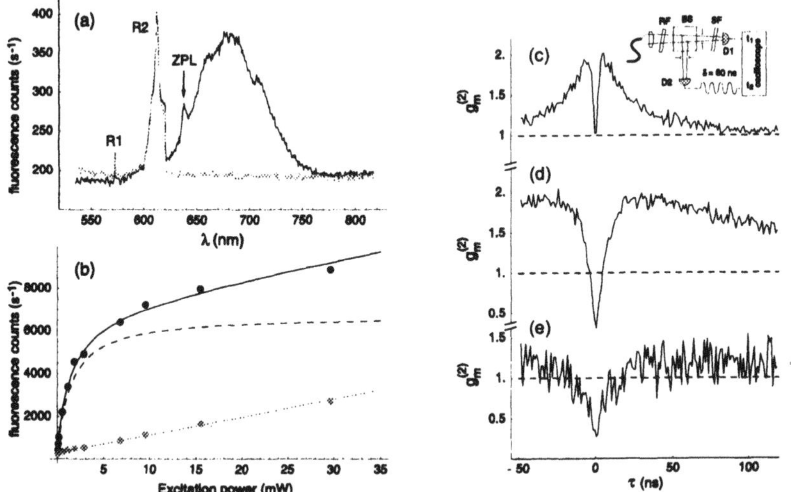

Figure 2 (a) Fluorescence spectrum of a single N–V center (black) and reference spectrum in type Ib diamond (gray). Besides the zero phonon line (ZPL) at $\lambda = 637$ nm and the vibrationally broadened spectrum of the N–V center, we find single- (R1) and two-phonon Raman scattering (R2) from bulk diamond. (b) Fluorescence as a function of the excitation power from a single N–V center (black) and from bulk diamond (grey). (c) Measured pair correlation function $g_m^{(2)}(\tau)$ for excitation powers of 30 P_{sat}, (d) 1.6 P_{sat} and (e) 0.16 P_{sat}: The fluorescence light is detected by two photodiodes D1, D2 behind a beam splitter BS and two filters RF, SF. A clear signature of photon antibunching is seen in $g_m^{(2)}(\tau)$ around $\tau = 0$.

teristic spectral shape with an overall width of about 120 nm [4]. The observed fluorescence light shows a clear saturation behaviour with a saturation intensity of $I_{\text{sat}} = 3.6 \times 10^9$ Wm^{-2} ($P_{\text{sat}} = 1.3$ mW)(figure2b).

We analyzed the photon statistics of the fluorescence light of our single photon source in a Hanbury-Brown–Twiss configuration. Figures 2c-e exemplarily show $g_m^{(2)}(\tau)$ for different excitation powers. Most prominently, the minimum at zero delay clearly proves the single photon character of the emitted fluorescence [5].

3. OTHER DEFECTS

One of the few drawbacks using N–V centers as single photon sources is the broad spectral emission. To increase the spectral yield in narrow wavelength bands, microcavities have been considered [6].

Alternatively, the narrow emission spectrum of other luminescence centers could be used. Particularly, we observed defects emitting at 695 nm (figure 3a) and at 800 nm (figure 3b). These systems exhibit saturation behaviour but

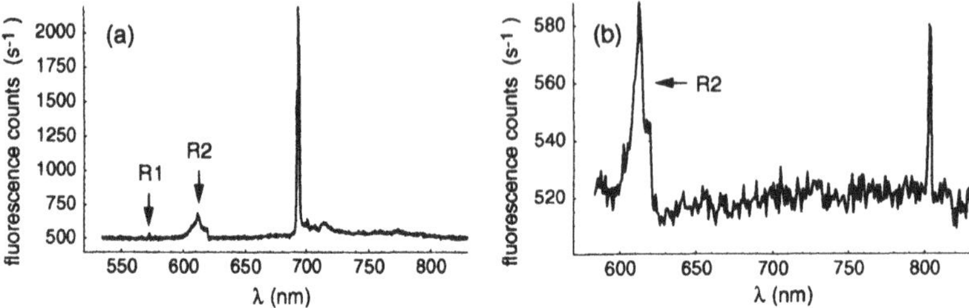

Figure 3 (a) Spectra of other color centers emitting light at 695 nm and (b) 800 nm with Raman background (R1, R2).

seem to suffer from photo bleaching after a few hours. Up to now, we could not observe antibunching from these centers.

4. SUMMARY

We have shown that fluorescence of N–V centers in synthetic Ib diamond shows photon antibunching. The emission spectrum lies in the conveniently detectable red to near infrared region, decay times are short, and the radiative quantum efficiency is close to one. The robustness against photo bleaching and the simplicity of the all-solid state setup distinguishes N–V centers from other fluorescing quantum systems. The potential for miniaturizing the setup and the superior stability makes N–V centers very attractive as practical single photon sources for quantum communication.

References

[1] H.J. Kimble et al., Phys. Rev. Lett. **39**, 691 (1977); D. Diedrich and H. Walther, Phys. Rev. Lett. **58**, 203 (1987).

[2] P. Michler et al., Nature **406**, 968 (2000).

[3] W.E. Moerner and L. Kador, Phys. Rev. Lett. **62**, 2535 (1989); X.S. Xie and J.K. Trautmann, Annu. Rev. Phys. Chem. **49**, 441 (1998); B. Lounis and W. E. Moerner, Nature **407**, 491 (2000).

[4] *The Properties of Natural and Synthetic Diamond*, edited by J.E. Field (Academic Press, London, 1992); A. Gruber et al., Sience **276**, 2021 (1997); G. Davies and M.F. Hamer, Proc. R. Soc. Lond. A **348**, 285-298 (1976).

[5] C. Kurtsiefer et al., Phys. Rev. Lett. **85**, 290 (2000).

[6] F. De Martini et al., Phys. Rev. Lett. **76**, 900 (1996); S.C. Kitson et al., Phys. Rev. A **58**, 620 (1998).

[7] R. Brouri et al., Opt. Lett. **25(17)**, 1294 (2000).

V

ENTANGLEMENT AND TELEPORTATION

NON-LOCALITY AND QUANTUM THEORY: NEW EXPERIMENTAL EVIDENCE

Luigi Accardi

Centro Vito Volterra

Università degli Studi di Roma "Tor Vergata"

Roma, Italy

WEB page: http://volterra.mat.uniroma2.it

accardi@volterra.mat.uniroma2.it

Massimo Regoli

Centro Vito Volterra

Università degli Studi di Roma "Tor Vergata"

Roma, Italy

regoli@volterra.mat.uniroma2.it

Abstract Starting from the late 60's many experiments have been performed to verify the violation Bell's inequality by Einstein–Podolsky–Rosen (EPR) type correlations. The idea of these experiments being that: (i) Bell's inequality is a consequence of locality, hence its experimental violation is an indication of non locality ; (ii) this violation is a typical quantum phenomenon because any classical system making local choices (either deterministic or random) will produce correlations satisfying this inequality. Both statements (i) and (ii) have been criticized by quantum probability on theoretical grounds (not discussed in the present paper) and the experiment discussed below has been devised to support these theoretical arguments. We emphasize that the goal of our experiment is not to reproduce classically the EPR correlations but to prove that there exist perfectly local classical dynamical systems violating Bell's inequality. The conclusions of the present experiment are:

(I) no contradiction between quantum theory and locality can be deduced from the violation of Bell's inequality.

(II) The Copenhagen interpretation of quantum theory becomes quite reasonable and not metaphyisic if interpreted at the light of the chameleon effect.

(III) One can realize quantum entanglement by classical computers.

In section (7) we prove that our experiment also provides a classical analogue of the type of logical (i.e. independent of statistics) incompatibilities pointed out by Greenberger, Horne and Zeilinger.

1. INTRODUCTION

The experiment described in the present paper has been performed in the attempt to clarify a question that has accompanied quantum theory since its origins: *is it true that the theory of relativity, quantum mechanics and a realistic interpretation of natural phenomena are mutually contradictory*?

In the past 30 years the answer to this question, accepted by the majority of physicists has been: *yes they are and this can be proved by theory and confirmed by experiment.* Furthermore, since the experimentally confirmed EPR type correlations are ... *necessarily nonlocal in character...* [GrHoZe93], the experiments also solve the contradiction in favor of quantum theory by showing that the basic pillar of relativity theory is violated in nature. Usually this statement is tempered by the clause that quantum nonlocality cannot be used to send superluminal signals. This amounts to downgrade the locality principle from a law of nature (no action at distance exists) to a principle of telecommunications theory (the existing actions at distance cannot be used to build up a superluminal television). Of course, if this downgrade is the only reasonable way to interpret the experimental data, we have to accept it. However, given the relevance of the issue, the question whether it effectively is the only reasonable way to interpret the experimental data, naturally arises. In several papers starting from [Ac81], (cf. [Ac97] for a general presentation, [Ac99], [AcRe99a] for more recent results) the arguments, relating Bell's inequality to locality, have been criticised on a theoretical ground. In the present paper we will substantiate these theoretical arguments with an experiment.

The usual *proof* of the contradiction between relativity, quantum mechanics and realism is based on a combination of Einstein, Podolsky, Rosen (EPR) type arguments with Bell's inequality and goes as follows. In the EPR type experiments a source emits pairs of systems with the following properties:

i) the members of each pair are distinguishable after separation and we denote them 1 and 2. After emission the two become spatially separated, say: 1 goes to the left, 2 goes to the right.

ii) for each member $j = 1, 2$ of each pair we can measure a family of observables (spin): $\hat{S}_a^{(j)}$ $j = 1, 2$, parametrized by an index set T ($a \in T$). To fix the ideas let us say that T is the unit circle so that each a is a unit vector in the plane.

iii) each observable $\hat{S}_a^{(j)}$ $(j = 1, 2; a \in T)$ can take only two values: ± 1 and, if $a \neq b$, the two observables $\hat{S}_a^{(j)}$, $\hat{S}_b^{(j)}$ $j = 1, 2)$ cannot be simultaneously measured on the same system

iv) The *singlet condition* is satisfied, i.e. if the same observable is measured on both particles, then the results are opposite.

According to EPR the values of the spins must be *pre–determined* otherwise, if one assumes that *these values are created* by a choice of nature at the act of measurement, the only way to explain the strict validity of the singlet condition would be to postulate a mechanism of instantaneous action at distance through which a particle *instantaneously knows* which choice nature is going to do on its distant partner so that the two choices can be *matched* into the singlet law. The pre–determination of the result is called a *realism condition*. Thus the negation of the realism condition in the sense of EPR, would imply a nonlocality effect. In our experiment, a crucial role will be played by a distinction between the EPR realism , or *ballot box realism*, on which the whole of classical statistics is based, from what might be called the *chameleon realism* which seems to be more appropriate when dealing with quantum systems.

According to Bell, given the realism condition (as defined above), the *vital assumption* in the deduction of his inequality is a locality condition:

... *The vital assumption [2] [i.e. the EPR paper is that the result B for particle 2 does not depend on the setting $\vec{a}$ of the magnet for particle 1, nor A on $\vec{b}$*... [Be64].

A theory which satisfies both assumptions is called a *local realistic theory*. With this terminology Bell's result is often reformulated in the form: *a theory that violates Bell's inequality must be either non–local or non–realistic.*

This is Bell's original thesis and, even if balanced by a multiplicity of subtle differences on minor points, from the huge literature on this topic a substantial consensus emerges with its main points which one may summarize as follows:

(a) locality implies Bell's inequality; (b) quantum mechanical (EPR type) experimental correlations violate Bell's inequality; (c) therefore quantum mechanics is non–local; (d) furthermore, no classical system can violate Bell's inequality by local choices.

Therefore if we can construct a classical system (even deteministic and macroscopic) which, by means of purely local choices produces a set of statistical correlations violating Bell's inequality, this will be an experimental proof of the fact that: (i) locality does not imply Bell's inequality; (ii) the experimental violation of Bell's inequality achieved by the quantum mechanical (EPR type) correlations is not an indication of non–locality; (iii) the experimental violation of Bell's inequality by purely local experiments is not a typically quantum phenomenon.

The goal of the present experiment is not to build a hidden variable model for the quantum mechanical singlet correlations (such a model has now [October 2000] been constructed by us and will be discussed elsewhere), but to construct a classical system showing that the three conditions of pre–determination, locality and singlet condition are not incompatible with a violation of Bell's inequality.

2. DESCRIPTION OF THE EXPERIMENT

Our experiment describes a classical system made of three computers:

(2.1) one, playing the role of the source of the singlet pairs, will produce pseudo–random points in the unit disk $p_1, p_2, \ldots, p_N$. Each point p_j plays the role of a singlet pair, therefore in the following we shall speak indifferently of *the point* p_j or of *the singlet pair* p_j. The choice of the pseudo–random generator is not relevant for the results, provided the algorithm has a reasonably good performance. The algorithm we use has been taken from [PrTe93]. An order of $N = 50.000$ points is already sufficient to obtain acceptable results. Increasing the number of points gives a higher precision in the computation of the areas, but does not change the order of magnitude of the violation of the inequality. This also shows that the violation we obtain cannot be attributed to round–off errors in the measurements of the areas.

(2.2) the other two computers play the role of the two measurement apparata. The three computers are separated (e.g. they may be in different room's or different countries, ...).

(2.3) The role of the experimentalists will be played by two persons, one for each of the "measuring computers". Given a point p_j (a singlet pair), operator 1 makes a local independent choice of a unit vector a in the plane (analogue of the direction of the magnetic field) and activates a programme in computer 1 which computes the value of a function $S_a^{(1)}(p_j)$, depending only on the local choice a and on the point (singlet pair) p_j. Similarly operator 2 computes the value $S_b^{(2)}(p_j)$, depending only on her own local choice b, but (as in all EPR experiments) on the same singlet pair p_j. We shall say that each operator *asks a (local) question to the system (computer)*

(2.4) the values of the functions $S_a^{(1)}$, $S_b^{(2)}$ can only be $+1$ or -1. A priori, for any given point (singlet pair) p_j, experimenter 1 (resp. 2) can choose among infinitely many questions to be asked to system 1 (resp. 2). However only one question at a time can be asked to any single system.

(2.5) The calculation of the values $S_a^{(1)}(p_j)$, $S_b^{(2)}(p_j)$ is purely local, however the programme has been devised so that if, by chance, the same question is asked to both systems then they will give opposite answers (singlet condition).

(2.6) Finally we introduce, in our classical model, a strong form of the disturbance effect that a measurement of an observable induces on the other observables of the same system, incompatible with the given one, by requiring that, if the observable $\hat{S}_x$ is measured, then both types of particles instantaneously change the value of all the other observables from $\hat{S}_y$ $(y \neq x)$ to $-\hat{S}_y$ (if I measure the weight of a chameleon in a closed box, its color need not to be the same I would have found if I would have measured it on a leaf). This additional prescription plays a role in the deduction of the Bell inequality (section (5)), but is not necessary for the Zeilinger–Horne–Greenberger type contradiction of section (7).

We want to study the statistics of these answers. In order to do so we have to ask the same pair of questions to many, say N, pairs: $p_1, \ldots, p_N$. In particular, denoting

$$S_a^{(\nu)}(p_j)\ ; \quad \nu = 1,2\ ; \quad j = 1,\ldots,N \tag{1}$$

the answer given by system $\nu(= 1,2)$ of the j–th pair to the local questions $S_a^{(\nu)}$, we can define the *empirical correlation* for any pair $(S_a^{(1)}, S_b^{(2)})$ of local questions

$$\langle S_a^{(1)} S_b^{(2)} \rangle = \frac{1}{N} \sum_{j=1}^{N} S_a^{(1)}(p_j) S_b^{(2)}(p_j) \tag{2}$$

The question we want to answer with our experiment is the following: *can the members of each pair p_j make an agreement on their answers so that the inequality*

$$|\langle S_a^{(1)} S_b^{(2)} \rangle - \langle S_c^{(1)} S_b^{(2)} \rangle| \leq 1 + \langle S_a^{(1)} S_c^{(2)} \rangle \tag{3}$$

is violated for some choices of the indices a, b, c?

3. INTUITIVE INTERPRETATION OF THE EXPERIMENT

The present experiment describes an ensemble of pairs of classical particles which can interact with a vector observable $\vec{B}$, that we call *magnetic field*. We assume that there are two types of particles, type I (left handed) and type II (right handed) and that each pair contains exactly one particle of type I and one particle of type II.

The pairs decay, i.e. split, sequentially, one after the other, and we suppose that left handed particles always go on the left and right handed particles always go on the right, so that after a short time they become spatially separated. On the path of each type of particles there is an experimentalist and the two have coordinated their space–time reference frames so that it makes sense to say that they make simultaneous measurements and that they orient a physical vector quantity in a common direction. Without loss of generality we can assume that: (i) all the experiments take place in the same plane; (ii) in this plane we have fixed an oriented reference frame $e = (e_1, e_2)$. In particular, with respect to this frame, we can speak of upper and lower half–plane and of counterclockwise or clockwise measurement of the angles. Moreover for any point c in the disk,we denote Rc its reflection with respect to the origin

Denote D the unit disk in the plane: $D := \{p = (x, y) \in \mathbf{R}^2 : x^2 + y^2 \leq 1\}$; ∂D the unit circle; $D_+ =$ (resp. ∂D_+) the upper semi–disk (resp. semi–circle); $D_- =$ (resp. ∂D_-) the lower semi–disk (resp. semi–circle).

To both types of particles we associate a family $\{\hat{S}_a : a \in \partial D\}$ of observables (that we call *spin*) parametrized by the points on the unit circle in the plane and with values ± 1 in appropriate units. The two types of particles differ because a magnetic field of unit strength in direction $a \in \partial D$ splits a beam of particles of type I according to the values of the observable $\hat{S}_a$, and a beam of particles of type II according to the values of the observable $\hat{S}_{Ra}$. More precisely: if a is any unit vector in the upper half plane (which includes the vector e_1, but not the vector Re_1) then, under the action of a magnetic field $B(a)$ in direction a, a particle of type I will deviate in the upward direction if $\hat{S}_a = +1$, in the downward direction if $\hat{S}_a = -1$. Under the action of the same magnetic field $B(a)$, a particle of type II will deviate in the upward direction if $\hat{S}_{Ra} = +1$, in the downward direction if $\hat{S}_{Ra} = -1$. For a as above, a magnetic field in direction Ra will have the symmetric effect: i.e. particles of type I will go up if $\hat{S}_{Ra} = -1$, down if $\hat{S}_{Ra} = +1$ and particles of type II will go up if $\hat{S}_a = -1$, down if $\hat{S}_a = +1$. In this sense the observable $\hat{S}_a$ can be understood as measuring the response of a particle (either of Type or of I Type II) to a magnetic field in direction a, *being understood that no other interaction has to take place with the same particle in the same time.*

From this definition it is clear that for any two different vectors $a, b \in \partial D$ the corresponding observables $\hat{S}_a$, $\hat{S}_b$ are incompatible because it is impossible to place two different magnetic fields $B(a), B(b)$ in the same point and, even if it were possible, a single particle cannot interact *only with* $B(a)$ and, at the same time, *only with* $B(b)$. (if $\hat{S}_a$ denotes the color of a chameleon on a leaf and $\hat{S}_b$ its color on a log, it is self–contradictory to say that the color *is simultaneously measured only on a leaf and only on a log*). This expression of logical impossibility has to be distinguished from the usual statement of the Heisenberg principle, expressing the *physical* (but by no means logical) impossibility of simultaneously measuring, with arbitrary precision and on the same system, position and momentum.

REMARK. In standard quantum mechanics it is well known that the spin in direction a is a pseudo–scalar quantity, i.e. under proper rotations it transforms as a scalar but, under parity, according to the rule ([Sak85], chap. 4) $S_{Ra} = -S_a$ and obviously the two hermitean matrices S_a and $-S_a$ commute. However the above considerations on the impossibility, for a single particle, to interact simultaneously with two magnetic fields oriented in opposite directions and placed in the same point also apply to spin variables and show that some care is needed when identifying the simultaneous physical measurability of two observables with the commutativity of the associated operators, at least when pseudo–scalar quantities are involved.

On two different particles any pair of observables can be simultaneously measured. Accordingly an ordered pair $(B(a), B(b))$ of magnetic fields in directions $a, b \in \partial D_+$ in the upper half plane will describe a simultaneous measurement of observable $\hat{S}_a$, for particle I, and of observable $\hat{S}_{Rb}$, for particle II, of the pair. We will also use the notation $(S_a^{(1)}, S_b^{(2)})$ for such a measurement.

4. PRESCRIPTIONS FOR THE LOCAL ANSWERS

The state space of our classical particles is the unit disk D and, for c in the unit circle, we denote $S_c^{(j)}$ the response of particle j to the magnetic field $B(c)$ as described in section (3). We will describe these responses by means of functions $S_x : p \in D \to S_x(p) = \pm 1$, parametrized by points p in the unit circle, through the following rule:

$$S_c^{(1)} = S_c = -S_{Rc} = S_c^{(2)} \tag{1}$$

As explained in the previous section, $S_c^{(1)}$ corresponds to the measurement of S_c on particle 1 and $S_c^{(2)}$ to the measurement of S_{Rc} on particle 2 while, to measure S_c on both particles we have to place a magnetic field oriented in direction c for particle 1 and in direction Rc for particle 2. This gives, according to (1):

$$S_{Rc}^{(2)} = -S_{R^2c} = -S_c = -S_c^{(1)} \tag{2}$$

which is the singlet law. For $a \in \partial D_+$, making a angle $\alpha \in [0, \pi)$ (measured counterclockwise) with the x–axis, the calculation of the value of $S_a =: S_a^{(1)}$ in the point p of the disk D is done as follows:

– one rotates the point p, counterclockwise, of an angle α

$$R_{-\alpha}p \quad ; \quad R_{-\alpha} = \begin{pmatrix} \cos\alpha & \sin\alpha \\ -\sin\alpha & \cos\alpha \end{pmatrix} \tag{3}$$

– one chooses the value: $S_a(p) = \pm 1$ if $R_{-\alpha}p \in D_\pm$.

– The value of $S_{Ra}(p)$ is determined by the prescription: $S_{Ra} := -S_a$.

5. THE CHAMELEON EFFECT

In this section we show that, if one pretends to apply to our experiment the same type of arguments which are applied to the EPR type experiments, then one arrives to the conclusion that Bell's inequality should be satisfied and this leads to a contradiction with the experimental data. Recall that the chameleon effect means that *the dynamical evolution of a system depends on the observable A one is going to measure*. In particular, if we measure A, the evolution of another observable B, hence its value at the time of measurement, might be quite different from what it would have been if B, and not A, had been measured. As explained in item (2.6) of section (2), in our classical model, this *disturbance effect* is not only deterministic but also known, so that we can use it in our calculations. For a given pair (point in the unit disk) $p \in D$, we do not know a priori the values assumed by $S_x^{(j)}(p)$ (because p is a hidden parameter for the experimentalists). However whatever these values are they must satisfy the inequality

$$|S_a^{(1)}(p)S_b^{(2)}(p) - S_c^{(1)}(p)S_b^{(2)}(p)| \leq 1 - S_a^{(1)}(p)S_c^{(1)}(p) \tag{1}$$

Since, in the second experiment, $S_b^{(2)}$ has been measured and $b \neq c$, we now, from item (2.6) of section (2), that particle 2 of the pair p, has changed the value of $S_c^{(2)}$ from $S_c^{(2)}(p)$ to $-S_c^{(2)}(p)$. Moreover, from section (4), we know that $S_c^{(2)}(p) = S_c^{(1)}(p)$. Therefore, if we want to insert the value of $S_c^{(2)}(p)$ *that we would have found if* $S_c^{(2)}$ *would have been measured instead of* $S_b^{(2)}$, then we have to replace in (1), $S_c^{(1)}(p)$ by $-S_c^{(2)}(p)$ thus obtaining

$$|S_a^{(1)}(p)S_b^{(2)}(p) - S_c^{(1)}(p)S_b^{(2)}(p)| \leq 1 + S_a^{(1)}(p)S_b^{(2)}(p)$$

From which Bell's inequality (2.2) is easily deduced by taking averages. In the following section we show that the experimental results contradict this argument.

6. VIOLATION OF THE BELL'S INEQUALITY

We will now prove that there exist three directions a, b, c in the plane such that the correlations (2.3) violate the Bell inequality. To this goal we fix a to be the x–axis and consider a generic choice of both vectors c, b in the upper semi–circle. Denoting $(+,+)$, $(-,-)$ the probabilities of concordance, and $(-,+)$, $(+,-)$, those of discordance, one has

$$\langle S_c^{(1)} S_b^{(2)} \rangle = 2(-,-) - 2(+,-) = 2(-,-) - 2(\frac{1}{2} - (-,-)) = -1 + 4(-,-)$$

Denoting $\widehat{cb}$ the probability of the $(-,-)$–concordance for the choice $(S_c^{(1)}, S_b^{(2)})$, this gives the general formula

$$\langle S_c^{(1)} S_b^{(2)} \rangle = -1 + 4\widehat{cb} \tag{1}$$

valid for any choice of the vectors c, b in the upper semi–circle.

Let us now choose the three directions a, b, c in the plane as indicated in Figure (1). Then, according to equation (1), the two sides of the Bell inequality are respectively:

$$|\langle S_a^{(1)} S_b^{(2)} \rangle - \langle S_c^{(1)} S_b^{(2)} \rangle| = 4|\widehat{ab} - \widehat{cb}| = 4(\widehat{cb} - \widehat{ab})$$

$$1 + \langle S_a^{(1)} S_c^{(2)} \rangle = 4\widehat{ac}$$

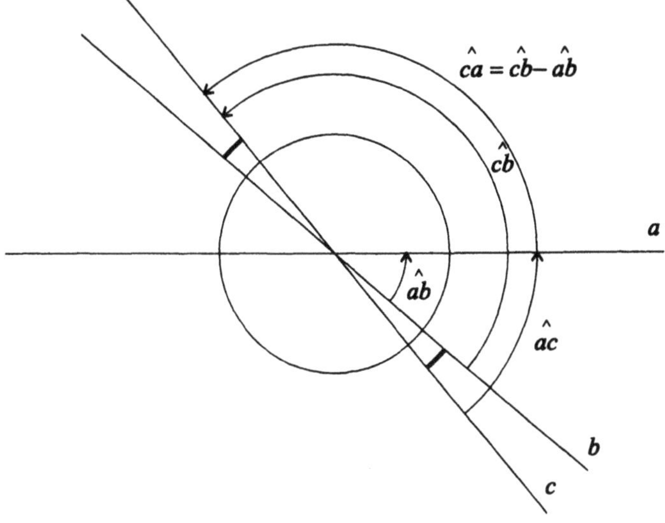

Figure 1

and we are reduced to compare $\widehat{cb} - \widehat{ab} = \widehat{ca}$ and $\widehat{ac}$. But, because of our choice of the axes (cf. Figure 1), one has $\widehat{ca} > \widehat{ac}$. Thus we conclude that

$$|\langle S_a^{(1)} S_b^{(2)} \rangle - \langle S_c^{(1)} S_b^{(2)} \rangle| > 1 + \langle S_a^{(1)} S_c^{(2)} \rangle$$

which violates the Bell inequality.

7. A GREENBERGER–HORNE–ZEILINGER TYPE CONTRADICTION

Greenberger, Horne and Zeilinger [GrHoZe93] have constructed an example showing that the attempt to attribute simultaneous values to 2–valued observables represented by non commuting operators may lead to a contradiction independently of any statistical consideration. In this section we show that our experiment also provides a classical analogue of the construction of these authors. From (3.1) and (3.3) we know that the answers of a particle of type I to the two (mutually exclusive) measurements $B(a)$ and $B(Ra)$ will be opposite:

$$S_a^{(1)} = -S_{Ra}^{(1)} \quad ; \qquad S_a^{(2)} = -S_{Ra}^{(2)} \tag{1}$$

Moreover we know that

$$S_a^{(1)} = S_a^{(2)} \quad ; \qquad S_a^{(1)} = -S_{Ra}^{(2)} \tag{2}$$

Notice that, if we interpret S_a (resp. S_{Ra}) as the response of a type *I* particle to the only action of the magnetic field $B(a)$ (resp. $B(Ra)$) the joint event

$[S_a = +]$ and $[S_{Ra} = -]$ makes no sense as a simultaneous statement on the same particle. Now let us show that, if we pretend (as done in the original Bell's argument and in all discussions of the EPR type experiments) that all these relations hold simultaneously, in the sense of joint events, then we arrive to a contradiction. Interpreting the diagrams below as explained in section (3) (both columns in each diagram are referred to the same particle), we see that relations (1) and (3.1), (3.2) imply that, for any vector a in the upper half plane:

$$I \to B(a) \to \begin{cases} S_a = + \, ; & S_{Ra} = - \\ S_a = - \, ; & S_{Ra} = + \end{cases} \tag{3}$$

$$II \to B(a) \to \begin{cases} S_{Ra} = + \, ; & S_a = - \\ S_{Ra} = - \, ; & S_a = + \end{cases} \tag{4}$$

Relations (1) and (3.3), (3.4) imply:

$$I \to B(Ra) \to \begin{cases} S_{Ra} = - \, ; & S_a = + \\ S_{Ra} = + \, ; & S_a = - \end{cases} \tag{5}$$

$$II \to B(Ra) \to \begin{cases} S_a = - \, ; & S_{Ra} = + \\ S_a = + \, ; & S_{Ra} = - \end{cases} \tag{6}$$

But, if we pretend to attribute to $S_a^{(2)}$ the values given by the second column of (4) then, because of (3) this would contradict (2) because it gives $S_a^{(1)} = -S_a^{(2)}$. Similarly also (5) and (6) contradict (2) because they give $S_{Ra}^{(1)} = -S_{Ra}^{(2)}$. Summing up: the functions of the "hidden parameter" $p \in D$ correctly describe the behavior of **pairs of observables referred to different particles.** In fact this behavior can be simulated on the computer with arbitrary precision. However, if we pretend to extend this descriptions to triples (or quadruples) of observables by including **pairs of observables referred to the same particle**, then we arrive to a logical contradiction, independent of any statistics.

References

[Ac81] Luigi Accardi: "Topics in quantum probability", Phys. Rep. 77 (1981) 169–192.

[Ac97] Luigi Accardi: Urne e camaleonti. Dialogo sulla realtà, le leggi del caso e la teoria quantistica. Il Saggiatore (1997).

[Ac99] Luigi Accardi: On the EPR paradox and the Bell inequality Volterra Preprint (1998) N. 350.

[AcRe99a] Luigi Accardi, Massimo Regoli: Quantum probability and the interpretation of quantum mechanics: a crucial experiment, Invited talk at the workshop: *"The applications of mathematics to the sciences of nature: critical moments and aspetcs"*, Arcidosso June 28-July 1 (1999). To appear in the proceedings of the workshop, Preprint Volterra N. 399.

[Be64] J.S. Bell: On the Einstein Podolsky Rosen Paradox Physics 1 no. 3 (1964) 195–200.

[GrHoZe93] A. Zeilinger, M.A. Horne, D.M. Greenberger: Multiparticle interferometry and the superposition principle. American Institute of Physics (1993).

[PrTe93] Press H. William, A. Teukolsky Saul: Numerical recepees in C. The art of scientific computing, Cambridge University Press, 1993.

[Sak85] Sakurai Jun John: Modern Quantum Mechanics, Benjamin (1985).

ON ENTANGLED QUANTUM CAPACITY

Viacheslav P. Belavkin
Department of Mathematics, University of Nottingham, Nottingham, NG7 2RD UK
vpb@maths.nott.ac.uk

Keywords: Entanglements, Compound States, Quantum Information.

Abstract The pure quantum entanglement is generalized to the case of mixed compoundstates to include the classical-quantum encoding as a particular case. The mutual information of the entangled states leads to a different type of entropy which doubles the von Neumann entropy for a given state and is the true quantum entropy. The conditional q-entropy is positive, and q-information and q-capacity of a quantum channel are additive.

Introduction

The entanglements [1] as specifically quantum (q-) correlations, are used for quantum information processes in quantum computation, quantum teleportation, and quantum cryptography [2, 3, 4]. In this paper, we develope the operational approach to entanglement in order to provide a comparison of different quantum information processings.

We show that any compound state can be achieved by a generalized entanglement, and the classically (c-) entangled states of c-q encodings can be achieved by d-entanglements, the diagonal c-entanglements for these disentangled states. The pure d-entanglements are most informative among the c-entanglements in the sense that they achieve the c-capacity $C = \log \operatorname{rank}\mathcal{B}$, where $\operatorname{rank}\mathcal{B}$ is the dimensionality of a maximal Abelian subalgebra $\mathcal{A} \subset \mathcal{B}$.

We prove that the truly entangled states are most informative in the sense that the maximum of mutual entropy over all entanglements to the quantum system algebra $\mathcal{B}$ achieves the q-capacity $C_q = \log \dim \mathcal{B}$.

Here we consider the case of a simple algebra $\mathcal{B}$ for which the proofs are given in the paper [1].

1. MATHEMATICS OF ENTANGLEMENT

Let $\mathcal{H}$ denote a separable Hilbert space of quantum system, and $\mathcal{B} = \mathcal{L}(\mathcal{H})$ be the algebra of all linear operators $B : \mathcal{H} \to \mathcal{H}$ with the Hermitian adjoints $B^\dagger$ on $\mathcal{H}$. A normal state can be expressed as

$$\varrho(B) = \mathrm{tr}_{\mathcal{G}} \chi^\dagger B \chi = \mathrm{tr} B \rho, \quad B \in \mathcal{B}. \tag{1}$$

Here $\mathcal{G}$ is another separable Hilbert space, χ is a Hilbert-Schmidt operator from $\mathcal{G}$ to $\mathcal{H}$ and $\chi^\dagger$ is its adjoint $\mathcal{H} \to \mathcal{G}$. This χ is called the amplitude operator (just amplitude if $\mathcal{G}$ is one dimensional space $\mathbb{C}$, $\chi = \psi \in \mathcal{H}$ in which case $\chi^\dagger$ is the functional $\psi^\dagger$ from $\mathcal{H}$ to $\mathbb{C}$).

In general, $\mathcal{G}$ is not one dimensional, the dimensionality $\dim \mathcal{G}$ must be not less than $\mathrm{rank}\rho$, the dimensionality of the range $\mathrm{ran}\rho \subseteq \mathcal{H}$ of the density operator ρ. We shall always equip $\mathcal{G}$ with an isometric involution $J = J^\dagger$, $J^2 = I$, having the properties of complex conjugation on $\mathcal{G}$ with respect to which $J\sigma = \sigma J$ for the positive and so self-adjoint operator $\sigma = \chi^\dagger \chi = \sigma^\dagger$ on $\mathcal{G}$. The latter can also be expressed as the symmetricity property $\tilde{\varsigma} = \varsigma$ of the state $\varsigma(A) = \mathrm{tr} A \sigma$ given by the real and so symmetric density operator $\bar{\sigma} = \sigma = \tilde{\sigma}$ on $\mathcal{G}$ with respect to the complex conjugation $\bar{A} = JAJ$ and the tilda operation ($\mathcal{G}$-transponation) $\tilde{A} = JA^\dagger J$ on the auxiliary algebra $\mathcal{A} = \mathcal{L}(\mathcal{G})$.

Given the amplitude operator χ, one can define not only the states ϱ by $\rho = \chi\chi^\dagger$and ς by $\sigma = \chi^\dagger \chi$ on the algebras $\mathcal{B} = \mathcal{L}(\mathcal{H})$ and $\mathcal{A} = \mathcal{L}(\mathcal{G})$ but also a pure entanglement state ϖ on the algebra $\mathcal{A} \otimes \mathcal{B}$ of all bounded operators on the tensor product Hilbert space $\mathcal{G} \otimes \mathcal{H}$ by

$$\varpi(A \otimes B) = \mathrm{tr}_{\mathcal{H}} B \chi \tilde{A} \chi^\dagger = \mathrm{tr}_{\mathcal{G}} \tilde{A} \chi^\dagger B \chi. \tag{2}$$

This state is pure as it is given by an amplitude $\psi \in \mathcal{G} \otimes \mathcal{H}$ defined as

$$(\zeta \otimes \eta)^\dagger \psi = \eta^\dagger \chi J \zeta, \quad \forall \zeta \in \mathcal{G}, \eta \in \mathcal{H}, \tag{3}$$

and ϱ and ς are the marginals of ϖ:

$$\varpi(I \otimes B) = \mathrm{tr}_{\mathcal{H}} B \rho, \quad \varpi(A \otimes I) = \mathrm{tr}_{\mathcal{G}} A \sigma. \tag{4}$$

The next theorem generalizes this to any compound state

$$\varpi(A \otimes B) = \psi^\dagger (A \otimes B) \psi, \quad A \in \mathcal{A}, B \in \mathcal{B}. \tag{5}$$

Theorem 1 *Let $\varpi : \mathcal{A} \otimes \mathcal{B} \to \mathbb{C}$ be a compound state*

$$\varpi(A \otimes B) = \mathrm{tr}_{\mathcal{F}} \upsilon^\dagger (A \otimes B) \upsilon, \tag{6}$$

defined by an amplitude operator $\upsilon : \mathcal{F} \to \mathcal{G} \otimes \mathcal{H}$ on a separable Hilbert space $\mathcal{F}$ into the tensor product Hilbert space $\mathcal{G} \otimes \mathcal{H}$ with $\mathrm{tr}\upsilon^\dagger \upsilon = 1$. *Then this state is achieved by an entanglement*

$$\varpi(A \otimes B) = \mathrm{tr}_{\mathcal{F} \otimes \mathcal{H}} (I \otimes B) \chi \tilde{A} \chi^\dagger = \mathrm{tr}_{\mathcal{G}} \tilde{A} \chi^\dagger (I \otimes B) \chi \tag{7}$$

of the states (4) with $\sigma = \chi^{\dagger}\chi$ and $\rho = \mathrm{tr}_{\mathcal{F}}\chi\chi^{\dagger}$, where χ is an operator $\mathcal{G} \to \mathcal{F} \otimes \mathcal{H}$. The entangling operator χ is uniquely defined up to a unitary transformation U of the minimal space $\mathcal{F} = \mathrm{ran}\upsilon^{\dagger}$ by $\tilde{\chi}U = \upsilon$, where

$$(\zeta \otimes \eta)^{\dagger} \tilde{\chi}\xi = (J\xi \otimes \eta)^{\dagger} \chi J\zeta, \quad \forall \xi \in \mathcal{F}, \zeta \in \mathcal{G}, \eta \in \mathcal{H}, \tag{8}$$

Note that the entangled state (7) is written as

$$\varpi (A \otimes B) = \mathrm{tr}_{\mathcal{H}} B\pi (A) = \mathrm{tr}_{\mathcal{G}} A\pi^{\intercal} (B), \tag{9}$$

where the operator $\pi (A) = \mathrm{tr}_{\mathcal{F}}\chi\tilde{A}\chi^{\dagger} \in \mathcal{B}$, bounded by $\|A\| \rho \in \mathcal{B}_*$, is in the predual space $\mathcal{B}_* = \mathcal{T}(\mathcal{H})$ of $\mathcal{B}$ for any $A \in \mathcal{L}(\mathcal{G})$, and

$$\pi^{\intercal} (B) = J\chi^{\dagger} \left(I \otimes B^{\dagger}\right) \chi J = \chi^{\dagger} \widetilde{(I \otimes B)} \chi \tag{10}$$

is in $\mathcal{A}_*$ as a trace-class operator in $\mathcal{G}$, bounded by $\|B\| \sigma \in \mathcal{A}_*$. The linear map $\pi^{\intercal}$ is written in the Steinspring form [5] of the normal completely positive map $B \mapsto \widetilde{\pi^{\intercal} (B)}$, while $\pi : \mathcal{A} \to \mathcal{B}_*$ is written in the Kraus form [6] of the normal completely positive map $A \mapsto \pi\left(\tilde{A}\right)$. Thus the linear maps $\pi^{\intercal}$ and π are dual to each other, $\pi^{\intercal\intercal} = \pi$, both are positive, but they are not *completely* positive but *co-positive* as the compositions of the transponation $A \mapsto \tilde{A}$ and the completely positive (CP) maps. In terms of the compound density operator $\omega = \upsilon\upsilon^{\dagger}$ for the entangled state $\varpi (A \otimes B) = \mathrm{tr} (A \otimes B) \omega$ they can be written simply as

$$\pi (A) = \mathrm{tr}_{\mathcal{G}} (A \otimes I) \omega, \quad \pi^{\intercal} (B) = \mathrm{tr}_{\mathcal{H}} (I \otimes B) \omega. \tag{11}$$

Definition 1.1. The completely co-positive map $\pi^{\intercal} : \mathcal{B} \to \mathcal{A}_*$ (or its dual map $\pi : \mathcal{A} \to \mathcal{B}_*$), normalized as $\mathrm{tr}_{\mathcal{G}}\pi^{\intercal} (I) = 1$ (or $\mathrm{tr}_{\mathcal{H}}\pi (I) = 1$) is called the quantum entanglement of the state ϱ on $\mathcal{B}$ to $\sigma = \pi^{\intercal} (I)$ (or of ς on $\mathcal{A}$ to $\rho = \pi (I)$). The entanglement $\pi = \pi_q = \pi^{\intercal}$ by

$$\pi_q (B) = \rho^{1/2} \tilde{B} \rho^{1/2}, \quad B \in \mathcal{B} \tag{12}$$

of the state $\varsigma = \varrho$ on the algebra $\mathcal{A} = \mathcal{B}$ is called standard for $(\mathcal{B}, \varrho)$.

The standard entanglement defines the standard compound state

$$\varpi_q (A \otimes B) = \mathrm{tr}_{\mathcal{H}} B\rho^{1/2} \tilde{A} \rho^{1/2} = \mathrm{tr}_{\mathcal{H}} A\rho^{1/2} \tilde{B} \rho^{1/2}. \tag{13}$$

2. ENTANGLED QUANTUM ENTROPY

As we shall prove in this section, the most informative for a quantum system $(\mathcal{B}, \varrho)$ is the standard entanglement $\pi_0 = \pi_q = \pi_0^\intercal$ of the probe system $(\mathcal{A}^0, \varsigma_0) = (\mathcal{B}, \varrho)$, described in (12).

Let us consider the entangled mutual information and quantum entropies of states by means of the different types of compound states. To define the quantum mutual information, we need to apply a quantum version of the relative entropy to compound state on the algebra $\mathcal{A} \otimes \mathcal{B}$, called also the informational divergency of the state ϖ with respect to a reference state φ. It can be defined by the density operators ω, ϕ of these states as

$$\mathfrak{I}(\varpi; \varphi) = \mathrm{tr}\omega(\ln \omega - \ln \phi). \tag{14}$$

The most important property of the information divergence $\mathfrak{I}$ is its monotonicity property [7, 8], i. e. nonincrease of the divergency $\mathfrak{I}(\varpi_0; \varphi_0)$ on the application of the pre-dual K_* of a normal CP unital map $\mathrm{K} : \mathcal{M} \to \mathcal{M}^0$ to the states ϖ_0 and φ_0 on a von Neumann algebra $\mathcal{M}^0$:

$$\varpi = \varpi_0 \mathrm{K}, \varphi = \varphi_0 \mathrm{K} \Rightarrow \mathfrak{I}(\varpi; \varphi) \leq \mathfrak{I}(\varpi_0; \varphi_0). \tag{15}$$

The *mutual information* $\mathfrak{I}_{\mathcal{A},\mathcal{B}}(\pi) = \mathfrak{I}_{\mathcal{B},\mathcal{A}}(\pi^\intercal)$ in a compound state ϖ achieved by an entanglement $\pi : \mathcal{B} \to \mathcal{A}_*$, or by $\pi : \mathcal{A} \to \mathcal{B}_*$ with

$$\varsigma(A) = \varpi(A \otimes I) = \mathrm{tr}_{\mathcal{G}} A\sigma, \ \varrho(B) = \varpi(I \otimes B) = \mathrm{tr}_{\mathcal{H}} B\rho \tag{16}$$

is defined as the relative entropy (14) of the state ϖ on $\mathcal{M} = \mathcal{A} \otimes \mathcal{B}$ with respect to the product state $\varphi = \varsigma \otimes \varrho$:

$$\mathfrak{I}_{\mathcal{A},\mathcal{B}}(\pi) = \mathrm{tr}\, \omega (\ln \omega - \ln(\sigma \otimes I) - \ln(I \otimes \rho)). \tag{17}$$

The next proposition follows from the monotonicity property (15) of the relative entropy on $\mathcal{M} = \mathcal{A} \otimes \mathcal{B}$ with respect to the predual $\mathrm{K}_*(\omega_0) = \varpi_0(\mathrm{K} \otimes \mathrm{I})$ to the ampliation $\mathrm{K} \otimes \mathrm{I}$ of a normal completely positive unital map $\mathrm{K} : \mathcal{A} \to \mathcal{A}^0$.

Proposition 2.1. Let $\pi_0 : \mathcal{A}^0 \to \mathcal{B}_*$ be an entanglement of a state $\varsigma_0(A) = \mathrm{tr}\pi_0(A)$, $A \in \mathcal{A}^0 = \mathcal{L}(\mathcal{G}_0)$, to $(\mathcal{B}, \varrho)$ with $\rho = \pi^0(I)$ on $\mathcal{B}$, where $\pi^0 = \pi_0^\intercal$, and $\pi = \pi_0 \mathrm{K}$ be an entanglement defined as the composition of π_0 with a normal CP unital map $\mathrm{K} : \mathcal{A} \to \mathcal{A}^0$. Then $\mathfrak{I}_{\mathcal{A},\mathcal{B}}(\pi) \leq \mathfrak{I}_{\mathcal{A}^0,\mathcal{B}}(\pi^0)$. In particular, for any c-entanglement π to $(\mathcal{B}, \varrho)$ there exists a not less informative d-entanglement $\pi_d(A) = \sum \langle n|A|n\rangle \rho(n)$ with the same $\rho = \sum \rho(n)$, and the standard entanglement $\pi_q(A) = \rho^{1/2}\tilde{A}\rho^{1/2}$ of $\varsigma_0 = \varrho$ on $\mathcal{A}^0 = \mathcal{B}$ is the maximal one in this sense.

Note that any extreme d-entanglement decomposed into pure normalized states $\rho_n = \rho(n)/\mathrm{tr}\rho(n)$ is maximal among all c-entanglements in the sense $\mathfrak{I}_{\mathcal{A},\mathcal{B}}(\pi_d) \geq \mathfrak{I}_{\mathcal{A},\mathcal{B}}(\pi^c)$. The supremum of the information gain (17) over all c-entanglements to the system $(\mathcal{B}, \varrho)$ is the von Neumann entropy

$$S_{\mathcal{B}}(\varrho) = -\mathrm{tr}_{\mathcal{H}}\rho \ln \rho. \tag{18}$$

It is achieved on any extreme π_d, for example given by a Schatten decomposition $\rho = \sum_n |n\rangle\langle n|\lambda(n)$. The maximal value $\ln \mathrm{rank}\mathcal{B}$ of the von Neumann entropy is defined by $\mathrm{rank}\mathcal{B} = (\dim \mathcal{B})^{1/2}$, i.e. by $\dim \mathcal{H}$.

Definition 2.1. The maximal mutual entropy

$$H_{\mathcal{B}}(\varrho) = \sup_{\pi(I)=\rho} \mathfrak{I}_{\mathcal{A},\mathcal{B}}(\pi) = \mathfrak{I}_{\mathcal{B},\mathcal{B}}(\pi_q), \tag{19}$$

achieved on $\mathcal{A}^0 = \mathcal{B}$ by the standard q-entanglement $\pi_q(B) = \rho^{1/2}\tilde{B}\rho^{1/2}$ for a fixed $\varrho(B) = \mathrm{tr}_{\mathcal{H}}B\rho$ is called q-entropy of the state ϱ. The difference

$$H_{\mathcal{B}|\mathcal{A}}(\varpi) = H_{\mathcal{B}}(\varrho) - \mathfrak{I}_{\mathcal{A},\mathcal{B}}(\varpi) \tag{20}$$

called the q-conditional entropy on $\mathcal{B}$ with respect to $\mathcal{A}$.

Obviously, $H_{\mathcal{B}|\mathcal{A}}(\varpi)$ is positive in contrast to $S_{\mathcal{B}}(\varrho) - \mathfrak{I}_{\mathcal{A},\mathcal{B}}(\varpi)$ which can achieve also the negative value

$$S_{\mathcal{B}}(\varrho) - H_{\mathcal{B}}(\varrho) = \mathrm{tr}\rho \ln \rho \tag{21}$$

in the case the standard entanglement as the following theorem states.

Theorem 2 *The q-entropy for the simple algebra $\mathcal{B} = \mathcal{L}(\mathcal{H})$ is given by*

$$H_{\mathcal{B}}(\varrho) = -2\mathrm{tr}_{\mathcal{H}}\rho \ln \rho = 2S_{\mathcal{B}}(\rho), \tag{22}$$

It is positive, $H_{\mathcal{B}}(\varrho) \in [0, \infty]$, and if $\mathcal{B}$ is finite dimensional, it is bounded, with the maximal value $H_{\mathcal{B}}(\varrho^\circ) = \ln \dim \mathcal{B}$ which is achieved on the tracial $\rho^\circ = (\dim \mathcal{H})^{-1} I$, where $\dim \mathcal{B} = (\dim \mathcal{H})^2$.

3. ENTAGLED QUANTUM CAPACITY

A noisy quantum channel sends pure input states ϱ_1 on the algebra $\mathcal{B}^1 = \mathcal{L}(\mathcal{H}_1)$ into mixed ones $\varrho = \Lambda^*(\varrho_1)$ given by the predual $\Lambda_* = \Lambda^*|\mathcal{B}^1_*$ to a normal completely positive unital map $\Lambda : \mathcal{B} \to \mathcal{B}^1$,

$$\Lambda(B) = \mathrm{tr}_{\mathcal{F}_1} Y^\dagger B Y, \quad B \in \mathcal{B} \tag{23}$$

where Y is a linear operator from $\mathcal{H}_1 \otimes \mathcal{F}_+$ to $\mathcal{H}$ with $\mathrm{tr}_{\mathcal{F}_+} Y^\dagger Y = I$, and $\mathcal{F}_+$ is a separable Hilbert space of quantum noise in the channel. Each input mixed state ϱ_1 is transmitted into an output state $\varrho = \varrho_1\Lambda$ given by the density operator

$$\Lambda^* (\rho_1) = Y \left(\rho_1 \otimes I^+\right) Y^\dagger \in \mathcal{B}_* \tag{24}$$

for each density operator $\rho_1 \in \mathcal{B}_*^1$ (I^+ is the identity operator in $\mathcal{F}_+$).

The input entanglements $\pi^1 = \mathcal{A} \to \mathcal{B}_*^1$ dual to $\pi_1 : \mathcal{B}^1 \to \mathcal{A}$ will be denoted as $\pi^1 = \kappa = \pi_1^\intercal$. They define the quantum-quantum correspondences (q-encodings) of probe systems $(\mathcal{A}, \varsigma)$ with the density operator $\sigma = \kappa^\intercal\left(I^1\right)$, to the input $\left(\mathcal{B}^1, \varrho_1\right)$ of the channel Λ with $\rho = \kappa(I)$.

Let $\mathcal{K}_q^1$ be the set of all normal completely positive maps $\kappa : \mathcal{A} \to \mathcal{B}_*^1$ with any probe algebra $\mathcal{A}$, normalized as $\mathrm{tr}\kappa(I) = 1$, and $\mathcal{K}_q(\varrho_1)$ be the subset of $\kappa \in \mathcal{K}_q^1$ with $\kappa(I) = \varrho_1$. Each $\kappa \in \mathcal{K}_q^1$ can be decomposed as $\pi_q^1\mathrm{K}$, where $\pi_q^1 = \kappa_0$ is the standard entanglement on $\left(\mathcal{A}^0, \varsigma_0\right) = \left(\mathcal{B}^1, \varrho_1\right)$, and K is a normal unital CP map $\mathcal{A} \to \mathcal{B}^1$. Further let $\mathcal{K}_c^1$ be the set of all c-entanglements κ described by the combinations $\kappa(A) = \sum_n \varsigma_n(A)\rho_1(n)$of the primitive maps $A \mapsto \varsigma_n(A)\rho_1(n)$, and $\mathcal{K}_d^1$ be the subset of the diagonalizing entanglements κ, i.e. the decompositions $\kappa(A) = \sum_n \langle n|A|n\rangle \rho_1(n)$. As in the first case $\mathcal{K}_c(\varrho_1)$ and $\mathcal{K}_d(\rho_1)$ denote the subsets corresponding to a fixed $\kappa(I) = \varrho_1$, and each $\mathcal{K}_c(\varrho_1)$ can be represented as $\kappa = \pi_d^1\mathrm{K}$, where $\pi_d^1 = \kappa_0$ is a pure d-entanglement $\mathcal{A}^0 = \mathcal{B}^1 \to \mathcal{B}_*^1$ normalized as $\mathrm{tr}\pi_d^1(B) = \varrho_1(B)$ for all $B \in \mathcal{B}^1$, by a proper choice of the CP map $\mathrm{K} : \mathcal{A} \to \mathcal{B}^1$. Furthermore let $\mathcal{K}_o^1$ and $\mathcal{K}_o(\varrho_1)$ be the subsets with orthogonal $\rho_1(n)$ (and fixed $\sum_n \rho_1(n) = \rho_1$). Each $\kappa \in \mathcal{K}_o(\varrho_1)$ can also be represented as $\kappa = \kappa_o^1\mathrm{K}$, where $\kappa_o^1 = \kappa_0$ is a pure o-entanglement on $\mathcal{A}^0 = \mathcal{B}^1$ corresponding to the same ϱ_1.

Now, let us maximize the entangled mutual entropy for a given quantum channel Λ (and a fixed input state ϱ_1) by means of the above four types entanglements κ.

Proposition 3.1. The entangled mutual informations achieve the following maximal values

$$\sup_{\kappa \in \mathcal{K}_q(\varrho_1)} \mathfrak{I}_{\mathcal{A},\mathcal{B}}\left(\Lambda^*\kappa\right) = \mathfrak{I}_q(\varrho_1, \Lambda) := \mathfrak{I}_{\mathcal{B}^1,\mathcal{B}}\left(\Lambda^*\pi_q^1\right), \tag{25}$$

$$\mathfrak{I}_c(\varrho_1, \Lambda) = \sup_{\kappa \in \mathcal{K}^c(\varrho_1)} \mathfrak{I}_{\mathcal{A},\mathcal{B}}\left(\Lambda^*\kappa\right) = \sup_{\pi_d^1} \mathfrak{I}_{\mathcal{B}^1,\mathcal{B}}\left(\Lambda^*\pi_d^1\right) = \mathfrak{I}_d(\varrho_1, \Lambda), \tag{26}$$

$$\sup_{\kappa \in \mathcal{K}_o(\varrho_1)} \mathfrak{I}_{\mathcal{A},\mathcal{B}}\left(\Lambda^*\kappa\right) = \mathfrak{I}_o(\varrho_1, \Lambda) := \sup_{\pi_o^1} \mathfrak{I}_{\mathcal{B}^1,\mathcal{B}}\left(\Lambda^*\pi_o^1\right), \tag{27}$$

where $\pi_\cdot^1$ are the corresponding extremal input entangled states on $\mathcal{B}^1 \otimes \mathcal{B}^1$ with $\mathrm{tr}\pi_\cdot^1(B) = \varrho_1(B)$ for all $B \in \mathcal{B}^1$. They are ordered as

$$\mathfrak{I}_q(\varrho_1, \Lambda) \geq \mathfrak{I}_c(\varrho_1, \Lambda) = \mathfrak{I}_d(\varrho_1, \Lambda) \geq \mathfrak{I}_o(\varrho_1, \Lambda). \tag{28}$$

We shall denote the information $\mathfrak{I}_c = \mathfrak{I}_d$ simply as $\mathfrak{I}_1(\varrho_1, \Lambda)$.

Definition 3.1. The suprema

$$C_q(\Lambda) = \sup_{\kappa \in \mathcal{K}_q^1} \mathfrak{I}_{\mathcal{A},\mathcal{B}}(\Lambda^* \kappa) = \sup_{\varrho_1} \mathfrak{I}_q(\varrho_1, \Lambda), \tag{29}$$

$$\sup_{\kappa \in \mathcal{K}_d^1} \mathfrak{I}_{\mathcal{A},\mathcal{B}}(\Lambda^* \kappa) = C_1(\Lambda) := \sup_{\varrho_1} \mathfrak{I}_1(\varrho_1, \Lambda), \tag{30}$$

$$C_o(\Lambda) = \sup_{\kappa \in \mathcal{K}_o^1} \mathfrak{I}_{\mathcal{A},\mathcal{B}}(\Lambda^* \kappa) = \sup_{\varrho_1} \mathfrak{I}_o(\varrho_1, \Lambda), \tag{31}$$

are called the q-, c- or d-, and o-capacities respectively for the quantum channel defined by a normal unital CP map $\Lambda : \mathcal{B} \to \mathcal{B}^1$.

Obviously the capacities (30) satisfy the inequalities

$$C_o(\Lambda) \leq C_1(\Lambda) \leq C_q(\Lambda). \tag{32}$$

Theorem 3 *Let $\Lambda(B) = Y^\dagger B Y$ be a unital CP map $\mathcal{B} \to \mathcal{B}^1$ describing a quantum deterministic channel. Then*

$$\mathfrak{I}_1(\varrho_1, \Lambda) = \mathfrak{I}_o(\varrho_1, \Lambda) = S(\varrho_1), \quad \mathfrak{I}_q(\varrho_1, \Lambda) = H(\varrho_1), \tag{33}$$

and thus in this case

$$C_1(\Lambda) = C_o(\Lambda) = \ln \operatorname{rank} \mathcal{B}^1, \quad C_q(\Lambda) = \ln \dim \mathcal{B}^1 \tag{34}$$

4. ADDITIVITY OF ENTANGLED CAPACITY

In order to consider block entanglements let us introduce the product systems $(\mathcal{B}^{\otimes n}, \varrho^{\otimes n})$ in the tensor product $\mathcal{H}^{\otimes n} = \otimes_{l=1}^n \mathcal{H}_l$ of identical spaces $\mathcal{H}_l = \mathcal{H}$, and the product channels $\Lambda_*^{\otimes n} : \mathcal{B}_*^n \mapsto \mathcal{B}_*^{\otimes n}$, the preduals of the normal unital CP product maps

$$\Lambda^{\otimes n} : \mathcal{B}^{\otimes n} \to \mathcal{B}^n = \otimes_{l=1}^n \mathcal{B}_l^1, \quad \Lambda^{\otimes n}(B^{\otimes n}) = \Lambda(B)^{\otimes n}, \tag{35}$$

where $\mathcal{B}_l = \mathcal{B}^1$ for all l. Let us denote

$$\mathfrak{I}_q^n(\varrho_1, \Lambda) = \mathfrak{I}_q(\varrho_1^{\otimes n}, \Lambda^{\otimes n}), \quad C_q^n(\Lambda) = C_q(\Lambda^{\otimes n}) \tag{36}$$

the suprema of the mutual information $\mathfrak{I}_{\mathcal{A},\mathcal{B}^{\otimes n}}(\Lambda_*^{\otimes n}\kappa)$ over input entanglements $\kappa : \mathcal{A} \to \mathcal{B}_*^n$ with any probe algebra $\mathcal{A}$, normalized as $\kappa(I) = \rho_1^{\otimes n}$, and with any such normalization respectively It is easily seen, by applying the monotonicity property (15) with respect to the normal unital CP map $\mathrm{K} : \mathcal{A} \to \mathcal{A}^0 = \mathcal{B}^n$ in $\kappa = \pi_q^n \mathrm{K}$, where

$$\pi_q^n(A^0) = (\rho_1^{\otimes n})^{1/2} \tilde{A}^0 (\rho_1^{\otimes n})^{1/2}, \quad A^0 \in \mathcal{B}^n, \tag{37}$$

that the quantities $\mathfrak{I}_q^n(\varrho_1, \Lambda)$, $C_q^n(\Lambda)$ are additive:

$$\mathfrak{I}_q^n(\varrho_1, \Lambda) = n\mathfrak{I}_q(\varrho_1, \Lambda), \quad C_q^n(\Lambda) = nC_q(\Lambda). \tag{38}$$

This additivity cannot be proved but super-additivity for the quantities

$$\mathfrak{I}_1^n(\varrho_1, \Lambda) = \mathfrak{I}_1\left(\varrho_1^{\otimes n}, \Lambda^{\otimes n}\right), \quad C_1^n(\Lambda) = C_1\left(\Lambda^{\otimes n}\right) \tag{39}$$

corresponding to the c-entanglements (or d-entanglements) which are usually used to describe the classical-quantum encodings. This implies that the quantities $\mathfrak{I}_n = \mathfrak{I}_1^n/n, C_n = C_1^n/n$ have the limits

$$\mathfrak{I}(\varrho_1, \Lambda) = \lim_{n\to\infty} \mathfrak{I}_n((\varrho_1, \Lambda)) \geq \mathfrak{I}_1(\varrho_1, \Lambda), \quad C(\Lambda) = \lim C_n(\Lambda) \geq C_1(\Lambda) \tag{40}$$

which are usually taken as the bounds of classical information and capacity [9, 10] for a quantum channel Λ. As it has been recently proved in [11], the upper bound $C(\Lambda)$ is indeed asymptotically achievable by long block classical-quantum encodings. Note that the c-quantities $\mathfrak{I}_n$, C_n are not easy to evaluate for each n, but they all are bounded by the corresponding q-quantities $\mathfrak{I}_q$, C_q, and

$$\mathfrak{I}(\varrho_1, \Lambda) \leq \mathfrak{I}_q(\varrho_1, \Lambda), \quad C(\Lambda) \leq C_q(\Lambda). \tag{41}$$

In order to measure the "real" entangled information of quantum channels "without the classical part", another quantity, the "coherent information" was introduced in [3]. It is defined in our notations as

$$\mathfrak{I}_s(\varrho_1, \Lambda) = \mathfrak{I}_q(\varrho_1, \Lambda) - S(\varrho_1), \quad C_s^1(\Lambda) = \sup_{\varrho_1} \mathfrak{I}_s(\varrho_1, \Lambda) \tag{42}$$

if these quantities are positive, otherwise they are taken to be zero. The supremum $C_s^n(\Lambda) = C_s(\Lambda^{\otimes n})$ of $\mathfrak{I}_s(\varrho_1^n, \Lambda^{\otimes n})$ over all states $\varrho_1^n \in \otimes_1^n \mathcal{B}_*^1$ in general is not additive but superadditive, and the coherent capacity is defined as the limit

$$C_s(\Lambda) = \lim_{n\to\infty} C_s^n(\Lambda)/n. \tag{43}$$

Obviously this capacity has the bounds

$$C_s^1(\Lambda) \leq C_s(\Lambda) \leq C_q(\Lambda) \tag{44}$$

as $\mathfrak{I}_s(\varrho_1^n, \Lambda^{\otimes n}) \leq \mathfrak{I}_q(\varrho_1^n, \Lambda^{\otimes n}) \geq 0$ for each input state ϱ_1.

Thus the entangled information $\mathfrak{I}_q(\varrho_1, \Lambda)$ for a single channel corresponding to the standard entanglement κ_1 and the entangled capacity $C_q(\Lambda) = C_q^1(\Lambda)$ give upper bounds of all other capacities and are good additive analogs of the corresponding classical quantities.

References

[1] Belavkin, V.P., and M. Ohya, Los Alamos Archive Quant–Ph/9812082, 1-16, 1998.

[2] Bennett, C.H. and G. Brassard, C. Crépeau, R. Jozsa, A. Peres, W.K. Wootters, Phys. Rev. Lett., **70**, pp.1895-1899, 1993.

[3] Schumacher, B., Phys. Rev. A, **51**, pp.2614-2628, 1993; Phys. Rev. A, **51**, pp.2738-2747, 1993, Phys. Rev. A, **54**, p.2614, 1996.

[4] Jozsa, R. and B. Schumacher, J. Mod. Opt., **41**, pp.2343-2350, 1994.

[5] Stinespring, W. F., Proc. Amer. Math. Soc. **6**, p. 211 (1955).

[6] Kraus, K., Ann. Phys. **64**, p. 311 (1971).

[7] Lindblad, G., Comm. in Math. Phys. **33**, p305–322 (1973).

[8] Uhlmann, A., Commun. Math. Phys., **54**, pp.21-32, 1977.

[9] Holevo, A.S, Probl. Peredachi Inform., **9**, no. 3, pp3-11, 1973.

[10] Stratonovich, R.S. and A.G. Vancian, Probl. Control Inform. Theory, **7**, no. 3, pp.161-174, 1978.

[11] Hausladen P., R. Jozsa, B. Schumacher, M. Westmoreland, W. Wootters, Phys. Rev. A, **54**, no. 3, pp. 1869-1876, 1996.

CONTROL OF SQUEEZED LIGHT PULSE SPECTRUM IN THE KERR MEDIUM WITH AN INERTIAL NONLINEARITY

A. S. Chirkin, F. Popescu
Moscow State University,
119899, Moscow, Russia
chirkin@foton.ilc.msu.su florentin_p@hotmail.com

Keywords: Self-Phase Modulation, Cross-Phase Modulation, Squeezed Light

Abstract The consistent quantum theory of self-phase modulation (**SPM**) and cross-phase modulation (**XPM**) for ultrashort light pulses (**USPs**) in medium with electronic Kerr-nonlinearity are developed. The approach makes use of momentum operator of electrical field which takes account of the inertial behaviour of the nonlinearity. The spectrum of quantum fluctuations of squeezed-quadrature component as a function of response time of nonlinearity and values of nonlinear phase shifts due to the **SPM** and **XPM** effects, is investigated.

1. INTRODUCTION

There are some methods to obtain USPs in a nonclassical state. One is the degenerate three-frequency parametric amplification which has high phase sensitivity. Other is SPM in a Kerr-nonlinear medium which in comparison with parametrical processes, does not require phase-matching. This property is a real advantage for the pulsed nonclassical state formation. The both participation of SPM and dispersion in the nonlinear medium leads to the optical soliton formation [1]. The quantum theory of SPM and XPM for USPs developed during last 15 years have difficulties produced, in particular, by the multi-frequency structure of USPs and nonlinear kind of their interaction.

The aim of our study is to develop the consistent quantum theory of nonlinear propagation of USPs. An account is taken of the role of a response time of the nonlinearity. The first attempt has been made in [2] to give quantum analysis of SPM of USPs for the case of relaxation nonlinearity of medium, and has been noted that the correct quantum theory should take account of additional noise sources connected with nonlinear absorption. However, the theory developed in [2] does not take account of noise sources, so that the commutation relation for annihilation and creation photon operators is not fulfilled. This was carried out in the developed in [3] quantum theory of SPM, where noise sources are considered as a fluctuation addition to the relaxation nonlinearity of medium. Therefore, the results of [3] are connected with the range of carrier frequencies of the pulse, where the nonlinear absorption is important. Actually the authors [3] have developed the quantum theory of USP propagation in a Raman active medium. In [3] the electronic Kerr nonlinearity with finite response time is modeled as the Raman active medium. Note, if we deal with the USPs propagation, for example, through fused-silica fibres only about 0.2 of the Kerr effect is attributable to the Raman oscillators and about 0.8 of the Kerr effect is due to electronic motion [4]. The case when the USP carrier frequency is far enough from some resonances (one- and two-photon and the Raman resonances), and therefore absorption is absent, was investigated by us in [5] where the finite response time is considered in the interaction Hamiltonian, and the commutation relation is exactly fulfilled.

In the present work the results of the consistent quantum theory of SPM and XPM of USPs based on the use of the momentum operator (quantity of movement) are presented. The developed approach can be used when the response time is much shorter than the pulse duration and dispersion in nonlinear medium is neglected. There are no restrictions on the pulse intensity in our theory.

2. QUANTUM THEORY OF SPM OF USPs

The traditional way to describe the SPM of USP is based on the interaction Hamiltonian use, at that usually we solve the time-evolution equation. The transition to the spatial-cvolution equation is realized substituting t with z/u, where z is the distance passed in medium and u is the group velocity. This approach seems to be good enough for single-mode radiation. If we deal with nonlinear propagation of USP then both t and z are present in analytical description. Thus, we use the momentum operator of pulse field related to the space-evolution [6]. We describe the SPM of USP using the momentum operator (cf. [5])

$$\hat{G}_{spm}(z) = \frac{1}{2}\hbar\beta \int_{-\infty}^{\infty} dt \int_{-\infty}^{t} H(t-t_1)\,\hat{\mathbf{N}}\left[\hat{n}(t,z)\hat{n}(t_1,z)\right] dt_1, \quad (1)$$

where $\hat{n}(t,z) = \hat{A}^+(t,z)\hat{A}(t,z)$, $\hat{A}^+(t,z)$ $(\hat{A}(t,z))$ is the Bose operator creating (annihilating) photons in a given cross-section z of the medium at a given time t, $\hbar$ is the Planck's constant and $\hat{\mathbf{N}}$ is the operator of normal ordering. The coefficient β is defined by the Kerr nonlinearity of a medium at stationary conditions. $H(t)$ is the function of the nonlinear response of a medium; $H(t) \neq 0$ at $t \geq 0$ and $H(t) = 0$ at $t < 0$. The expression under the first integral in

(1) can be interpreted as generalized force acting in a defined cross-section z, which at the moment of time t depends only on the previous ones ($t_1 \leq t$), i.e. it satisfies the causality principle. The nonlinear response function of a medium should be introduced as :

$$H(t) = (1/\tau_r)e^{-t/\tau_r} \qquad (t \geq 0), \tag{2}$$

where τ_r is the response time of nonlinear medium, that we assume to be much shorter than the pulse duration τ_p. This nonlinearity takes place in absence of the one- and two- photons absorption and Raman resonances [1]. Besides, as it will be shown below, also in this limit case the account of finite nonlinear response time plays an important role. The space evolution equation for $\hat{A}(t,z)$ follows from (see [6])

$$-i\hbar\frac{\partial\hat{A}(t,z)}{\partial z} = \left[\hat{A}(t,z), \hat{G}_{spm}(z)\right], \tag{3}$$

and making use of (1), we finally get

$$\frac{\partial\hat{A}(t,z)}{\partial\, z} - i\frac{1}{2}\beta q\left[\hat{n}(t,z)\right]\hat{A}(t,z) = 0, \tag{4}$$

where

$$q\left[\hat{n}(t,z)\right] = \int_0^{\infty} H(t_1)\left[\hat{n}(t-t_1,z) + \hat{n}(t+t_1,z)\right] dt_1. \tag{5}$$

Here $\hat{n}(t,z) = \hat{A}^+(t,z)\hat{A}(t,z)$ is the photon number "density". The Eq. (4) describes SPM in the moving coordinate frame: $z = z^{'}, t = t^{'} - z/u$, where $t^{'}$ is the running time. According to (4) the operator $\hat{n}(t,z)$ does not depend on z, i.e. $\hat{n}(t,z) = \hat{n}(t,z=0) = \hat{n}_0(t) = \hat{A}_0^+(t)\hat{A}_0(t)$ ($z=0$ corresponds to the entrance into the medium). Taking account of this fact $q[\hat{n}(t,z)] = q[\hat{n}_0(t)]$, and (5) can be rewritten as:

$$q\left[\hat{n}_0(t)\right] = \int_{-\infty}^{\infty} h(t_1)\hat{n}_0(t-t_1)\, dt_1, \qquad (h(t) = H(|t|)). \tag{6}$$

Taking account of (6), we get the solution of (4)

$$\hat{A}(t,z) = \exp\left\{i\gamma q\left[\hat{n}_0(t)\right]\right\}\hat{A}_0(t). \tag{7}$$

For hermitian conjugate operator of $\hat{A}(t,z)$ we have

$$\hat{A}^+(t,z) = \hat{A}_0^+(t)\exp\left\{-i\gamma q\left[\hat{n}_0(t)\right]\right\}. \tag{8}$$

In expressions (7), (8) $\gamma = \beta z/2$. For time-independent operator $\hat{n}_0$, Eqs. (7), (8) give us the results for single mode radiation [7]. In the case of the instantaneous response of the nonlinearity $H(t) = \delta(t)$ we get [2]

$$\hat{A}(t,z) = e^{i2\gamma\hat{n}_0(t)}\hat{A}_0(t), \qquad \hat{A}^+(t,z) = \hat{A}_0^+(t)e^{-i2\gamma\hat{n}_0(t)}. \tag{9}$$

At the input to the medium Bose operators satisfy the commutation relation $[\hat{A}_0(t_1), \hat{A}_0^+(t_2)] = \delta(t_1 - t_2)$. In a correct quantum methodology the analogical commutation relation must also be satisfied in nonlinear medium, i.e. for any z

$$[\hat{A}(t_1, z), \hat{A}^+(t_2, z)] = \delta(t_1 - t_2). \tag{10}$$

The calculation of average values of (9) and the normal ordering are followed by the non-integrable singularities appearance (for instance, $\delta(t)$ appears in the exponent). These shortcomings are absent using (7) and (8). Besides, their use requires knowledge of an algebra of time-dependent Bose operators. The later has been developed in [2, 8]. The following permutation relations [8]

$$\hat{A}_0(t_1)e^{\hat{O}(t_2)} = e^{\hat{O}(t_2)+i\gamma h(t_2-t_1)}\hat{A}_0(t_1), \tag{11}$$
$$e^{\hat{O}(t_2)}\hat{A}_0^+(t_1) = \hat{A}_0^+(t_1)e^{\hat{O}(t_2)+i\gamma h(t_2-t_1)}, \tag{12}$$

(where $\hat{O}(t) = i\gamma q[\hat{n}_0(t)]$) and the theorem of normal ordering [9]

$$e^{\hat{O}(t)} = \hat{\mathbf{N}} \exp\left\{\int_{-\infty}^{\infty} \left[e^{i\gamma\tilde{h}(\theta)} - 1\right] \hat{n}_0(t - \tau_r\theta)\, d\theta\right\}, \tag{13}$$

are valid, where $\tilde{h}(\theta) = e^{-|\theta|}$, $\theta = t/\tau_r$. As $[\hat{O}(t_1), \hat{O}(t_2)] = 0$, then $e^{\hat{O}(t_1)}e^{\hat{O}(t_2)} = e^{\hat{O}(t_1)+\hat{O}(t_2)}$. Making use of relations (11) and (12), one can verify that the commutation relation (10) is exactly fulfilled.

As we analyse the SPM of an initial coherent USP, the equation on eigenvalues $\hat{A}_0(t)|\alpha(t)\rangle = \alpha(t)|\alpha(t)\rangle$ is satisfied by $\hat{A}_0(t)$ (see [1]), where $|\alpha(t)\rangle$ is the initial coherent state, $\alpha(t)$ is the eigenvalue of $\hat{A}_0(t)$, and $|\alpha(t)|^2 = \bar{n}_0(t)$ is the average photon density. Particular attention is given to the quantum fluctuations behaviour of X-quadrature; $\hat{X}(t, z) = [\hat{A}(t, z) + \hat{A}^+(t, z)]/2$. The behaviour of another quadrature component is shifted in phase with $\pi/2$. For correlation function (see [5])

$$R(t, \tau) = \langle\hat{X}(t, z)\hat{X}(t + \tau, z)\rangle - \langle\hat{X}(t, z)\rangle\langle\hat{X}(t + \tau, z)\rangle, \tag{14}$$

where the brackets denote averaging over the initial coherent state of the pulse, making use of (11)-(13) we get

$$R(t, \tau) = \frac{1}{4}\left[\delta(\tau) - \psi(t)h(\tau)\sin 2\Phi(t) + \psi^2(t)g(\tau)\sin^2\Phi(t)\right]. \tag{15}$$

In (15) the following notations have been introduced: $\Phi(t) = \psi(t) + \varphi(t)$, $\varphi(t) = arg\ \alpha(t)$ - the phase of USP or the heterodyne pulse's phase at balanced homodyne detection, $\psi(t) = 2\gamma\bar{n}_0(t) = \psi_0\rho^2(t)$ - the nonlinear phase shift resulted from SPM, $\psi_0 = \psi(0) = 2\gamma\bar{n}_0$ - the maximum nonlinear phase shift, $\rho(t)$ - the envelope of USP ($|\alpha(t)| = \alpha_0\rho(t)$, $\rho(0) = 1$), and

$g(\tau) = \int_{-\infty}^{\infty} \tilde{h}(\theta)\tilde{h}(\theta+\tau)\, d\theta$. The derivation of (15) took into consideration the fact that, in many experimental situations nonlinear phase shift per photon $\gamma \ll 1$. The instantaneous spectral density of the quantum fluctuations of X-quadrature component $S_X(\omega, t) = \int_{-\infty}^{\infty} R_X(t,\tau)e^{i\omega\tau}\, d\tau$, according to (15), takes the form

$$S_X(\Omega, t) = \frac{1}{4}\Big[1 - 2\psi(t)L(\Omega)\sin 2\Phi(t) + 4\psi^2(t)L^2(\Omega)\sin^2\Phi(t)\Big], \quad (16)$$

where $L(\Omega) = 1/[1+\Omega^2]$, $\Omega = \omega\tau_r$. From (16) it follows that the quantum fluctuations level bellow the short noise one depends on the nonlinear phase shift $\psi(t)$ and $\Phi(t)$. At the initial phase of the pulse chosen optimal for a frequency $\Omega_0 = \omega_0\tau_r$

$$\varphi_0(t) = \frac{1}{2}\arctan\left[\frac{1}{\psi(t)L(\Omega_0)}\right] - \psi(t), \quad (17)$$

the spectral density (16) is

$$S_X(\Omega_0, t) = \frac{1}{4}\Big\{\left[1 + \psi^2(t)L^2(\Omega_0)\right]^{1/2} - \psi(t)L(\Omega_0)\Big\}^2. \quad (18)$$

From (18) it follows that the spectral density adiabatically changes itself with the changing pulse's envelope and it is lower than $1/4$ which corresponds to the coherent state of the pulse. From (18) we can also see that at the optimal phase $\varphi_0(t)$ with the increasing of the nonlinear phase shift $\psi(t)$, the spectral density $S_X(\Omega_0, t)$ monotonously decreases. We point out that at $\tau_r = 0$ from (16) the results for single mode radiation can be obtained, at that the quantum fluctuation spectrum's level does not depend on the frequency. According to (18), increasing τ_r (increasing Ω_0) at fixed frequency ω_0, the quadrature fluctuations level increases.

3. QUANTUM THEORY OF XPM OF USPs

We analyse now two-pulse propagation in an inertial nonlinear medium. We will consider that pulses have orthogonal polarizations or/and different frequencies. Then, besides SPM of each pulse, the XPM effect takes place. Here we assume that the parametrical interaction of pulses can be neglected. In this case, the analysed process can be depicted making use of the following momentum operator [10].

$$\hat{G}(z) = \sum_{j=1}^{2} \hat{G}_{spm}^{(j)}(z) + \hat{G}_{xpm}^{(1,2)}(z). \quad (19)$$

Here $\hat{G}^{(j)}_{spm}(z)$ is due to the SPM of j-pulse ($j = 1, 2$) (see 1) and operator $\hat{G}^{(1,2)}_{xpm}(z)$ is connected with the XPM of pulses

$$\hat{G}^{(1,2)}_{xpm}(z) = \hbar\widetilde{\beta}\int_{-\infty}^{\infty} dt \int_{-\infty}^{t} H(t-t_1)[\hat{n}_1(t,z)\hat{n}_2(t_1,z)+\hat{n}_2(t,z)\hat{n}_1(t_1,z)]dt_1. \tag{20}$$

In (20) $\hat{n}_j(t,z)$ is the photon number operator of j-pulse and coefficients β_j, $\widetilde{\beta}$ are responsible for the SPM and XPM respectively. As stated earlier, the operator $\hat{n}_j(t,z)$ commutes with $\hat{G}(z)$. In consequence $\hat{n}_j(t,z) = \hat{n}_j(t, z = 0) = \hat{n}_{0,j}(t)$, which means that the photons statistics of each pulse remains unchanged in the nonlinear medium. In accordance with (3), the evolution of $\hat{A}_1(t,z)$ for the first pulse is given by (cf. (4))

$$\frac{\partial \hat{A}_1(t,z)}{\partial z} - i\left\{\frac{1}{2}\beta_1 q\left[\hat{n}_{0,1}(t)\right] + \widetilde{\beta} q\left[\hat{n}_{0,2}(t)\right]\right\}\hat{A}_1(t,z) = 0. \tag{21}$$

One can get for $\hat{A}_2(t,z)$ the similar equation changing the index $1 \mapsto 2$ in (21). The functions $q[\hat{n}_{0,j}(t)]$ is similar to (6). We remark that (4), (21), are written in the moving frame. The solution of (21) is

$$\hat{A}_1(t,z) = e^{i\gamma_1 q[\hat{n}_{0,1}(t)] + i\widetilde{\gamma} q[\hat{n}_{0,2}(t)]}\hat{A}_{0,1}(t), \tag{22}$$

where $\gamma_1 = \beta_1 z/2$, $\widetilde{\gamma} = \widetilde{\beta} z$. We define the correlation function of the investigated light pulse as (14). For the spectra of quantum fluctuations of X_1-quadrature we get

$$S_{X_1}(\Omega,t) = \frac{1}{4}\Big\{1 - 2\psi_1(t)L(\Omega)\sin 2\Phi_{1,2}(t) + 4[\psi_1^2(t)+\widetilde{\psi}_1(t)\widetilde{\psi}_2(t)]L^2(\Omega)\sin^2\Phi_{1,2}(t)\Big\}. \tag{23}$$

Here we denoted: $\Phi_{1,2}(t) = \Phi_1(t) + \widetilde{\psi}_2(t)$, $\Phi_1(t) = \psi_1(t) + \varphi_1(t)$, $\psi_1(t) = 2\gamma_1\bar{n}_{0,1}(t)$, $\widetilde{\psi}_j(t) = 2\widetilde{\gamma}\bar{n}_{0,j}(t)$ ($j = 1, 2$). As already mentioned, the phase $\Phi_1(t)$ is connected with own parameters of the investigated pulse, and the phase $\widetilde{\psi}_1(t)$ with the XPM. From comparison of (23) and (16) one can see that the XPM adds new terms in the multiplier and in the phase in (23). This circumstance allows us to control the fluctuations spectrum of the investigated pulse. At the initial phase

$$\varphi_{0,1}(t) = \frac{1}{2}\arctan\left[\frac{\psi_1(t)}{\psi^*_{1,2}(t)L(\Omega_0)}\right] - \psi_1(t) - \widetilde{\psi}_2(t), \tag{24}$$

chosen optimal for frequency $\Omega_0 = \omega_0\tau_r$, the spectral density (23) is

$$S_{X_1}(\Omega_0,t) = \frac{1}{4}\Big\{1 - 2L(\Omega_0)\left[\psi_1^2(t) + \psi^{*2}_{1,2}(t)L^2(\Omega_0)\right]^{1/2} - \psi^*_{1,2}(t)L(\Omega_0)\Big\}, \tag{25}$$

where $\psi^*_{1,2}(t) = \psi_1^2(t) + \widetilde{\psi}_1(t)\widetilde{\psi}_2(t)$. The influence of the second pulse to the spectrum of the first one is depicted in *Figure 1*. Increasing the photon number

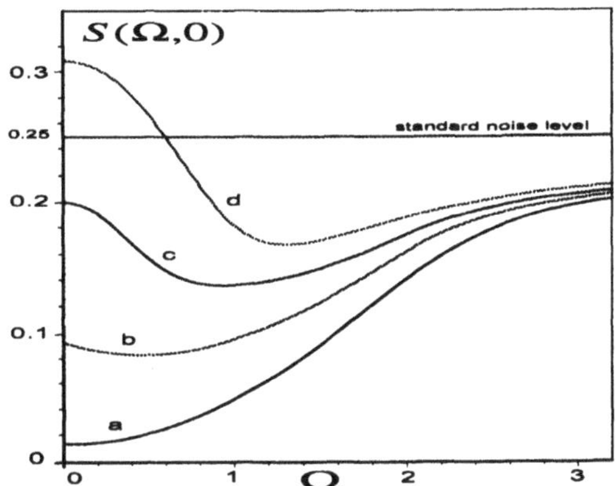

Figure 1 The spectrum of quantum fluctuations of the investigated light pulse at SPM and XPM as a function of reduced frequency $\Omega = \omega\tau_r$ for different relations between average photon numbers of pulses $n_{\bar{0},2}/n_{\bar{0},1}$: 0 (a), 3 (b), 5 (c), 8 (d). The graphics are depicted for the case $\gamma_1 = \gamma_2 = 2\tilde{\gamma}$, and nonlinear phase shift $\psi_{0,1}(0) = 2$ at $t = 0$. At SPM only (curve (a)), the maximum squeezing is realized for $\Omega = 0$.

(intensity) of the second pulse, their quantum fluctuations can substantially increase the level of this ones for the investigated pulse. Changing the intensity of the second pulse, one can control the spectrum of quantum fluctuations of the the investigated pulse's quadrature.

4. CONCLUSIONS

The main result of our work is represented by the development of the consistent quantum theory of SPM and XPM of USPs in non-absorption nonlinear media, when the carrier frequencies of the pulse is far enough from any resonances. However, the consideration of the finite response time of nonlinearity is necessary to get correct solutions ((7), (8), (22)). Our results are valid for response time of nonlinearity much shorter than duration of pulses but for unspecified intensities of USPs. In some sense, our results are complementary with the ones presented in [3] which treats the situation when the Raman resonance is important. The quantum theory of XPM of USPs in the developed approach here is presented for the first time. We have shown that the form and the level of the fluctuation spectrum can be controlled with the change of the pulse's phase and the another pulse's intensity in the presence of XPM. This fact can be used for quantum non-demolition measurements of parameters of the pulse [11].

Acknowledgments

The authors are grateful to K.N. Drabovich (MSU, Moscow) for useful discussions and to S. Codoban (JINR, Dubna) for rendered help. This work has been partial supported by Programme "Fundamental Metrology".

References

[1] S.A. Akhmanov, V.A. Vysloukh, and A.S. Chirkin, *Optics of Femtosecond Laser Pulses*, N.Y.: AIP, 1992.

[2] K.J. Blow, R. Loudon, and S.J.D. Phoenix, J. Opt. Soc. Am. B, 8:1750, 1991.

[3] L. Boivin, F.X. Kärtner, and H.A. Haus, Phys. Rev. Lett., 73:240, 1994; L. Boivin, Phys. Rev. A., 52:754, 1994.

[4] L.G. Joneckis, and J.H. Shapiro, J. Opt. Soc. Am. B, 10:1102, 1993.

[5] F. Popescu, A.S. Chirkin, Pis'ma Zh. Eksp. Teor. Fiz., 69:481, 1999. [JETP Lett. 69:516, 1999].

[6] M. Toren, and Y. Ben-Aryeh, Quantum Opt., 9:425, 1994.

[7] S.A. Akhmanov, A.V. Belinskii, and A.S. Chirkin, in *New Physical Principles of Optical Information Processing* (in Russian). Eds: S.A. Akhmanov, M.A. Vorontsov, Nauka, Moscow, p. 83, 1990.

[8] F. Popescu, and A.S. Chirkin, Phys. Rev. A. (to be published) (see. LANL E-print: quant-ph/0003028), 2000.

[9] K.J. Blow, R. Loudon, S.J.D. Phoenix, and T.J. Shepherd, Phys. Rev. A., 42:4102, 1990.

[10] F. Popescu, and A.S. Chirkin, LANL E-print: quant-ph/0004049 (to be published), 2000.

[11] D.F. Walls, and G.J. Milburn, *Quantum Optics*, Springer, Berlin, 1995.

MACROSCOPIC QUANTUM SUPERPOSITION BY AMPLIFICATION OF ENTANGLED STATES

Giovanni Di Giuseppe
Dipartimento di Fisica and Istituto Nazionale di Fisica della Materia
Universitá "La Sapienza", Roma 00185, Italy
digiuseppe@uniroma1.it

Francesco De Martini
Dipartimento di Fisica and Istituto Nazionale di Fisica della Materia
Universitá "La Sapienza", Roma 00185, Italy
f.demartini@caspur.it

Abstract A novel phase sensitive amplifier/squeezer of entangled two-photon states is adopted to demonstrate the first realization of an all optical multiphoton *Schroedinger-Cat* state with a large, virtually infinite, signal-to-noise ratio. Because of its intrinsic nonlocal structure the new system is expected to play a relevant role in quantum information and in the basic physics of entanglement.

Since the golden years of quantum mechanics the interference of classically distinguishable quantum states, first introduced by the famous "Schroedinger Cat" apologue [1], has been the object of extensive theoretical studies and recognized as a major conceptual paradigm of physics [2, 3]. In modern times the sciences of quantum information and quantum computation deal precisely with collective processes involving a multiplicity of interfering states, generally mutually "entangled" and rapidly de-phased by decoherence [4]. For many respects the experimental implementation of this intriguing classical-quantum condition represents today an open problem in spite of recent successful studies carried out mostly with atoms [5, 6, 7]. Recently one of us proposed the process of the *single-photon quantum injected* parametric amplification (QIOPA) in a nonlinear (NL) crystal as a viable solution towards the realization of an efficient S-Cat scheme with a small decoherence rate [8]. The injected photon was associated with one of the input modes of the squeezer/amplifier in a superposition of *linear* polarization (π) states, viz. a *qubit*. Unexpectedly the implementation of that scheme, albeit successfully leading to the first experimental realization of an "optimum universal quantum cloning machine" [9],

Quantum Communication, Computing, and Measurement 3
Edited by P. Tombesi and O. Hirota, Kluwer Academic/Plenum Publishers, 2001

was found too difficult to represent a practical solution. In facts the high level of the *unfiltered squeezed-vacuum* noise emitted over the output modes of the amplifier prevented the attainment of a suitable signal-to-noise ratio (S/N) at the output [10]. In the present work we present a novel quantum parametric amplifier/squeezer that transforms an input 2 photon π-*entangled* state *(ebit)* into a multiphoton S-Cat π-*entangled* state having the same input phase [11]. Most important, the new system implements an efficient nonlocal interferometric scheme which, on the basis of the *parity* of the input state, selectively conveys on different output channels the squeezed vacuum "*noise*" and the "*signal*", viz. the amplified ebit state. This results in the generation of a multiphoton entangled S-Cat state with a (S/N) which may be large, virtually infinite. Consider the experimental arrangement shown in Figure 1. A NL crystals, BBO (beta barium borate) cut for Type II phase-matching and 1.5 mm thick, was excited in both "*left*" (L-) and "*right*" (R-) directions in virtue of a spherical UV coated rear mirror M_{UV} with curvature radius $(cr_{UV}) = 30$ cm, by two counterpropagating beams derived from a common UV laser.

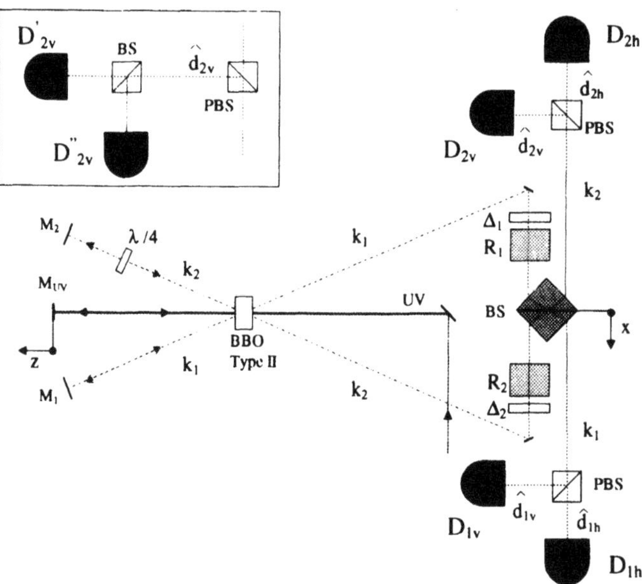

Figure 1 Experimental apparatus.

A computer controlled mount allowed micrometric displacements of M_{UV} along the axis **Z**. The primary UV beam was created by second-harmonic-generation of the output of a Ti:Sa Coherent MIRA mode-locked laser emitting pulses at the wavelength (wl) $\lambda_p = 397.5nm$ with a coherence time $\delta_t = 150fs$ at a 76 Mhz rep-rate and with an average power of 0.3 W. By the L-amplification of the input vacuum, viz. by a Spontaneous Parametric Down Conversion (SPDC) process excited by the primary UV beam focused by a lens with focal length $f_{UV} = 1m$, two photon entangled π-states with a wl $\lambda = 795nm$ were generated with a parity determined by the spatial orientation of the NL crystal [12]. Each emitted photon pair was selected by a couple of pinholes placed in front of two equal spherical mirrors M_j ($j = 1, 2$) with reflectivity 100% at λ and cr_λ =30 cm. Both M_j were placed at an adjustable distance l $\simeq$ 30 cm from the source crystal.

The L-generated two-particle maximally entangled state (Bell-state) may be expressed as:

$$|\Phi\rangle = \frac{1}{\sqrt{2}}\left(|1100\rangle + e^{i\Phi}\,|0011\rangle\right) \tag{1}$$

where the state $|n_1 n_2 n_3 n_4\rangle \equiv |n_{1h}\rangle|n_{2v}\rangle|n_{1v}\rangle|n_{2h}\rangle$ expresses the particle occupancies of the Fock states associated with the relevant $\mathbf{k}$-modes, $\mathbf{k}_j$ ($j = 1, 2$) with horizontal or vertical $\pi-$polarizations. During the M_j back-reflections a $\lambda/4$ plate, with an angular orientation (φ) introduced a change of the phase Φ of the L-generated state before *quantum injection* into the NL crystal. In the present experiment the originally L-generated totally *symmetric triplet* state $|\Phi\rangle_t$=$2^{-\frac{1}{2}}(|1100\rangle + |0011\rangle)$ was transformed before re-injection into the totally *anti-symmetric singlet:* $|\Phi\rangle_s$=$2^{-\frac{1}{2}}(|1100\rangle - |0011\rangle)$ by setting the optical axes of the $\lambda/4$ plate in the horizontal/vertical directions: $\varphi = 0$. In general, by adoption of different $\varphi's$ and of an optional $\pi-$analyzer, all possible two particle maximally- or non maximally entangled states, with *any* parity could be generated. All these states could be amplified by the system albeit the S/N discrimination at the output is dependent on the phase difference $\Delta_\Phi = |\Phi - \Psi|$, being the phase Ψ an intrinsic property of the amplifier, to be defined shortly. Precisely, the condition of maximum S/N, viz. $\Delta_\Phi = \pi$ was chosen for the experiment. Of course the parity selectivity of the system is based on the fact that the symmetry of the state obtained by R-amplification of the *input vacuum* state $|vac\rangle \equiv |0000\rangle$ only depends on the NL crystal orientation, i.e. on the phase Ψ, and is unaffected by the $\lambda/4$ plate placed outside the crystal.

Let us analyze the R-amplification process. For a Type II NL crystal operating in noncollinear configuration the overall amplification process taking place over $\mathbf{k}_j$ is contributed by two equal and independent amplifiers OPA_A and OPA_B inducing unitary transformations respectively on two couples bosonic field operators: $\hat{a}_1(t) \equiv \hat{a}(t)_{1h}$, $\hat{a}_2(t) \equiv \hat{a}(t)_{2v}$ and $\hat{b}_1(t) \equiv \hat{a}(t)_{1v}$, $\hat{b}_2(t) \equiv \hat{a}(t)_{2h}$ for which, at the initial interaction time $t = 0$ is $[\hat{a}(0)_i, \hat{a}(0)_j^\dagger] = [\hat{b}(0)_i, \hat{b}(0)_j^\dagger] = \delta_{ij}$ and $[\hat{a}(0)_i, \hat{b}(0)_j^\dagger] = 0$ [8]. The Hamiltonian of the interaction is expressed in the form: $H_I = i\hbar g[\hat{A}^\dagger + e^{i\Psi}\hat{B}^\dagger] + h.c.$ where: $\hat{A}^\dagger \equiv \hat{a}_1(t)^\dagger \hat{a}_2(t)^\dagger$, $\hat{B}^\dagger \equiv \hat{b}_1(t)^\dagger \hat{b}_2(t)^\dagger$, $g \equiv \chi t$ is a real number expressing the *amplification gain*: χ is the coupling term proportional to the product of the 2nd-order NL susceptibility of the crystal and of the *pump* field, assumed classical and undepleted; t is the interaction time. The quantum dynamics of OPA_A and OPA_B is expressed by the mutually commuting, unitary *squeeze operators*: $U_A(t) = \exp[g(\hat{A}^\dagger - \hat{A})]$ and $U_B(t) = \exp[ge^{i\Psi}(\hat{B}^\dagger - \hat{B})]$ implying the following Bogoliubov transformations for the field operators: $\hat{a}_i(t) = C\hat{a}(0)_i + S\hat{a}(0)_j{}^\dagger$; $\hat{b}_i(t) = C\hat{b}(0)_i + \tilde{S}\hat{b}(0)_j{}^\dagger$ with $i, j = 1, 2$ and $i \neq j$. Here: $C \equiv \cosh g$, $S \equiv \sinh g$, $\Gamma = \tanh g$, $\tilde{S} \equiv e^{i\Psi}S$. Of course the same dynamics holds for the L-amplification, i.e. SPDC, generally with a different value of the gain: $g' = \eta g$, $\Gamma' \simeq \eta\Gamma$ for $\Gamma \ll 1$, being the scaling parameter: $\eta \simeq (cr_{UV}/f_{UV})$. Hereafter, primed and umprimed parameters will refer to the processes of L- and R-amplifications, respectively. The entangled state L-generated by SPDC is found [8]:

$$|\Phi\rangle_L = -\frac{1}{C'} \sum_{n,m:0}^{\infty} e^{im\Psi}\Gamma'^{(n+m)} |nn; mm\rangle \tag{2}$$

with $\Gamma' < 1$. Assume for instance $\Psi = 0$, i.e. a *totally symmetric* Hamiltonian, determined by a suitable spatial orientation of the NL crystal [12]. In this case, corresponding to the experimental conditions, the L-generated state is expressed in increasing powers of $\Gamma' \ll 1$ by: $|\Psi\rangle_L \simeq C'^{-1}[|0000\rangle + \Gamma'(|1100\rangle + |0011\rangle) + \Gamma'^2|1111\rangle + ...]$. The lowest order significant term is a totally symmetric *triplet* state which during back reflection is $\Phi-$phase transformed by the $\lambda/4$ plate, as said. By use of the evolution operator $U_{AB}(t) = U_A(t)U_B(t)$ and of the *disentangling theorem* the quantum injection of the general input state (1) leads to the output state [9]:

$$|\Phi\rangle_{OUT} = U_{AB}(t)\,|\Phi\rangle \simeq \frac{\eta}{C^5}[\sum_{n,m:0}^{\infty} (n + e^{i\Phi}m)\Gamma^{(n+m)}\,|nn; mm\rangle] \tag{3}$$

This expression may be cast in the Schroedinger Cat form: $|\Phi\rangle_{OUT} = 2^{-\frac{1}{2}}[|\Psi_A\rangle + e^{i\Phi}|\Psi_B\rangle]$ being $|\Psi_A\rangle$ and $|\Psi_B\rangle$ *pure* quantum macrostates [8]. Note that the phase Φ of the input state (1) is reproduced into the *output* multiparticle state and determines the quantum superposition character of the S-Cat: this is a common feature of all parametric amplification and squeezing transformations of entangled quantum states [8, 9].

Turn now the attention to the *parity-selective* interferometric part of our system which operates over the output beams $\mathbf{k}_j$ emerging from the QIOPA amplifier. This may be considered an improved version of a previously reported system introduced some years ago as a basic precursor of the modern chapter of *entangled nonlocal interferometry* [12]. Consider the field emitted by the NL crystal after R-amplification. The two beams associated with modes

$\mathbf{k}_j$ $(j = 1, 2)$ were phase shifted $\Delta_j = (\psi_{jh} - \psi_{jv})$ by two equal birefringent plates Δ_j and the π−polarizations were rotated by two equal Fresnel-Rhomb π−rotators $R_j(\theta)$ by angles θ_j respect to directions taken at 45^0 with the horizontal. The beams were then linearly superimposed by a beam splitter (BS) and coupled by two polarizing beam splitters (PBS) to equal EGG SPCM-AQR14 Si-avalanche detectors D_{1h} , D_{1v}, D_{2h}, D_{2v} which measured the horizontal and vertical π−polarizations on the output modes associated with the field operators: $\widehat{d}_{1h}$, $\widehat{d}_{1v}$, $\widehat{d}_{2h}$, $\widehat{d}_{2v}$. A computer controlled mount allowed micrometric displacements of BS along the axis **X**. Consider now the probability of multiple coincidences detected by the $D's$. For instance, the rate of *double* coincidences $(D_{1h}D_{2v}) \equiv \langle\Phi|\widehat{N}_{1h}\widehat{N}_{2v}|\Phi\rangle = (D_{2h}D_{1v})$, where $\widehat{N}_{1h} \equiv \widehat{d}^{\dagger}_{1h}\widehat{d}_{1h}$, is found up to the order n=2 of the gain parameter $S \equiv \sinh g$ and by setting $\Delta \equiv (\Delta_1 - \Delta_2) = 0$, $(\theta_1 + \theta_2) = \frac{1}{2}\pi$, e.g. by eliminating the devices Δ_j and $R_j(\theta)$ from the optical circuit:

$$(D_{1h}D_{2v}) = \frac{1}{4}(1 + \cos\Phi) + S^2(5 + \cos\Phi) \tag{4}$$

The *phase* Φ of the input state critically determines the value of this quantity (This is shown by the data given in Figure 2 as function of the position **X** of the BS).

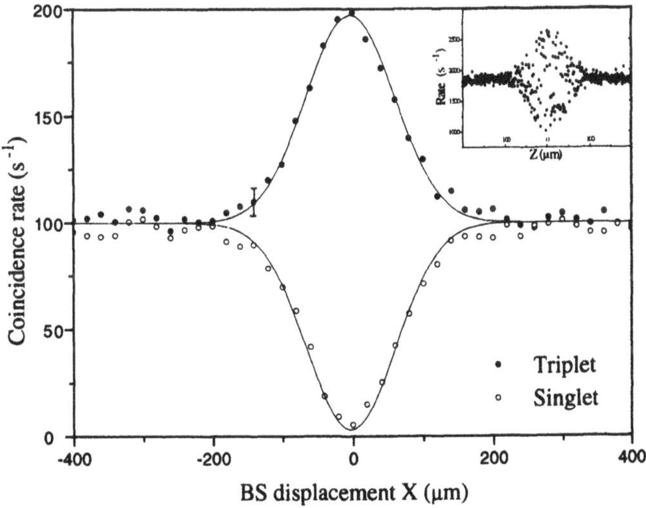

Figure 2 Parity selectivity by the double coincidence $(D_{1h}D_{2v})$.

There the width of the resonance expresses the coherence time $(225fs)$ of the detected photons which is determined in this case by the passband $(\Delta\lambda = 2nm)$ of the equal gaussian IF filters placed in front of the $D's$. Of course the expected behavior is realized when the value of **X** realizes the perfect *in principle* indistinguishability of the Feynman paths affecting the dynamics of each correlated photon couple before detection [12]. A detailed analysis of the system given in [12] shows that the phase selectivity properties expressed by Equation (4) can be *inverted* by adoption of the complementary coincidences $(D_{1h}D_{1v})$=$(D_{2h}D_{2v})$: i.e., there an input *singlet* leads to a resonance *peak*, an input *triplet* to a *dip* etc. Of course changes of the selectivity properties can also be realized by appropriate settings of Δ_i and θ_i.

Let us turn the attention to the R-amplification process by first reporting on an interesting $1^{st} - order$ quantum interference phenomenon. We found that when the spatial position **Z** of the mirror M_{UV} was set at a value realizing the *time superposition* of the back reflected UV pulse wave-packet (wp) with wl λ_p and of the back reflected SPDC generated wp's with wl λ, a sinusoidal interference fringe pattern with periodicity $= \lambda$ and *visibility* V up to 40%, was revealed by the $D's$ within either single detector and multiple coincidence measurements [13] (See Figure 2, inset).

Apart from its intrinsic relevance and novelty, because of its first realization with an entangled state, this result was helpful to determine the value of **Z** corresponding to maximum R-amplification. The main R-amplification was carried out with $\Delta = 0$, $(\theta_1 + \theta_2) = \frac{1}{2}\pi$ by adopting closely the measurement configuration $(D_{2v}D_{2h})$ according to the arguments just given. In fact we found convenient to adopt a related more complex measurement scheme expressed by the concidence rate: $\Re = (D'_{2v}D''_{2v}D_{2h})XOR(D_{1h} + D_{1v})$. This scheme consists of: (a) A *triple coincidence* involving 2 detectors D'_{2v}, D''_{2v} coupled to the output field $\widehat{d}_{2v}$ by a normal 50/50 BS: Figure 1, inset. Furthermore: (b) The triple coincidence was taken in *anti-coincidence* with either D_{1h} or D_{1v}. The reasons for the new options are: (a) In Equation (4) the amplified contribution $\propto S^2$ had to be discriminated against the dominant first term arising from the L-generated, back reflected, *non amplified single* photon couples: this can indeed be obtained by the present BS technique as shown by [14]. (b) The "*noise*" coincidence rate determined by the *input vacuum* field $(D'_{2v}D''_{2v}D_{2h})_{vac}$=$S^4$+$\frac{1}{2}S^2(1 - \cos\Delta)\cos^2(\theta_1 - \theta_2)$ contains a term S^4 arising from *double* detections of output noise photon-couples. These detections are carried out by $D-$couples involving either D_{1h} and D_{1v}. Then the S^4 term was eliminated by the given XOR operation. All this leads, for $\Delta = 0$ to $(D_{2v}D_{2h})_{vac}$= 0, viz. virtually to $S/Nrightarrow\infty$. Of course this ideal condition implies an *ideally perfect* alignment of the nonlocal interferometer, i.e. leading to a 100% visibility (V) of the patterns shown in Figure 2. In practice a value not better than $V \simeq 98\%$ could be attained so far. Another parameter that needs to be improved is the value of the QIOPA gain which is presently small: $g \approx 0.02$.

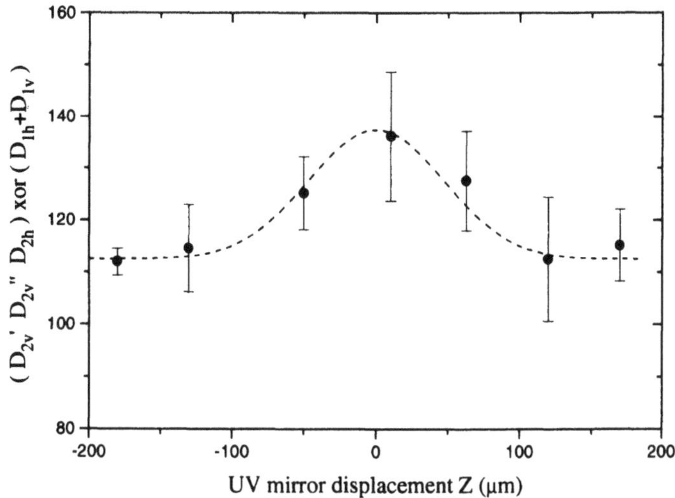

Figure 3 Coincidence rate showing the amplification of an injected 2-photon entangled singlet state as function of the superposition time of the injected and pump wavepackets.

By the adoption within the UV laser system of a Ti:Sa Regenerative Amplifier Coherent REGA9000, already installed in our laboratory, a gain increase up to the value: $g \simeq 0.40$ is expected. The result given in Figure 3 shows the effect of the R-amplification of a quantum-injected 2-photon *entangled "singlet"* state as function of the position Z of the UV mirror M_{UV}. As shown by the above theory this corresponds to the realization of the Schroedinger-Cat state given by Equation (3). The average number of parametrically generated couples has been determined to be ≥ 3. A more detailed account of the experiment and of the results will be published elsewhere. In conclusion we have reported the realization of an all-optical Schroedinger Cat state. Of course the beast is presently not very big. However the present results can be *linearly scaled* by adoption of a more powerful laser apparatus and of larger and more efficient NL crystals. This will be done in the future.

We acknowledge enlightening discussions and very useful collaboration with M. Lucamarini, S. Pirandola, D. Boschi, S. Branca, V. Mussi, F. Bovino, M. D'Ariano. We thank MURST and INFM (Contract PRA97-cat) for funding.

References

[1] E. Schroedinger, *Naturwissenshaften* **23**, 807; 823 (1935).

[2] A. Caldeira, A. Leggett, *Physica A* **121**, 587 (1983);

[3] A. Leggett, A. Garg, *Phys. Rev. Lett.* **54**, 857 (1985).

[4] W. Zurek, *Physics Today*, October 1991, pag.36.

[5] C. Monroe, D.Meekhof, B. King, D.Wineland, *Science* **272**, 1131 (1996).

[6] M. Brune, E. Hagley, J. Dreyer, X. Maistre, A. Maali, C. Wunderlich, J. M. Raimond, S. Haroche, *Phys. Rev. Lett.***77**, 4887 (1996).

[7] M. W. Noel, C. R. Stroud, *Phys. Rev. Lett.***77**, 1913 (1996).

[8] F. De Martini, *Phys. Rev. Lett.* **81**, 2842 (1998); *Phys.Lett.A*, **250**, 15 (1998).

[9] F. De Martini, V. Mussi, F. Bovino, *Optics Comm.* **179**, 581 (2000).

[10] D. Walls, G. Milburn, *Quantum Optics (*Springer, *N.Y. 1994).*

[11] S. Popescu in *Introduction to Quantum Computation and Information* (World Scientific, N.Y. 1998).

[12] G. Di Giuseppe, F. De Martini, D. Boschi, *Fortschritte der Physik*, **46**, 643 (1998).

[13] T. Herzog, J.Rarity, H.Weinfurter, A. Zeilinger, *Phys. Rev. Lett.* **72**, 629 (1994).

[14] Z. Y. Ou, J. K. Rhee and L. J. Wang, *Phys. Rev. A* **60** 593 (1999)

QUANTUM TELEPORTATION WITH ATOMIC ENSEMBLES AND COHERENT LIGHT

Lu-Ming Duan, J. I. Cirac, P. Zoller
Institute for Theoretical Physics, University of Innsbruck, Austria

E. S. Polzik
Institute of Physics and Astronomy, Aarhus University, Denmark

Abstract We show that an unknown quantum state can be teleported from one free-space atomic ensemble to another by using only coherent light. Neither non-classical light nor cavities are needed in the scheme, which greatly simplifies its experimental implementation.

Quantum teleportation [1] of an unknown state from a photon to a photon [2, 3], or from a single-mode beam of light to another [4] has been demonstrated experimentally. A desired goal is to perform quantum teleportation of the state of massive particles, since the massive particles are ideal for storage of quantum information, and they play an important role in local quantum information processing, such as quantum computation. At the same time, the information should be transferred from one location to another via optical states, since light is the best long distance carrier of information. There have been several proposals for quantum teleportation of atomic motional or internal states, by transmitting single-photon or non-classical light [5, 6, 7]. Most of these proposals are based on the assumption that atoms are trapped inside high-Q optical cavities, which is difficult to achieve experimentally [5, 6]. The recent proposal [7] eliminates this requirement, however it still requires an external source of entanglement (non-classical light). Here, we propose and analyze a quantum communication scheme, which teleports an *unknown collective internal state* from one *free-space atomic ensemble* to another only using *coherent light*. This result is indeed surprising, since strong coherent light (light from an ordinary laser) is usually thought to be 'purely classical', but via it unknown quantum states of free-space atomic ensembles can nonetheless be teleported from one location to another!

The system we are considering is a cloud of identical atoms with the relevant level structure shown in Fig. 1. Each atom has two degenerate ground states

Quantum Communication, Computing, and Measurement 3
Edited by P. Tombesi and O. Hirota, Kluwer Academic/Plenum Publishers, 2001

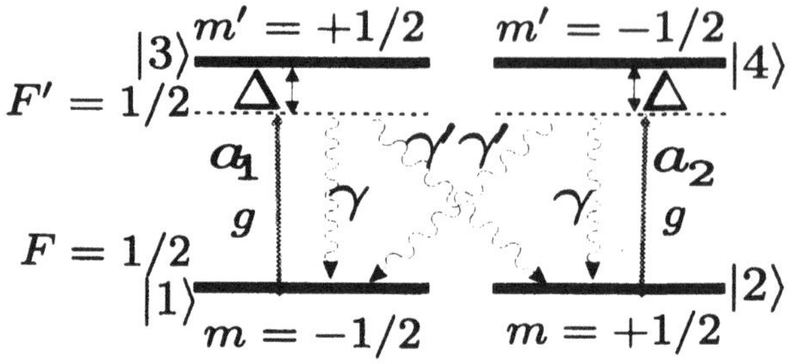

Figure 1 Level structure of the atoms.

and two degenerate excited states. The transitions $|1\rangle \rightarrow |3\rangle$ and $|2\rangle \rightarrow |4\rangle$ are coupled with a large detuning Δ to propagating light fields with different circular polarizations. This kind of interaction has been analyzed semiclassically in [8], and recently shown to be applicable for quantum non-demolition measurements [9, 10] and teleportation with non-classical light [7], with an adiabatic Hamiltonian.

We create entanglement between two atomic ensembles through a nonlocal Bell measurement with the schematic setup shown by Fig. 2. We assume a one-dimensional model for the propagating light field. The three-dimensional description can be shown to exactly reduce to the one-dimensional model [11] by introducing in the latter the spontaneous emissions to account for contributions from the other vacuum modes as we will do below. The input laser pulse is linearly polarized and expressed as $E^{(+)}(z,t) = \sqrt{\frac{\hbar\omega_0}{4\pi\epsilon_0 A}} \sum_{i=1,2} a_i(z,t)\, e^{i(k_0 z-\omega_0 t)}$, where $\omega_0 = k_0 c = 2\pi c/\lambda_0$ is the carrier frequency (λ_0 is the optical wave length), and i denotes two orthogonal circular polarizations, with the standard commutation relations $[a_i(z,t), a_j(z',t)] = \delta_{ij}\delta(z-z')$. The light is weakly focused with cross area A to match the atomic ensemble. For a strong coherent input with linear polarization, the initial condition is expressed as $\langle a_i(0,t)\rangle = \alpha_t$, with the total photon number over the pulse duration T satisfies $2N_p = 2c\int_0^T |\alpha_t|^2 dt \gg 1$. The Stokes operators are introduced for the free-space input and output light (light before entering or after leaving the atomic ensemble) by $S_x^p = \frac{c}{2}\int_0^T \left(a_1^\dagger a_2 + a_2^\dagger a_1\right) d\tau$, $S_y^p = \frac{c}{2i}\int_0^T \left(a_1^\dagger a_2 - a_2^\dagger a_1\right) d\tau$, $S_z^p = \frac{c}{2}\int_0^T \left(a_1^\dagger a_1 - a_2^\dagger a_2\right) d\tau$. In free space, $a_i(z,t)$ only depends on $\tau = t - z/c$, and then the Stokes operators satisfy the spin commutation relations $[S_y^p, S_z^p] = iS_x^p$. For our coherent input, we have $\langle S_x^p\rangle = N_p$ and $\langle S_y^p\rangle = \langle S_z^p\rangle = 0$. With a very large N_p, the off-resonant interaction with atoms is only a small perturbation to S_x^p, and we can treat S_x^p classically by replacing it with its mean value $\langle S_x^p \rangle$. Then, we define two canonical observables for light by $X^p = S_y^p/\sqrt{\langle S_x^p \rangle}$, $P^p = S_z^p/\sqrt{\langle S_x^p \rangle}$ with a standard commutator $[X^p, P^p] = i$. These operators are the quantum variables we are interested in. Similar operators can be introduced for atoms. For an atomic ensemble with many atoms, it is convenient to define the continuous atomic operators $\sigma_{\mu\nu}(z,t) = \lim_{\delta z \to 0} \frac{1}{\rho A \delta z} \sum_i^{z \le z_i < z+\delta z} |\mu\rangle_i \langle \nu|$ $(\mu, \nu = 1,2,3,4)$ with the commutation relations $[\sigma_{\mu\nu}(z,t), \sigma_{\nu'\mu'}(z',t)] = (1/\rho A)\,\delta(z-z')\left(\delta_{\nu\nu'}\sigma_{\mu\mu'} - \delta_{\mu\mu'}\sigma_{\nu'\nu}\right)$. In the definition, z_i is the position of the i atom, and ρ is the number density of the atomic ensemble with the total atom number $2N_a = \rho AL \gg 1$, where L is the length of the ensemble. The collective spin operators are introduced for the ground states of the atomic ensemble by $S_x^a = \frac{\rho A}{2} \int_0^L \left(\sigma_{12} + \sigma_{12}^\dagger\right) dz$, $S_y^a = \frac{\rho A}{2i} \int_0^L \left(\sigma_{12} - \sigma_{12}^\dagger\right) dz$, $S_z^a = \frac{\rho A}{2} \int_0^L (\sigma_{11} - \sigma_{22})\, dz$. All the atoms are initially prepared in the superposition of the two ground states $(|1\rangle + |2\rangle)/\sqrt{2}$ (this can be obtained with negligible noise by applying classical laser pulses with detuning $\Delta \gg \gamma$), which is an eigenstate of S_x^a with a very large eigenvalue N_a. As before, we treat S_x^a classically, and define the canonical operators for atoms by $X^a = S_y^a/\sqrt{\langle S_x^a \rangle}$, $P^a = S_z^a/\sqrt{\langle S_x^a \rangle}$ with $[X^a, P^a] = i$ and an initial vacuum state. As we have shown in detail in Ref. [12], after the laser pulse passes through the atomic ensemble, the off-resonant interaction changes the canonical operators according to

$$\begin{aligned} X^{p\prime} &= \sqrt{1-\varepsilon_p}\,(X^p - \kappa P^a) + \sqrt{\varepsilon_p} X_s^p, \\ X^{a\prime} &= \sqrt{1-\varepsilon_a}\,(X^a - \kappa P^p) + \sqrt{\varepsilon_a} X_s^a, \\ P^{\beta\prime} &= \sqrt{1-\varepsilon_\beta}\,P^\beta + \sqrt{\varepsilon_\beta} P_s^\beta, (\beta = a,p), \end{aligned} \tag{1}$$

where the symbols with (without) a prime denote the operators after (before) the interaction, and X_s^a, P_s^a and X_s^p, P_s^p are the standard vacuum noise operators with variance $1/2$. The interaction and damping coefficients $\kappa, \varepsilon_p, \varepsilon_a$ are given respectively by $\kappa = -\frac{2\sqrt{N_p N_a}|g|^2}{\Delta c}$, $\varepsilon_p = \frac{N_a |g|^2 \gamma}{\Delta^2 c}$, $\varepsilon_a = \frac{N_p |g|^2 \gamma'}{\Delta^2 c}$, where g is the coupling constant and γ, γ' are spontaneous emission rates (see Fig. 1). Equation (1) is obtained under the conditions $\varepsilon_{p,a} \ll 1$ and $\kappa \ll \sqrt{N_{p,a}}$. For our applications, we would like to have $\kappa > 1$. This is possible if we choose $N_p \sim N_a \gg 1$ and $\Delta \gg \gamma$. The number matching condition $N_p \sim N_a$ is an important requirement obtained here to minimize the noise effect, since we have $\kappa = 2\sqrt{\varepsilon_p \varepsilon_a}\Delta/\sqrt{\gamma\gamma'}$ and the best choice is $\varepsilon_p \sim \varepsilon_a$ to increase the signal-to-noise ratio. It is helpful to give an estimation of the relevant parameters for typical experiments. The interaction parameter κ can be rewritten as $\kappa = \left(3\rho\lambda_0^2 L\gamma\right) / \left(8\pi^2\Delta\right)$ with $N_p \approx N_a$ and $\gamma \approx \gamma'$. For a atomic sample of density $\rho \sim 5 \times 10^{12}\text{cm}^{-3}$ and of length $L \sim 2$cm, $\kappa \sim 5$ is obtainable with the choice $\Delta \sim 300\gamma$, and at the same time the loss $\varepsilon_p \sim \varepsilon_a < 1\%$.

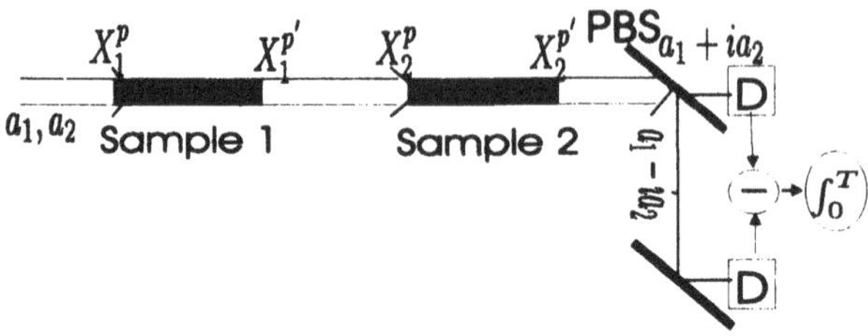

Figure 2 Schematic setup for Bell measurements. A linearly polarized strong laser pulse (decomposed into two circular polarization modes a_1, a_2) propagates successively through the two atomic samples. The two polarization modes $(a_1 + ia_2)/\sqrt{2}$ and $(a_1 - ia_2)/\sqrt{2}$ are then split by a polarizing beam splitter (PBS), and finally the difference of the two photon currents (integrated over the pulse duration T) is measured.

Now we show that the transformation (1) allows us to generate entanglement, and to achieve quantum communication between atomic ensembles using only coherent light. Entanglement is generated through a nonlocal Bell measurement of the EPR operators $X_1^a - X_2^a$ and $P_1^a + P_2^a$ with the setup depicted by Fig. 2. This setup measures the Stokes operator $X_2^{p\prime}$ of the output light. Using Eq. (1) and neglecting the small loss terms, we have $X_2^{p\prime} = X_1^p + \kappa\left(P_1^a + P_2^a\right)$, so we get a collective measurement of $P_1^a + P_2^a$ with some inherent vacuum noise X_1^p. The efficiency $1 - \eta$ of this measurement is determined by the parameter κ with $\eta = 1/\left(1 + 2\kappa^2\right)$. After this round of measurements, we rotate the collective atomic spins around the x axis to get the transformations $X_1^a \to -P_1^a$, $P_1^a \to X_1^a$ and $X_2^a \to P_2^a$, $P_2^a \to -X_2^a$. The rotation of the atomic spin can be easily obtained with negligible noise by applying classical laser pulses with detuning $\Delta \gg \gamma$. After the rotation, the measured observable of the first round of measurement is changed to $X_1^a - X_2^a$ in the new variables. We then make another round of collective measurement of the new variable $P_1^a + P_2^a$. In this way, both the EPR operators $X_1^a - X_2^a$ and $P_1^a + P_2^a$ are measured, and the final state of the two atomic ensembles is collapsed into a two-mode squeezed state with variance $\delta\left(X_1^a - X_2^a\right)^2 = \delta\left(P_1^a + P_2^a\right)^2 = e^{-2r}$, where the squeezing parameter r is given by

$$r = \frac{1}{2}\ln\left(1 + 2\kappa^2\right). \tag{2}$$

Thus, using only coherent light, we generate continuous variable entanglement [13] between two nonlocal atomic ensembles. With the interaction parameter $\kappa \approx 5$, a high squeezing (and thus a large entanglement) $r \approx 2.0$ is obtainable. Note that entanglement generation is the key step for many quantum protocols, and is the basis of quantum communication, quantum cryptography , and tests of Bell inequality. In the following, we show as an example how to achieve indirect quantum communication, i.e., quantum teleportation, between distant atomic ensembles using only coherent light.

We consider unconditional quantum teleportation of continuous variables [14, 15, 4] from one atomic ensemble to the other since we have continuous variable entanglement. To achieve quantum teleportation, first two distant atomic samples 1 and 2 are prepared in a continuously entangled state using the nonlocal Bell measurement described above. Then, a Bell measurement with the same setup as shown by Fig. 2 on the two local samples 1 and 3, together with a straightforward displacement of X_3^a, P_3^a on the sample 3, will teleport an unknown collective spin state from the atomic sample 3 to 2. The teleported state on the sample 2 has the same form as that in the original proposal of continuous variable teleportation using squeezing light [15], with the squeezing parameter r replaced by Eq. (2) and with an inherent Bell detection inefficiency $\eta = 1/\left(1+2\kappa^2\right)$. The teleportation quality is best described by the fidelity, which, for a pure input state, is defined as the overlap of the teleported state and the input state. For any coherent input state of the sample 3, the teleportation fidelity is given by

$$F = 1/\left(1 + \frac{1}{1+2\kappa^2} + \frac{1}{2\kappa^2}\right). \tag{3}$$

Equation (3) shows, if there is no extra noise, a high fidelity $F \approx 96\%$ would be possible for the teleportation of the collective atomic spin state with the interaction parameter $\kappa \approx 5$.

As our last point, let us analyze the influence of some important noise terms on the teleportation fidelity. The noise includes the spontaneous emission noise described by Eq. (1), the detector inefficiency, and the transmission loss of the light from the first sample to the second sample. The spontaneous emission noise can be included partly in the transmission loss and partly in the detector efficiency, so we do not analyze it separately. The effect of the detector inefficiency η_d is to replace κ^2 in Eqs. (2) and (3) with $\kappa^2\left(1-\eta_d\right)$, and the teleportation fidelity is decreased by a term η_d/κ^2, which is very small and can be safely ignored. The most important noise comes from the transmission loss. The transmission loss is described by $X_2^p = \sqrt{1-\eta_t}X_1^{p'} + \sqrt{\eta_t}X_s^t$ (see Fig. 2), where η_t is the loss rate and X_s^t is the standard vacuum noise. The transmission loss changes the measured observables to be $\sqrt{1-\eta_t}X_1^a - X_2^a$ and $\sqrt{1-\eta_t}P_1^a + P_2^a$. These two observables do not commute, and the two rounds of measurements influence each other. To minimize the influence on the teleportation fidelity, we choose the following configuration (for simplicity, we assume we have the same loss rate η_t from the sample 1 to 2 and from 1 to 3): In the nonlocal Bell measurements on the samples 1 and 2 (the entanglement generation process), we choose a suitable interaction coefficient κ_2 (where its optimal value will be determined below) for the second round measurement, whereas κ_1 for the first round of measurement is large with $\kappa_1^2 \gg \kappa_2^2$ (the interaction coefficient can be easily adjusted, for instance, by changing the detuning). In the local Bell measurement, we choose the same κ_2 for the first round of measurement and the large κ_1 for the second round of measurement. For a coherent input state of the sample 3, the teleported state on the sample 2 is still Gaussian, and the teleportation fidelity F' is found to be

$$F' \approx 2/\left(2 + \frac{1}{\kappa_2^2} + \kappa_2^2 \eta_t\right) \leq 1/\left(1 + \sqrt{\eta_t}\right), \tag{4}$$

which is still independent of the coherent input state with suitable gain for the displacements [15, 4]. The optimal value for κ_2 is thus given by $\kappa_2 = 1/\sqrt[4]{\eta_t}$. Even with a notable transmission loss rate $\eta_t \sim 0.2$, quantum teleportation with a remarkable high fidelity $F \sim 0.7$ is still achievable. It is known that for coherent inputs a fidelity exceeding $1/2$ has ensured quantum teleportation [17].

In summary, we have shown that quantum communication between free space atomic ensembles can be achieved using only coherent laser beams. Quantum teleportation of the atomic spin state is observable even in the presence of significant noise. This result, together with the much simplified experimental setup proposed here, suggests that efficient quantum communication between atomic samples is within reach of present experimental conditions.

Acknowledgments

ESP acknowledges fruitful discussions with A. Kuzmich. LMD and JIC acknowledge discussions with A. Sorensen. This work was supported by the Austrian Science Foundation, the Europe Union project EQUIP, the Danish Research Council, the ESF and the European TMR network Quantum Information.

References

[1] C. H. Bennett *et al.*, Phys. Rev. Lett. 70, 1895 (1996).

[2] D. Bouwmeester *et al.*, Nature 390, 595 (1997).

[3] D. Boschi *et al.*, Phys. Rev. Lett. 80, 1121 (1998).

[4] A. Furusawa, *et al.*, Science 282, 706 (1998).

[5] A. S. Parkins and H. J. Kimble, J. Opt. B 1, 496 (1999).

[6] S. Bose *et al.*, Phys. Rev. Lett. 83, 5158 (1999); C. S. Maierle,*et al.*, Phys. Rev. Lett. **81**, 5928 (1998).

[7] A. Kuzmich and E. S. Polzik, quant-ph/0003015.

[8] W. Happer, Rev. Mod. Phys. 44, 169 (1972); W. Happer and B. S. Mathur, Phys. Rev. Lett. 18, 577 (1967).

[9] A. Kuzmich, N. P. Bigelow, and L. Mandel, Europhys. Lett. A 42, 481 (1998); A. Kuzmich *et al.*, Phys. Rev. A 60, 2346 (1999); Y. Takahashi *et al.*, Phys. Rev. A **60**, 4974 (1999).

[10] K. Molmer, Eur. Phys. J. D 5, 301 (1999).

[11] A. Sorensen *et al.*, to be published.

[12] L. M. Duan *et al.*, quant-ph/0003111.

[13] L. M. Duan *et al.*, Phys. Rev. Lett. 84, 2722 (2000).

[14] L. Vaidman, Phys. Rev. A 49, 1473 (1994).

[15] S. L. Braunstein and H. J. Kimble, Phys. Rev. Lett. 80, 869 (1998).

[16] C. W. Gardiner and P. Zoller, Quantum Noise (Springer–Verlag, Berlin, 1999).

[17] S. L. Braunstein, C. A. Fuchs, and H. J. Kimble, quant-ph/9910030.

ENTANGLED STATE BASED ON NONORTHOGONAL STATE

Osamu Hirota
Research Center for Quantum Communications
Tamagawa University, Tokyo, Japan
hirota@lab.tamagawa.ac.jp

Masahide Sasaki
Communication Research Laboratory
Ministry of Post and Telecommunication, Tokyo, Japan
psasaki@crl.go.jp

Keywords: entanglement, nonorthogonal state, quantum teleportation

Abstract Properties of entangled states based on nonorthogonal states are clarified. Especially, it is shown that they can have complete degree of entanglement.

INTRODUCTION

Entanglement and its information theoretic aspects have been studied by many authors [1, 2, 3, 4, 5]. For a pure entangled state of a bipartite system $|\rho_{AB}\rangle$, a measure of entanglement is defined as [1, 6]

$$E(|\rho_{AB}\rangle) = -\mathrm{Tr}_A \rho_A \log \rho_A, \quad \rho_A = \mathrm{Tr}_B |\rho_{AB}\rangle\langle\rho_{AB}|, \tag{1}$$

which is called as "entropy of entanglement". This quantity enjoys two kinds of information theoretic interpretations. One of them is entanglement of formation which means the asymptotic number k of standard singlet required to locally prepare faithfully n identical copies of a system in bipartite state $|\rho_{AB}\rangle$ for very large k and n. Other is distillable entanglement which means the asymptotic number of singlets k that can be distilled from n identical copies of $|\rho_{AB}\rangle$. In particular, it satisfies

$$\lim_{n,k\to\infty} \frac{k}{n} = E(|\rho_{AB}\rangle). \tag{2}$$

Explicit expressions for $E(|\rho_{AB}\rangle)$ is only known in the case of two qubit systems [4]. In fact, it is given as

$$E(|\rho_{AB}\rangle) = H[\frac{1}{2}(1 + \sqrt{1 - C(\rho_{AB})^2}) \qquad (3)$$

where $H[x]$ is the entropy function and $C(\rho_{AB})$ is the "concurrence" defined by $C(\rho_{AB}) = |\langle\rho_{AB}|\tilde{\rho}_{AB}\rangle|$ with $|\tilde{\rho}_{AB}\rangle = \sigma|\rho_{AB}\rangle^*$. The similar analytic formulas for mixed states of qubits is also obtained for the properly defined entanglement of formation [5]. In this paper, we study properties of entangled states based on nonorthogonal states such as coherent states. An implementation scheme for manipulating such states are also discussed.

QUASI BELL STATE

General definition

Let us define the entangled state based on nonorthogonal states $|\psi_1\rangle$ and $|\psi_2\rangle$ such as $\langle\psi_1|\psi_2\rangle = \kappa$ and $\langle\psi_2|\psi_1\rangle = \kappa^*$. They can be described by

$$\left\{ \begin{array}{lcl} |\Psi_1\rangle & = & h_1(|\psi_1\rangle_A|\psi_2\rangle_B + |\psi_2\rangle_A|\psi_1\rangle_B) \\ |\Psi_2\rangle & = & h_2(|\psi_1\rangle_A|\psi_2\rangle_B - |\psi_2\rangle_A|\psi_1\rangle_B) \\ |\Psi_3\rangle & = & h_3(|\psi_1\rangle_A|\psi_1\rangle_B + |\psi_2\rangle_A|\psi_2\rangle_B) \\ |\Psi_4\rangle & = & h_4(|\psi_1\rangle_A|\psi_1\rangle_B - |\psi_2\rangle_A|\psi_2\rangle_B) \end{array} \right. \qquad (4)$$

where $\{h_i\}$ are normalized constant:$h_1 = h_3 = 1/\sqrt{2(1+\kappa^2)}$, $h_2 = h_4 = 1/\sqrt{2(1-\kappa^2)}$. They are not orthogonal each other. Here, if $\kappa = \kappa^*$, then the Gram matrix of them becomes very simple as follows:

$$G = \begin{pmatrix} 1 & 0 & D & 0 \\ 0 & 1 & 0 & 0 \\ D & 0 & 1 & 0 \\ 0 & 0 & 0 & 1 \end{pmatrix} \qquad (5)$$

where $D = \frac{2\kappa}{1+\kappa^2}$. If the basic states are orthogonal, then they are Bell states. Let us discuss the entropy of entanglement for the above states. We, first, calculate the reduced density operators of the quasi Bell state. They are ${\rho_A}^{(1)} = {\rho_A}^{(3)}$ and ${\rho_A}^{(2)} = {\rho_A}^{(4)}$. Their concrete forms are

$${\rho_A}^{(1)} = \frac{1}{2(1+\kappa^2)}\{|\psi_1\rangle_A\langle\psi_1| + \kappa|\psi_1\rangle_A\langle\psi_2| + \kappa|\psi_2\rangle_A\langle\psi_1| + |\psi_2\rangle_A\langle\psi_2\} \quad (6)$$

$${\rho_A}^{(2)} = \frac{1}{2(1-\kappa^2)}\{|\psi_1\rangle_A\langle\psi_1| - \kappa|\psi_1\rangle_A\langle\psi_2| - \kappa|\psi_2\rangle_A\langle\psi_1| + |\psi_2\rangle_A\langle\psi_2|\} \quad (7)$$

The eigenvalues of the above density operators ${\rho_A}^{(1)}$(or ${\rho_A}^{(3)}$) are given as follows by using the Gram matrix elements $C_{ij} = |\langle\Psi_i|\Psi_j\rangle|$ of Eq. (5):

$$\lambda_{1/1} = \frac{(1+\kappa)^2}{2(1+\kappa^2)} = \frac{1+C_{13}}{2}, \quad \lambda_{2/1} = \frac{(1-\kappa)^2}{2(1+\kappa^2)} = \frac{1-C_{13}}{2} \tag{8}$$

and for ${\rho_A}^{(2)}$(or ${\rho_A}^{(4)}$), we have

$$\lambda_{1/2} = \frac{1}{2} = \frac{1+C_{24}}{2}, \quad \lambda_{2/2} = \frac{1}{2} = \frac{1-C_{24}}{2}. \tag{9}$$

The entropy of entanglement is then

$$E(|\Psi_1\rangle) = E(|\Psi_3\rangle) = -\frac{1+C_{13}}{2}\log\frac{1+C_{13}}{2} - \frac{1-C_{13}}{2}\log\frac{1-C_{13}}{2} \tag{10}$$

and $E(|\Psi_2\rangle) = E(|\Psi_4\rangle) = 1$. Thus $|\Psi_2\rangle$ and $|\Psi_4\rangle$ have perfect entanglement, even though the enatangled states consist of nonorthogonal state in each subsystem. These results are true for arbitrary nonorthogonal states with $\langle\psi_1|\psi_2\rangle = \langle\psi_2|\psi_1\rangle = \kappa$ and do not depend on the physical dimension of the systems.

Generation of quasi Bell states

It is well known that an entangled state can be generated by Walsh-Hadamard gate and CN gate. That is, when the input state for control bit is a superposition state generated by Walsh-Hadamard gate, the output state of the CN gate is an entangled state. The W-H gate(Walsh-Hadamard transformation) is described by

$$U_{WH} = \exp\{\theta(|0\rangle\langle 1| - |1\rangle\langle 0|)\} \tag{11}$$

and the unitary operator for the control NOT is

$$U_{CN} = |0\rangle_C\langle 0| \otimes I_T + |1\rangle_C\langle 1| \otimes (|0\rangle_T\langle 1| + |1\rangle_T\langle 0|) \tag{12}$$

The function of the control NOT(CN gate) is as follows:

$$\left\{\begin{array}{lcl} |0\rangle_C|0\rangle_T & \rightarrow & U_{CN}|0\rangle_C|0\rangle_T = |0\rangle_C|0\rangle_T \\ |0\rangle_C|1\rangle_T & \rightarrow & U_{CN}|0\rangle_C|1\rangle_T = |0\rangle_C|1\rangle_T \\ |1\rangle_C|0\rangle_T & \rightarrow & U_{CN}|1\rangle_C|0\rangle_T = |1\rangle_C|1\rangle_T \\ |1\rangle_C|1\rangle_T & \rightarrow & U_{CN}|1\rangle_C|1\rangle_T = |1\rangle_C|0\rangle_T \end{array}\right. \tag{13}$$

where C and T mean control mode, and target mode, respectively.

On the two state space spanned by nonorthogonal states: $|\psi_1\rangle$ and $|\psi_2\rangle$, we can consider general scheme to manipulate the quasi Bell states. Let us define the orthonormal basis

$$|\psi_e\rangle = (|\psi_1\rangle + |\psi_2\rangle)/\sqrt{2(1+\kappa)}, \tag{14}$$

$$|\psi_o\rangle = (|\psi_1\rangle - |\psi_2\rangle)/\sqrt{2(1-\kappa)}. \tag{15}$$

They play a role of qubit basis $\{|0\rangle, |1\rangle\}$. We can then go along with quantum logic operations on qubit systems. In terms of the basis defined by Eq. (14), and Eq. (15), the required gates are

$$U_{WH} = \exp\{\theta(|\psi_e\rangle\langle\psi_o| - |\psi_o\rangle\langle\psi_e|)\} \tag{16}$$

$$\begin{aligned} U_{CN} &= |\psi_e\rangle_C\langle\psi_e| \otimes I_T \\ &+ |\psi_o\rangle_C\langle\psi_o| \otimes \Big(|\psi_e\rangle_T\langle\psi_o| + |\psi_o\rangle_T\langle\psi_e|\Big) \end{aligned} \tag{17}$$

The W-H gate acts on the input superposition state as follows:

$$\begin{aligned} |\psi_e\rangle_C &\rightarrow U_{WH}|\psi_e\rangle_C \\ &= |\psi_e\rangle_C + |\psi_o\rangle_C \end{aligned} \tag{18}$$

Thus we have

$$\begin{aligned} K\Big(|\psi_e\rangle_C &+ |\psi_o\rangle_C\Big)|\psi_e\rangle_T \\ &\rightarrow U_{CN}K\Big(|\psi_e\rangle_C + |\psi_o\rangle_C\Big)|\psi_e\rangle_T \\ &= K\Big(|\psi_e\rangle_C|\psi_e\rangle_T + |\psi_o\rangle_C|\psi_o\rangle_T\Big) \\ &= h_3\Big(|\psi_1\rangle_C|\psi_1\rangle_T + |\psi_2\rangle_C|\psi_2\rangle_T\Big) \end{aligned} \tag{19}$$

where K is the normalized constant. The final state is one of quasi Bell state. Thus we have quasi Bell states based on such operations. If we use the coherent states as the basic states, then the above gates correspond to bosonic gates whose realization is discussed in the later section.

General case

Here let us consider the general pure entangled state of nonorthogonal state. They can be described as follows:

$$\left\{ \begin{aligned} |\Psi_1\rangle &= g_1(\beta|\psi_1\rangle_A|\psi_2\rangle_B + \sqrt{1-\beta^2}|\psi_2\rangle_A|\psi_1\rangle_B) \\ |\Psi_2\rangle &= g_2(\beta|\psi_1\rangle_A|\psi_2\rangle_B - \sqrt{1-\beta^2}|\psi_2\rangle_A|\psi_1\rangle_B) \\ |\Psi_3\rangle &= g_3(\beta|\psi_1\rangle_A|\psi_1\rangle_B + \sqrt{1-\beta^2}|\psi_2\rangle_A|\psi_2\rangle_B) \\ |\Psi_4\rangle &= g_4(\beta|\psi_1\rangle_A|\psi_1\rangle_B - \sqrt{1-\beta^2}|\psi_2\rangle_A|\psi_2\rangle_B) \end{aligned} \right. \tag{20}$$

where g_i is normalized constant, and β is real number. Since all the elements of the Gram matrix is not zero, they are not orthogonal states. The reduced density operators in this case become ${\rho_A}^{(1)} = {\rho_A}^{(3)}$ and ${\rho_A}^{(2)} = {\rho_A}^{(4)}$, and

$$\rho_A^{(1)} = k_1\{\beta^2|\psi_1\rangle_A\langle\psi_1| + \kappa\beta\sqrt{1-\beta^2}|\psi_1\rangle_A\langle\psi_2$$

$$
\begin{aligned}
&+ \quad \kappa\beta\sqrt{1-\beta^2}|\psi_2\rangle_A\langle\psi_1| + (1-\beta^2)|\psi_2\rangle_A\langle\psi_2|\} \qquad (21)\\
\rho_A^{(2)} &= k_2\{\beta^2|\psi_1\rangle_A\langle\psi_1| - \kappa\beta\sqrt{1-\beta^2}|\psi_1\rangle_A\langle\psi_2| \\
&- \quad \kappa\beta\sqrt{1-\beta^2}|\psi_2\rangle_A\langle\psi_1| + (1-\beta^2)|\psi_2\rangle_A\langle\psi_2|\} \qquad (22)
\end{aligned}
$$

where $k_i = \frac{1}{1\pm 2\kappa^2\beta\sqrt{1-\beta^2}}$ is normalized constant. The analysis in this case is also easy. Here we again assumed that $\langle\psi_1|\psi_2\rangle = \langle\psi_2|\psi_1\rangle = \kappa$.

QUASI BELL STATES OF COHERENT STATES

Let us consider the binary coherent states of a bosonic mode $\{|\alpha\rangle, |-\alpha\rangle\}$, where $(\kappa = \langle\alpha|-\alpha\rangle = e^{-2|\alpha|^2})$. Then the quasi Bell states are

$$
\left\{
\begin{aligned}
|\Psi_1\rangle &= h_1(|\alpha\rangle_A|-\alpha\rangle_B + |-\alpha\rangle_A|\alpha\rangle_B)\\
|\Psi_2\rangle &= h_2(|\alpha\rangle_A|-\alpha\rangle_B - |-\alpha\rangle_A|\alpha\rangle_B)\\
|\Psi_3\rangle &= h_3(|\alpha\rangle_A|\alpha\rangle_B + |-\alpha\rangle_A|-\alpha\rangle_B)\\
|\Psi_4\rangle &= h_4(|\alpha\rangle_A|\alpha\rangle_B - |-\alpha\rangle_A|-\alpha\rangle_B)
\end{aligned}
\right. \qquad (23)
$$

where α is coherent amplitude of light field. The average photon numbers of the reduced states are

$$
\langle n_A^{(1)}\rangle = \frac{(1-\kappa^2)}{(1+\kappa^2)}|\alpha|^2, \quad \langle n_A^{(2)}\rangle = \frac{(1+\kappa^2)}{(1-\kappa^2)}|\alpha|^2 \qquad (24)
$$

Thus the quasi Bell states can have arbitrary photon, and approach to the Bell states as $|\alpha| \to \infty$. We mention the characteristic function of quasi Bell states defined as

$$
\begin{aligned}
C(\xi,\eta) &= \mathrm{Tr}[|\Psi\rangle\langle\Psi|\exp(\xi a_A^\dagger)\exp(-\xi^* a_A)\exp(\eta a_B^\dagger)\exp(-\eta^* a_B)]\\
&\times \exp\{-(|\xi|^2+|\eta|^2)/2\} \qquad (25)
\end{aligned}
$$

where a and $a^\dagger$ are the annihilation and creation operators, respectively. They are actually

$$
\begin{aligned}
C(\xi,\eta|i=1,2) &= h_i{}^2\exp\{-(|\xi|^2+|\eta|^2)/2\}\{\exp(A_1-B_1)\alpha\\
&+ \exp(-A_1+B_1)\alpha \pm \exp(A_2-B_2)\alpha\\
&\pm \exp(-A_2+B_2)\alpha\} \qquad (26)\\
C(\xi,\eta|i=3,4) &= h_i{}^2\exp\{-(|\xi|^2+|\eta|^2)/2\}\{\exp(A_1+B_1)\alpha\\
&+ \exp(-A_1-B_1)\alpha \pm \exp(A_2+B_2)\alpha\\
&\pm \exp(-A_2-B_2)\alpha\} \qquad (27)
\end{aligned}
$$

where $A_1 = (\xi - \xi^*), A_2 = (\xi + \xi^*), B_1 = (\eta - \eta^*), B_2 = (\eta + \eta^*)$. It is worthy to mention that the quasi Bell states do not belong the Gaussian state in contrast to that two mode squeezed state does so.

PHYSICAL REALIZATION

In order to manipulate the quasi Bell states of bosonic coherent states, one needs quantum gates acting on a state space spanned by the relevant coherent states. A convenient basis is the even and odd coherent states. Let us denote them as $\{|e\rangle, |o\rangle\}$ hereafter. It would be much difficult to realize quantum gates for these *macroscopic qubits*. Cochrane, Milburn, and Munro proposed a physical model for such gates [7]. In their model, a CN gate is made by applying an H gate to the mode a, the target bit, then coupling the target and the control (the mode b), and finally applying another H gate to the target again. This CN gate operation is actually valid in a certain limited case of the coherent state amplitude α and the classical field amplitude γ. However, their model is indeed indicative. If the H gate is capable of generating the Schrödinger cat state of coherent state, that is, includes an appropriate nonlinear Hamiltonian instead of the linear interaction of Eq. [7], then their scheme provides the universal bosonic gates. (This was also recognized by Tatsuta et al. [8]) Concerning the two bit interaction, the cross Kerr effect always suffices. Thus the problem is to synthesize the appropriate nonlinear Hamiltonian.

One bit gate operations for macroscopic bosonic qubits can be represented by rotations on the two-state space. For example, the Hadamard gate is represented by

$$\hat{U}_H = -i\exp[i\frac{\pi}{2}\hat{Q}]\exp[\frac{\pi}{4}\hat{P}], \tag{28}$$

$$\hat{P} = |e\rangle\langle o| - |o\rangle\langle e|, \quad \hat{Q} = |e\rangle\langle e| - |o\rangle\langle o|, \tag{29}$$

The physical process corresponding to $\hat{U}_H$ includes essentially multiphoton nonlinear precess. The corresponding Hamiltonians were studied by Sasaki and Hirota [9].

The case of our interest is when the amplitude $|\alpha|$ is small, in which the assumption used in [7]. Then the Hamiltonian $\hat{P}$ and $i\hat{Q}$ can be effected by the nonlinear Hamiltonian including finite number of nonlinearity. For mathematical convenience, we consider the Hamiltonian for $\tilde{U}_H = \hat{D}(-\alpha)\hat{U}_H\hat{D}(\alpha)$. First we introduce a cut off photon number M for a weak coherent state such that its photon number distribution in $n > M$ becomes negligibly small. Second define

$$\hat{P}_M = -2\sqrt{1-c_0^2}\left(\sum_{l=0}^{M}\frac{(-\hat{a}^\dagger)^l\hat{a}^l}{l!}\sum_{n=1}^{M}d_n\frac{\hat{a}^n}{\sqrt{n!}} - \text{h.c.}\right), \tag{30}$$

$$\hat{Q}_M = 4c_0\sum_{l=0}^{M}\frac{(-\hat{a}^\dagger)^l\hat{a}^l}{l!} + 2\sqrt{1-c_0^2}\left(\sum_{l=0}^{M}\frac{(-\hat{a}^\dagger)^l\hat{a}^l}{l!}\sum_{n=1}^{M}d_n\frac{\hat{a}^n}{\sqrt{n!}} + \text{h.c.}\right), \tag{31}$$

where

$$c_n = \mathrm{e}^{-2\alpha^2}\frac{(-2\alpha)^2}{\sqrt{n!}}, \quad d_n = \frac{c_n}{\sqrt{\sum_{n=1}^{M} c_n^2}}. \tag{32}$$

Then $\tilde{U}_H$ can be represented

$$\tilde{U}_H = -i\exp[i\frac{\pi}{2}\hat{Q}_M]\exp[\frac{\pi}{4}\hat{P}_M] + O(\delta_M), \tag{33}$$

where

$$\delta_M = 1 - \frac{\sum_{n=1}^{M} c_n^2}{1 - c_0^2}. \tag{34}$$

The Hamiltonian of $\hat{P}_M$ and $\hat{Q}_M$ still seem to be unrealistic. One possible way to make realistic is to decompose them into a cascade process of lower order nonlinear processes. In fact, as suggested by Harel and Akulin [10], and Lloyd and Braunstein[11], it is possible in principle to synthesize the required unitary dynamics by lower order nonlinear Hamiltonians. In particular, it can be shown that the nonlinearity up to third order and the cross Kerr nonlinearity suffice to implement the universal gates for macroscopic bosonic qubits.

CONCLUSION

We investigated properties of entangled states based on nonorthogonal state such as coherent states. Implementation of quantum gates for such macroscopic qubits was suggested. We would like to find more simple generation method of such a quasi Bell states.

Acknowledgments

We are grateful to M.Ban, S.Barnett, C.Bennett, S. J. van Enk, C.Fuchs, A.Holevo, K.Kato, R.Jozsa, M.Osaki, and P.Shor for helpful discussions.

References

[1] C.H. Bennett, H.J. Bernstein, S. Popescu, and B. Schumacher, Phys. Rev., **A-53**, 2046, (1996).

[2] C.H. Bennett, D.P. DiVincenzo, J.A. Smolin, and W.K. Wootters, Phys. Rev., **A-54**, 3824, (1996).

[3] V. Vedral and M.B. Plenio, Phys. Rev., **A-57**, 1619, (1998).

[4] S. Hill and W.K. Wootters, Phys. Rev. Lett., **78**, 5022, (1997).

[5] W.K. Wootters, Phys. Rev. Lett., **80**, 2245, (1998).

[6] S.M. Barnett and S.J. Phoenix, Phys. Rev., **A-40**, 2404, (1989).

[7] P. T. Cochrane, G. J. Milburn, and W. J. Munro, Phys. Rev., **A-59**, 2631 (1999)

[8] S.Tatsuta, T.S.Usuda, Techinical Report in IEICE of Japan, **QIT-3**, no-22, (2000).

[9] M. Sasaki and O. Hirota, Phys. Rev., **A-54**, 2728 (1996).

[10] G. Harel and V. M. Akulin, Phys. Rev. Lett., **82**, 1, (1999).

[11] S. Lloyd and S. L. Braunstein, Phys. Rev. Lett., **82**, 1784, (1999).

LONG-DISTANCE HIGH-FIDELITY TELEPORTATION USING SINGLET STATES

Jeffrey H. Shapiro
Department of Electrical Engineering and Computer Science
Massachusetts Institute of Technology
Cambridge, MA 02139
jhs@mit.edu

Keywords: Teleportation, entanglement, parametric amplifier, quantum memory

Abstract A quantum communication system is proposed that uses polarization-entangled photons and trapped-atom quantum memories. This system is capable of long-distance, high-fidelity teleportation, and long-duration quantum storage.

Introduction

This paper proposes a singlet-based approach to quantum communication that uses a novel ultrabright narrowband source of polarization-entangled photon pairs [1], and a trapped-atom quantum memory [2] whose loading can be nondestructively verified and whose structure permits all four Bell-state measurements to be performed. The system is designed to operate with standard telecommunication fiber as its transmission medium. It can achieve a loss-limited throughput as high as 200 entangled-pairs/sec with a 97.5% fidelity over a 50 km path when there is 10 dB of fixed loss in the overall system and 0.2 dB/km propagation loss in the fiber.

1. LONG-DISTANCE TELEPORTATION

The notion that singlet states could be used to achieve teleportation is due to Bennett *et al* [3]. The transmitter and receiver stations share the entangled qubits of a singlet state and the transmitter then accepts an input-mode qubit. Making the Bell-state measurements on the joint input-mode/transmitter system then yields the two bits of classical information that the receiver needs to reconstruct the input state. An initial experimental demonstration of teleportation using singlet states was performed by the Innsbruck group [4],[5], but only one of the Bell states was measured, the demonstration was a table-top experiment, and it did not include a quantum memory. Our proposal for a singlet-based quantum communication system, which is shown in Fig. 1, remedies all of these limitations.

Figure 1 Schematic of long-distance quantum communication system: P = ultrabright narrowband source of polarization-entangled photon pairs; L = L km of standard telecommunication fiber; M = trapped-atom quantum memory.

An ultrabright narrowband source of polarization-entangled photon pairs [1] launches the entangled qubits from a singlet state into two L-km-long standard telecommunication fibers. The photons emerging from the fibers are then loaded into trapped-atom quantum memories [2]. These memories store the photon-polarization qubits in long-lived hyperfine levels. Because it is compatible with fiber-optic transmission, this configuration is capable of long-distance teleportation. Because of the long decoherence times that can be realized with trapped atoms, this configuration supports long-duration quantum storage. We devote the rest of this section to summarizing the basic features of our proposal.

Each M block in Fig. 1 is a quantum memory in which a single ultra-cold ^{87}Rb atom ($\sim$6 MHz linewidth) is confined by a CO_2-laser trap in an ultra-high vacuum chamber with cryogenic walls within a high-finesse ($\sim$15 MHz linewidth) single-ended optical cavity. This memory can absorb a 795 nm photon, in an arbitrary polarization state, transferring the qubit from the photon to the degenerate B levels of Fig. 2a and thence to long-lived storage levels, by coherently driving the B-to-D transitions. (We are using abstract symbols here for the hyperfine levels of rubidium, see [2] for the actual atomic levels involved as well as a complete description of the memory and its operation.) With a liquid helium cryostat, so that the background pressure is less than 10^{-14} Torr, the expected lifetime of the trapped rubidium atom will be more than an hour. Moreover, the decoherence time can be expected to be about the same as this lifetime for the levels we have chosen to use for storage. By using optically off-resonant Raman (OOR) transitions, the Bell states of two atoms in a single vacuum-chamber trap can be converted to superposition states of one of the atoms. All four Bell measurements can then be made, sequentially, by detecting the presence (or absence) of fluorescence as an appropriate sequence of OOR laser pulses is applied to the latter atom. The Bell-measurement results in one memory can be sent to a distant memory, where at most two additional OOR pulses are needed to complete the Bennett *et al.* state transformation. The qubit stored in a trapped rubidium atom can be converted back into a photon by reversing the Raman excitation process that occurs during memory loading.

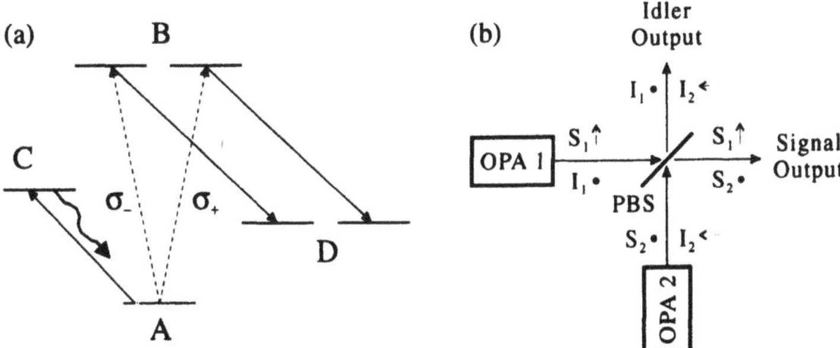

Figure 2 Essential components of singlet-state quantum communication system from Fig. 1. Left (a), simplified atomic-level schematic of the trapped rubidium atom quantum memory: A-to-B transition occurs when one photon from an entangled pair is absorbed; B-to-D transition is coherently driven to enable storage in the long-lived D levels; A-to-C cycling transition permits nondestructive determination of when a photon has been absorbed. Right (b), ultrabright narrowband source of polarization-entangled photon pairs: each optical parametric amplifier (OPA1 and OPA2) is type-II phase matched; for each optical beam the propagation direction is $\hat{z}$, and $\hat{x}$ and $\hat{y}$ polarizations are denoted by arrows and bullets, respectively; PBS, polarizing beam splitter.

The P-block in Fig. 1 is an ultrabright narrowband source of polariz-ation-entangled photon pairs, capable of producing $\sim 10^6$ pairs/sec in ~ 30 MHz bandwidth by appropriately combining the signal and idler output beams from two doubly-resonant type-II phase-matched optical parametric amplifiers (OPAs), as sketched in Fig. 2b [1]. The fluorescence spectrum of the signal and idler beams is controlled by the doubly-resonant OPA cavities. These can be advantageously and easily tailored to produce the desired (factor-of-two broader than the memory-cavity's) bandwidth. By using periodically-poled potassium titanyl phosphate (PPKTP), a quasi-phase-matched type-II nonlinear material, we can produce $\sim 10^6$ pairs/sec at the 795 nm wavelength of the rubidium memory for direct memory-loading (i.e., local-storage) applications. For long-distance transmission to remotely-located memories, we use a different PPKTP crystal and pump wavelength to generate $\sim 10^6$ pairs/sec in the 1.55-μm-wavelength low-loss fiber transmission window. After fiber propagation we then shift the entanglement to the 795 nm wavelength needed for the rubidium-atom memory via quantum-state frequency translation [6],[7].

Successful singlet transmission requires that polarization not be degraded by the propagation process. Our scheme for polarization maintenance relies on time-division multiplexing. Time slices from the signal beams from our two

OPAs are sent down one fiber in the same linear polarization but in nonoverlapping time slots, accompanied by a strong out-of-band laser pulse. By tracking and restoring the linear polarization of the strong pulse, we can restore the linear polarization of the signal-beam time slices at the far end of the fiber. After this linear-polarization restoration, we then reassemble a time-epoch of the full vector signal beam by delaying the first time slot and combining it on a polarizing beam splitter with the second time slot after the latter has had its linear polarization rotated by 90°. A similar procedure is performed to reassemble idler time-slices after they have propagated down the other fiber in Fig. 1. This approach, which is inspired by the Bergman *et al.* two-pulse fiber-squeezing experiment [8], common-modes out the vast majority of the phase fluctuations and the polarization birefringence incurred in the fiber, permitting standard telecommunication fiber to be used in lieu of the lossier and much more expensive polarization-maintaining fiber.

2. LOSS-LIMITED PERFORMANCE

Quantum communication is carried out in the Fig. 1 configuration via the following protocol. The entire system is clocked. Time slots of signal and idler (say 400 ns long) are transmitted down optical fibers to the quantum memories. These slots are gated into the memory cavities—with their respective atoms either physically displaced or optically detuned so that no A-to-B (i.e., no 795 nm) absorptions occur. After a short loading interval (a few cold-cavity lifetimes, say 400 ns), each atom is moved (or tuned) into the absorbing position and B-to-D coherent pumping is initiated. After about 100 ns, coherent pumping ceases and the A-to-C cycling transition (shown in Fig. 2a) is repeatedly driven (say 30 times, taking nearly 1 μs). By monitoring a cavity for the fluorescence from this cycling transition, we can reliably detect whether or not a 795 nm photon has been absorbed by the atom in that cavity. If neither atom or if only one atom has absorbed such a photon, then we cycle both atoms back to their A states and start anew. If no cycling-transition fluorescence is detected in either cavity, then, because we have employed enough cycles to ensure very high probability of detecting that the atom is in its A state, it must be that both atoms have absorbed 795 nm photons and stored the respective qubit information in their long-lived degenerate D levels. These levels are not resonant with the laser driving the cycling transition, and so the loading of our quantum memory is nondestructively verified in this manner.

We expect that the preceding memory-loading protocol can be run at rates as high as $R = 500$ kHz, i.e., we can get an independent try at loading an entangled photon pair into the two memory elements of Fig. 1 every 2 μs. With a high probability, P_{erasure}, any particular memory-loading trial will result in an erasure, i.e., propagation loss and other inefficiencies combine to preclude both atoms from absorbing photons in the same time epoch. With a small probability, P_{success}, the two atoms will absorb the photons from a single polarization-entangled pair, viz., we have a memory-loading success. With a much smaller probability, P_{error}, both atoms will have absorbed photons but these photons will not have come from a single polarization-entangled pair; this is the error event.

There are two key figures-of-merit for the Fig. 1 configuration: throughput and fidelity. Propagation losses and other inefficiencies merely increase P_{erasure} and hence reduce the throughput, i.e., the number of successful entanglement-loadings/sec, $N_{\text{success}} \equiv RP_{\text{success}}$, that could be achieved if the quantum memories each contained a lattice of trapped atoms for sequential loading of many pairs. Loading errors, which occur with probability P_{error}, provide the ultimate limit on the entanglement fidelity of the Fig. 1 configuration: $F_{\text{max}} = 1 - P_{\text{error}}/2(P_{\text{success}} + P_{\text{error}})$, where we have assumed that the error event loads independent, randomly-polarized photons into each memory.

OPA Statistics

Assume matched signal and idler cavities, each with linewidth Γ, zero detuning, and no excess loss. Also assume anti-phased pumping at a fraction, G^2, of oscillation threshold, with no pump depletion or excess noise. From [1] we then have that the output beams from OPAs 1 and 2 are in an entangled, zero-mean Gaussian pure state, which is completely characterized by the following normally-ordered and phase-sensitive correlation functions:

$$\langle \hat{A}^{\dagger}_{k_j}(t+\tau)\hat{A}_{k_j}(t)\rangle = \frac{G\Gamma}{2}\left[\frac{e^{-(1-G)\Gamma|\tau|}}{1-G} - \frac{e^{-(1+G)\Gamma|\tau|}}{1+G}\right], \tag{1}$$

$$\langle \hat{A}_{S_j}(t+\tau)\hat{A}_{I_j}(t)\rangle = \frac{(-1)^{j-1}G\Gamma}{2}\left[\frac{e^{-(1-G)\Gamma|\tau|}}{1-G} + \frac{e^{-(1+G)\Gamma|\tau|}}{1+G}\right], \tag{2}$$

where $\{\hat{A}_{k_j}(t)e^{-i\omega_k t} : k = S\,(\text{signal}), I\,(\text{idler}), j = 1,2\}$ are positive-frequency, photon-units OPA-output field operators. The presence of excess loss within the OPA cavities, and/or propagation loss along the fiber can be incorporated into this OPA analysis in a straightforward manner [9]. Assuming symmetric operation, in which the signal and idler encounter identical intra-cavity and fiber losses, then the correlation-function formulas, Eqs. 1 and 2, are merely multiplied by $\eta_L\gamma/\Gamma$, where $\eta_L < 1$ is the transmission through the fiber and $\gamma < \Gamma$ is the output-coupling rate of the OPA cavity.

Cavity-Loading Statistics

The internal annihilation operators of quantum memory cavities—over the T_c-sec-long loading interval—are related to the incoming signal and idler field operators as follows:

$$\hat{\vec{a}}_k(T_c) = \hat{\vec{a}}_k(0)e^{-\Gamma_c T_c} + \int_0^{T_c} dt\, e^{-\Gamma_c(T_c-t)} \left[\sqrt{2\gamma_c}\hat{\vec{A}}_k(t) + \sqrt{2(\Gamma_c-\gamma_c)}\hat{\vec{A}}_{k_v}(t)\right], \tag{3}$$

for $k = S, I$, where $\gamma_c < \Gamma_c$ is the input-coupling rate and Γ_c is the linewidth of the (assumed to be identical for signal and idler) memory cavities. The initial intracavity operators and the loss-operators, $\{\hat{\vec{a}}_k(0), \hat{\vec{A}}_{k_v}(t)\}$, are in vacuum states.

It is now easy to show that the joint density operator (state) for $\{\hat{\vec{a}}_S(T_c), \hat{\vec{a}}_I(T_c)\}$, takes the factored form, $\hat{\rho}_{\vec{S}\vec{I}} = \hat{\rho}_{S_xI_y}\hat{\rho}_{S_yI_x}$, where the two-mode density operators on the right-hand side are Gaussian mixed states given by the anti-normally ordered characteristic functions,

$$\begin{aligned} &\mathrm{tr}\left[\hat{\rho}_{S_xI_y} e^{-\zeta_S^*\hat{a}_{S_x} - \zeta_I^*\hat{a}_{I_y}} e^{\zeta_S\hat{a}_{S_x}^\dagger + \zeta_I\hat{a}_{I_y}^\dagger}\right] \\ &= \mathrm{tr}\left[\hat{\rho}_{S_yI_x} e^{-\zeta_S^*\hat{a}_{S_y} + \zeta_I^*\hat{a}_{I_x}} e^{\zeta_S\hat{a}_{S_y}^\dagger - \zeta_I\hat{a}_{I_x}^\dagger}\right] \\ &= e^{-(1+\bar{n})(|\zeta_S|^2+|\zeta_I|^2)+2\tilde{n}\mathrm{Re}(\zeta_S\zeta_I)}, \end{aligned} \tag{4}$$

where $\bar{n} \equiv I_- - I_+$ and $\tilde{n} \equiv I_- + I_+$, with $I_\mp \equiv \eta_L\gamma\gamma_c/\Gamma_c(1 \mp G)[(1 \mp G)\Gamma + \Gamma_c]$.

Throughput and Fidelity Calculations

From the loaded-cavity state we can find the erasure, success, and error probabilities via,

$$\begin{aligned} P_{\text{erasure}} &= \left({}_{S_x}\langle 0|\hat{\rho}_{S_x}|0\rangle_{S_x}\right)\left({}_{S_y}\langle 0|\hat{\rho}_{S_y}|0\rangle_{S_y}\right) + \left({}_{I_x}\langle 0|\hat{\rho}_{I_x}|0\rangle_{I_x}\right)\left({}_{I_y}\langle 0|\hat{\rho}_{I_y}|0\rangle_{I_y}\right) \\ &- \left({}_{S_x}\langle 0|{}_{I_y}\langle 0|\hat{\rho}_{S_xI_y}|0\rangle_{I_y}|0\rangle_{S_x}\right)\left({}_{S_y}\langle 0|{}_{I_x}\langle 0|\hat{\rho}_{S_yI_x}|0\rangle_{I_x}|0\rangle_{S_y}\right), \end{aligned} \tag{5}$$

$$P_{\text{success}} = {}_{\text{SI}}\langle\psi|\hat{\rho}_{\vec{S}\vec{I}}|\psi\rangle_{\text{SI}} \quad \text{and} \quad P_{\text{error}} = 1 - P_{\text{erase}} - P_{\text{success}}, \tag{6}$$

where $|\psi\rangle_{\text{SI}} \equiv \left(|1\rangle_{S_x}|1\rangle_{I_y}|0\rangle_{S_y}|0\rangle_{I_x} - |0\rangle_{S_x}|0\rangle_{I_y}|1\rangle_{S_y}|1\rangle_{I_x}\right)/\sqrt{2}$, is the singlet state.

In Fig. 3 we have plotted the throughput and loss-limited fidelity for our quantum communication system under the following assumptions: OPAs pumped at 1% of their oscillation thresholds ($G^2 = 0.01$); 5 dB of excess loss in each P-to-M block path in Fig. 1; 0.2 dB/km loss in each fiber; $\Gamma_c/\Gamma = 0.5$;

and $R = 500$ kHz memory cycling rate. We see from this figure that a throughput of 200 pairs/sec can be sustained out to an end-to-end path length ($2L$) of 50 km, with a loss-limited fidelity of 97.5%.

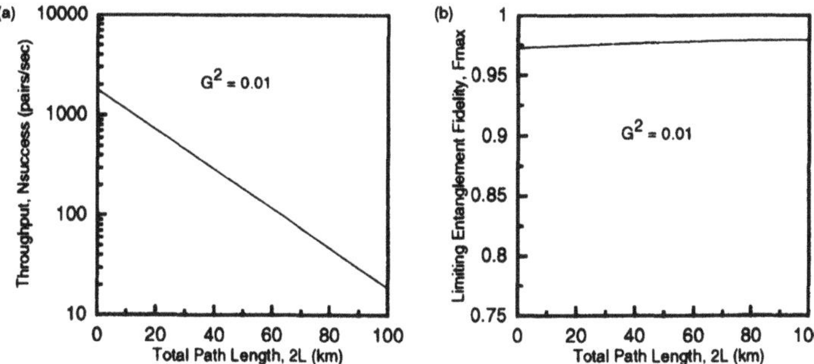

Figure 3 Figures of merit for Fig. 1 configuration. Left (a), throughput, $N_{success}$, vs. total path length, $2L$. Right (b), limiting entanglement fidelity, F_{max}, vs. total path length, $2L$. All curves assume OPAs operating at 1% of their power thresholds, 5 dB of excess loss per P-to-M block connection, and 0.2 dB/km fiber-propagation loss.

3. DISCUSSION

We have described a single-hop, long-distance, high-fidelity quantum communication system whose loss-limited operating range extends well beyond that of previous quantum repeater proposals . At $2L = 50$ km there is 20 dB end-to-end loss in our system example, yet, because of the nondestructive memory-loading verification, the ultrabright nature of our entanglement source, and our ability to employ the low-loss wavelength window in standard telecommunication fiber, we can sustain appreciable throughputs and high fidelity. Of course, this analysis has neglected additional degradations that may arise from residual phase errors in transmission, imperfect Bell-state measurements, etc., which will reduce the teleportation fidelity. Nevertheless, the Fig. 1 configuration offers substantial promise for bringing singlet-based teleportation from a conditional demonstration in the laboratory to a viable quantum communication system.

Acknowledgments

This research was supported in part by U.S. Army Research Office Grant DAAD19-00-1-0177. The author acknowledges fruitful technical discussions with Phil Hemmer, Prem Kumar, Seth Lloyd, Selim Shahriar, Franco Wong, and Horace Yuen.

References

[1] J. H. Shapiro and N. C. Wong, *J. Opt. B: Quantum Semiclass. Opt.* 2:L1 (2000).

[2] S. Lloyd, M. S. Shahriar, and P. R. Hemmer, Teleportation and the quantum Internet, submitted to *Phys. Rev. A* (quant-phy/003147).

[3] C. H. Bennett, G. Brassard, C. Crépeau, R. Josza, A. Peres, and W. K. Wootters, *Phys. Rev. Lett.* 70:1895 (1993).

[4] D. Bouwmeester, J.-W. Pan, K. Mattle, M. Eibl, H. Weinfurter, and A. Zeilinger, *Nature* 390:575 (1997).

[5] D. Bouwmeester, K. Mattle, J.-W. Pan, H. Weinfurter, A. Zeilinger, and M. Zukowski, *Appl. Phys. B* 67:749 (1998).

[6] P. Kumar, *Opt. Lett.* 15:1476 (1990).

[7] J. M. Huang and P. Kumar, *Phys. Rev. Lett.* 68:2153 (1992).

[8] K. Bergman, C. R. Doerr, H. A. Haus, and M. Shirasaki, *Opt. Lett.* 18:643 (1993).

[9] N. C. Wong, K. W. Leong, and J. H. Shapiro, *Opt. Lett.* 15:891 (1990).

QUANTUM TELEPORTATION WITH COMPLETE BELL STATE MEASUREMENT

Yoon-Ho Kim
Department of Physics
University of Maryland, Baltimore County
Baltimore, Maryland 21250, USA

Sergei P. Kulik
Department of Physics, Moscow State University, Moscow, Russia

Yanhua Shih
Department of Physics
University of Maryland, Baltimore County
Baltimore, Maryland 21250, USA
yokim@umbc.edu and shih@umbc.edu

Abstract We report a quantum teleportation experiment in which nonlinear interactions are used for Bell state measurement. The experimental results demonstrate the working principle of irreversibly teleporting an unknown arbitrary quantum state from one system to another distant system by disassembling into and then later reconstructing from purely classical information and nonclassical EPR correlations. The distinct feature of this experiment is that *all* four Bell states can be distinguished in the Bell state measurement. Teleportation of a quantum state can thus occur with certainty in principle.

1. INTRODUCTION

The idea of quantum teleportation is to utilize the nonlocal correlations between an Einstein-Podolsky-Rosen pair of particles [1] to prepare a quantum system in some state, which is the exact replica of an arbitrary unknown state of a distant individual system [2]. Three experiments in this direction were published recently [3, 4, 5].

The following conditions must be satisfied in any claim for quantum teleportation: (i) the input quantum state, which is teleported in the experiment

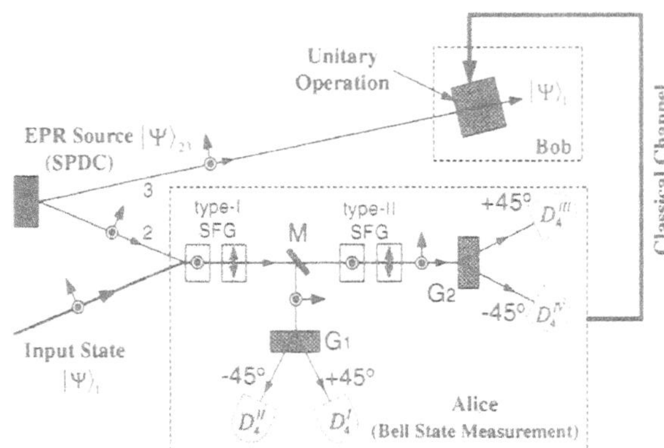

Figure 1 Schematic of quantum teleportation with complete BSM. Nonlinear interactions (SFG) are used to perform the BSM. ⊙ and ↕ represent the respective horizontal and vertical orientations of the optic axes of the crystals.

must be an *arbitrary* state, (ii) there must be an output quantum state which is an "instantaneous copy" of the input quantum state, (iii) the Bell state measurement (BSM) must be able to distinguish the complete set of the orthogonal Bell states so that the input state can be teleported with certainty, and (iv) for any input state, the teleportation must be deterministic, not "statistical".

In this paper, we experimentally demonstrate a quantum teleportation scheme which satisfies all four of the above conditions. The input state is an *arbitrary polarization state* and the BSM can distinguish *all* four orthogonal Bell states so that the state has a 100% certainty to be teleported in principle. This is because the BSM is based on nonlinear interactions which are *necessary* and *non-trivial* physical processes for correlating the input state and the entangled EPR pair [6, 7].

2. QUANTUM TELEPORTATION

The basic elements of the experiment are schematically shown in Fig. 1. Just as the original proposal of quantum teleportation [2], it consists of four essential parts: (a) the input state, (b) the EPR pair, (c) Alice (who performs the BSM of the input state and her EPR particle), and (d) Bob (who carries out unitary operations on his EPR particle). The input quantum state is an *arbitrary polarization state* given by,

$$|\Psi_1\rangle = \alpha|0_1\rangle + \beta|1_1\rangle, \tag{1}$$

where $|0\rangle$ and $|1\rangle$ represent the two orthogonal linear polarization bases (specifically in this paper) $|H\rangle$ (horizontal) and $|V\rangle$ (vertical) respectively. α and β are two arbitrary complex amplitudes with respect to the $|0\rangle$ and $|1\rangle$ bases and they satisfy the condition $|\alpha|^2 + |\beta|^2 = 1$. The EPR pair shared by Alice and Bob is prepared by spontaneous parametric down conversion (SPDC) as,

$$|\Psi_{23}\rangle = \frac{1}{\sqrt{2}}\{|0_2 0_3\rangle - |1_2 1_3\rangle\}, \tag{2}$$

with the subscripts 2 and 3 as labeled in Fig.1 [8]. The complete state of the three particles before Alice's measurement is then,

$$|\Psi_{123}\rangle = \frac{\alpha}{\sqrt{2}}\{|0_1 0_2 0_3\rangle - |0_1 1_2 1_3\rangle\} + \frac{\beta}{\sqrt{2}}\{|1_1 0_2 0_3\rangle - |1_1 1_2 1_3\rangle\}. \quad (3)$$

The four Bell states which form a complete orthonormal basis for both particle 1 and particle 2 are usually represented as,

$$\begin{aligned} |\Phi_{12}^{(\pm)}\rangle &= \frac{1}{\sqrt{2}}\{|0_1 0_2\rangle \pm |1_1 1_2\rangle\}, \\ |\Psi_{12}^{(\pm)}\rangle &= \frac{1}{\sqrt{2}}\{|0_1 1_2\rangle \pm |1_1 0_2\rangle\}. \end{aligned}$$

State (3) can now be re-written in the following form based on the above orthonormal Bell states,

$$\begin{aligned} |\Psi_{123}\rangle = \tfrac{1}{2}\{ \quad & |\Phi_{12}^{(+)}\rangle \; (\; \alpha|0_3\rangle - \beta|1_3\rangle \;) \\ + \; & |\Phi_{12}^{(-)}\rangle \; (\; \alpha|0_3\rangle + \beta|1_3\rangle \;) \\ + \; & |\Psi_{12}^{(+)}\rangle \; (\; -\alpha|1_3\rangle + \beta|0_3\rangle \;) \\ + \; & |\Psi_{12}^{(-)}\rangle \; (\; -\alpha|1_3\rangle - \beta|0_3\rangle \;) \quad \}. \end{aligned} \quad (4)$$

To teleport the state of particle 1 to particle 3 reliably, Alice must be able to distinguish her four Bell states by means of the BSM performed on particle 1 and her EPR particle (particle 2). She then tells Bob through a classical channel to perform a corresponding linear unitary operation on his EPR particle (particle 3) to obtain an exact replica of the state of particle 1. This completes the process of quantum teleportation.

3. COMPLETE BSM USING NONLINEAR INTERACTIONS

The distinct feature of the scheme shown in Fig.1 is that the BSM is based on nonlinear interactions: optical Sum Frequency Generation (SFG) (or "up-conversion"). Four SFG nonlinear crystals are used for "measuring" and "distinguishing" the complete set of the four Bell states. Photon 1 and photon 2 may interact either in the two type-I crystals or in the two type-II crystals to generate a higher frequency photon (labeled as photon 4). The projection measurements on photon 4 (either the 45° or 135° direction) correspond to the four Bell states of photon 1 and photon 2, $|\Phi_{12}^{(\pm)}\rangle$ and $|\Psi_{12}^{(\pm)}\rangle$.

Let us now discuss the BSM in detail (see Fig.1). The first type-I SFG crystal converts two $|V\rangle$ polarized photons $|1_1 1_2\rangle$ into a single horizontal polarized photon $|H_4\rangle$. Likewise, the second type-I SFG crystal converts two $|H\rangle$ polarized photons $|0_1 0_2\rangle$ into a single vertical polarized photon $|V_4\rangle$. The first and the last terms on the right-hand side in Eq.(3) thus become,

$$|\Psi_{43}\rangle = \alpha|V_4 0_3\rangle - \beta|H_4 1_3\rangle.$$

Dichroic beamsplitter M reflects only SFG photons to the 45° polarization projector G_1. Two detectors D_4^I and D_4^{II} are placed at the 45° and 135° output ports of G_1 respectively. Denoting the 45° and 135° polarization bases by $|45°\rangle$ and $|135°\rangle$, the state $|\Psi_{43}\rangle$ may be re-written as,

$$|\Psi_{43}\rangle = \frac{1}{\sqrt{2}}\{|45°\rangle_4(\alpha|0_3\rangle - \beta|1_3\rangle) + |135°\rangle_4(\alpha|0_3\rangle + \beta|1_3\rangle)\}, \quad (5)$$

which gives,

$$\begin{aligned} |\Psi\rangle_{3|D_4^I} &= \alpha|0_3\rangle - \beta|1_3\rangle, \\ |\Psi\rangle_{3|D_4^{II}} &= \alpha|0_3\rangle + \beta|1_3\rangle, \end{aligned} \quad (6)$$

i.e., if detector D_4^I (45°) is triggered, the quantum state of Bob's EPR photon (photon 3) is: $|\Psi_3\rangle = \alpha|0_3\rangle - \beta|1_3\rangle$ and if detector D_4^{II} (135°) is triggered, the quantum state of Bob's photon is: $|\Psi_3\rangle = \alpha|0_3\rangle + \beta|1_3\rangle$.

As we have analyzed above, the 45° and the 135° polarized type-I SFG components in Eq.(5) correspond to the superposition of $|0_1 0_2\rangle$ and $|1_1 1_2\rangle$ which are the respective Bell states $|\Phi_{12}^{(+)}\rangle$ and $|\Phi_{12}^{(-)}\rangle$.

Similarly, the other two Bell states are distinguished by the type-II SFG's. The states $|0_1 1_2\rangle$ and $|1_1 0_2\rangle$ are made to interact in the first and the second type-II SFG crystals respectively to generate a higher frequency photon with either horizontal (the first type-II SFG) or vertical (the second type-II SFG) polarization. A 45° polarization projector G_2 is used after the type-II SFG crystals and two detectors D_4^{III} and D_4^{IV} are placed at the 45° and the 135° output ports of G_2 respectively. On the new bases of 45° and 135° for the SFG photon, the second and the third terms on the right-hand side in Eq.(3) thus become,

$$|\Psi_{43}\rangle = \frac{1}{\sqrt{2}}\{|45°\rangle_4(-\alpha|1_3\rangle + \beta|0_3\rangle) + |135°\rangle_4(-\alpha|1_3\rangle - \beta|0_3\rangle)\}, \quad (7)$$

which gives,

$$\begin{aligned} |\Psi\rangle_{3|D_4^{III}} &= -\alpha|1_3\rangle + \beta|0_3\rangle, \\ |\Psi\rangle_{3|D_4^{IV}} &= -\alpha|1_3\rangle - \beta|0_3\rangle, \end{aligned} \quad (8)$$

i.e., if detector D_4^{III} (45°) is triggered, the quantum state of Bob's photon is: $|\Psi_3\rangle = -\alpha|1_3\rangle + \beta|0_3\rangle$ and if detector D_4^{IV} (135°) is triggered, the quantum state of Bob's photon is: $|\Psi_3\rangle = -\alpha|1_3\rangle - \beta|0_3\rangle$.

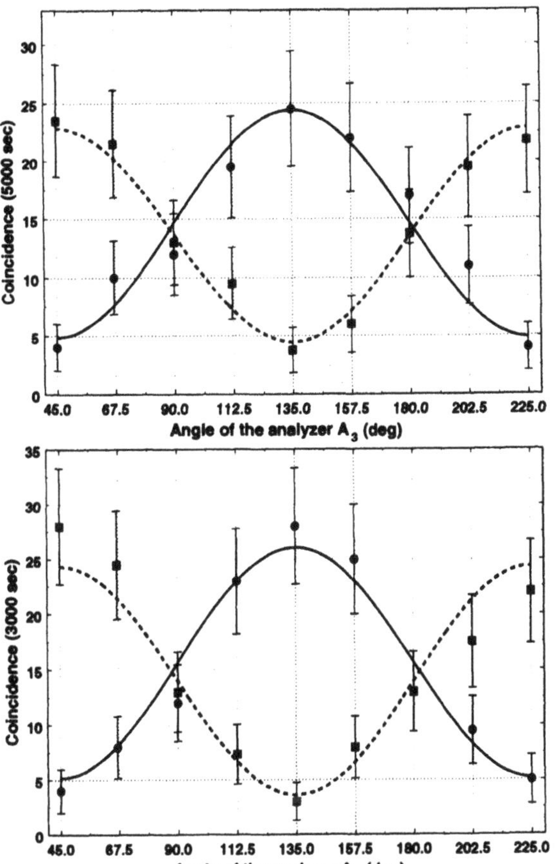

Figure 2 Solid line (circled data points) is the joint detection rate D_4^I-D_3 for 45° linear polarization as an input state. Dashed line (square data points) is for D_4^{II}-D_3 for the same input state. The expected π phase shift is clearly demonstrated.

Figure 3 Solid line (circled data points) is the joint detection rate D_4^{III}-D_3 and dashed line (square data points) is for D_4^{IV}-D_3. Again, the expected π phase shift is clearly demonstrated.

The 45° and the 135° polarized type-II SFG components correspond to the superposition of $|0_1 1_2\rangle$ and $|1_1 0_2\rangle$ which are the Bell states $|\Psi_{12}^{(+)}\rangle$ and $|\Psi_{12}^{(-)}\rangle$ respectively.

To obtain the exact replica of the state of Eq.(1), Bob needs simply to perform a corresponding unitary transformation after learning from Alice which of her four detectors, D_4^I, D_4^{II}, D_4^{III}, or D_4^{IV}, has triggered [9].

To demonstrate the working principle of this scheme, we measure the joint detection rates between detectors D_4^I-D_3, D_4^{II}-D_3, D_4^{III}-D_3 and D_4^{IV}-D_3, where D_3 is Bob's detector (see Fig.4). In these measurements we choose the input state $|\Psi_1\rangle$ as a linear polarization state. For a fixed input polarization state, the angle of the polarization analyzer A_3 which is placed in front of Bob's detector is rotated and the joint detection rates are recorded. Figure 2 show two typical data sets for D_4^I-D_3 and D_4^{II}-D_3. The input polarization state is 45°. Clearly, these data curves confirm Eq.(6). The different phases of the two curves reflect the phase difference between the two states in Eq.(6). As expected, experimental data for D_4^{III}-D_3 and D_4^{IV}-D_3 show similar behavior, see Fig.3.

4. EXPERIMENT

We now discuss the details of the experimental setup. The schematic of the experimental setup is shown in Fig. 4. The input polarization state is prepared by using a $\lambda/2$ plate from a femtosecond laser pulse (pulse width $\approx$ 100fsec and central wavelength = 800nm) [10]. The EPR pair (730nm-885nm photon pair) is generated by two non-degenerate type-I SPDC's. The optical axes of the first and the second SPDC crystals are oriented in the respective horizontal ($\odot$) and vertical ($\updownarrow$) directions. The SPDC crystals are pumped by a 45° polarized 100fsec laser pulse with 400nm central wavelength. The BBO crystals (each with thickness 3.4mm) are cut for collinear non-degenerate phase matching. Since the two crystals are pumped equally, the SPDC pair can be generated either in the first BBO as $|V_{885}\rangle_2|V_{730}\rangle_3$ ($|1_2 1_3\rangle$) or in the second BBO as $|H_{885}\rangle_2|H_{730}\rangle_3$ ($|0_2 0_3\rangle$) with equal probability (885 and 730 refer to the wavelengths in nanometer).

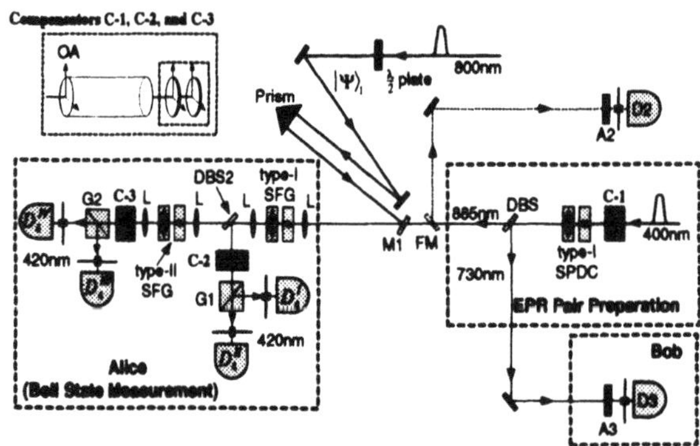

Figure 4 Schematic of the Experimental Setup. The inset shows the details of the compensators. See text for details.

In order to prepare an EPR state in the form of Eq.(2) (a Bell state), these two amplitudes have to be quantum mechanically "indistinguishable" and have the expected relative phase. A Compensator (C-1) is used for this purpose and it consists of two parts: a thick quartz rod and two thin plates. The thick quartz rod is used to compensate the time delay between the two amplitudes $|1_2 1_3\rangle$ and the $|0_2 0_3\rangle$, and two thin quartz plates are used to adjust the relative phase between them by angular tilting. A dichroic beamsplitter DBS is placed behind the SPDC crystals to separate and send the photon 2 (885nm) and photon 3 (730nm) to Alice and Bob respectively. To check the EPR state, a flipper mirror FM is used to send the photon 2 (885nm) to a photon-counting detector D_2 for EPR correlation measurement. Both the space-time and polarization correlations must be checked before teleportation measurements, in order to be certain of having high degree EPR entanglement and the expected relative phase between the $|1_2 1_3\rangle$ and the $|0_2 0_3\rangle$ amplitudes (see Ref.[11] for details). Once the EPR state in Eq.(2) is prepared, FM is flipped-down and photon 2 (885nm) is given to Alice for BSM with photon 1.

The BSM consists of four SFG nonlinear crystals, two 45° projectors (G_1 and G_2), four single photon counting detectors (D_4^I, D_4^{II}, D_4^{III}, and D_4^{IV}) and two compensators as well as other necessary optical components. The input photon (800nm) and photon 2 (885nm) may either interact in the two type-I or in the two type-II SFG crystals. Two pairs of lenses (L) are used as telescopes to focus the input beams onto the crystals. The vertical (horizontal) polarized amplitudes of the input photon (800nm) and the vertical (horizontal) polarized photon 2 (885nm) interact in the first (second) type-I SFG to generate a 420nm horizontal (vertical) polarized photon. The horizontal (vertical) polarized amplitudes of the input photon and the vertical (horizontal) polarized photon 2 interact in the first (second) type-II SFG to generate a 420nm horizontal (vertical) polarized photon. The 420nm photons generated in the type-I SFG process is reflected to detectors D_4^I and D_4^{II} (after passing through C-2 and a 45° polarization projector G_1) by a dichroic beamsplitter DBS_2 and similarly for the 420nm photons created in two type-II SFG processes. It is very important to design and adjust the Compensators (C-2 and C-3) correctly in order to make the horizontal and the vertical components of the 420nm SFG quantum mechanically indistinguishable and to attain the expected relative phase. These two compensators are similar to C-1.

Since the input state (photon 1) and photon 2 should overlap inside the SFG crystals exactly, a prism is used to adjust the path-length of the input pulse. M_1 is a dichroic mirror which reflects the 800nm photons while transmitting the 885nm ones. By scanning the prism, we observed the SFG process between the input pulse (photon 1) and photon 2 (single-photon created by the SPDC process) when they overlap perfectly inside the SFG crystals.

In summary, we have shown a proof-of-principle experimental demonstration of quantum teleportation with a complete set of Bell state measurement. The two main features lie at the heart of our scheme: (i) EPR-Bohm type quantum correlation and (ii) the BSM using nonlinear interactions. Single photon SFG is used as the BSM and the working principle is demonstrated by observing correlations between the joint measurement of Alice and Bob. In the current experiment, femtosecond laser pulses are used to prepare the input polarization state to reduce data collection time. Recent research on nonlinear optics at low light levels may enable high-efficiency SFG at single-photon level in the near future [12].

Acknowledgments

We would like to thank C.H. Bennett and M.H. Rubin for helpful discussions. This work was supported in part by the Office of Naval Research, ARDA, and the National Security Agency.

References

[1] A. Einstein, B. Podolsky, and N. Rosen, Phys. Rev. **47**, 777 (1935).

[2] C.H. Bennett *et al.*, Phys. Rev. Lett. **70**, 1895 (1993).

[3] D. Bouwmeester *et al.*, Nature **390**, 575 (1997).

[4] D. Boschi *et al.*, Phys. Rev. Lett. **80**, 1121 (1998).

[5] A. Furusawa *et al.*, Science **282**, 706 (1998).

[6] Complete BSM cannot be performed with only linear elements. See, L. Vaidman and N. Yoran, Phys. Rev. A **59**, 116 (1999); N. Lütkenhaus, J. Calsamiglia, and K.-A. Suominen, *ibid.* **59**, 3295 (1999).

[7] D.N. Klyshko, JETP **87**, 639 (1998).

[8] Note that any one of the four Bell states can be used for this purpose.

[9] Complete BSM using SFG is also useful for other applications, see C.H. Bennett and S.J. Wiesner, Phys. Rev. Lett. **69**, 2881 (1992).

[10] In this experiment, the input state is a polarization state of a femtosecond laser pulse. Only one out of approximately 10^{10} photons, all in the same polarization state, in each laser pulse is actually "upconverted" in the SFG process. What is being teleported is the state or qubit associated with that photon. (Note, quantum teleportation does not teleport the "quantum" but rather, the state of the quantum). In principle, it does not prevent one to use a single-photon qubit as the input state in this experiment. Due to the low efficiency of SFG, one needs to wait a much longer time for teleportation to occur.

[11] Y.-H. Kim, S.P. Kulik, and Y.H. Shih, Phys. Rev. A **62**, 011802(R), (2000); Y.-H. Kim, S.P. Kulik, and Y.H. Shih, quant-ph/0007067.

[12] S.E. Harris and L.V. Hau, Phys. Rev. Lett. **82**, 4611 (1999).

COMPLETE QUANTUM TELEPORTATION WITH A CROSSED-KERR NONLINEARITY

D. Vitali, M. Fortunato, P. Tombesi
Dipartimento di Matematica e Fisica, Università di Camerino,
INFM, Unità di Camerino, via Madonna delle Carceri 62032, Camerino, Italy

Keywords: Entanglement, Bell-state measurement, crossed-Kerr effect, electromagnetic induced transparency, quantum teleportation

Abstract We show how perfect Bell-state discrimination and complete quantum teleportation of the polarization state of a photon can be implemented employing a cross-Kerr medium. We show that, using the recently demonstrated ultraslow light propagation in coherent atomic media, the proposal can be realized with presently available technology.

Quantum teleportation [1] is the "reconstruction", with 100% success, of an unknown state given to one station (Alice), performed at another remote station (Bob), on the basis of two bits of classical information sent by Alice to Bob. Perfect teleportation is possible only if the two parties share a maximally entangled state. The most delicate part needed for the effective realization of teleportation is the Bell-state measurement, *i.e.* the discrimination between the four, maximally entangled, Bell states which has to be performed by Alice and whose result is communicated to Bob through the classical channel. There have been numerous proposals for its realization in different systems and recently successful, pioneering experiments [2, 3, 4] have provided convincing experimental proof-of-principle of the correctness of the teleportation concept.

These experiments differ by the degrees of freedom used as qubits and for the different ways in which the Bell-state measurement is performed. The Innsbruck experiment [2] is the conceptually simplest one, since each qubit is represented by the polarization state of a single photon pulse. In this experiment, however, only two out of the four Bell states can be discriminated and therefore the success rate cannot be larger than 50% [5]. The Rome experiment [3] employs the entanglement between the spatial and the polarization degrees of freedom of a photon and it is able to distinguish all the corresponding four Bell states completely. However in this scheme the state to be teleported is generated within the apparatus (it cannot come from the outside) and therefore the scheme cannot be used as a computational primitive in a larger quantum network for further information processing, as it has been recently proposed in Ref. [6]. Finally the Caltech experiment [4] is conceptually completely different since it implies the teleportation of the state of a continuous degree of freedom, the mode of an electromagnetic field, employing the entangled two-mode squeezed states at the output of a parametric amplifier. In this case, the Bell-state measurement is replaced by two homodyne measurements and a direct comparison with the original quantum teleportation scheme of Ref. [1] cannot be made. Up to now, only coherent states of the electromagnetic field have been successfully teleported using this scheme.

It is therefore desirable to have a scheme for a Bell-state measurement that can be used in the simplest case of the Innsbruck scheme. This would imply the possibility of realizing the first *complete* verification of the original quantum teleportation scheme [1] and also of having a device useful for other quantum protocols, as quantum dense coding [7]. What we need is a device able to discriminate among the four Bell states that can be realized with the polarization-entangled photon pairs produced in Type-II phase matched parametric down conversion, that is

$$|\psi^{\pm}\rangle = \frac{|V_1, H_2\rangle \pm |H_1, V_2\rangle}{\sqrt{2}} \qquad |\phi^{\pm}\rangle = \frac{|V_1, V_2\rangle \pm |H_1, H_2\rangle}{\sqrt{2}} \tag{1}$$

where $|H\rangle$ and $|V\rangle$ denote the horizontally and vertically polarized one-photon states, respectively, and $1, 2$ refer to two different spatial modes.

It has been recently shown that it is impossible to perform a complete Bell measurement on two-mode polarization states using only linear passive elements [8] (unless the two photons are entangled in more than one degree of freedom [9]), and for this reason schemes involving some effective nonlinearities, such as resonant atomic interactions [10], or the Kerr effect [11], have been proposed. Here we propose a scheme for a perfect Bell-state discrimination based on a nonlinear optical effect, the cross-phase modulation taking place in Kerr media. In this respect, our scheme is based on a $\chi^{(3)}$ medium as the "Fock-filter" proposal of Ref. [11]. However, our scheme is different and simpler and, above all, is feasible using available technology, since we shall show that the needed crossed-Kerr nonlinearity can be obtained using the recently demonstrated ultraslow light propagation [12], achieved via electromagnetic induced transparency (EIT) [13] in ensembles of cold atoms.

Our "Bell box" is described in Fig. 1 and can be divided into two parts: the left part is composed by three polarization rotators (R, $R^{\dagger}$) and by the "quantum phase gate" (QPG) which will be described below, and can be called "the disentangler", since it realizes the unitary transformation changing each

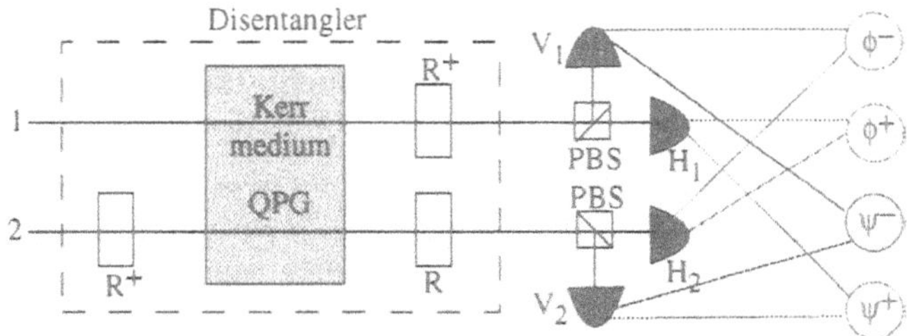

Figure 1 Scheme of the Bell-state measurement: QPG is the quantum phase gate of Eqs. (6)-(7) with $\varphi = \pi$; R ($R^{\dagger}$) rotates the polarization by $\pi/4$ ($-\pi/4$), and PBS are polarizing beam splitters. The "disentangler" performs the unitary transformation of Eqs. (2)-(3).

Bell state of Eq. (1) into one of the four factorized polarization states , *i.e.*

$$|\psi^+\rangle \rightarrow |H_1, V_2\rangle \qquad |\psi^-\rangle \rightarrow |V_1, V_2\rangle \tag{2}$$

$$|\phi^+\rangle \rightarrow |H_1, H_2\rangle \qquad |\phi^-\rangle \rightarrow |V_1, H_2\rangle \;. \tag{3}$$

The right part of the scheme is composed by two polarizing beam-splitters (PBSs) and by four detectors with single-photon sensitivity, and simply serves the purpose of detecting the four states of the factorized polarization basis

$$\{|e_1\rangle, |e_2\rangle, |e_3\rangle, |e_4\rangle\} = \{|H_1, H_2\rangle, |H_1, V_2\rangle, |V_1, H_2\rangle, |V_1, V_2\rangle\} \;, \tag{4}$$

where $\{|e_i\rangle\}$ are the tensor product of the single-photon polarization basis states,

$$|H_i\rangle = \begin{pmatrix} 1 \\ 0 \end{pmatrix}_i \qquad |V_i\rangle = \begin{pmatrix} 0 \\ 1 \end{pmatrix}_i \;. \tag{5}$$

Due to the one-to-one correspondence of Eqs. (2)-(3), it is clear that the detection of each Bell state corresponds to a different pair of detector clicks, so that they are unambiguously distinguishable. The disentangler, and in particular the QPG, is the most delicate part as concerns the experimental implementation, since it involves a two-qubit operation, *i.e.*, an effective photon-photon interaction. In fact, if $\tilde{R}_i$ is a simple polarization rotation by $\pi/4$ radians for mode i (and $\tilde{R}_i^{\dagger}$ its inverse), *i.e.*, $|H_i\rangle \rightarrow (|H_i\rangle + |V_i\rangle)/\sqrt{2}$, $|V_i\rangle \rightarrow (|V_i\rangle - |H_i\rangle)/\sqrt{2}$, we have $R_1 = \tilde{R}_1 \otimes I_2$, $R_2 = I_1 \otimes \tilde{R}_2$, which can be obtained using a $\lambda/2$ retardation plate at a $\pi/8$ angle. (I_i is the 2×2 unit matrix for mode i). The general QPG $P(\varphi)$ is a universal two-qubit gate as long as $\varphi \neq 0$ [14, 15], and in the two-photon polarization basis (4) we are considering here, it can be written as

$$|H_1, H_2\rangle \rightarrow |H_1, H_2\rangle \qquad |H_1, V_2\rangle \rightarrow |H_1, V_2\rangle \tag{6}$$

$$|V_1, H_2\rangle \rightarrow |V_1, H_2\rangle \qquad |V_1, V_2\rangle \rightarrow e^{i\varphi}|V_1, V_2\rangle \;. \tag{7}$$

The experimental realization of this gate has been reported in Ref. [16], in the case when one qubit is given by the internal state of a trapped ion and the other qubit by its two lowest vibrational states, and recently in Ref. [17], where the two qubits are represented by two circular Rydberg states of a Rb atom and by the two lowest Fock states of a microwave cavity. In the optical case we are interested in, the QPG between two frequency-distinct cavity modes has been experimentally investigated in Ref. [15], using however weak coherent states instead of single photon pulses, demonstrating therefore only conditional quantum dynamics and not the full quantum transformation of Eqs. (6)-(7). As it can be easily checked, the QPG can be realized using a crossed-Kerr interaction involving the vertically polarized modes only, i.e., $H_K = \hbar\chi a_{V_1}^{\dagger} a_{V_1} a_{V_2}^{\dagger} a_{V_2}$, so that the conditional phase shift is $\varphi = \chi t_{int}$, where t_{int} is the interaction time within the Kerr medium.

The disentangler of Fig. 1 realizes the transformation (2)-(3) when the conditional phase shift is $\varphi = \pi$, as it can be checked in a straightforward way by writing the matrix form of the transformation $R_1^{\dagger} R_2 P(\pi) R_2^{\dagger}$ of Fig. 1 in the factorized polarization basis (4), which is just the matrix form of Eqs. (2)-(3) in the chosen basis. The proposed Bell box is therefore extremely simple and also robust against detector inefficiencies. This is due to the fact that in our scheme, only one photon at most impinges on each of the four detectors. First of all this means that only single photon *sensitivity* and not single photon *resolution* is needed, and in this case solid-state photomultipliers can provide up to 90% efficiency [9]. Moreover, this implies that the detection scheme is reliable, *i.e.*, it always discriminates the correct Bell state, whenever it answers. In the case of detectors with the same efficiency η, our Bell box gives the (always correct) output with probability η^2 and it does not give any output (only zero or one photon is detected) with probability $1 - \eta^2$.

As we have already remarked, the most difficult part for the experimental implementation of the scheme is the QPG with a conditional phase shift $\varphi = \pi$. In fact, realizing the transformation (6)-(7) means having a large cross-phase modulation at the single photon level between two traveling-wave pulses, with negligible absorption, which is very demanding. For example, in the experiment of Ref. [15], a conditional phase-shift $\varphi = 16°$ has been measured, which however involved two frequency-distinct *cavity* modes in a high-finesse cavity. However, the recent demonstration of ultraslow light propagation in a cold gas of sodium atoms [12] and with hot Rb atoms [18], opens the way for the realization of significant conditional phase shifts also between two traveling single photon pulses. In fact, the extremely slow group velocity is obtained as a consequence of EIT [13], which however, as originally suggested by Schmidt and Imamoğlu in Ref. [19], can also be used to achieve giant crossed-Kerr nonlinearities. In particular Harris and Hau [20] have showed that in the limit of very small group velocity, a conditional phase shift per photon between the slowed-down pulse and a detuned probe pulse not subject to EIT, given by $\varphi = \gamma_{24}\Delta\omega_{24}/\left(4\gamma_{24}^2 + 4\Delta\omega_{24}^2\right)$, is realized, where $\Delta\omega_{24}$ is the detuning of the probe pulse and γ_{24} the associated linewidth (see Eq. (10) of Ref. [20]). In this case one can reach $\varphi = 1/8$ at most, and therefore one has a conditional phase shift far from the optimal value $\varphi = \pi$, because the two pulses travel at very different group velocities and therefore the interaction time within the Kerr medium is limited. The situation can be greatly improved if *both* light pulses are subject to EIT and propagate with slow but equal group velocities. In this case the interaction time becomes much longer and conditional phase shifts of the order of π could be in principle achieved (see for example [21]). In our case of polarization-encoded photonic qubits, this could be achieved using EIT between Zeeman sublevels of the ground and excited states, in which only pulses with a given polarization are subject to EIT. Anyhow, it is easy to understand that realizing the condition $\varphi = \pi$ *exactly* is highly non trivial and this could represent a problem since, when $\varphi \neq \pi$, the scheme of Fig. 1 is no longer perfect and it does not discriminate the four Bell states with 100% success (in general the probability of a successful discrimination is equal to $\cos^2 \varphi/2$). However, as mentioned above, the QPG represented by $P(\varphi)$ is a universal two-qubit gate, capable of entangling and disentangling qubits as soon as $\varphi \neq 0$. Moreover, even though different from π, the conditional phase shift φ is a given and measurable property, and it is reasonable to expect that, using the knowledge of the actual value of φ, it is possible to adapt and optimize the teleportation protocol in order to achieve a truly *quantum* teleportation (*i.e.*, that cannot be achieved with only classical means), even in the presence of an imperfect Bell-state measurement. Optimization means that Bob has to suitably modify the four local unitary transformations he has to perform on the received qubit according to the Bell measurement result communicated by Alice. In the optimized protocol, Bob's local unitary transformations will now depend on the phase φ of the QPG and will reduce to those of the original proposal [1] in the ideal case of perfect Bell-state discrimination $\varphi = \pi$. We expect that, $\forall\varphi \neq 0$, the average fidelity of the teleported state will be always larger than 2/3, as it must be for any truly quantum teleportation of a qubit state [22].

Let us therefore consider a generic one-photon state $|\psi\rangle_1 = \alpha|H_1\rangle + \beta|V_1\rangle$, which is given to Alice and has to be teleported to Bob, and let us assume that Alice and Bob share the Bell state $|\psi^+\rangle_{23} = (|H_2V_3\rangle + |V_2H_3\rangle)/\sqrt{2}$, so that the input state for the teleportation process is $|\psi\rangle_1 \otimes |\psi^+\rangle_{23}$. Alice is provided with the "imperfect" Bell box with a QPG $P(\varphi)$, so that the disentangler of Fig. 1 will now be described by the transformation $R_1^\dagger R_2 P(\varphi) R_2^\dagger$. It is easy to check that when $\varphi \neq \pi$, the four Bell states are no longer completely disentangled and therefore no longer discriminated with 100% success.

Alice has to perform the Bell-state measurement on modes 1 and 2, and the resulting joint state of the three modes just before the photodetections is

$$|\tilde{\psi}\rangle_{123} = \sum_{i=1}^{4} |e_i\rangle_{12} \hat{G}_i(\varphi)|\psi\rangle_3 \ , \tag{8}$$

where $|e_i\rangle_{12}$ are the factorized basis states (4) and

$$\hat{G}_1(\varphi) = \frac{1}{2}\begin{pmatrix} 0 & -ie^{i\varphi/2}\sin\frac{\varphi}{2} \\ 1 & e^{i\varphi/2}\cos\frac{\varphi}{2} \end{pmatrix} \tag{9}$$

$$\hat{G}_2(\varphi) = \frac{1}{2}\begin{pmatrix} 1 & e^{i\varphi/2}\cos\frac{\varphi}{2} \\ 0 & -ie^{i\varphi/2}\sin\frac{\varphi}{2} \end{pmatrix} \tag{10}$$

$$\hat{G}_3(\varphi) = \frac{1}{2}\begin{pmatrix} 0 & -ie^{i\varphi/2}\sin\frac{\varphi}{2} \\ -1 & e^{i\varphi/2}\cos\frac{\varphi}{2} \end{pmatrix} \tag{11}$$

$$\hat{G}_4(\varphi) = \frac{1}{2}\begin{pmatrix} -1 & e^{i\varphi/2}\cos\frac{\varphi}{2} \\ 0 & -ie^{i\varphi/2}\sin\frac{\varphi}{2} \end{pmatrix} . \tag{12}$$

When the photons are detected, Alice sends the results through the classical channel to Bob. Bob is left with the photon of mode 3, and applies a local unitary transformation $\hat{U}_i(\varphi)$ in correspondence to the i-th result of the Bell-state measurement. As a consequence, the output state of the teleportation process is

$$\rho_{out} = \sum_{i=1}^{4} \hat{U}_i(\varphi)\hat{G}_i(\varphi)|\psi\rangle_3\langle\psi|\hat{G}_i(\varphi)^\dagger \hat{U}_i(\varphi)^\dagger \ . \tag{13}$$

Since the output state has to reproduce the unknown input state $|\psi\rangle$ as much as possible, it is evident that to optimize the local unitary transformations $\hat{U}_i(\varphi)$, one should "invert" $\hat{G}_i(\varphi)$. The best strategy is suggested by the use of the polar decomposition of the matrices $\hat{G}_i(\varphi)$,

$$\hat{G}_i(\varphi) = \hat{T}_i(\varphi)\hat{R}_i(\varphi) \ , \tag{14}$$

where $\hat{R}_i(\varphi) = \sqrt{\hat{G}_i(\varphi)^\dagger \hat{G}_i(\varphi)}$ is Hermitian and $\hat{T}_i(\varphi)$ unitary, so that Bob's optimal local unitary transformations will be

$$\hat{U}_i(\varphi) = \hat{T}_i(\varphi)^{-1} = \hat{R}_i(\varphi)\hat{G}_i(\varphi)^{-1} \ . \tag{15}$$

Using Eqs. (9)-(12), (14) and (15), one finds the following Bob's optimal unitary transformations

$$\hat{U}_1(\varphi) = \begin{pmatrix} -i\cos\frac{\pi+\varphi}{4} & \sin\frac{\pi+\varphi}{4} \\ ie^{-i\varphi/2}\sin\frac{\pi+\varphi}{4} & e^{-i\varphi/2}\cos\frac{\pi+\varphi}{4} \end{pmatrix} \tag{16}$$

$$\hat{U}_2(\varphi) = \begin{pmatrix} \cos\frac{\pi-\varphi}{4} & -i\sin\frac{\pi-\varphi}{4} \\ e^{-i\varphi/2}\sin\frac{\pi-\varphi}{4} & ie^{-i\varphi/2}\cos\frac{\pi-\varphi}{4} \end{pmatrix} \quad (17)$$

$$\hat{U}_3(\varphi) = \begin{pmatrix} i\cos\frac{\pi+\varphi}{4} & -\sin\frac{\pi+\varphi}{4} \\ ie^{-i\varphi/2}\sin\frac{\pi+\varphi}{4} & e^{-i\varphi/2}\cos\frac{\pi+\varphi}{4} \end{pmatrix} \quad (18)$$

$$\hat{U}_4(\varphi) = \begin{pmatrix} -\cos\frac{\pi-\varphi}{4} & i\sin\frac{\pi-\varphi}{4} \\ e^{-i\varphi/2}\sin\frac{\pi-\varphi}{4} & ie^{-i\varphi/2}\cos\frac{\pi-\varphi}{4} \end{pmatrix}, \quad (19)$$

which (once the conditional phase shift is known) can be easily implemented using appropriate birefringent plates and polarization rotators. It can be checked that, in the special case $\varphi = \pi$, the above optimized teleportation protocol coincides with the original one [1], since one has $\hat{U}_1(\pi) = \sigma_x$, $\hat{U}_2(\pi) = 1$, $\hat{U}_3(\pi) = -i\sigma_y$, and $\hat{U}_4(\pi) = -\sigma_z$.

Finally, we have to check that the proposed teleportation protocol, even though no longer with 100% success when $\varphi \neq \pi$, always implies the realization of a true quantum teleportation, that cannot be achieved with only classical means. This amounts to check that the average fidelity of the output state is larger than $2/3$ for $0 < \varphi < 2\pi$. For pure qubit states, the average fidelity F_{av} is defined as

$$F_{av} = \frac{1}{4\pi}\int d\Omega \langle\psi|\rho_{out}|\psi\rangle \,, \quad (20)$$

where the integral is over the Bloch sphere and $|\psi\rangle$ is the generic input state. Using Eqs. (1) and (15) one has

$$\langle\psi|\rho_{out}|\psi\rangle = \sum_{i=1}^{4} \left|\langle\psi|\hat{R}_i(\varphi)|\psi\rangle\right|^2 \,, \quad (21)$$

so that, using the explicit expressions for $\hat{R}_i(\varphi)$ that can be obtained from Eqs. (9)-(12), and performing the average over the Bloch sphere, one finally finds [23]

$$F_{av}(\varphi) = \frac{2}{3} + \frac{1}{3}\sin\frac{\varphi}{2} \,, \quad (22)$$

which is larger than the upper classical bound $F_{av} = 2/3$ [22] for $0 < \varphi < 2\pi$, as expected.

References

[1] C.H. Bennett *et al.*, Phys. Rev. Lett. **70**, 1895 (1993).

[2] D. Bouwmeester *et al.*, Nature (London) **390**, 575 (1997).

[3] D. Boschi *et al.*, Phys. Rev. Lett. **80**, 1121 (1998).

[4] A. Furusawa *et al.*, Science, **282** 706 (1998).

[5] S.L. Braunstein and H.J. Kimble, Nature (London) **394**, 47 (1998).

[6] D. Gottesman, and I.L. Chuang, Nature (London) **402**, 390 (1999).

[7] C.H. Bennett and S.J. Wiesner, Phys. Rev. Lett. **69**, 2881 (1992).

[8] N. Lütkenhaus *et al.*, Phys. Rev. A **59**, 3259 (1999).

[9] P.G. Kwiat and H. Weinfurter, Phys. Rev. A **58**, R2623 (1998).

[10] M.O. Scully *et al.*, Phys. Rev. Lett. **83**, 4433 (1999); E. DelRe *et al.*, Phys. Rev. Lett. **84**, 2989 (2000).

[11] M.G.A. Paris *et al.*, LANL e-print quant-ph/9911036.

[12] L.V. Hau *et al.*, Nature (London) **397**, 594 (1999).

[13] E. Arimondo, in *Progress in Optics* XXXV, ed. by E. Wolf, (Elsevier, Amsterdam, 1996); S.E. Harris, Phys. Today **50**, 36 (1997).

[14] S. Lloyd, Phys. Rev. Lett. **75**, 346 (1995).

[15] Q.A. Turchette *et al.*, Phys. Rev. Lett. **75**, 4710 (1995).

[16] C. Monroe *et al.*, Phys. Rev. Lett. **75**, 4714 (1995).

[17] A. Rauschenbeutel *et al.*, Phys. Rev. Lett. **83**, 5166 (1999).

[18] M.M. Kash *et al.*, Phys. Rev. Lett. **82**, 5229 (1999).

[19] H. Schmidt and A. Imamoğlu, Opt. Lett. **21**, 1936 (1996).

[20] S.E. Harris and L.V. Hau, Phys. Rev. Lett. **82**, 4611 (1999).

[21] M. D. Lukin and A. Imamoğlu, Phys. Rev. Lett. **84**, 1419 (2000).

[22] H. Barnum, Ph. D. thesis, University of New Mexico (1998).

[23] D. Vitali, M. Fortunato, and P. Tombesi, Phys. Rev. Lett. **85**, 445 (2000).

QUANTUM LITHOGRAPHY

Pieter Kok, Samuel L. Braunstein
Informatics, University of Wales, Bangor LL57 1UT, UK
pieter@sees.bangor.ac.uk

Agedi N. Boto, Daniel S. Abrams,
Colin P. Williams, Jonathan P. Dowling
Jet Propulsion Laboratory, California Institute of Technology,
Mail Stop 126-347, 4800 Oak Grove Drive, Pasadena, California 91109

Abstract As demonstrated in Boto *et al.* [quant-ph/9912052], quantum lithography offers an increase in resolution without an upper bound. This procedure makes use of the entangled state $(|n,0\rangle + e^{in\varphi}|0,n\rangle)/\sqrt{2}$. It allows us to write evenly spaced lines with sub-wavelength resolution. Here we generalise this procedure in such a way that it enables us to create arbitrary patterns in one dimension. We distinguish two methods: the 'Fourier method' and 'the superposition method'. The Fourier method is conceptually easier since it depends on Fourier series, but it also involves a minimal finite amount of unwanted exposure of the substrate. The superposition method gets around this problem and gives generally better results, but lacks the intuitive clarity of the Fourier method.

Optical lithography is a widely used printing method. In this process light is used to etch a substrate. The (un)exposed areas on the substrate then define the pattern. In particular, the micro-chip industry uses lithography to produce smaller and smaller processors. However, classical optical lithography can only achieve a resolution comparable to the wavelength of the used light [1, 2, 3]. It therefore limits the scale of the patterns. To create smaller patterns we need to venture beyond this classical boundary [4]. In this paper we study how *quantum* lithography can be used to increase the resolution indefinitely, and how it allows us to create arbitrary patterns in one dimension. The two-dimensional case will be studied in a forthcoming paper [5].

1. INTRODUCTION TO QUANTUM LITHOGRAPHY

Suppose we have two crossing light beams a and b. Where the two beams meet we place some suitable substrate, such that the interference pattern is recorded [6]. We consider the grazing limit in which the angle θ between the two beams is π (see Fig. 1). Classically, the interference pattern on the substrate has a resolution of the order of $\lambda/2$, where λ is the wavelength of the light used. However, by using photon-number states (i.e., inherently non-classical states) we can increase the resolution to the sub-wavelength regime.

As in Ref. [6], let the two counter-propagating light beams a and b be in the combined (entangled) state

$$|\psi_n\rangle_{ab} = \left(|n,0\rangle_{ab} + e^{in\varphi}|0,n\rangle_{ab}\right)/\sqrt{2}\,. \tag{1}$$

Now define the mode operator $e = (a + b)/\sqrt{2}$ and its adjoint $e^\dagger = (a^\dagger + b^\dagger)/\sqrt{2}$. The deposition rate Δ on the substrate is then given by

$$\Delta_n = \langle\psi_n|\delta_n|\psi_n\rangle \qquad \text{with} \quad \delta_n = \frac{(e^\dagger)^n e^n}{n!}\,, \tag{2}$$

i.e., we look at the higher moments of the electric field operator [7, 8, 9]. The deposition rate Δ is measured in units of intensity. Leaving the substrate exposed for a time t to the light source will result in an intensity pattern $p(\varphi) = \Delta_n t$. After a straightforward calculation we see that

$$\Delta_n \propto (1 + \cos n\varphi)\,. \tag{3}$$

We interpret this as follows. A phase-shift φ in light beam b results in a displacement x of the interference pattern on the substrate. Using two classical waves, a phase-shift of 2π will return the pattern to its original position. However, according to Eq. (3) one cycle is completed after a shift of $2\pi/n$. This means that a shift of 2π will displace the pattern n times. In other words, we have n times more maxima in the interference pattern. These need to be closely spaced, yielding an effective resolution proportional to λ/n.

In order for quantum lithography to become a mature technology, we have to solve two problems. First, we have to create the entangled states given by Eq. (1). One possibility which needs to be studied further is to use optical components like parametric down-converters. Contrary to Ref. [10], we are not concerned with the usually large vacuum contribution of these processes, since the vacuum will not contribute to the deposition rate [see Eqs. (1) and (2)]. Secondly, we need substrates which are sensitive to predominantly one higher moment of the electric field operator.

This paper is organised as follows: in section 2 we consider a generalised version of the state given in Eq. (1). We show how, given this state we can tailor arbitrary one-dimensional patterns.

2. ONE-DIMENSIONAL GENERALISATION

So far, we have only described a method to print a simple pattern of evenly spaced lines of sub-wavelength resolution. However, for any application we

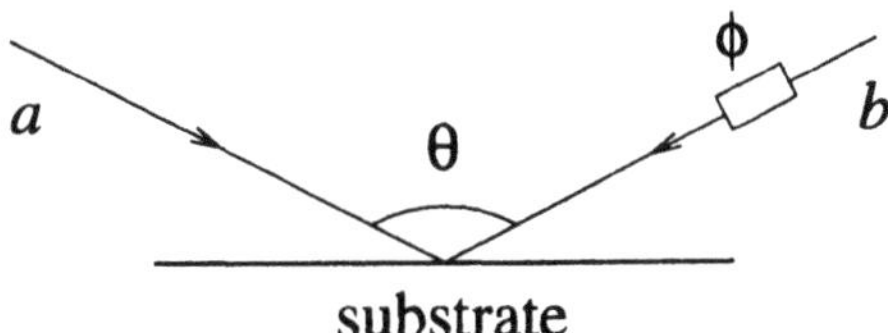

Figure 1 Two light beams a and b cross each other at the surface of a photosensitive substrate. The angle between them is θ and they have a relative phase difference φ. We consider the limit case of $\theta \to \pi$.

need the ability to produce more complicated patterns. To this end, we introduce the state

$$|\psi_{nm}\rangle_{ab} = \left(e^{im\varphi}|n-m,m\rangle_{ab} + e^{i(n-m)\varphi}|m,n-m\rangle_{ab}\right)/\sqrt{2}\,. \quad (4)$$

This is the generalised version of Eq. (1). In particular, Eq. (4) reduces to Eq. (1) when $m = 0$. Again, we calculate the deposition rate

$$\begin{aligned} \Delta_{nm}^{n'm'} &= \langle\psi_{nm}|\delta_n|\psi_{n'm'}\rangle \\ &\propto \delta_{n'n}\sqrt{\binom{n}{m}\binom{n}{m'}}\cos\left(\frac{n-2m}{2}\varphi\right)\cos\left(\frac{n-2m'}{2}\varphi\right). \end{aligned} \quad (5)$$

Obviously, $\langle\psi_{nm}|\delta_k|\psi_{n'm'}\rangle = 0$ when $k \neq n, n'$. For $m = m'$, the deposition rate takes on the form

$$\Delta_{nm}^{n'm} \propto \delta_{n'n}\binom{n}{m}\left[1+\cos(n-2m)\varphi\right]\,, \quad (6)$$

which, in the case of $m = 0$ coincides with Eq. (3). Note that we added extra subscripts and superscripts to Δ_n to accommodate for different n and m. This does not look like an improvement over Eq. (3), since $n - 2m < n$, which means that the resolution decreases. However, we will show later how these states *can* help us to produce non-trivial patterns.

First, note that so far we have deposition rates of the form $1 + \cos n\varphi$. We can achieve 'shifted' deposition rates ($1 - \cos n\varphi$ and $1 \pm \sin n\varphi$) by including a phase shift which is independent of the photon number. That is, we include a factor $e^{i\xi}$ in the second term in Eq. (4) In particular, as illustrated by Fig. 1, the state $|\psi_n\rangle$ is obtained by running the state $(|n,0\rangle + |0,n\rangle)/\sqrt{2}$ through a phase shifter in mode b. Alternatively, we can prepare a state

$$|\phi_n\rangle = (|n,0\rangle + i|0,n\rangle)/\sqrt{2}\,. \quad (7)$$

The factor i is independent of the number of photons (i.e., it is not affected by the phase shift φ), and the deposition rate Δ'_n will now be

$$\Delta'_n = \langle\phi_n|\delta_n|\phi_n\rangle \propto (1+\sin n\varphi)\,. \quad (8)$$

This allows us to tailor odd patterns as well.

Similarly, we can take the state to be

$$|\psi_n\rangle = \left(|n,0\rangle - |0,n\rangle\right)/\sqrt{2}\,, \tag{9}$$

which will yield a deposition rate

$$\Delta_n = \langle\psi_n|\delta_n|\psi_n\rangle \propto (1 - \cos n\varphi)\,. \tag{10}$$

A similar procedure using the state $(|n,0\rangle - i|0,n\rangle)/\sqrt{2}$ will provide us with $\Delta_n \propto (1 - \sin n\varphi)$.

Now, *there are two methods we can use to tailor non-trivial patterns in one dimension.* Note that there are two fundamentally different ways we can superpose states given by Eq. (4). We can superpose states with different photon numbers n and a fixed distribution m over the two modes:

$$|\Psi_m\rangle = \sum_{n=1}^{N} \alpha_n|\psi_{nm}\rangle\,. \tag{11}$$

Alternatively, we can superpose states with a fixed photon number n, but with different distributions m:

$$|\Psi_n\rangle = \begin{cases} \sum_{m=0}^{n/2} \alpha_m|\psi_{nm}\rangle & n \text{ even,} \\ \sum_{m=0}^{(n-1)/2} \alpha_m|\psi_{nm}\rangle & n \text{ odd.} \end{cases} \tag{12}$$

In the first case [Eq. (11)], there will be *no* interference between the different branches of the superposition, because every branch has a different total photon number n and is thus in principle distinguishable. We can therefore simply tailor an intensity pattern $p(\varphi) = \Delta t$ according to $(m = 0)$

$$p(\varphi) = t\sum_{n=0}^{\infty} c_n(1 + \cos n\varphi) \tag{13}$$

where t is the exposure time. This is a Fourier series up to a constant, which we see from Eq. (3) is just a sum over the deposition rates for different n. Needless to say, this is a very simple and attractive way of tailoring complicated (even) patterns up to a constant. As we have seen above, we can also use the 'Fourier components' $1 - \cos n\varphi$ and $1 \pm \sin n\varphi$.

However, there is a serious drawback with this procedure. The deposition rate Δ is a positive definite quantity, which means that once the substrate is exposed at a particular Fourier component, there is no way this can be undone. Technically, Eq. (13) can be written as

$$p(\varphi) = P \cdot t + \sum_{n=0}^{\infty} c_n \cdot t \cos n\varphi\,, \tag{14}$$

where P is the uniform background 'penalty exposure' $P = \sum_{n=0}^{\infty} c_n$. The second term on the right-hand side is a true Fourier series with coefficients $c_n t$. Thus in the 'Fourier method' there is always a minimum exposure of the substrate. Ultimately, this penalty can be traced to the absence of interference between the terms with different photon number in Eq. (11).

The second method of tailoring patterns [Eq. (12)] *does* exhibit interference between the different branches in the superposition. Take for instance a superposition of two terms

$$|\Psi_n\rangle = \alpha_1|\psi_{nm}\rangle + \alpha_2|\psi_{nm'}\rangle \,, \tag{15}$$

with $|\alpha_1|^2 + |\alpha_2|^2 = 1$. Using Eq. (5), the deposition rate can be easily calculated to be

$$\Delta_{nm}^{nm'} = (\alpha_1, \alpha_2) \propto |\alpha_1|^2 \binom{n}{m} [1 + \cos(n - 2m)\varphi]$$
$$+|\alpha_2|^2 \binom{n}{m'} [1 + \cos(n - 2m')\varphi] + 2(\alpha_1\alpha_2^* + \alpha_1^*\alpha_2)$$
$$\times\sqrt{\binom{n}{m}\binom{n}{m'}} \cos\left(\frac{n-2m}{2}\varphi\right) \cos\left(\frac{n-2m'}{2}\varphi\right).$$

Where we have now written the deposition rate Δ as a function of α_1 and α_2.

For more than two branches in the superposition this becomes a very complicated function, which is not nearly as well understood as Fourier series. If we want to tailor a pattern $p(\varphi)$, however, it might be the case that this type of superposition will also converge to the required pattern.

In order to investigate this we study the test function

$$p(\varphi) = \begin{cases} h \text{ if } -\frac{\pi}{2} < \varphi < \frac{\pi}{2} \,, \\ 0 \text{ otherwise.} \end{cases} \tag{16}$$

For example, suppose we have a superposition

$$|\Psi_n\rangle = \sum_{m=0}^{n/2} \alpha_m |\psi_{nm}\rangle \,, \tag{17}$$

where we assume n even (we can do this without loss of generality). To find the best fit with the test pattern we have to minimise the absolute difference between the deposition rate $\Delta_n(\vec{\alpha})$ times the exposure time t and the test function $p(\varphi)$. We have chosen $\vec{\alpha} = (\alpha_0, \ldots, \alpha_{n/2})$. Mathematically, we have to evaluate the $\vec{\alpha}$ and t which minimize

$$\int_0^{2\pi} |p(\varphi) - \Delta_n(\vec{\alpha})t|^2 d\varphi \,, \tag{18}$$

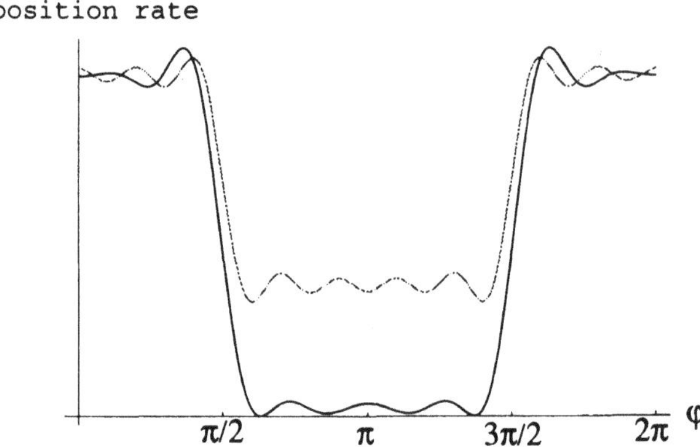

Figure 2 The deposition rate on the substrate resulting from a superposition of states with $n = 10$ and different m (black curve) and resulting from a superposition of states with different n and $m = 0$ (grey curve). The coefficients of the superposition yielding the black curve are optimised using a genetic algorithm [11], while the grey curve is a truncated Fourier series. Notice the 'penalty' (displaced from zero) deposition rate of the Fourier series between $\pi/2$ and $3\pi/2$.

with

$$\Delta_n(\vec{\alpha}) = \langle \Psi_n | \delta_n | \Psi_n \rangle \ . \tag{19}$$

We have to fit both t and $\vec{\alpha}$. Using a genetic optimalisation algorithm [11] (with $h = 1$, a normalised height of the test function) we found that the deposition rate is actually very close to zero in the interval $\pi/2 \leq \varphi \leq 3\pi/2$, unlike the 'Fourier method', where we have to pay a uniform background penalty. This means that in this case a superposition of different photon distributions m given a fixed total number of photons n works better than a superposition of different photon number states (see Fig. 2). In particular, *the second method (with fixed photon number) allows for the substrate to remain virtually unexposed in certain areas.*

Besides the ability to fit a pattern, another criterion of comparison between the 'Fourier method' and the 'superposition method' is the time needed to create the n-photon entangled states. We can make a crude estimate without going into the actual details of the state preparation. Note that both methods involve n-photon states. Given current technologies (like, e.g., parametric down-conversion) the probability of making such a state will be proportional to p^n, where p is the probability of making a single-photon state. Since both methods need n-photon states the exposire time for both methods is proportional to p^{-n}. This first, and crude, estimate therefore suggests that from this perspective there is no preferred method.

3. CONCLUSIONS

In this paper we have generalised the theory of quantum lithography presented in Ref. [6]. In particular, we have shown how we can create arbitrary patterns in one dimension.

We distinguish two methods: the 'Fourier method' and 'the superposition method'. The Fourier method is conceptually easier since it depends on Fourier analysis, but it also involves a finite amount of unwanted exposure of the substrate. More specifically, the deposition rate equals the pattern in its Fourier basis *plus* a term yielding unwanted background exposure. The superposition method gets around this problem and gives generally better results, but lacks the intuitive clarity of the Fourier method.

There are several issues to be addressed in the future. First, we have to show that the superposition method can reasonably tailor any arbitrary pattern in one and two dimensions. In order to do this, we need to define what we mean by 'reasonable'. Secondly, we need to study the specific restrictions on the substrate and how we can physically realise them. Finally, we need to create the various entangled states involved in the quantum lithography protocol.

Acknowledgments

This research was supported by the National Aeronautics and Space Administration and the British Engineering and Physical Sciences Research Council.

References

[1] S.R.J. Brück, *et al.*, Microelectron. Eng. **42**, 145 (1998).

[2] C.A. Mack, Opt. Phot. News **7**, 29 (1996).

[3] M. Mansuripur and R. Liang, Opt. Phot. News **11**, 36 (2000).

[4] E. Yablonovich and R.B. Vrijen, Opt. Eng. **38**, 334 (1999).

[5] P. Kok, A.N. Boto, D.S. Abrams, C.P. Williams, S.L. Braunstein and J.P. Dowling, in preparation.

[6] A.N. Boto, P. Kok, D.S. Abrams, S.L. Braunstein, C.P. Williams and J.P. Dowling, Phys. Rev. Lett. **85**, 2733 (2000).

[7] M. Göppert-Mayer, Ann. Phys. **5**, 273 (1931).

[8] J. Javanainen and P.L. Gould, Phys. Rev. A **41**, 5088 (1990).

[9] J. Perina jr., B.E.A. Saleh and M.C. Teich, Phys. Rev. A **57**, 3972 (1998).

[10] P. Kok and S.L. Braunstein, Phys. Rev. A, to appear.

[11] K. Price and R. Storn, *Dr. Dobb's Journal*, April, p. 18ff, (1997).

EXPERIMENTAL TEST OF LOCAL REALISM USING NON-MAXIMALLY ENTANGLED STATES

M. Genovese, G. Brida, C. Novero
Istituto Elettrotecnico Nazionale Galileo Ferraris, Str. delle Cacce 91, I-10135 Torino
genovese@ien.it

E. Predazzi
Dip. Fisica Teorica Univ. Torino and INFN, via P. Giuria 1, I- 10125 Torino

Entanglement is the main resource of Quantum Information Processing and thus deserves to be widely studied. One of the most characteristic properties of entanglement are the correlations between space-like separated entangled systems studied by Einstein-Podolsky-Rosen, for which Bell showed that no Local Hidden Variable Theory can reproduce all the results of Quantum Mechanics and, in particular, that some inequality, valid for every LHV theory, is violated by QM.

Many interesting experiments have been devoted to a test of Bell inequalities, the most interesting of them using photon pairs [1, 2, 3, 4], leading to a substantial agreement with quantum mechanics and disfavouring LHV theories. But, up to now, no experiment has yet been able to exclude definitively such theories [6], because of the low total detection efficiency, which requires the hypothesys that the observed sample of particle pairs is a faithful subsample of the initial set of pairs. In the 90's a big progress in the direction of eliminating this loophole has been obtained by using parametric down conversion (PDC) processes, which overcomes some former limitations, as the poor angular correlation of atomic cascade photons [2]. The first experiments were performed with type I PDC, which gives phase and momentum entanglement and can be used for a test of Bell inequalities using two spatially separated interferometers [3]. The use of beam splitters, however, strongly reduces the total quantum efficiency. In alternative, a polarisation entangled state can be generated [5], but in most of the used configurations, half of the initial photon flux is lost.

Recently, an experiment where a polarisation entangled state is directly generated, has been realised using Type II PDC [4]. This scheme has permitted, at the price of delicate compensations for having identical arrival time of the ordinary and extraordinary photon, a much higher total efficiency than the previous ones, which is, however, still far from the value of 0.81 required for eliminating the detection loophole for a maximally entangled state. A large interest remains therefore for new experiments increasing total quantum efficiency in order to reduce and finally overcome the efficiency loophole.

For this purpose, we have considered [8] the possibility of generating a polarisation entangled state via the superposition of the spontaneous fluorescence emitted by two non-linear crystals (rotated for having orthogonal polarisation) driven by the same pumping laser [9]. The crystals are put in cascade along the propagation direction of the pumping laser and the superposition is obtained by using an appropriate optics. If the path between the two crystal is smaller than the coherence length of the laser, the two photons path are indistinguishable and a polarisation entangled state is created. The possibility of easily obtaining a non maximally entangled state is very important, for it has been recognised that in this case the lower limit on the total detection efficiency for eliminating the detection loophole is as small as 0.67 [7].

The experimental set-up is composed of two crystals of $LiIO_3$ placed along the pump laser propagation, 250 mm apart, a distance smaller than the coherence length of the pumping laser. This guarantees indistinguishability in the creation of a couple of photons in the first or in the second crystal. A couple of planoconvex lenses of 120 mm focal length centred in between focalises the spontaneous emission from the first crystal into the second one. A hole of 4 mm diameter is drilled into the centre of the lenses to allow transmission of the pump. A small quartz plate compensates the displacement of the pump due to birefringence in the first crystal. Finally, a half-wavelength plate immediately after the condenser rotates the polarisation of the Argon beam. We have used as photo-detectors two avalanche photodiodes with active quenching (EG&G SPCM-AQ) coupled to an optical fiber.

A very interesting degree of freedom of this configuration is given by the fact that by tuning the pump intensity between the two crystals, one can easily tune the ratio f between the two component of the entangled state ($f = 1$ for a maximally entangled state). Furthermore, every Bell state can be generated (for example for applications to quantum information processing) changing the phase between the two components (moving the second crystal) and/or changing the polarisation on one of the branches after the second crystal. The main difficulty of this configuration is in the alignment, which is of fundamental importance for having a high visibility. This problem has been solved using [10] an optical amplifier scheme, where a solid state laser is injected into the crystals together with the pumping laser. We think that the proposed scheme is well suited for leading to a further step toward a conclusive experimental test of non-locality in quantum mechanics. The main advantage of the proposed configuration with respect to most of the previous experimental set-ups is that all the entangled pairs are selected (and not only $< 50\%$ as with beams splitters), furthermore it does not require delicate compensations for the optical paths of the ordinary and extraordinary rays after the crystal.

At the moment, the results which we are going to present are still far from a definitive solution of the detection loophole; nevertheless, being the first test of Bell inequalities using a non-maximally entangled state, they represent an important step in this direction. Furthermore, this configuration permits to use any pair of correlated frequencies and not only the degenerate ones. We have thus realised this test using for a first time two different wave lengths (at 633 and 789 nm).

It must be acknowledged that a set-up for generating polarisation entangled pairs of photons, which presents analogies with our, has been realised recently in Ref. [11] The main difference between the two experiments is that in [11] the two crystals are very thin and in contact with orthogonal optical axes: this permits a "partial" [12] superposition of the two emissions with opposite polarisation. This overlapping is mainly due to the finite dimension of the pump laser beam, which reflects into a finite width of each wavelength emission. A much better superposition can be obtained with the present scheme. More recently [13], they have also performed a test about local realism using a particular kind of Hardy equalities [14]. Their result is in agreement with quantum mechanics modulo the detection loophole, no discussion concerning the elimination of loopholes for this equality is presented.

As a first check of our apparatus, we have measured the interference fringes, varying the setting of one of the polarisers, leaving the other fixed. We have found a high visibility, $V = 0.973 \pm 0.038$.

Our results are summarised by the value obtained for the Clauser-Horne sum

$$CH = N(\theta_1, \theta_2) - N(\theta_1, \theta_2') + N(\theta_1', \theta_2) + N(\theta_1', \theta_2') - N(\theta_1', \infty) - N(\infty, \theta_2) \tag{1}$$

which is strictly negative for local realistic theory. In (1), $N(\theta_1, \theta_2)$ is the number of coincidences between channels 1 and 2 when the two polarisers are rotated to an angle θ_1 and θ_2 respectively (∞ denotes the absence of selection of polarisation for that channel). On the other hand, quantum mechanics predictions for CH can be larger than zero. In our case we have generated a state with $f \simeq 0.4$: in this case the largest violation of the inequality is reached for $\theta_1 = 72^o.24$, $\theta_2 = 45^o$, $\theta_1' = 17^o.76$ and $\theta_2' = 0^o$. Our experimental result is $CH = 512 \pm 135$ coincidences per second, which is almost four standard deviations different from zero and compatible with the theoretical value predicted by quantum mechanics.

References

[1] see L. Mandel, and E. Wolf, Optical Coherence and Quantum Optics, Cambridge University Press, 1995 and references therein.

[2] A. Aspect et al., Phys. Rev. Lett. 49 , 1804, (1982).

[3] J. P. Franson, Phys. Rev. Lett. 62, 2205 (1989); J. G. Rarity, and P. R. Tapster, Phys. Rev. Lett. 64, 2495 (1990); J. Brendel et al. Eur.Phys.Lett. 20, 275 (1992); P. G. Kwiat el al, Phys. Rev. A 41, 2910 (1990); W. Tittel et al, quant-ph 9806043

[4] T.E. Kiess et al., Phys. Rev. Lett. 71, 3893, (1993); P.G. Kwiat et al., Phys. Rev. Lett. 75, 4337, (1995).

[5] Z.J. Ou and L. Mandel, Phys. Rev. Lett. 61, 50, (1988); Y.H.Shih et al., Phys. Rev. A, 1288, (1993); J.R.Torgerson et al., Phys. Lett. A 204, 323 (1995); G.Di Giuseppe et al., Physical Review A 56, 176 (1997); D.Boschi et al., Phys. Rev. Lett. 79, 2755 (1998).

[6] L. De Caro and A. Garuccio Phys. Rev. A 54, 174 (1996) and references therein.

[7] P. H. Eberhard, Phys. Rev. A 47, R747 (1993).

[8] M. Genovese, G. Brida, C. Novero and E. Predazzi, proceeding of ICSSUR, Napoli may 1999.

[9] L. Hardy, Phys. Lett. A 161, 326 (1992).

[10] G. Brida, M. Genovese and C. Novero quant-ph 9911032 and ref.s therein (Journ. Mod. Opt. in press).

[11] P.G. Kwiat et al., Phys. Rev. A 60 (1999) R773 and thcsc proceedings.

[12] A. Sergienko, these proceedings.

[13] A.G. White et al., Phys. Rev. Lett. 83 (1999) 3103 and these proceedings.

[14] L. Hardy, Phys. Rev. Lett. 71 (1993) 1665.

NEW SCHEMES FOR MANIPULATING QUANTUM STATES WITH A KERR CELL

M. Genovese, C. Novero
Istituto Elettrotecnico Nazionale Galileo Ferraris, Str. delle Cacce 91, I-10135 Torino
genovese@ien.it

Recently, Quantum Non Demolition (QND) detection of a single photon has become possible [1].This is due to the discovery of new materials with very high Kerr coupling as the Quantum Coherent Atomic Systems (QCAS) [1, 2] and the Bose-Einstein condensate [3]. These great technical improvements could allow the realisation of small Kerr cells, capable of large phase shift, even with a low-intensity probe. This recent development has a large relevance for application to quantum information theory and foundations of Quantum Mechanics. For example, it has lead to the proposal of schemes for complete teleportation [4] and for realising quantum gates [4, 5]. In this proceeding we describe some other proposals of application of this device, as the fast modulation of quantum interference [6], the generation of optical Schrödinger cats [7] and of GHZ states and the realisation of translucent eavesdropping.

Let us consider first a scheme for realising a fast modulation of quantum interference. A "signal" photon enters a Mach-Zender interferometer through a Beam-Splitter (BS I) from port 1; assuming a 50% BS (the treatment of the non 50% case is a trivial extension) one has after the BS

$$a_2 = \frac{a_1 + i a_0}{\sqrt{2}} \quad a_3 = \frac{a_0 + i a_1}{\sqrt{2}} \tag{1}$$

A probe laser crosses the Kerr cell on the third arm acquiring a phase which in principle will be measurable, providing *welcher Weg* information. The usual treatment of Kerr interaction gives after recombination on Beam Splitter II:

$$\begin{aligned} \langle \Psi | n_4 | \Psi \rangle &= \frac{1}{2} \left[1 - \exp[-2|\nu|^2 \sin^2(\chi T)] \cos[(\omega_s + \chi_s)T + \Theta \right. \\ &\quad \left. + |\nu|^2 \sin(2\chi T)] \right] \end{aligned} \tag{2}$$

where Θ takes into account different lengths of arms 2 and 3 and could be varied interposing a variable phase shift on one of the interferometer arms. χ_s and χ denote the self- and the cross- Kerr couplings respectively and n_3 and

Quantum Communication, Computing, and Measurement 3
Edited by P. Tombesi and O. Hirota, Kluwer Academic/Plenum Publishers, 2001

n_p are the photon number operators on the arm 3 and for the probe, which is assumed to be described by a coherent field $|\nu\rangle$.

To this point, the usual treatment [1] of QND ideal *welcher Weg* experiment has been considered, with the well known result that the probe laser acquires a phase which permits, by a homodyne measurement, the identification of the path followed by the signal photon. The signal-to-noise ratio in the homodyne measurement, given by $R = 4|\nu|\sin(\chi T)$, is directly related to the suppression of interference, whose visibility is $\exp(-R^2/8)$.

Let us now consider the insertion of a second Kerr cell on the 2nd arm and let us assume that the interaction time between the probe and signal fields in this cell is T'. We now have:

$$\begin{aligned} \langle\Psi|n_4|\Psi\rangle &= \frac{1}{4}\langle\Psi|n_1[2-(\exp\left[-i\left(\Theta+\beta(T-T')\right)\right] \\ &+\exp\left[i\left(\Theta+\beta(T-T')\right)\right])]|\Psi\rangle \end{aligned} \tag{3}$$

where $\beta = (\omega_s + \chi_s/2 + \chi_s n_s + 2\chi n_p)$. If we choose the Kerr cells so that $T = T'$, the phase into the probe due to the photon in path 3 or 2 would be the same and the interference pattern $\frac{1-\cos(\Theta)}{2}$ is recovered for the signal field. On the other hand, if one considers the case where the distance between the two Kerr cells is larger than the coherence length of the probe laser the two paths will still be distinguishable and interference will be lost. Quantum interference can thus be regulated, changing the coherence length of the laser before injecting it into the first Kerr cell. The observation of this effect represents a very good and illustrative example of the effect of disappearance of quantum interference when *welcher Weg* information is obtained and of the effect of erasing this information.

Let us now substitute the first beam splitter with a Polarising-Beam-Splitter I (PBS I) and let us suppose that the entering photon is in a superposition of vertical (V) and horizontal (H) polarisation, which will take different paths, for example the vertical one will follow path 2 and the horizontal path 3. As before, a probe laser crosses the Kerr cell on the arm 3 acquiring a phase or not according if the photon crosses or not the cell. Thus the entangled state:

$$|\Psi\rangle = \frac{|H\rangle|\nu'\rangle + |V\rangle|\nu\rangle}{\sqrt{2}} \tag{4}$$

is generated, where the coherent state $|\nu'\rangle$ differs in phase from $|\nu\rangle$. The two signal photon paths are then recombined on a second beam splitter (BSII) and a polarisation measurement is performed on this photon on the base at 45°. This is the conditional measurement producing the Schrödinger cat: if the signal photon is found to have a 45° (135°) polarisation, the coherent state is projected into the superposition

$$|\psi_{+(-)}\rangle = \frac{|\nu'\rangle + (-)|\nu\rangle}{\sqrt{2}} \tag{5}$$

A superposition of two "many photons" states is thus obtained.

The one photon signal state can be easily produced, for example, using parametric down conversion (PDC) in a non-linear crystal. In this case the second photon of the down-converted pair can be used as trigger. Furthermore, if the same pulsed laser is used both for pumping the crystal and for the Kerr effect, one can easily obtain a good timing for the crossing of the Kerr cell for the signal photon and the coherent state. A main advantage of the present configuration respect to the Schrödinger cat generation using cavities is that the coherent states superposition travels in air between the Kerr cell and the detection apparatus allowing a much longer time before decoherence takes over.

For identifying the macroscopic superposition one can look for a negative part of the Wigner function, reconstructed by tomographic techniques [9].We have performed a numerical simulation keeping into account the deterioration due to errors: our results show that even with 25%, or larger, errors on the homodyne measurement the cat can be easily identified.

The use as input of a photon from PDC in a Kerr cell allows also the creation of a three photons GHZ entangled state, which is one of the three elements necessary for realising an optical quantum computer, together with single qubit operations, which are easily implemented, and teleportation. For what concerns this last , a description of a scheme performing this operation using a Kerr cell appears in Ref. [4]. Using the same polarisation dependence of the Kerr interaction of Ref. [4], where there is no effect except when both the photons interacting in the Kerr cell have vertical polarization ($|V\rangle|V\rangle \rightarrow |V\rangle|V\rangle e^{i\phi}$), a GHZ state can be generated by the interaction of an entangled pair of photons with a third one. Let us assume, for example, of having generated the entangled state [8]: $|\Phi^+\rangle = |H\rangle|H\rangle + |V\rangle|V\rangle/\sqrt{2}$. A simple calculation shows that the interaction in the Kerr medium of the second photon of the pair with a third photon polarised at 45° (denoted by $|45\rangle$, whilst its orthogonal state is $|135\rangle$), is, for a phase shift $\phi = \pi/2$, the GHZ state:

$$|\Psi_{GHZ}\rangle = \frac{|H\rangle|H\rangle|45\rangle + |V\rangle|V\rangle|135\rangle}{\sqrt{2}}. \tag{6}$$

Analogous results are easily derived for the other three Bell states.

Finally, let us notice that such an apparatus can also be used for performing translucent eavesdropping on a quantum channel where polarisation is used for distinguishing qubits. In this case the input port of the first beam splitter is fed with a single photon of vertical polarisation, which splits on the two interferometer arms. On arm 1 it interacts with the transmitted qubit inside the

Kerr cell (with the same polarisation dependence as before). Let us suppose that the transmitted qubit is in the general form $|u\rangle = cos(\theta)|H\rangle + sin(\theta)|V\rangle$, denoting the probe photon with $|p\rangle$, the final state is:

$$|\Psi\rangle = \frac{1}{\sqrt{2}} \left[cos(\theta)|H\rangle|p\rangle_1 + sin(\theta)|V\rangle|p\rangle_1 e^{(i\phi)} + i|u\rangle|p\rangle_2 \right] \quad (7)$$

where the suffixes after the probe $|p\rangle$ denotes the path followed and where we have considered a 50 % BS (which gives Eve the largest information on Alice-Bob transmission). We have thus obtained the desired entanglement between the probe and the transmitted qubit, which can be used for translucent or coherent eavesdropping.

References

[1] B.C. Sanders and G.J. Milburn, Phys. Rev. A 39 (1989) 694. Q.A. Turchette et al., Phys. Rev. Lett 75 (1995) 4710; H. Schmidt and A. Imamoglu, Opt. Lett. 21 (1996) 1936; A. Imamoglu et al., Phys. Rev. Lett. 79 (1997) 1467.

[2] U. Rathe et al., Phys. Rev. A 47 (1993) 4994; M.M. Kash et al., Phys. Rev. Lett. 82 (1999) 5229.

[3] L. Vestergaard Hau et al., Nature 397 (1999) 594; S.E. Harris and L.V. Hau, Phys. Rev. Lett. 82 (1999) 4611.

[4] D. Vitali et al., Phys. Rev. Lett. 85 (2000) 445.

[5] G. M. D'Ariano et al., Fort. Phys. **48**, 573 (2000).

[6] M. Genovese and C. Novero, Phys. Rev. A 61 032102 (2000).

[7] M. Genovese and C. Novero, Phys. Lett. A 271 (2000) 48.

[8] G. Brida, M. Genovese, C. Novero and E. Predazzi, Phys. Lett. A 268 (2000) 12 and these proceedings and ref.s therein.

[9] G.M. D'Ariano et al., Phys. Rev. A 50 (1994) 4298; Nuov. Cim. 110B (1995) 237.

LOCAL AND NONLOCAL PROPERTIES OF WERNER STATES

Tohya Hiroshima
Fundamental Research Laboratories
System Devices and Fundamental Research
NEC Corporation, 34 Miyukigaoka, Tsukuba 305-8501, Japan
tohya@frl.cl.nec.co.jp

Satoshi Ishizaka
Fundamental Research Laboratories
System Devices and Fundamental Research
NEC Corporation, 34 Miyukigaoka, Tsukuba 305-8501, Japan
isizaka@frl.cl.nec.co.jp

Keywords: Entanglement, Werner state

Abstract We consider a special kind of mixed states: a *Werner derivative*, which is a state transformed from a Werner state by either local or nonlocal unitary operations. We show that the amount of entanglement of Werner derivatives cannot exceed that of the original Werner state and that although it is generally possible to increase the entanglement of a single copy of a Werner derivative by LQCC, the maximal entanglement cannot exceed the entanglement of the original Werner state.

Quantum entanglement plays an essential role in various types of quantum information processing. Since the best performance of such tasks requires maximally entangled states, one of the most important entanglement manipulations is entanglement purification or distillation [1, 2, 3, 4]. Most of protocols proposed for entanglement purification (distillation) utilize the collective operations on many copies of a given state ρ [3, 4]. If we have only a single copy of a given mixed state, the only thing we can do by local quantum operations and classical communications (LQCC) is to increase the amount of entanglement. There are, however, entangled mixed states for which this cannot be done [5, 6, 7]. It is, therefore, of fundamental importance to clarify the limit of entanglement manipulations of a single copy of a given mixed state.

We consider in this paper a special kind of mixed states, a *Werner derivative*. This is a state transformed from a Werner state[8] by unitary operations – local or nonlocal. We show the following. (i) The amount of entanglement of Werner derivatives cannot exceed that of the original Werner state. (ii) Although it is generally possible to increase the entanglement of a single copy of a Werner derivative by LQCC (single-state LQCC), the maximal entanglement cannot exceed the entanglement of the original Werner density matrix.

The degree of entanglement of mixed states of two qubits is customarily measured by the entanglement of formation (EOF) [4]. In 2×2 systems the EOF is given by

$$E_F(\rho) = H\left(\frac{1+\sqrt{1-C^2}}{2}\right), \tag{1}$$

where $H(x) = -x\log_2 x - (1-x)\log_2(1-x)$ [9]. The nonnegative real number $C = \max\{0, \lambda_1 - \lambda_2 - \lambda_3 - \lambda_4\}$ is called a concurrence, where λ_i are the square roots of eigenvalues of positive matrix $\rho\widetilde{\rho}$ in descending order. The density matrix $\widetilde{\rho}$ is defined as $\widetilde{\rho} = \sigma_2 \otimes \sigma_2 \rho^* \sigma_2 \otimes \sigma_2$, where the asterisk denotes complex conjugation in the standard basis $\{|00\rangle, |01\rangle, |10\rangle, |11\rangle\}$ and σ_i are usual Pauli matrices. Since E_F is a monotonic function of C, the concurrence C is also a measure of entanglement.

A Werner state in 2×2 systems takes the following form:

$$\rho_W = \frac{1-F}{3}\mathbf{I}_4 + \frac{4F-1}{3}\left|\Psi^-\right\rangle\left\langle\Psi^-\right|, \tag{2}$$

where $\mathbf{I}_n$ denotes the $n \times n$ identity matrix and $|\Psi^-\rangle = (|01\rangle - |10\rangle)/\sqrt{2}$. The Werner state ρ_W is characterized by a single real parameter F called fidelity. In the following, we assume $1/2 < F \leq 1$ so that $C(\rho_W) = \max\{0, 2F-1\} = 2F-1$

The nonlocal unitary transformation brings ρ_W to a new density matrix of the form

$$\rho = \frac{1-F}{3}\mathbf{I}_4 + \frac{4F-1}{3}\left|\psi\right\rangle\left\langle\psi\right|, \tag{3}$$

where $|\psi\rangle$ is written in a Schmidt decomposed form, $|\psi\rangle = \sqrt{a}\,|00\rangle + \sqrt{1-a}\,|11\rangle$ with $1/2 \leq a \leq 1$. The Werner derivative of Eq. (3) is shown to be entangled if and only if

$$\frac{1}{2} \leq a < \frac{1}{2}\left(1 + \frac{\sqrt{3(4F^2-1)}}{4F-1}\right). \tag{4}$$

The range of parameter a is assumed to be limited by the above inequalities so that ρ is always entangled. The square roots of eigenvalues of $\rho\widetilde{\rho}$ are calculated as

$$\{\lambda_i\}_{i=1}^4 = \left\{\frac{(4F-1)}{3}G_+, \frac{(4F-1)}{3}G_-, \frac{1-F}{3}, \frac{1-F}{3}\right\}, \tag{5}$$

where

$$G_{\pm} = \left[2a(1-a) + G \pm 2\sqrt{a(1-a)(a(1-a)+G)}\right]^{\frac{1}{2}}, \tag{6}$$

with $G = 3F(1-F)/(4F-1)^2$. By using Eqs. (5) and (6), we calculate the concurrence $C(\rho) = \lambda_1 - \lambda_2 - \lambda_3 - \lambda_4$ as a function of a. The maximum value of $C(\rho)$ is found to be $2F-1$ [10]. Hence $E_F(\rho) \leq E_F(\rho_W)$.

It is possible to increase the EOF of ρ by a single-state LQCC [7]. As shown below, however, the maximum EOF thus obtained is still less than or equal to the EOF of the original Werner state. According to Theorem 3 in [7], there exist a single-state LQCC mapping ρ to a Bell diagonal state ρ' with maximal possible EOF of the form,

$$\rho' = \frac{1}{4}\left(\mathbf{I}_4 + \sum_{i=1}^{3} r_i \sigma_i \otimes \sigma_i\right), \tag{7}$$

with $r_1 \leq r_2 \leq r_3 \leq 0$. The square roots of eigenvalues of $\rho'\widetilde{\rho'}$ in descending order are $\lambda_1' = (1 - r_1 - r_2 - r_3)/4$, $\lambda_2' = (1 - r_1 + r_2 + r_3)/4$, $\lambda_3' = (1 + r_1 - r_2 + r_3)/4$, and $\lambda_4' = (1 + r_1 + r_2 - r_3)/4$. Since the ratios λ_i'/λ_j' are invariant under LQCC, $\lambda_i'/\lambda_4' = \lambda_i/\lambda_4$ $(i = 1, 2, 3)$, where λ_i are given by Eq. (5). Therefore the concurrence $C(\rho')$ can be expressed in terms of λ_i:

$$C(\rho') = \frac{\lambda_1 - \lambda_2 - \lambda_3 - \lambda_4}{\lambda_1 + \lambda_2 + \lambda_3 + \lambda_4}. \tag{8}$$

Substituting the explicit forms of λ_i into this equation, we obtain

$$C(\rho') - C(\rho_W) = 2\frac{(1-F)G_+ - FG_- - 2\frac{1-F}{4F-1}}{G_+ + G_- + 2\frac{1-F}{4F-1}}. \tag{9}$$

The straightforward calculations show that the right-hand side of Eq. (9) is less than or equal to zero [10]. Hence $E_F(\rho') \leq E_F(\rho_W)$.

Our results above can be also stated in other words: the EOF of a Werner state cannot be increased by a single-state LQCC followed by nonlocal unitary transformations. This property is unique to Werner states, as shown below. If another state ρ which does not belong to a family of Werner states has the property stated above, it must be one of the maximally entangled mixed states [11]. Otherwise $E_F(\rho)$ could be increased by nonlocal unitary transformation. The maximally entangled mixed state takes the following form [11]:

$$\rho = p_1 \left|\Psi^-\right\rangle \left\langle\Psi^-\right| + p_2 \left|00\right\rangle \left\langle 00\right| + p_3 \left|\Psi^+\right\rangle \left\langle\Psi^+\right| + p_4 \left|11\right\rangle \left\langle 11\right|, \tag{10}$$

where $|\Psi^+\rangle = (|01\rangle + |10\rangle)/\sqrt{2}$ and p_i are eigenvalues of ρ in decreasing order ($p_1 \geq p_2 \geq p_3 \geq p_4 \geq 0$). According to the arguments in [7], the

EOF of ρ can be increased further by a single-state LQCC. This contradicts the assumed property of ρ. Therefore the equality $p_2 = p_4$ must hold, which implies $p_2 = p_3 = p_4 = (1 - p_1)/3$. It follows that the state ρ takes the form of a Werner state of fidelity p_1.

In summary, the present results combined with previously obtained ones [6, 11] reveal the following peculiar property of a Werner state of two qubits: its EOF cannot be increased (i) by a single-state LQCC, (ii) by a nonlocal unitary transformation, or (iii) by a single-state LQCC followed by nonlocal unitary transformations.

References

[1] N. Gisin, Phys. Lett. A **210**, 151 (1996).

[2] M. Horodecki, P. Horodecki, and R. Horodecki, Phys. Rev. Lett. **78**, 574 (1997).

[3] C. H. Bennett, G. Brassard, S. Popescu, B. Schumacher, J. Smolin, and W. K. Wootters, Phys. Rev. Lett. **76**, 722 (1996).

[4] C. H. Bennett, D. P. DiVincenzo, J. Smolin, and W. K. Wootters, Phys. Rev. A **54**, 3824 (1996).

[5] A. Kent, Phys. Rev. Lett. **81**, 2839 (1998).

[6] N. Linden, S. Massar, and S. Popescu, Phys. Rev. Lett. **81**, 3279 (1998).

[7] A. Kent, N. Linden, and S. Massar, Phys. Rev. Lett. **83**, 2656 (1999).

[8] R. Werner, Phys. Rev. A **40**, 4277 (1989).

[9] W. K. Wootters, Phys. Rev. Lett. **80**, 2245 (1998).

[10] T. Hiroshima and S. Ishizaka, Phys. Rev. A **62**, 044302 (2000).

[11] S. Ishizaka and T. Hiroshima, Phys. Rev. A **62**, 022310 (2000).

MAXIMALLY ENTANGLED MIXED STATES IN TWO QUBITS

Satoshi Ishizaka, Tohya Hiroshima
Fundamental Research Laboratories
System Device and Fundamental Research, NEC Corporation
isizaka@frl.cl.nec.co.jp, tohya@frl.cl.nec.co.jp

Abstract We propose novel mixed states in two qubits, which have a property that the amount of entanglement of these states cannot be increased by any unitary transformation. The corresponding entanglement of formation specified by its eigenvalues gives an upper bound of that for density matrices with the same eigenvalues.

Entanglement is an important resource for most applications of quantum information. In recent years, quantification of the amount of entanglement has attracted much attention, and a number of measures, such as the entanglement of formation (EOF) [1] and negativity [2], have been proposed. The amount of entanglement of the Bell states is maximum and defined as unity in these measures. However, the maximum is achieved only when the system is in a pure state. What is the maximally entangled states when the system is in a mixed state?

Recently, a class of mixed states has been proposed [3], in which the probabilistic increase of the EOF cannot be achieved by any local quantum operations and classical communication (LQCC) protocol. Bell diagonal states, for example, belong to this class. In this sense, the Bell diagonal states can be considered as *maximally entangled mixed states* in LQCC protocols. In this paper, however, we consider the other operation than LQCC. The maximally entangled mixed states in two qubits we present in this paper have a property that the entanglement cannot be increased by any (local or nonlocal) unitary transformation. The corresponding entanglement of formation gives an upper bound when the eigenvalues of the density matrix are fixed. These states are truly regarded as *maximally entangled mixed states*, since the measure of mixture, such as von Neumann entropy and purity, is usually determined by the eigenvalues of the density matrix.

Quantum Communication, Computing, and Measurement 3
Edited by P. Tombesi and O. Hirota, Kluwer Academic/Plenum Publishers, 2001

The states we propose are those obtained by applying any *local* unitary transformation to

$$M = p_1|\Psi^-\rangle\langle\Psi^-| + p_2|00\rangle\langle 00| + p_3|\Psi^+\rangle\langle\Psi^+| + p_4|11\rangle\langle 11|, \qquad (1)$$

where $|\Psi^{\pm}\rangle = (|01\rangle \pm |10\rangle)/\sqrt{2}$ are Bell states. Here, p_i's are eigenvalues of M in decreasing order ($p_1 \geq p_2 \geq p_3 \geq p_4$), and $p_1+p_2+p_3+p_4=1$. These include states such as

$$\rho = p_1|\Phi^-\rangle\langle\Phi^-| + p_2|01\rangle\langle 01| + p_3|\Phi^+\rangle\langle\Phi^+| + p_4|10\rangle\langle 10|, \qquad (2)$$

where $|\Phi^{\pm}\rangle = (|00\rangle \pm |11\rangle)/\sqrt{2}$ are also Bell states, and include those that are obtained by exchanging $|\Psi^-\rangle \leftrightarrow |\Psi^+\rangle$, $|00\rangle \leftrightarrow |11\rangle$ in Eq. (1), or $|\Phi^-\rangle \leftrightarrow |\Phi^+\rangle$, $|01\rangle \leftrightarrow |10\rangle$ in Eq. (2).

The analytical form for EOF of a mixed states in two qubits is given by Hill and Wootters [4]. Since the EOF is a monotonic function of the concurrence, the maximum of the concurrence corresponds to the maximum of the EOF. Applying the analytical form to the above states, we obtain the concurrence of

$$\begin{aligned} C^* &= \max\{0, C^*(p_i)\}, \\ C^*(p_i) &\equiv p_1 - p_3 - 2\sqrt{p_2 p_4}. \end{aligned} \qquad (3)$$

The concurrence C^* is maximum among the density matrices with the same eigenvalues, when the density matrices have a rank less than 4 ($p_4=0$) or when the eigenvalues satisfy a relation even for $p_4 \neq 0$. The proof is as follows:

(1) Rank 1 and rank 2 case ($p_3=p_4=0$). Any density matrices of two qubits can be decomposed into Lewenstein-Sanpera (L-S) form [5]:

$$\rho = q|\psi\rangle\langle\psi| + (1-q)\rho_{\text{sep}}, \qquad (4)$$

where $|\psi\rangle$ is an entangled state and ρ_{sep} is a separable density matrix. The convexity of the concurrence [6] implies that

$$C(\rho) \leq qC(|\psi\rangle\langle\psi|) + (1-q)C(\rho_{\text{sep}}) = qC(|\psi\rangle\langle\psi|). \qquad (5)$$

Since ρ_{sep} is a positive operator, q is equal to or less than the maximum eigenvalue of ρ, and thus

$$C(\rho) \leq p_1. \qquad (6)$$

The upper bound in Eq. (6) coincides with C^* for $p_3=p_4=0$.

(2) Rank 3 case ($p_4=0$). Any rank 3 density matrices can be decomposed into two density matrices as

$$\rho = (1-3p_3)\rho_2 + 3p_3\rho_3, \qquad (7)$$

where the eigenvalues of ρ_2 are

$$\left\{\frac{p_1-p_3}{1-3p_3}, \frac{p_2-p_3}{1-3p_3}, 0, 0\right\}, \qquad (8)$$

and eigenvalues of ρ_3 are $\{1/3, 1/3, 1/3, 0\}$. According to Lemma 3 in Ref. [7], ρ_3 is always separable, since the purity of ρ_3 is equal to 1/3. Therefore, convexity of the concurrence implies that

$$C(\rho) \leq (1 - 3p_3)C(\rho_2) \leq p_1 - p_3. \tag{9}$$

Here, we have used that, as shown above, the maximum concurrence of rank 2 density matrices is its maximum eigenvalue. The upper bound in Eq. (9) again coincides with C^* for $p_4 = 0$.

(3) $p_4 \neq 0$ case. The rank 4 density matrices satisfying $C^*(p_i) \geq 0$ are decomposed as

$$\rho = C^*(p_i)|1\rangle\langle 1| + \rho_4, \tag{10}$$

where $|1\rangle$ is an eigenvector of ρ, and the eigenvalues of ρ_4 (not normalized) are $\{p_3 + 2\sqrt{p_2 p_4}, p_2, p_3, p_4\}$. When the eigenvalues of ρ satisfy

$$p_3 = p_2 + p_4 - \sqrt{p_2 p_4}, \tag{11}$$

the purity of (normalized) ρ_4 is equal to 1/3 and ρ_4 becomes always separable. Therefore, convexity of the concurrence again implies that the upper bound of the concurrence is C^* for density matrices satisfying Eq. (11) and $C^*(p_i) \geq 0$.

It should be noted that, when $p_2 = p_3 = p_4 (\leq 1/4)$, M is reduced to the Werner state [8], whose eigenvalues satisfy Eq. (11). Therefore, the EOF of the Werner state cannot be increased by any unitary transformation.

At this stage, we conjectured that the C^* is maximum for any general density matrices in two qubits. In order to check the correctness of our conjecture, we have performed a numerical calculation whose result was published elsewhere [9]. The numerical result strongly supports the truth of our conjecture.

Accepting our conjecture implies that all the states satisfying $C^*(p_i) \leq 0$ become automatically separable. This condition of separability is looser than $\mathrm{Tr}\rho^2 \leq 1/3$. In fact, $C^*(p_i) \leq 0$ is only a necessary condition of $\mathrm{Tr}\rho^2 \leq 1/3$. The difficulty with the rigorous proof of our conjecture might relate to the difficulty in completely understanding the separable-inseparable boundary in the 15-dimensional space of the density matrices due to its complex structure.

It is worth testing whether M is maximally entangled in the other entanglement measures. The negativity [2] based on the partial transposition is a kind of entanglement measures, and we have checked numerically that the negativity of M

$$\begin{aligned} 2E_N^* &= \max\{0, 2E_N^*(p_i)\}, \\ 2E_N^*(p_i) &\equiv -p_2 - p_4 + \sqrt{(p_1 - p_3)^2 + (p_2 - p_4)^2}, \end{aligned} \tag{12}$$

is indeed maximum [9], in spite that the EOF and negativity generally do not induce the same ordering of density matrices with respect to the amount of entanglement [10].

It should be noted finally that M is decomposed as

$$M = C^*(p_i)|\Psi^-\rangle\langle\Psi^-| + (1 - C^*(p_i))\rho_{\rm sep}, \quad (13)$$

where the separability of $\rho_{\rm sep}$ is easily confirmed by Peres criteria. q in the optimum L-S decomposition Eq. (4) is also a kind of entanglement measures [5]. The above decomposition is an optimum L-S decomposition for M, since the convexity of the concurrence implies $C \leq q$ and equality is satisfied in the above decomposition.

It will be extremely important to seek out the *maximally entangled mixed states* as well as the measure in systems with a larger dimension, for understanding the nature of entanglement of general mixed states.

References

[1] C. H. Bennett, D. P. DiVincenzo, J. A. Smolin, and W. K. Wootters, Phys. Rev. A **54**, 3824 (1996).

[2] A. Peres, Phys. Rev. Lett. **76**, 1413 (1996), K. Życzkowski and P. Horodecki, Phys. Rev. A **58**, 883 (1998).

[3] N. Linden, S. Massar, and S. Popescu, Phys. Rev. Lett. **81**, 3279 (1998), A. Kent, N. Linden, and S. Massar, Phys. Rev. Lett. **83**, 2656 (1999).

[4] S. Hill and W. K. Wootters, Phys. Rev. Lett. **78**, 5022 (1997), W. K. Wootters, Phys. Rev. Lett. **80**, 2245 (1998).

[5] M. Lewenstein and A. Sanpera, Phys. Rev. Lett. **80**, 2261 (1998).

[6] A. Uhlmann, quant-ph/9909060 .

[7] K. Życzkowski and P. Horodecki, Phys. Rev. A **58**, 883 (1998).

[8] R. Werner, Phys. Rev. A **40**, 4277 (1989).

[9] S. Ishizaka and T. Hiroshima, Phys. Rev. A **62**, 22310 (2000).

[10] J. Eisert, J. Mod. Opt. **46**, 145 (1999).

ENGINEERING BELL-LIKE STATES OF TWO HIGH-Q CAVITY FIELDS

A. Napoli, A. Messina, S. Maniscalco
Dipartimento di Scienze Fisiche ed Astronomiche
via Archirafi 36, 90123 Palermo

Abstract A simple conditional single-atom based method for generating Bell-like states of two high-Q single mode cavity fields is presented. The practical feasibility of the scheme is briefly discussed.

Recent developments in cavity quantum electrodynamics as well as the possibility of exciting and manipulating single Rydberg atoms , have provided favourable conditions for shading light on some puzzling quantum predictions. A great deal of attention has for example been devoted to the possibility of generating coherent superpositions of macroscopically distinguishable states as well as to the fundamental issue of nonlocality that is the existence of entanglement of distant systems. A pure quantum state of two subsystems is said to be entangled if it is not a product of states of each subsystem. >From an applicative point of view, reliable sources of entangled states are required for the implementation of many quantum communication and computation protocols [1]. They are, for example, essential in quantum teleportation procedure which is, on the other hand, an extremely useful tool for understanding many properties of quantum entanglement itself. The puzzling implications of entanglement as well as its applicability, have spurred an intense theoretical and experimental research work [2, 3, 4, 5, 6]. In this paper we present a simple method for generating Bell-like states of two high-Q single mode cavity fields. The implementation of our scheme requires a single effective three-level atom and two spatially separeted high-Q cavities. Indicate by ω_1 and ω_2 ($\omega_1 \sim \omega_2 \sim 10^{10} Hz$) the fundamental frequencies of the two resonators and denote by $|0\rangle$ (E_0^A), $|1\rangle$ (E_1^A) and $|2\rangle$ (E_2^A) the three relevant atomic circular Rydberg states and their relative energies with $E_0^A < E_1^A < E_2^A$. We will show that it is possible to generate two different classes of Bell-like states in accordance with two resonance conditions. We begin imposing as first condition $E_1^A - E_0^A \sim \omega_1$ and $E_2^A - E_1^A \sim \omega_2$ ($\hbar = 1$). Under these hypotesis the interaction between the atom and each cavity can be cast in the form of

JC-like model so that the effective RWA Hamiltonian describing the system under scrutiny can be written down as $H = H_0 + H_1 + H_2$ where

$$H_0 = \sum_{i=1}^{2} \omega_i a_i^\dagger a_i + \sum_{j=0}^{2} E_j^A |j\rangle\langle j| \tag{1}$$

$$H_1 = g_1 a_1 |1\rangle\langle 0| + h.c. \qquad H_2 = g_2 a_2 |2\rangle\langle 1| + h.c. \tag{2}$$

In eqs. (1.1) - (1.2) $a_i(a_i^\dagger)$ $(i = 1, 2)$ is the annihilation (creation) operator relative to the $i-$th cavity whereas g_1 and g_2 measure the strenghts of the energy exchanges between the Rydberg atom and the cavity 1 and 2 respectively. Suppose now that our single three-level Rydberg atom, prepared in the state $\frac{1}{\sqrt{2}}(|0\rangle + |2\rangle)$ by means an appropriate Ramsey zone, is injected into the first cavity where the mode of frequency ω_1 has been previously excited in its Fock state $|p\rangle_1$. After this interaction the atom enters the second resonator whose fundamental mode has been prepared in the Fock state with p photons, $|p\rangle_2$. Then the state of the atom-field system at $t = 0$ is given by

$$|\psi(0)\rangle = \frac{1}{\sqrt{2}}(|0\rangle + |2\rangle)|p\rangle_1|p\rangle_2 \tag{3}$$

Let's indicate by t_1 and t_2 the interaction times between the atom and the cavity 1 and 2 respectively and suppose that the condition

$$g_j t_j = \frac{\pi}{4\sqrt{p}} \quad (j = 1, 2) \tag{4}$$

is satisfied. We have exactly proved that after the atom leaves the second resonator, the state of the system, apart from an overall phase factor, can be cast in the entangled atom-field form

$$\begin{aligned} |\ \psi\ \rangle &= \frac{1}{2}\{[|p\rangle_1|p\rangle_2 - e^{-i\phi_1}\sqrt{2}sin(\frac{\pi}{4}\sqrt{\frac{p+1}{p}})|p+1\rangle_1|p\rangle_2]|0\rangle \\ &- i[|p-1\rangle_1|p\rangle_2 + e^{-i\phi_2}\sqrt{2}sin(\frac{\pi}{4}\sqrt{\frac{p+1}{p}})|p\rangle_1|p+1\rangle_2]|1\rangle \\ &+ [\frac{1}{\sqrt{2}}|p-1\rangle_1|p-1\rangle_2 - e^{-i\phi_2}\sqrt{2}cos^2(\frac{\pi}{4}\sqrt{\frac{p+1}{p}})|p\rangle_1|p\rangle_2]|2\rangle\} \end{aligned} \tag{5}$$

where $\phi_1 = [\frac{\omega_1+\omega_2}{g_1} + \frac{\omega_2}{g_2}]\frac{\pi}{4\sqrt{p}}$ and $\phi_2 = \frac{\pi(\omega_1+\omega_2)}{4\sqrt{p}}(\frac{1}{g_1} + \frac{1}{g_2})$. Eq. (1.5) clearly suggests that if immediately after the atom leaves the second cavity, an appropriate atomic state detector finds it in its excited state $|2\rangle$, then the cavity fields collapse onto an entangled state well-approximated by a Bell-like state $|\chi\rangle$ having the form

$$|\chi\rangle = N[|p-1\rangle_1|p-1\rangle_2 + e^{-i\phi_2}|p\rangle_1|p\rangle_2] \tag{6}$$

where N is an appropriate normalization constant. Taking into account eq. (1.5) it is easy to convince oneself that the probability of finding the atom in its excited state $|2\rangle$, that is the probability of success of the experimental scheme under scrutiny, can be written down as

$$P = \frac{1}{8} + \frac{1}{2}cos^4(\frac{\pi}{4}\sqrt{\frac{p+1}{p}}) > \frac{1}{8} \tag{7}$$

Consider now a different resonance condition, namely $E_2^A - E_0^A \sim \omega_1$ and $E_1^A - E_0^A \sim \omega_2$. The effective Hamiltonian model now becomes $H = H_0 + H_1' + H_2'$ where

$$H_1' = g_1 a_1 |2\rangle\langle 0| + h.c. \quad H_2' = g_2 a_2 |1\rangle\langle 0| + h.c. \tag{8}$$

In this case we prepare the atom in the state $\frac{1}{\sqrt{2}}(|1\rangle + |2\rangle)$ and the two cavity fields as before. It is easily and exactly demonstrable that, manipulating the atomic velocity in order to satisfy the condition

$$g_1 t_1 = \frac{2\pi}{\sqrt{p+1}} \quad g_2 t_2 = \frac{\pi}{4\sqrt{p}} \tag{9}$$

when the atom leaves the second resonator, the state of the combined system neglecting an overall phase factor, assumes the form:

$$|\psi\rangle = \frac{1}{\sqrt{2}}\{[-isin(\frac{\pi}{4}\sqrt{\frac{p+1}{p}})|p\rangle_1|p+1\rangle_2 + \frac{1}{\sqrt{2}}e^{-i\chi_1}|p+1\rangle_1|p\rangle_2]|0\rangle \tag{10}$$
$$+[e^{-i\chi_2}cos(\frac{\pi}{4}\sqrt{\frac{p+1}{p}})|p\rangle_1|p\rangle_2 + e^{-i\chi_3}|p+1\rangle_1|p-1\rangle_2]|1\rangle\}$$

where $\chi_1 = \frac{\pi\omega_2}{4g_2\sqrt{p}}, \chi_2 = \omega_2(\frac{1}{g_1\sqrt{p+1}} + \frac{1}{2g_2\sqrt{p}})$ and $\chi_3 = \frac{\pi\omega_1}{4g_2\sqrt{p}}$. In view of equation (1.10) it is immediate to understand that a conditional measurement of the atomic internal state projects the two cavity fields in a superposition of $|p\rangle_1|p+1\rangle_2$ and $|p+1\rangle_1|p\rangle_2$ if the atom is found in its ground state $|0\rangle$. It is possible to see that the state so obtained can be well-approximated by the following Bell-like state

$$|\eta\rangle = \tilde{N}[|p+1\rangle_1|p\rangle_2 + e^{i\eta}|p\rangle_1|p+1\rangle_2] \tag{11}$$

with $\eta = -\frac{\pi}{2}(1 - \frac{\omega_2}{2g_2\sqrt{p}})$ and $\tilde{N}$ normalization constant. Even in this second case, the pertinent probability of success assumes experimentally interesting values being, from eq. (1.10)

$$P = \frac{1}{4} + \frac{1}{2}sin^2(\frac{\pi}{4}\sqrt{\frac{p+1}{p}}) > \frac{1}{4} \tag{12}$$

Before concluding it seems appropriate to point out some aspects related with the feasibility of the procedures discussed in this paper. First of all, let's observe that, being $g_j \sim (10^5 \div 10^4) sec^{-1}$ and realistically assuming $p < 10$, conditions (1.4) and (1.9) give values of the interaction times compatible with the atomic spontaneous lifetime ($\sim 30ms$). Moreover, considering that the availability of microwave superconducting resonator having quality factors Q up to 10^{10} are today within the reach of experimentalists, the cavity damping time, $\tau_c \sim 10^{-1} sec$, is much longer than the total duration of the experiment. Finally, we wish to stress that our conditional method is based on the passage through the two cavities of a single atom only and it is characterized by experimentally interesting probability of success as given by eq. (1.7) and (1.12). This circumstance is of particular relevance in view of the efficiency of the atomic velocity selectors ($1\% \div 2\%$) and, above all, of the low performance of the atomic state detectors.

Acknowledgments

This research was supported by Comitato Regionale di Ricerche Nucleari e di Struttura della Materia. One of the authors (A.N.) is indebted to MURST and FSE for partially supporting this work in the context of Programma Operativo Multiregionale 1994/99 940023I1.

References

[1] Ekert, E., Josza, R., 1996, Review of Mod. Phys. 68, 733.

[2] Gerry, C.C., 1996, Phys. Rev. A 53, 1179.

[3] Bergou, J.A., 1997, J. Mod. Opt. 44, 1957.

[4] Gerry, C.C., Grobe, R., 1999, J. Mod. Opt. 46, 1053.

[5] Banaszek K., 2000, Phys. Rev. A 62, 024301.

[6] Buzek V. and Hillery M., 2000, Phys. Rev. A 62, 022303.

NONDISSIPATIVE DECOHERENCE AND ENTANGLEMENT IN THE DYNAMICS OF A TRAPPED ION

S. Maniscalco, A. Messina, A. Napoli

INFM Unità di Palermo and Dipartimento di Scienze Fisiche ed Astronomiche
Università di Palermo, via Archirafi 36, 90123 Palermo, Italy

D. Vitali

Dipartimento di Matematica e Fisica, Università di Camerino,
INFM, Unità di Camerino, via Madonna delle Carceri 62032, Camerino, Italy

Keywords: Entanglement, nondissipative decoherence, trapped ions

Abstract We study the robustness of the entanglement between the 2D vibrational motion and two ground state hyperfine levels of a trapped ion with respect to the presence of non-dissipative sources of decoherence.

We study the motion of an ion isotropically confined in the radial plane of a Paul microtrap when it is irradiated by a properly chosen configuration of external laser beams. In Ref. [1] it has been shown that the center of mass vibrational motion and the electronic degrees of freedom of the ion may become entangled in a way which is very sensitive to the parity of the initial number of vibrational quanta. Here we focus on the robustness of this quantum effect with respect to various experimental imperfections. In fact, even if dissipation of the vibrational motion is completely negligible, non-dissipative sources of decoherence associated with fluctuations of external parameters may be important. We study these effects, and those of laser intensity fluctuations in particular, using the model-independent formalism of Ref. [2].

The system under study is a two-level ion confined in a bidimensional isotropic harmonic potential characterized by a trap frequency ν. The operators $\hat{a}$ and $\hat{b}$ are the annihilation operators of vibrational quanta in the X and Y directions respectively. It has been shown [3] that, irradiating the trapped ion with an appropriate configuration of laser beams, the physical system under scrutiny can be studied, in the Lamb-Dicke limit and in the interaction picture, by the following Hamiltonian model

$$\hat{H} = \hbar\nu(\hat{a}^\dagger\hat{a} + \hat{b}^\dagger\hat{b}) + \frac{\hbar\delta}{2}\hat{\sigma}_z + \hbar g\left[(\hat{a}\hat{b})\hat{\sigma}_+ + (\hat{a}^\dagger\hat{b}^\dagger)\hat{\sigma}_-\right] \quad (1)$$

where δ is the detuning of the laser beams from the atomic transition and $\hat{\sigma}_z = |+\rangle\langle+| - |-\rangle\langle-|$, $\hat{\sigma}_+ = |+\rangle\langle-|$, $\hat{\sigma}_- = |-\rangle\langle+|$ describe the internal degrees of freedom, $|+\rangle$ and $|-\rangle$ being the ionic excited and ground states respectively. Let's denote with $|n_a, n_b\rangle = |n_a\rangle|n_b\rangle$ the product of the vibrational Fock states along the X and Y directions and suppose that the initial state of the ion has the form

$$|\Psi(0)\rangle = \frac{1}{2^{N/2}}\sum_{k=0}^{N}\begin{pmatrix} N \\ k \end{pmatrix}^{1/2}|N-k,k\rangle|-\rangle = \sum_{k=0}^{N} P_k|N-k,k\rangle|-\rangle. \quad (2)$$

As pointed out by Gou and Knight [4], this state coincides with the vibrational Fock state with N quanta along the direction X' forming an angle $\pi/4$ relative to the X axis, and it can therefore be easily prepared experimentally (see [5, 6]). The time evolution of the system can be written as $|\Psi(t)\rangle = |\varphi_-(t)\rangle|-\rangle + |\varphi_+(t)\rangle|+\rangle$ with $|\varphi_-(t)\rangle = \sum_{k=0}^{N} P_k\cos(f_k t)|N-k,k\rangle$ and $|\varphi_+(t)\rangle = -i\sum_{k=1}^{N-1} P_k\sin(f_k t)|N-k-1,k-1,+\rangle$, where $f_k = 2g\sqrt{(N-k)k}$ are the Rabi frequencies. This shows that, starting from the factorized initial state of Eq. (2), the Hamiltonian model (1) generates entanglement between the external and internal degrees of freedom of the trapped ion. The dynamical quantity of interest, which can be directly measured, is the probability to find the ion in the ground state

$$P_-(t) = c(t) = \sum_{k=0}^{N}|P_k|^2\cos^2(f_k t) = \frac{1}{2}\left[1 + \sum_{k=0}^{N}|P_k|^2\cos(2f_k t)\right]. \quad (3)$$

It has been shown in Ref. [1] that, starting from a total vibrational excitation number N, there exist an N-dependent time instant at which the internal and external degrees of freedom of the trapped ion are disentangled ($c(t) = 1$ or $c(t) = 0$) or maximally entangled ($c(t) = 1/2$). More in detail, if $N \gg 1$ is odd, at the time instant $\bar{t}_o = \frac{\pi(N-1)}{4g}$ one has $c(\bar{t}_o) = 1$ ($c(\bar{t}_o) = 0$) if $(N-1)/2$ is even (odd). On the contrary, if $N \gg 1$ is even, at the time instant $\bar{t}_e = \frac{\pi N}{4g}$, one has $c(\bar{t}_e) = 1/2$ and therefore the vibrational and electronic degrees of freedom become maximally entangled. Therefore one has a totally different behavior of the system depending on the parity of the initial vibrational number.

The above parity-dependent entanglement effect is obtained if perfect unitary evolution is assumed. However, decoherence effects have been observed in the motion of a laser driven trapped ion [5, 6]. In fact, even when the entanglement with the environment is negligible, fluctuations of some classical parameter of the system may cause non-dissipative, phase-destroying effects. In the trapped ion case, phase decoherence is mainly caused by the fluctuations of the Rabi frequency, which are induced by the laser intensity fluctuations, making the coupling constant g fluctuate. A quantitative explanation of the decoherence observed in the Rabi oscillations of Ref. [5], in terms of a fluctuating laser pulse area, has been recently provided by Bonifacio *et al.* in Ref. [2], using a model-independent formalism able to describe non-dissipative decoherence phenomena due to the fluctuations of classical parameters or internal variables of a system. In this approach, the dynamical quantities can be obtained by averaging over an appropriate probability distribution of the fluctuating parameter. In the trapped ion case, the random parameter is the positive dimensionless random variable $A(t) = \int_0^t d\xi g(\xi)$, which is proportional to the laser pulse area. The time evolution of a generic dynamical quantity $O(t)$ becomes therefore the averaged quantity $\bar{O}(t) = \int_0^\infty dAP(t,A)O(A)$. The probability distribution $P(t,A)$ is the Gamma distribution function $P(t,A) = e^{-A/g\tau}(A/g\tau)^{(t/\tau)-1}/\left(g\tau\Gamma(t/\tau)\right)$, obtained in Ref. [2] by imposing the semigroup condition for the averaged density matrix of the system. The physical meaning of the parameters g and τ can be easily understood by considering the mean and the variance of the above probability distribution, $\langle A\rangle = gt$, $\sigma^2(A) = \langle A^2\rangle - \langle A\rangle^2 = g^2t\tau$, implying that g has now to be meant as a *mean* coupling constant, and that τ quantifies the strength of A fluctuations. In Ref. [2] the decay of the Rabi oscillations observed in Ref. [5] is well fitted assuming $\tau \simeq 1.5 \cdot 10^{-8}$ sec. In this case, the relevant experimental timescale t is much larger than τ and in this limit the Gamma distribution function can be well approximated by the Gaussian distribution $P(t,A) \simeq \{2\pi g^2t\tau\}^{-1/2}\exp\left\{-(A-gt)^2/2g^2t\tau\right\}$. Performing the average and using Eq. (3) we find

$$\bar{P}_-(gt) = \frac{1}{2}\left[1 + \sum_{k=0}^{N} e^{-8g^2\tau(N-k)kt}\cos(4\sqrt{(N-k)k}gt)\right], \qquad (4)$$

where the quantum coherent oscillations at frequencies f_k are now exponentially decaying because of the non-dissipative decoherence.

To quantify the robustness of the parity-dependent entanglement with respect to the various experimental imperfections, we have considered the quantity ΔP^N, defined as the difference between the values assumed by $\bar{P}_-(g\bar{t})$ in correspondence to N (odd) and $N+1$ total initial vibrational excitations, at the "intermediate" time $\bar{t} = \frac{\bar{t}_e+\bar{t}_o}{2}$. With no imperfections and $N \gg 1$, one has $\bar{t} \sim \bar{t}_e \sim \bar{t}_o$, and $\bar{P}_-^{N=odd}(g\bar{t}) \sim 1$, $\bar{P}_-^{N+1}(g\bar{t}) \sim 1/2$, so that $\Delta P^N = 1/2$. As long as the effect of phase decoherence increases, the difference between the

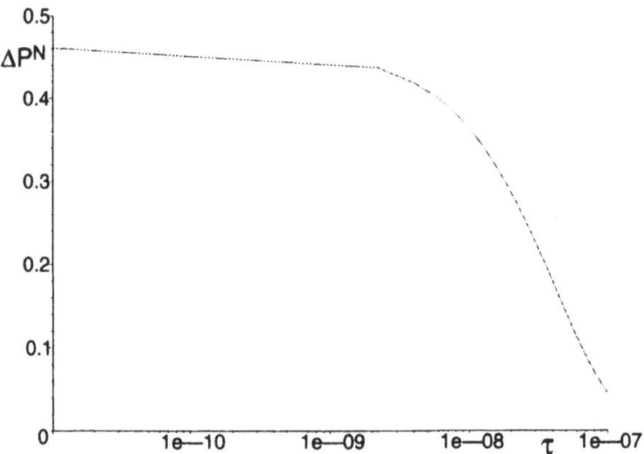

Figure 1 The probability difference ΔP^N defined in the text, at the time instant $\bar{t}$, as a function of the laser pulse area fluctuation strength parameter τ (expressed in seconds), in the case of a total initial vibrational quanta $N = 9$ and for $g = 10^5$ Hz.

odd N case and the even N case tends to zero, because in both cases the system state becomes the same statistical mixture, and ΔP^N tends to zero. In Fig. 1 we show the dependence of the probability difference ΔP^N as a function of the laser pulse area fluctuation strength τ. In particular one has a well visible transition from a quantum behaviour at small τ to a completely decoherent behaviour at larger τ, taking place at a threshold value $\tau_{th} \sim 3 \cdot 10^{-8}$ sec. Since current experimental situations correspond to $\tau \leq 10^{-8}$ sec, this means that the parity-dependent entanglement effect described in the preceding section can be experimentally detected even in the presence of laser intensity fluctuations.

References

[1] S. Maniscalco *et al.*, Phys. Rev A **61** 053806 (2000).

[2] R. Bonifacio *et al.*, Phys. Rev. A **61** 053802 (2000).

[3] S.-C. Gou *et al.*, Phys. Rev A **54** R1014 (1996).

[4] S.-C. Gou and P.L. Knight, Phys. Rev A **54** 1682 (1996).

[5] D.M. Meekhof *et al.*, Phys. Rev. Lett. **76** 1796 (1996).

[6] Ch. Roos *et al.*, Phys. Rev. Lett. **83** 4713 (1999).

ENTANGLEMENT MANIPULATION AND CONCENTRATION IN MIXED STATES

R. T. Thew, K. Nemoto, W. J. Munro
Special Research Centre for Quantum Computer Technology,
University of Queensland, Brisbane, Australia
billm@physics.uq.edu.au

Keywords: Entanglement Manipulation, Concentration, Mixed States

Abstract We introduce a simple experimentally realisable protocol for manipulating entanglement in mixed states of two polarisation-entangled qubits and show how concentration is achievable. We discuss a second protocol based on entanglement swapping with mixed states and show how the shared entanglement between distant parties can be improved.

1. INTRODUCTION

There has been a great deal of discussion recently concerning entanglement, motivated chiefly by the interest in quantum information, computing, teleportation and cryptography, with there reliance on entanglement. Much of this discussion has revolved around pure states though much current work is being put into understanding mixed state entanglement. In this article we introduce a simple experimentally realisable protocol to manipulate and explore mixed-state entanglement. We show how entanglement swapping via a teleportation protocol can be used to recover entanglement lost between distant parties in a noisy quantum channel.

Our motivation comes from focusing several ideas of various groups into a simple realisation of mixed state entanglement manipulation using controlled loss-elements[1, 2]. In conjunction with this are the significant recent advances in the experimental quantum state synthesis of pure[3] and mixed[4] polarisation-entangled photons. This allows the experimental exploration of the two qubit entanglement manipulation. For an arbitrary two qubit state the degree of entanglement can be analytically determined by the Tangle[6], τ, while the purity of the state which can be determined by the linear entropy, S_L [7].

2. TWO QUBIT CONCENTRATION

The general aim of our first protocol is to manipulate mixed states and explore the behaviour of such processes in terms of two properties, their degree of entanglement and their degree of purity. Our protocol, depicted schematically in Figure(1) (scheme(a)), is a variation of the Procrustean method introduced by Bennett *et.al* [5] to examine pure state distillation.

Consider a general two qubit polarisation-entangled density matrix (with the qubits labelled A and B). The polarisation modes for each qubit are separated

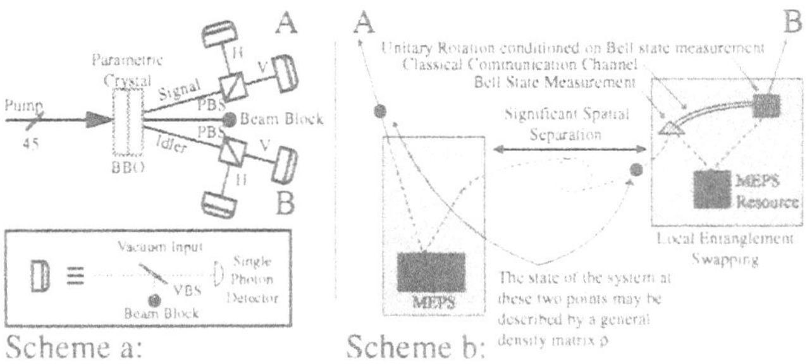

Figure 1 Schematic of two schemes for the manipulation of entanglement in mixed states. The schematic for scheme(a) illustrates how an entangled state shared between A and B can be concentrated. The second schematic (scheme(b)) considers an entangled state transferred from a Central Node to distant locations. A Repeater is present at one of these locations. By employing an entanglement swapping protocol between the mixed state and the repeater a state with enhanced entanglement and entropy characteristics is realised.

using a polarising beam splitter(labelling the four distinct modes by $|h\rangle_a$, $|v\rangle_a$, $|h\rangle_b$ and $|v\rangle_b$). Each polarisation mode is then incident onto a separate individually controllable variable-reflectivity beam splitter (BS) with transmission coefficients $\eta_{v,h}$. The second input port of these BSs are vacuum. We examine the transmitted modes of the beamsplitter and consider the situations where joint coincidences are registered at the detectors of the two subsystems A and B and discard the information present at the second output of each BS. As such the whole process we describe here is non-unitary. For an arbitrary two qubit input density matrix (whose elements are represented by ρ_{ij}) we obtain, in the polarisation coincidence basis, an reduced output density matrix of the form

$$\hat{\rho}_a = \mathcal{N}\begin{pmatrix} \rho_{11}\eta_{va}^2\eta_{vb}^2 & \rho_{12}\eta_{va}^2\eta_{vb}\eta_{hb} & \rho_{13}\eta_{va}\eta_{ha}\eta_{vb}^2 & \rho_{14}\eta \\ \rho_{12}^*\eta_{va}^2\eta_{vb}\eta_{hb} & \rho_{22}\eta_{va}^2\eta_{hb}^2 & \rho_{23}\eta & \rho_{24}\eta_{va}\eta_{ha}\eta_{hb}^2 \\ \rho_{13}^*\eta_{va}\eta_{ha} & \rho_{23}^*\eta & \rho_{33}\eta_{ha}^2\eta_{vb}^2 & \rho_{34}\eta_{ha}^2\eta_{vb}\eta_{hb} \\ \rho_{14}^*\eta & \rho_{24}^*\eta_{va}\eta_{ha}\eta_{hb}^2 & \rho_{34}^*\eta_{ha}^2\eta_{vb}\eta_{hb} & \rho_{44}\eta_{ha}^2\eta_{hb}^2 \end{pmatrix}$$

where $\eta = \eta_{va}\eta_{ha}\eta_{vb}\eta_{hb}$. Here $\mathcal{N} = [\rho_{11}\eta_{va}^2\eta_{vb}^2 + \rho_{22}\eta_{va}^2\eta_{hb}^2 + \rho_{33}\eta_{ha}^2\eta_{vb}^2 + \rho_{44}\eta_{ha}^2\eta_{hb}^2]^{-1}$ is the normalisation and is inversely proportional to the probability of obtaining our desired state.

To illustrate our protocol we introduce a class of mixed states that have a maximum amount of entanglement for the degree of linear entropy[8]. These maximally entangled mixed state, (MEMS), have the form

$$\hat{\rho}(\gamma) = \begin{pmatrix} g(\gamma) & 0 & 0 & \gamma/2 \\ 0 & 1-2g(\gamma) & 0 & 0 \\ 0 & 0 & 0 & 0 \\ \gamma/2 & 0 & 0 & g(\gamma) \end{pmatrix} \quad (1)$$

where $g(\gamma) = \gamma/2$ for $\gamma \geq 2/3$ and $g(\gamma) = 1/3$ for $\gamma < 2/3$. In figure(2a) the dashed curves represent the entanglement-purity bound for the MEMS. The solid curves represent a range of states that are obtainable, from the MEMS, as the BSs are tuned to optimise the output state characteristics. We see that if we start with a state, $\gamma < 2/3$ then the output can be significantly improved, however once we are above this value the states can be concentrated all the way up the dashed curve to a maximally entangled pure state (MEPS). This protocol can be shown to work for a wide range of states, distilling non-MEPS, and concentrating mixed states. Next we will outline our second protocol for the manipulation and concentration of mixed states.

3. REMOTE STATE MANIPULATION

In figure(1b) we present a protocol that allows certain mixed states to be concentrated via an entanglement swapping scheme. Consider a MEPS at some central node to be shared to distant locations. Both modes are transported to distant locations where the state becomes mixed due to environmental effects. At one of these distant locations we implement an entanglement swapping protocol using quantum teleportation[9]. By performing a Bell state measurement between one qubit of a local MEPS state and the incident central node qubit, one is able to transfer the entanglement from the incident qubit to the second qubit of the MEPS resource. When this Bell state measurement succeeds (probabilistic), the resulting state between A and B can be more entangled. We call this our *Quantum Repeater* protocol as it allows us to probabilistically recover some of the entanglement and entropy characteristics of the state that was lost. This repeater protocol could be applied successive times allowing the state from the central node to be distributed over even large distances while maintaining a degree of entanglement.

This protocol works for a wide variety of mixed states at the repeater location (not Werner states). For convenience we will again consider the MEMS of (1). In figure(2b) we observe the concentration characteristics for a range of parameter values for a MEMS that has been subject to one Repeater process. The entanglement-purity characteristics for a range of initial shared mixed states are labeled A. The characteristics for the shared state after the first repeater ap-

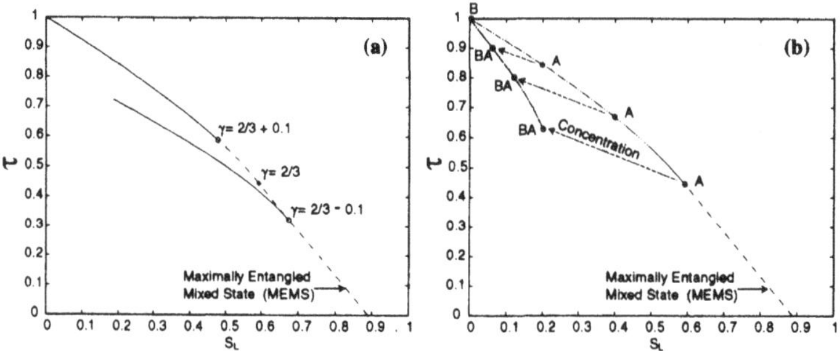

Figure 2 Improvement in entanglement and linear entropy characteristics for the two protocol presented in Figure (1). Both these figures have a dashed curve marking the bound on this plane for the MEMS. By manipulating the VBSs in scheme(a) an initial state, circles in figure(2a), can be manipulated producing a range of states, solid lines, with improved characteristics. For $\gamma > 2/3$ a MEPS can be recovered from this mixed state. Figure(2b) shows the entanglement swapping protocol taking a range of mixed states arriving from A and the local MEPS at B to produce a shared entanglement resource, AB, greatly improved over that initially realised.

plication with a local MEPS, B, are labeled BA. We observe that the protocol achieves a significant improvement for just one repeater application.

4. SUMMARY

In summary we have introduced two methods for manipulating the entanglement of two polarisation-entangled qubits. Both of these protocols would be of great benefit in the recovery and maintenance of an entanglement resource, either locally or between distant parties.

References

[1] S. Popescu, Phys. Rev. Lett. **74**,2619 (1995);

[2] N. Gisin, Phys. Lett. A **210**, 151 (1996).

[3] P. G. Kwiat *et. al*, Phys. Rev. A **60**, 773 (1999).

[4] A.G. White, D.F.V. James, W.J.Munro, and P.G. Kwiat, *In preparation.*

[5] C.H. Bennett *et. al* , Phys. Rev. A **53**, 2046 (1996).

[6] V. Coffman, J. Kundu and W.K. Wooters, Phys. Rev. A **61**, 052306 (2000).

[7] S. Bose and V. Vedral, Phys. Rev. A, **61**, 040101(R) (2000).

[8] W.J.Munro *et. al*, *In preparation*, (2000).

[9] J.W. Pan *et. al*, Phys.Rev.Lett. **80**, 3891 (1998).

RAMSEY INTERFEROMETRY WITH A SINGLE PHOTON FIELD IN CAVITY QED

S. Osnaghi, A. Rauschenbeutel, P. Bertet, G. Nogues, M. Brune,
J. M. Raimond, S. Haroche

Laboratoire Kastler Brossel, Département de Physique de l'Ecole Normale Supérieure,
24 rue Lhomond, F-15231 Paris Cedex 05, France

raimond@lkb.ens.fr

Keywords: Atom interferometry, cavity QED

Abstract We present the experimental realisation of a cavity QED atom interferometer based on two consecutive resonant interactions of a circular Rydberg atom with the field mode of a high Q microwave resonator that is initially prepared in the vacuum state. The interferometer scheme is analogous to the Ramsey method with the classical field replaced by the quantum field stored in the cavity mode. We demonstrate the possibility to use this technique to probe, by dispersive interaction, the field stored in a second mode of the resonator.

The measurement and manipulation of small fields stored in a high finesse resonator is an active domain of cavity QED. Due to the strong coupling between two-level atoms and the cavity mode, either the resonant or the dispersive atom-field interaction can be used to probe the field [1]. In the latter case, the light shift induced on the atom by the cavity mode is usually measured by means of a Ramsey interferometric scheme [2]: Before and after its interaction with the cavity mode, the atom is coupled to a quasi-resonant oscillatory field which mixes the atomic levels; the probability of detecting the atom in its initial state varies sinusoidally as a function of the detuning between the atom transition and the oscillatory field.

We present here an alternative method based on the set-up schematically depicted in Fig. 1. Our superconducting cavity sustains two orthogonally polarised modes at 51.1 GHz, a high frequency (HF) mode and a low frequency (LF) mode with frequency splitting $(\omega_{HF} - \omega_{LF})/2\pi$=130 kHz and damping time τ_c =1 ms. The atoms (^{85}Rb) are prepared in circular Rydberg states g and e (principal quantum number 50 and 51, lifetime 30 ms). The strong resonant coupling between single atoms and the vacuum stored in either one of the cavity modes exceeds all dissipative processes and results in a coherent and reversible energy exchange (quantum Rabi oscillation, with frequency at cavity centre $\Omega/2\pi$=47 kHz [3]). We are able to control the atomic frequency ω_{eg} at each point of the atom trajectory inside the cavity mode (waist 6 mm). As the preparation process of the atomic state (e or g) is pulsed and the atom velocity is selected [1], the atom position is known within ± 1 mm. Tuning of ω_{eg} inside the cavity is achieved through a weak dc voltage V_{cav} applied across the cavity mirrors: The $e \Rightarrow g$ transition experiences a quadratic Stark shift of -255kHz/$(V/cm)^2$. These features allow us to realise an atom interferometer in which the classical oscillating field of the Ramsey scheme is replaced by one of the modes of the resonator. The field stored in the other mode of the cavity can then be measured [4].

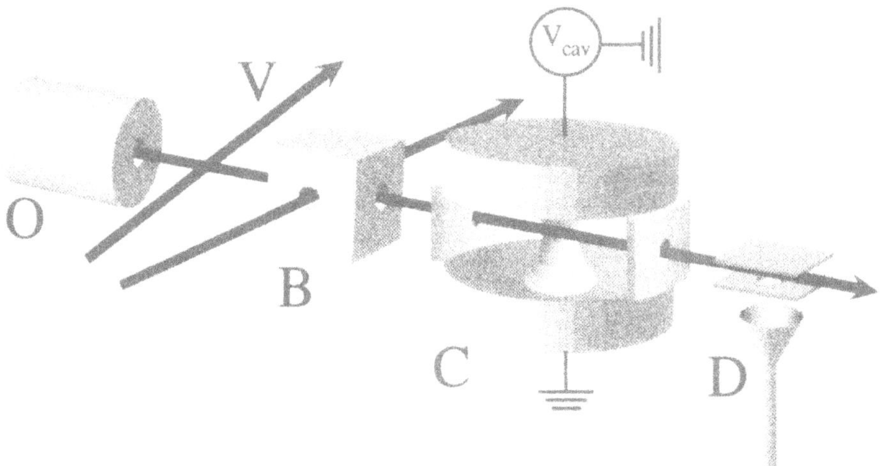

Figure 1 Experimental set-up. Atoms effuse from oven O; after velocity selection in V and state preparation in B, they cross the cavity C and are detected by ionisation in D.

The principle of the experiment is the following. An atom, prepared in e, enters the cavity at resonance with the HF mode prepared in vacuum. After a quarter period of Rabi oscillation ($\pi/2$ pulse), the atom-cavity state $|e,0\rangle$ (where 0 stands for the HF mode photon number) becomes:

$$|\Psi_I\rangle = |e,0\rangle - i|g,1\rangle \tag{1}$$

If we "freeze" the resonant evolution of the system at this point by detuning, at time $t = 0$, the atomic frequency from the HF mode ($|\omega_{HF} - \omega_{eg}| \gg \Omega$), no more energy exchange will take place. However the atom-cavity system is not in a stationary state and, in the frame rotating at frequency ω_{HF}, evolves into:

$$|\Psi(t)\rangle = |e,0\rangle - ie^{i(\omega_{HF}-\omega_{eg})\cdot t}|g,1\rangle \tag{2}$$

After a time $t = \delta T$ we tune the atom again into resonance with the HF mode in order to perform another $\pi/2$ pulse, and finally get:

$$|\Psi_{II}\rangle = (1 - e^{i\phi})|e, 0\rangle - i(1 + e^{i\phi})|g, 1\rangle \tag{3}$$

where $\phi = (\omega_{HF} - \omega_{eg}) \cdot \delta T$. The probability of detecting the atom in g is thus sinusoidally modulated as a function of ω_{eg}. We note that interference occurs in spite of the lack of an external reference phase, at variance with the classical Ramsey scheme where the phase reference is supplied by the macroscopic oscillating field. We replace here the classical field by vacuum fluctuations for which no definite phase exists. However, as the duration of the whole experiment is short compared to the cavity relaxation time, the coherence is preserved. We now focus on the low frequency (LF) mode of the cavity. During the non resonant evolution described in Eq. (2), the atom transition is tuned close to the LF mode and a dispersive interaction takes place. The corresponding light shift results in a phase ϕ_{LF} which adds to ϕ in Eq. (3). For a state with n_{LF} photons, this additional phase writes:

$$\phi_{LF} = \frac{\Omega^2}{\omega_{eg} - \omega_{LF}} \cdot (2n_{LF} + 1) \cdot \delta T \tag{4}$$

The LF photon number can therefore be inferred by measuring ϕ_{LF}. This is not straightforward because of the ϕ_{LF} dependence on ω_{eg}. We divide therefore the non-resonant evolution of the atom into two steps (Fig. 2). During a first short time interval δT_1, ω_{eg} is far detuned from both the cavity modes. In this situation ϕ still varies proportionally to the variation of ω_{eg}, whereas the contribution of ϕ_{LF} in Eq. (3) can be neglected, as $|\omega_{eg} - \omega_{LF}| \gg \Omega^2 \cdot \delta T_1$. The value of ω_{eg} during δT_1 can therefore be adjusted in order to scan ϕ and to get the fringes signal. In the second time interval, lasting δT_2, ω_{eg} is fixed and such that $|\omega_{eg} - \omega_{LF}|$ is of the order of $\Omega^2 \cdot \delta T_2$. A n_{LF}-dependent phase ϕ_{LF} is thus accumulated.

By this technique we have measured ϕ_{LF} with n_{LF}=1 for different δT_2 (i.e. for different atomic velocities). Each experimental sequence starts with a field-cooling procedure which prepares the two cavity modes in the vacuum state [1]. To get the fringes signal for n_{LF}=0, a "probe" atom is sent across the cavity and undergoes the interferometric procedure just described. The same experiment is repeated after having prepared a one photon field in the LF mode. This is achieved by means of a "source" atom. This atom, prepared in e, enters the cavity at resonance with the LF mode, undergoes a π Rabi pulse and emits a photon, being finally detected in g. The interference fringes displayed by the "probe" atom in this case are shifted with respect to the n_{LF}=0 signal. Fig. 3 reports the phase shift corresponding to two different atomic velocities (503 m/s and 200 m/s). For fast atoms the contrast is good (73%, compatible with the known imperfections [1]) whereas for slow atoms the signal shows a phase

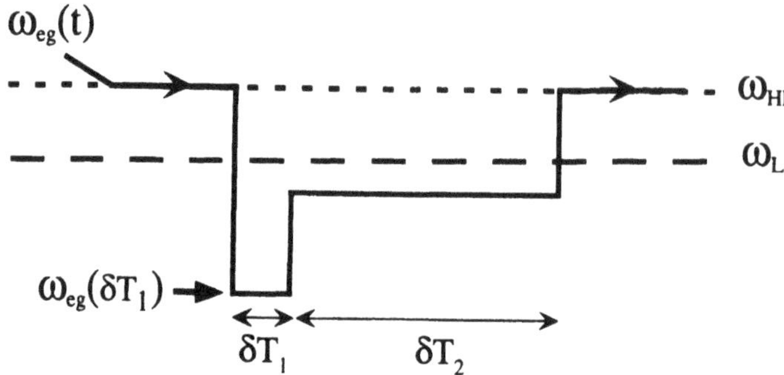

Figure 2 Scheme of the modulation of $\omega_{eg}(t)$ while the atom crosses the cavity mode. The value of $\omega_{eg}(\delta T_1)$ is scanned varying $V_{cav}(\delta T_1)$.

shift close to π (168°) but the contrast is lower, possibly due to inhomogeneous stray fields. This interferometric technique, which does not require any external field source, is a useful tool to explore the dispersive regime of cavity QED. We have observed light shifts of the order of π per photon, opening the way to a direct measurement of the parity of radiation states [5]. Applications would include Quantum Non Demolition measurements of any field state [5], the realisation of mesoscopic superpositions of radiation states with different field amplitudes and the full mapping of their Wigner function [6].

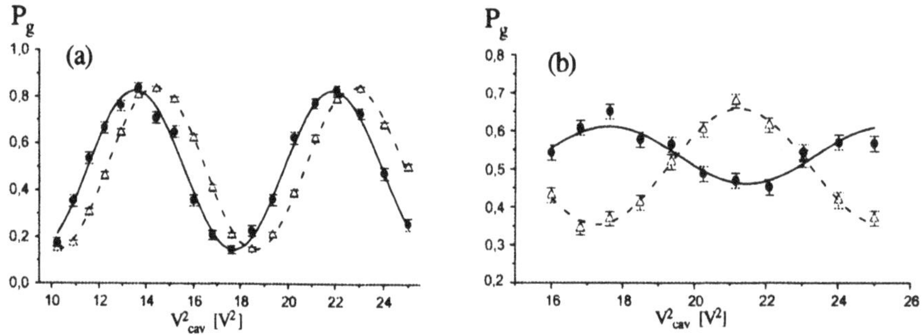

Figure 3 Probability P_g to detect the probe atom in g versus $V^2_{cav}(\delta T_1)$. Open triangles correspond to empty LF mode; solid circles to one LF photon. The measured phase shift between the two signals is 38° for atoms velocity 503 m/s (a) and 168° for atoms velocity 200 m/s (b).

References

[1] G. Nogues *et al.*, Nature (London) **400**, 239 (1999); A. Rauschenbeutel *et al.*, Science **288**, 2024 (2000) and references therein.

[2] N.F. Ramsey, *Molecular Beams* (Oxford University Press, New York, 1985).

[3] M. Brune *et al.*, Phys. Rev. Lett. **76**, 1800-1803 (1996).

[4] A. Rauschenbeutel *et al.*, to be submitted.

[5] M. Brune *et al.*, Phys. Rev. A **45**, 5193 (1992).

[6] L.G. Lutterbach and L. Davidovic, Phys. Rev. Lett. **78**, 2547 (1997).

ENTANGLEMENT TRANSFORMATION AT DIELECTRIC FOUR-PORT DEVICES

S. Scheel, L. Knöll
Theoretisch-Physikalisches Institut, Friedrich-Schiller-Universität Jena, Max-Wien-Platz 1, D-07743 Jena, Germany

T. Opatrný
Department of Theoretical Physics, Palacký University, Svobody 26, 771 46 Olomouc, Czech Republic

D.-G. Welsch
Theoretisch-Physikalisches Institut, Friedrich-Schiller-Universität Jena, Max-Wien-Platz 1, D-07743 Jena, Germany

Abstract Quantum communication schemes widely use dielectric four-port devices as basic elements for constructing optical quantum channels. Dielectrics with complex permittivity are typical examples of noisy quantum channels in which quantum coherence will not be preserved. Basing on quantization of phenomenological electrodynamics, we construct the transformation relating the output quantum state to the input quantum state without placing frequency restrictions. Knowledge of the full transformed quantum state enables us to compute the entanglement contained in the output state.

In order to study the problem of entanglement transformation in noisy quantum channels, quantization of the electromagnetic field in the presence of dielectric media is needed. A consistent formalism of QED in absorbing media is reviewed in [1]. It is based on the Green function expansion of the electromagnetic field with respect to the fundamental variables of the system composed of field and dielectric matter. The formalism is especially suited for deriving input-output relations of the field at dielectric slabs [2] on the basis of measurable quantities as transmission and absorption coefficients. From the input-output relations we can then derive closed formulas for calculating the output quantum state from the (known) input quantum state [3]. That is, the complete density matrix after the transformation is known, which makes the theory most suitable for studying entanglement properties of quantum states of light.

Quantum Communication, Computing, and Measurement 3
Edited by P. Tombesi and O. Hirota, Kluwer Academic/Plenum Publishers, 2001

We shortly review the basic formulas needed for the following considerations. Suppose the electromagnetic field has already been quantized and the Green function for a four-port device has been rewritten in the form of transmission and absorption matrices $\mathbf{T}(\omega)$ and $\mathbf{A}(\omega)$ [2]. The amplitude operators of the incoming and outgoing damped waves at frequency ω, $\hat{a}_i(\omega)$ and $\hat{b}_i(\omega)$, respectively, can then be connected by the quantum-optical input-output relations in the following way:

$$\hat{\mathbf{b}}(\omega) = \mathbf{T}(\omega)\hat{\mathbf{a}}(\omega) + \mathbf{A}(\omega)\hat{\mathbf{d}}(\omega) \tag{1}$$

with
$$\mathbf{T}(\omega)\mathbf{T}^{+}(\omega) + \sigma\mathbf{A}(\omega)\mathbf{A}^{+}(\omega) = \mathbf{I}, \tag{2}$$

where $\hat{d}_i(\omega)$ represent either device annihilation operators $\hat{g}_i(\omega)$ for absorbing devices ($\sigma = +1$) or creation operators $\hat{g}_i^{\dagger}(\omega)$ for amplifying devices ($\sigma = -1$). Eq. (2) is nothing but current (or energy) conservation. The formulas are valid for any chosen frequency.

By defining the "four-vector" operators $\hat{\boldsymbol{\alpha}}(\omega) = (\hat{\mathbf{a}}(\omega), \hat{\mathbf{d}}(\omega))^T$, $\hat{\boldsymbol{\beta}}(\omega) = (\hat{\mathbf{b}}(\omega), \hat{\mathbf{f}}(\omega))^T$, where $\hat{\mathbf{f}}(\omega) = \hat{\mathbf{h}}(\omega)$ for an absorbing device, and $\hat{\mathbf{f}}(\omega) = \hat{\mathbf{h}}^{\dagger}(\omega)$ for an amplifying device, with $\hat{\mathbf{h}}(\omega)$ being some auxiliary bosonic ("two-vector") operator, the input-output relation (2) can then be extended to the four-dimensional transformation

$$\hat{\boldsymbol{\beta}}(\omega) = \boldsymbol{\beta}\beta(\omega)\hat{\boldsymbol{\alpha}}(\omega) \tag{3}$$

with $\boldsymbol{\Lambda}(\omega)\mathbf{J}\boldsymbol{\Lambda}^{+}(\omega) = \mathbf{J}$, $\mathbf{J} = \mathrm{diag}(\mathbf{I}, \sigma\mathbf{I})$. Thus, $\boldsymbol{\Lambda}(\omega) \in$ SU(4) (for absorbing devices) or $\boldsymbol{\Lambda}(\omega) \in$ SU(2,2) (for amplifying devices). Using the (commuting) positive Hermitian matrices $\mathbf{C}(\omega) = \sqrt{\mathbf{T}(\omega)\mathbf{T}^{+}(\omega)}$, $\mathbf{S}(\omega) = \sqrt{\mathbf{A}(\omega)\mathbf{A}^{+}(\omega)}$ the matrix $\boldsymbol{\Lambda}(\omega)$ can be written in the form [3]

$$\boldsymbol{\Lambda}(\omega) = \begin{pmatrix} \mathbf{T}(\omega) & \mathbf{A}(\omega) \\ -\sigma\mathbf{S}(\omega)\mathbf{C}^{-1}(\omega)\mathbf{T}(\omega) & \mathbf{C}(\omega)\mathbf{S}^{-1}(\omega)\mathbf{A}(\omega) \end{pmatrix}. \tag{4}$$

It can then be shown that the density operator of the outgoing field is

$$\hat{\varrho}_{\text{out}}^{(\text{F})} = \mathrm{Tr}^{(\text{D})}\{\hat{\varrho}_{\text{in}}[\mathbf{J}\boldsymbol{\Lambda}^{+}(\omega)\mathbf{J}\hat{\boldsymbol{\alpha}}(\omega), \mathbf{J}\boldsymbol{\Lambda}^{\text{T}}(\omega)\mathbf{J}\hat{\boldsymbol{\alpha}}^{\dagger}(\omega)]\}, \tag{5}$$

where $\mathrm{Tr}^{(\text{D})}$ means trace with respect to the device. Equivalently, the Wigner function transforms as $W_{\text{out}}[\boldsymbol{\alpha}(\omega)] = W_{\text{in}}[\mathbf{J}\boldsymbol{\Lambda}^{+}(\omega)\mathbf{J}\,\boldsymbol{\alpha}(\omega)]$.

Now let us make first a general remark on entanglement transformation. One of the requirements to be satisfied by any entanglement measure E is that a CP map does not increase entanglement [5]. Looking at Eq. (5), we see that a quantum-state transformation is in fact a CP map for both absorbing and amplifying four-port devices because in both cases an ancilla (the device) is coupled to the Hilbert space of the field, a unitary transformation is performed in the product space $\mathcal{H}_{\text{field}} \otimes \mathcal{H}_{\text{device}}$, and the trace with respect to the device is taken at the end. An obvious consequence is that amplification does not help to increase entanglement.

Now we turn to the problem of transmitting two light beams prepared in an entangled quantum state through absorbing optical fibers (Fig. 1) represented by their transmission coefficients T_i. As an example let us consider the two types of Bell states $|\Psi^{\pm}\rangle = \frac{1}{\sqrt{2}}(|01\rangle \pm |10\rangle)$, $|\Phi^{\pm}\rangle = \frac{1}{\sqrt{2}}(|00\rangle \pm |11\rangle)$. Applying Eq. (5), after some algebra we get [4]

$$\hat{\varrho}_{\text{out}}^{(\text{F})}(|\Psi^{\pm}\rangle) = \tfrac{1}{2}[(2-|T_1|^2-|T_2|^2)|00\rangle\langle 00|] \\ +\tfrac{1}{2}(T_2|01\rangle \pm T_1|10\rangle)(T_2^*\langle 01| \pm T_1^*\langle 10|), \tag{6}$$

$$\hat{\varrho}_{\text{out}}^{(\text{F})}(|\Phi^{\pm}\rangle) = \tfrac{1}{2}[(1-|T_1|^2)(1-|T_2|^2)|00\rangle\langle 00| + |T_1|^2(1-|T_2|^2)|10\rangle\langle 10|+ \\ |T_2|^2(1-|T_1|^2)|01\rangle\langle 01|] + \tfrac{1}{2}(|00\rangle \pm T_1T_2|11\rangle)(\langle 00| \pm (T_1T_2)^*\langle 11|). \tag{7}$$

The density matrices (6) and (7) have been written as a sum of separable states and a single pure state such that the convexity property (see, e.g., [5]) $E[\lambda\hat{\varrho}_1 + (1-\lambda)\hat{\varrho}_2] \leq \lambda E(\hat{\varrho}_1) + (1-\lambda)E(\hat{\varrho}_2)$ can readily be used where the entanglement of a pure state is given by its reduced von Neumann entropy. The inequality reduces for equal fibers to

$$E[\hat{\varrho}_{\text{out}}^{(\text{F})}(|\Psi^{\pm}\rangle)] \leq |T|^2 \ln 2, \tag{8}$$

$$E[\hat{\varrho}_{\text{out}}^{(\text{F})}(|\Phi^{\pm}\rangle)] \leq \tfrac{1}{2}[(1+|T|^2)\ln(1+|T|^2) - |T|^2 \ln|T|^2]. \tag{9}$$

The numerically calculated relative entropies of the output quantum states (6) and (7) are shown in Fig. 2 for equal transmission coefficients of the fibers satisfying the Lambert-Beer law $|T| = e^{-l/L}$ with l and $L = c/(n_{\text{I}}\omega)$ being the propagation length through the fibers and the absorption length of the fibers with complex refractive index $n(\omega) = n_{\text{R}}(\omega) + in_{\text{I}}(\omega)$. It is seen that the entanglement of the states $\hat{\varrho}_{\text{out}}^{(\text{F})}(|\Phi^{\pm}\rangle)$ decays considerably faster than that of the states $\hat{\varrho}_{\text{out}}^{(\text{F})}(|\Psi^{\pm}\rangle)$, which can be understood from the argument that in the former case absorption acts on both photons simultaneously.

As a second example we consider the two-mode squeezed vacuum $|\text{TMSV}\rangle = \exp[\zeta(\hat{a}_1^{\dagger}\hat{a}_2^{\dagger} - \hat{a}_1\hat{a}_2)]\,|00\rangle = \sqrt{1-q^2}\sum q^n|nn\rangle$ with $q = \tanh\zeta$, which can be used in continuous-variable quantum teleportation [6]. Being difficult to follow the lines presented above for the Bell states, it is instructive to apply the separability criterion in [7] to this class of states. Since the criterion is based on properties of the Wigner function, we use the transformation of the Wigner function to obtain a relation for the bound between separability and inseparability. Assuming again equal optical fibers with some thermal photon

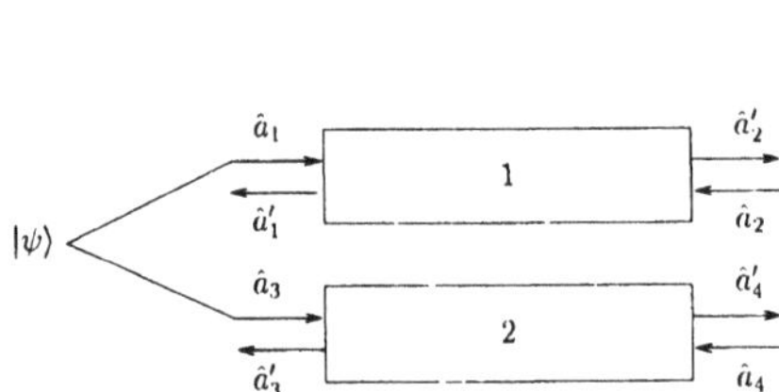

Figure 1 A two-mode input field in the state $|\psi\rangle$ is transmitted through two absorbing dielectric four-port devices with transmission coefficients T_1 and T_2.

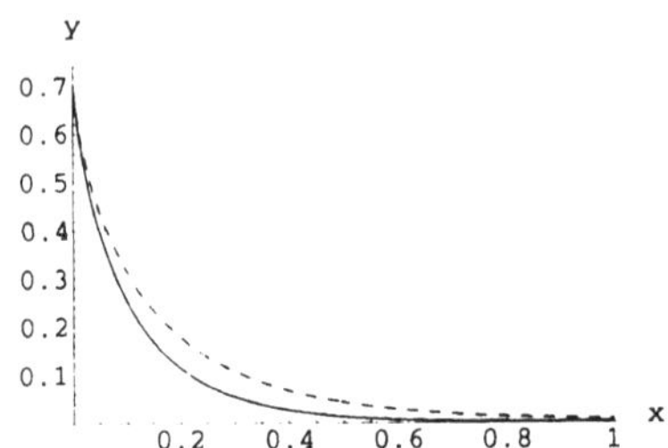

Figure 2 Entanglement degradation of Bell states $|\Phi^{\pm}\rangle$ (full curve) and $|\Psi^{\pm}\rangle$ (dashed curve).

number $n_{\rm th}$, we obtain [8]

$$n_{\rm th} \geq \frac{(1-\sigma)(1-|R|^2)+|T|^2(\sigma-\exp[-2|\zeta|])}{2\sigma(1-|R|^2-|T|^2)}\,. \tag{10}$$

Specifically in the absorbing case ($\sigma=+1$) we arrive at the formula for the maximal fiber length $l_{\rm max}$ after which the TMSV is still nonseparable

$$l_{\rm max}/L = \tfrac{1}{2}\ln\left[1+(1-\exp[-2|\zeta|])/2n_{\rm th}\right], \tag{11}$$

where we have set $R=0$ and, according to the Lambert-Beer law, $|T| = \exp[-l/L]$, with L being again the absorption length. On the other hand, from Eq. (10) it follows that for an amplifier in the low-temperature limit ($n_{\rm th}=0$) the maximal gain $g_{\rm max}=|T_{\rm max}|^2-1$ is given (for $R=0$) by $|T_{\rm max}|^2-1 = \tanh|\zeta| = |q|$. It essentially says that an amplifier that doubles the intensity of a signal destroys any initially given entanglement.

References

[1] L. Knöll, S. Scheel, and D.-G. Welsch, *QED in dispersing and absorbing media*, in *Coherence and Statistics of Photons and Atoms*, ed. J. Peřina (Wiley, to be published), arXiv: *quant-ph/0006121*.

[2] T. Gruner and D.-G. Welsch, Phys. Rev. A **54**, 1661 (1996).

[3] L. Knöll *et al.*, 4716 (1999).

[4] S. Scheel *et al.*, Phys. Rev. A **62**, 43803 (2000).

[5] V. Vedral and M.B. Plenio, Phys. Rev. A **57**, 1619 (1998); M. Horodecki, P. Horodecki, and R. Horodecki, Phys. Rev. Lett. **84**, 2014 (2000).

[6] S.L. Braunstein and H.J. Kimble, Phys. Rev. Lett. **80**, 869 (1998).

[7] L.-M. Duan, G. Giedke, J.I. Cirac, and P. Zoller, Phys. Rev. Lett. **84**, 2722 (2000); R. Simon, Phys. Rev. Lett. **84**, 2726 (2000).

[8] S. Scheel, T. Opatrný, and D.-G. Welsch, Contribution to Int. Conf. on Quantum Optics 2000, Raubichi, arXiv: *quant-ph/0006026*.

QUANTUM NOISE IN POLARIZATION MEASUREMENT AND POLARIZATION ENTANGLEMENT

Alexander S. Shumovsky
Physics Department, Bilkent University, Bilkent
Ankara, 06533 Turkey
shumo@fen.bilkent.edu.tr

Keywords: Zero-point oscillations, entanglement, polarization, nondemolition measurements

Abstract It is shown that the vacuum noise of electromagnetic field is concentrated near atoms. This effect can strongly influence the quantum measurements in the engineered entanglement in the system of trapped Ridberg atoms.
Keywords: Zero-point oscillations, entanglement, polarization, nondemolition measurements

The two main objectives of this note are on the one hand to show that the vacuum noise of electromagnetic field is concentrated near atoms and on the other hand to discuss how this noise can influence the precision of measurements in polarization entanglement in the system of trapped atoms and how to avoid it using a nondemolition quantum polarization measurement. The result is important for engineered entanglement in atomic systems as well as for experiments with single-atom masers (see [1, 2] and references therein).

It is well known that the field strengths continue to oscillate in the vacuum state (e.g., see [3]). These zero-point oscillations are usually described in terms of the vacuum contribution into the energy of free field

$$H^{(vac)}_{(plane)} = \sum_{k,\sigma} \hbar\omega_k/2 = \sum_k \hbar\omega_k, \tag{1}$$

in the case of plane waves of photons with wave number k and polarization $\sigma = 1, 2$ [4]. At the same time, it is well known that the atoms emit the multipole photons represented by the quantized spherical waves rather than plane waves [5]. In this case, the vacuum state energy is

$$H^{(vac)}_{(multi)} = \sum_k \sum_{\lambda,j,m} \hbar\omega_k/2 = \sum_k \left(\sum_{j=1}^{\infty} (2j+1)\hbar\omega_k \right). \tag{2}$$

Here $\lambda = E, M$ labels the type of field (either electric or magnetic), j is the angular momentum of photons, and $m = -j, \cdots j$. At first sight, the equations (1) and (2) are equivalent because both give infinite energy of the vacuum state. In fact, this infinity is inessential because any measurement implies an averaging over finite "volume of detection" and exposition time [3, 4]. This averaging is nothing but filtration leading to separation of a finite transmission frequency band in which the expressions (1) and (2) can be compared. It is seen that (2) always exceeds (1) because even in the simplest case of dipole radiation $H^{(vac)}_{(multi)}/H^{(vac)}_{(plane)} = 3$. Employing the spatial properties of multipole radiation, we can get a more interesting and important result. The point is that the multipole field is described by the positive frequency part of the operator vector potential with the components [6]

$$A^{(+)}_{\mu}(\vec{r}) = \sum_k \sum_{\lambda,j,m} V_{k\lambda jm\mu}(\vec{r}) a_{k\lambda jm}, \tag{3}$$

in the so-called circular polarization basis $\vec{e}_{\mu}$. The mode function $V(\cdot)$ here is represented by a certain combination of spherical Bessel functions, Clebsch-Gordon coefficients, and spherical harmonics. Then, the energy density of vacuum fluctuations can be defined by the following commutator

$$W^{(vac)}_{(multi)} = \sum_{\mu} [A^{(+)}_{\mu}, (A^{(+)}_{\mu})^{+}] = \sum_k \sum_{\lambda,j,m,\mu} |V_{k\lambda jm\mu}(\vec{r})|^2. \tag{4}$$

In view of the properties of spherical Bessel functions and spherical harmonics, it is easily seen that the vacuum noise (4) is independent of the angular variables θ and ϕ and concentrates near a local source (atom) while vanishes at $kr \gg 1$. More detailed investigation shows that (4) can strongly exceed the level (1) inside the region of the order of $r_0 \sim \lambda/3$, surrounding the atom. Here λ is the wave length of radiation.

Since the typical value of the interatomic distances in the experiments with trapped Ridberg atoms is of the same order, the above effect can strongly influence the quantum limit of precision of measurements in these experiments [1]. Consider as an example the two-atom Hertz-type experiment. The two identical atoms are separated by a distance d. The first atom is initially excited (source), while the second is in the ground state (detector). In the wave picture, the measurement can be represented in terms of a superposition of outgoing and incoming spherical waves focused on the source and detector respectively. Since the effect of concentration of the vacuum noise is valid for the outgoing, incoming, and standing spherical waves of photons, then there is an overlap of the vacuum fluctuations caused by the source and detector at $d \sim r_0$. It should lead to a strong change for the worse of the quantum limit of precision of measurement.

One of the promising ways in the engineered entanglement is represented by the so-called two-photon polarization entanglement (see Sec. 12.14 in [3]). In this case, a cascade decay of an atomic transition leads to the creation of two entangled photons with opposite helicities. The polarization of multipole atomic radiation is described by a (3×3) operator polarization matrix with the elements [7, 8]

$$P_{\mu,\mu'}(\vec{r}) = A_{\mu}^{(-)}(\vec{r}) A_{\mu'}^{(+)}(\vec{r}). \tag{5}$$

The vacuum fluctuations of the elements of (5) are described by the commutators $[A_{\mu}^{(+)}(\vec{r}), A_{\mu'}^{(-)}(\vec{r})]$, similar to (4). Therefore, the effect of concentration of vacuum noise near atoms seems to be important for the polarization measurements as well.

The standard polarization measurement implies the use of detection of the field variables done by photodetecting techniques, which are field destructive. As a result, successive measurements of the field variables yield different results. It seems to be quite tempting to use the measurement schemes avoiding back action of the detecting device on the detected variable. Such a nondemolition measurement can be done for the polarization through the use of the optical Aharonov-Bohm effect proposed in [9]. The geometry of the experiment assumes that the magnetic field oscillates along the axis orthogonal to the plane of a conducting loop used as a quantum interferometer. Then, the oscillations of conductance in the loop provide the measurement of the radial linearly polarized component of the multipole radiation. The estimation shows that the loop of diameter $\sim 1\mu m$ at temperature $\leq 1K$ can be used to measure the field soliton with approximately $1/\alpha \sim 137$ photons ($\alpha = e^2/\hbar c$) which corresponds to the oscillating power $\sim 10^{-3}w$. The radio-band superradiance represents a good local source for such a measurement [10].

References

[1] S. Haroche. (1999). *AIP Conf. Proc.* Vol. **464**, Issue 1, p. 45.

[2] M. Weidinger, B.T.H. Varcoe, R. Heerlein, and H. Walther. (1999). *Phys. Rev. Lett.* **82**, 3795.

[3] L. Mandel and E. Wolf. (1995). *Optical Coherence and Quantum Optics* (Cambridge University Press, New York).

[4] W.H. Louisell (1964). *Radiation and Noise in Quantum Electronics* (McGrow-Hill, New York).

[5] C. Cohen-Tannouji, J. Dupont-Roc, and G. Grinberg (1992). *Atom-Photon Interaction* (Wiley, New York).

[6] W. Heitler (1954). *The Quantum Theory of Radiation* (Oxford University Press, New York).

[7] A.S. Shumovsky and Ö.E. Müstecaplıoğlu (1998). *Phys. Rev. Lett.* **80**, 1202.

[8] A.S. Shumovsky and Müstecaplıoğlu (1998). *Optics Commun.* **146**, 124.

[9] I.O Kulik and A.S. Shumovsky (1996). *Appl. Phys. Lett.* **69**, 2779.

[10] Yu. Kiselev, A. Prutkoglyad, A. Shumovsky, and V. Yukalov (1998). *Sov. Phys. JETP* **94**, 344.

BRIGHT EPR-ENTANGLED BEAMS FOR QUANTUM COMMUNICATION

Ch. Silberhorn, P. K. Lam, N. Korolkova, G. Leuchs
Zentrum für Moderne Optik, Universität Erlangen - Nürnberg, Germany
korolkova@physik.uni-erlangen.de

Keywords: continuous EPR correlation, squeezing, Kerr nonlinearity, conditional variance

Abstract Bright Einstein-Podolsky-Rosen (EPR) entanglement is obtained in linear interference of two amplitude-squeezed beams produced in Kerr nonlinear interactions. The correlation of the amplitude (phase) quadratures are measured to be $4.0 \pm 0.2(4.0 \pm 0.4)$ dB below the quantum noise limit. Continuous variable EPR-paradox was demonstrated with measured product of the conditional variances of 0.64 ± 0.08, well below the quantum limit of unity.

We report on the generation of bright entangled beams with potential applications in quantum communication over continuous variables. This is an alternative approach which avoids some of the deficiencies of quantum information with discrete variables, i.e. low efficiency and the need for a nonlinear coupling between different modes. We use optical soliton pulses at the telecommunication wavelength in silica fibers as a reliable, efficient, and application-ready source of continuous variable entanglement. Optical solitons in fibers are stable particle-like pulses, propagating over large distances without distortion and exhibiting a complex internal quantum structure. In our experiment, an asymmetric fiber-Sagnac interferometer serves as a two-in-one squeezer for solitons delivering two independent amplitude-squeezed pulse trains (Fig. 1). EPR-like quantum correlations are generated by interference of these two amplitude-squeezed beams at a 50:50 beam splitter [1]. The amplitude quantum correlations were measured in direct detection and an indirect experimental test of the phase correlations was performed [2].

The experimental setup is as shown in Figure 1. A passively mode-locked Cr^{4+} : YAG laser is used to produce optical pulses at a centre wavelength of 1505 nm with a repetition rate of 163 MHz. The maximum average power of the laser is around 95 mW and the pulses have a bandwidth limited sech-shape with a FWHM of 130 fs. These pulses are injected into an asymmetric fibre Sagnac interferometer. The Sagnac loop consists of an 8 m long polarization maintaining fibre and a highly asymmetric beam splitter. The fibre is birefringent, thus supporting the s- and p- polarization states with negligible cross-talk. A strong and a weak counterpropagating pulses propagate within the Sagnac loop for each polarization. For certain input energies, the orthogonally polarized output beams $\hat{s}$ and $\hat{p}$ are amplitude squeezed [3, 4]. To attain a linear interference of these two squeezed pulse trains, the appropriate delay is used to compensate for the different propagation time along the fast and slow axes of the fibre and a $\lambda/2$ plate is introduced to obtain two optically coherent beams of the same polarization.

To model a superposition on the beam splitter, the classical parts of the input fields are both taken to be real with no relative phase difference. The beam splitter in Fig. 1 is chosen to introduce a phase shift of 90 ° for each reflection. This arrangement provides the optimal EPR-correlation between the output beams [1]. To detect the anti-correlation of the amplitude quadratures $\hat{X}_j$, the variance of the sum photo current $\delta X_a + \delta X_b$ have to be experimentally determined. For phase correlation, one measures the variance of difference photo current $\delta Y_a - \delta Y_b$, where Y_j denotes the phase quadratures.

The experimental data for the amplitude anti-correlation are recorded in direct detection [2]. The noise powers of the sum and difference photo currents and those of the single beams were measured as a function of the input power of the Sagnac interferometer (Fig. 2). The noise power of the sum photo currents was observed to drop up to 4 dB below the shot noise level, the quantum limit for the anti-correlation. As expected this quantum anti-correlation de-

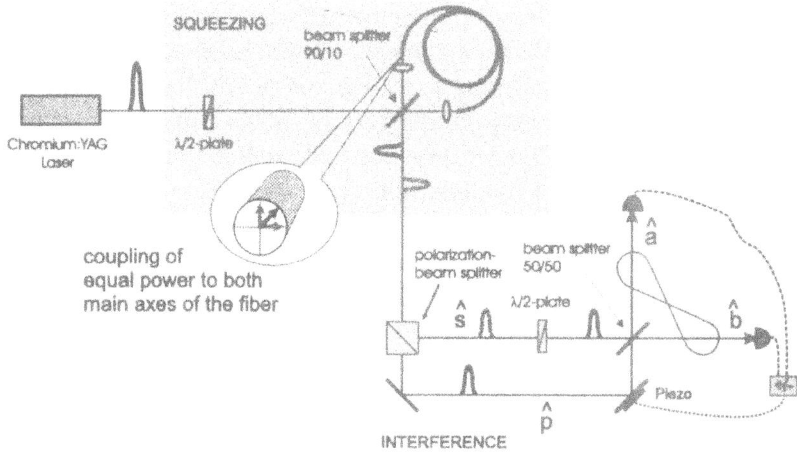

Figure 1 Schematic of the experimental setup. Insert shows the polarization direction of the input beam to get two squeezed output beams $\hat{s}$ and $\hat{p}$ of orthogonal polarization. Their interference produces the EPR entangled output beams $\hat{a}$ and $\hat{b}$. $\lambda/2$ is a half-wave plate.

pends on the squeezing of the input beams, as seen by the comparison of top and bottom graphs of Fig. 2.

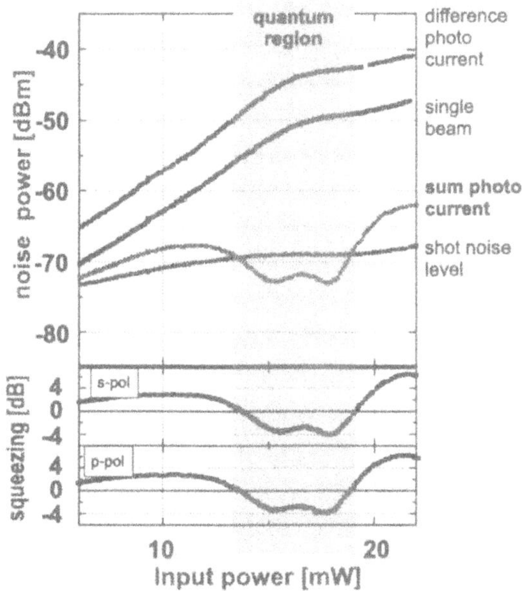

Figure 2 Amplitude anti-correlation of the output beams $\hat{a}$ and $\hat{b}$. In the quantum region where the noise power of the sum photo current drops below the shot noise the beams are quantum entangled. The respective squeezing curves of interfering beams $\hat{s}$ and $\hat{p}$ are plotted in the bottom part of the plot.

For the complete characterization of the EPR-entanglement the correlation of the phase quadrature was measured in direct detection as shown in Figure 3. The experiment described above is now assumed to be a black box with two output beams. These beams interfere at another 50:50 beam splitter. The phase correlation can be inferred from the amount of amplitude squeezing of $\hat{c}$ and $\hat{d}$, provided the variance of the sum of the amplitudes of beams $\hat{a}$ and $\hat{b}$ is already known and is substantially below shot noise. The relevant noise power was recorded by doubly balanced heterodyne detection whereas the relative phase of interference was scanned. The best results for the phase correlation reached up to 4 dB.

The notion of continuous EPR-like correlations of the amplitude and phase quadratures is related to the demonstration of the EPR-paradox for continuous variables [5]. If two spatially separated beams are correlated in their amplitude $\hat{X}_j$ and phase $\hat{Y}_j$ quadratures, it is possible to infer $\hat{X}_2$ by a measurement of $\hat{X}_1$. Due to the finite degree of correlation there will always be a limit in the accuracy of this inference. This was originally referred to as an inference

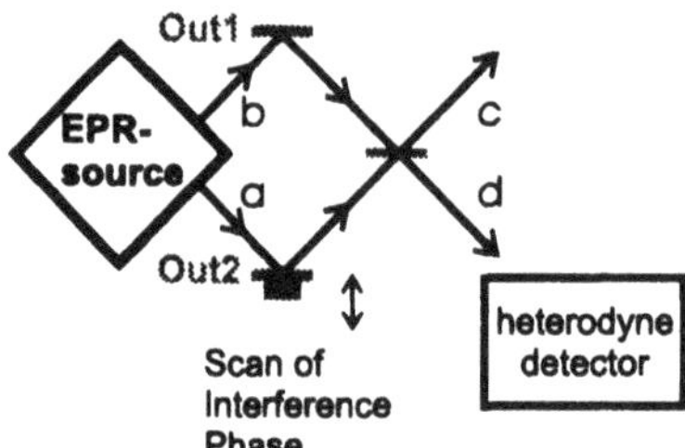

Figure 3 Schematic for the indirect interrogation of the phase quadrature correlation.

error [5] which can be minimized by introducing a variable gain g:

$$\Delta^2_{inf}X_1 = \langle(\delta X_1 \pm g_X \delta X_2)^2\rangle, \;\; \Delta^2_{inf}Y_1 = \langle(\delta Y_1 \mp g_Y \delta Y_2)^2\rangle \tag{1}$$

where the relevant variances are normalized to the shot noise level of a single beam. The demonstration of the EPR-paradox requires simultaneous non-classical values of these inference errors (1), which for the optimal gains correspond to non-classical respective conditional variances [5, 6]:

$$\Delta^2_{cond}X_{12} = \Delta^2_{inf}X_1 < 1, \qquad \Delta^2_{cond}Y_{12} = \Delta^2_{inf}Y_1 < 1; \tag{2}$$

This specifies the ability to infer "at a distance" either of the two non-commuting signal observables with a precision below the vacuum noise level of the signal beam [5]. For the bright entanglement source reported in this contribution the optimal gain is $g_{X,Y} = 1$ and the product of conditional variances (2) is equal to $\Delta^2_{cond}X_{ab}\, \Delta^2_{cond}Y_{ab} = 0.64 \pm 0.08$, well below the quantum limit. Hence the EPR entanglement of $\hat{a}$ and $\hat{b}$ is established. This bright entangled quantum state can be used as the basic building block for the implementation of quantum key distribution, quantum teleportation and long distance quantum communication.

References

[1] G. Leuchs, T. C. Ralph, C. Silberhorn, and N. Korolkova, *J. Mod. Opt.* **46,** 1927 (1999).

[2] Ch. Silberhorn, P. K. Lam, O. Weiß, F. König, N. Korolkova, and G. Leuchs, *Phys. Rev. Lett.,* submitted (2000).

[3] S. Schmitt, J. Ficker, M. Wolff, F. König, A. Sizmann, and G. Leuchs, *Phys. Rev. Lett.* **81,** 2446 (1998).

[4] N. Korolkova *et al. Nonlinear optics,* in print (2000).

[5] M. D. Reid, *Phys. Rev. A* **40,** 913 (1989).

[6] J.-P. Poizat, J.-F. Roch, and P. Grangier, *Ann. Phys. Fr.* **19,** 265 (1994).

POLARIZATION IN QUANTUM OPTICS: A NEW FORMALISM AND TWO EXPERIMENTS

T. Tsegaye, P. Usachev, J. Söderholm, A. Trifonov, G. Björk
Department of Electronics, Royal Institute of Technology (KTH)
Electrum 229, SE-164 40 Kista, Sweden
gunnarb@ele.kth.se

M. Atatüre, M. C. Teich
Quantum Imaging Laboratory, Department of Physics, Boston University
8 Saint Mary's Street, Boston, Massachusetts 02215 USA

A. V. Sergienko, B. E. A. Saleh
Quantum Imaging Laboratory, Department of Electrical and Computer Engineering
Boston University, 8 Saint Mary's Street, Boston, Massachusetts 02215 U.S.A.

Abstract We report on two quantum polarization experiments. The first one shows that three geometrical rotations can in principle be distinguished with certainty using two photons. The second experiment shows quantum polarization properties of classically unpolarized light. We outline a new treatment of polarization in quantum optics. Our formalism differs from the usual quantum description, which uses the Stokes operators.

1. THREE MUTUALLY ORTHOGONAL POLARIZATION STATES OF LIGHT

Using photon pairs, we have generated a state that is transformed into mutually orthogonal states when considered in a basis rotated by 60 or 120 degrees. Using such a state one could, in principle, distinguish three different rotation angles with certainty. Since the transformations describing geometric rotations of the basis are lossless, the treatment of quantum polarization properties is naturally divided into the different excitation manifolds. This is closely related to the quantum theory of the relative phase of two bosonic modes [1].

Quantum Communication, Computing, and Measurement 3
Edited by P. Tombesi and O. Hirota, Kluwer Academic/Plenum Publishers, 2001

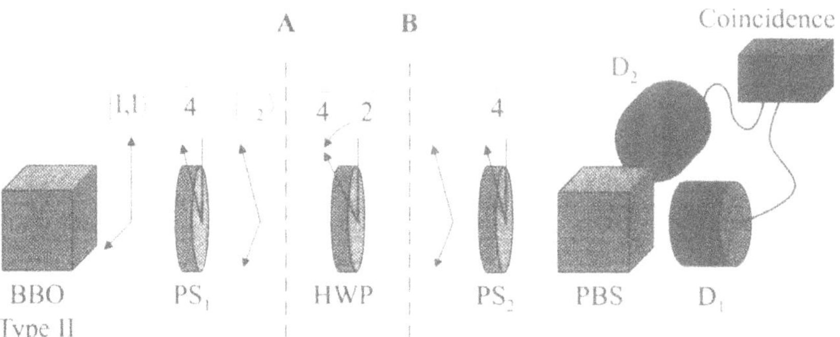

Figure 1 A simplified sketch of the experimental setup.

A simplified sketch of the experiment, which was performed in Boston, is shown in Fig. 1. Photon pairs consisting of colinearly propagating 702.2-nm photons in the horizontal-vertical basis were generated by SPDC of type II in a BBO crystal of length 0.5 mm. The pump source was a continuous-wave, 351.1 nm wavelength Argon-ion laser. As seen in Fig. 1, the state

$$|\zeta_2\rangle = \frac{i+\sqrt{2}}{\sqrt{6}}\,|0,2\rangle + \frac{i-\sqrt{2}}{\sqrt{6}}\,|2,0\rangle \tag{1}$$

is generated at A in the basis oriented at an angle 45 deg to the horizontal-vertical one by applying a phase shift (PS_1) of $-\arccos(1/3)$ in the rotated basis. The inverse transformation is realized on the right-hand side of B using a phase shift (PS_2) of $\arccos(1/3)$ oriented in parallel with PS_1. The state $|\zeta_2\rangle$ in the rotated basis at B will therefore be projected on the state $|1,1\rangle$ with one horizontally and one vertically polarized photon. Using a polarization beam splitter (PBS) these photons are detected by the actively quenched EG&G single-photon detectors D_1 and D_2, respectively. As the probability for generating more than two photons in the detected modes is very low, a simultaneous photon detection of D_1 and D_2, that is a coincidence count, can be seen as a successful projection onto the state $|\zeta_2\rangle$.

Instead of performing a geometric rotation of the generated state compared to the detected one, we rotated a half-wave plate (HWP) placed between A and B. In the considered case, the rotation of the half-wave plate by $\vartheta/2$ can be shown to result in an equivalent transformation as a geometric rotation by ϑ. The theoretical expression for the probability of projecting the state $|\zeta_2\rangle$ rotated by ϑ onto the state $|\zeta_2\rangle$ is

$$\left|\langle\zeta_2|e^{\vartheta(\hat{a}\hat{b}^\dagger-\hat{a}^\dagger\hat{b})}|\zeta_2\rangle\right|^2 = \frac{(1+2\cos 2\vartheta)^2}{9}, \tag{2}$$

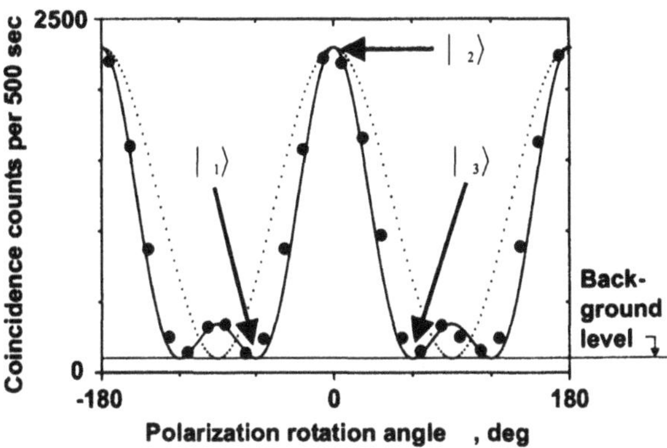

Figure 2 The measured coincidence counts (•) as a function of the rotation angle ϑ. The solid line is a fit of the theoretical form (2) apart from the background level. The sinusoidal dotted line shows the behavior expected from the classical theory. The states $|\zeta_1\rangle$, $|\zeta_2\rangle$ and $|\zeta_3\rangle$ are found to be mutually orthogonal.

where $\hat{a}$ ($\hat{b}$) is the annihilation operator of the mode corresponding to the left (right) index of the Dirac kets. The measured data shown in Fig. 2 is seen to agree with the theoretical curve except for a bias due to dark counts of the detectors. This is analogous to what we found for a similar experiment for the relative phase [2]. The expression (2) shows that $|\zeta_2\rangle$ is transformed into orthogonal states by rotating it by ± 60 deg. These three states ($|\zeta_1\rangle$, $|\zeta_2\rangle$ and $|\zeta_3\rangle$) can be taken as eigenstates in the second excitation manifold of the rotation-angle operator. Apart from the π-periodicity, this operator $\hat{\Theta}$ takes the same form as the relative phase operator [1]

$$\hat{\Theta} = \sum_{N=0}^{\infty} \sum_{k=0}^{N} \frac{\pi k}{N+1} |\zeta_k^{(N)}\rangle\langle\zeta_k^{(N)}|, \tag{3}$$

where N is the total photon number of the state, and

$$|\zeta_k^{(N)}\rangle = e^{\pi k(\hat{a}\hat{b}^\dagger - \hat{a}^\dagger\hat{b})/(N+1)}|\zeta^{(N)}\rangle. \tag{4}$$

Here $|\zeta^{(N)}\rangle$ is a superposition of all the N-photon eigenstates of the generator $\hat{a}\hat{b}^\dagger - \hat{a}^\dagger\hat{b}$ equally weighted.

2. QUANTUM POLARIZATION PROPERTIES OF A CLASSICALLY UNPOLARIZED STATE

In our second experiment, which was performed in Kista, we again used photon pairs generated by SPDC of type II in a BBO crystal. We also used the same type of detection as before. That is, the horizontally and vertically polarized modes were spatially separated by using a polarization beam splitter and subsequently impinged on different detectors.

The generated state $|1,1\rangle$ was imposed to pass through a half-wave plate or a quarter-wave plate. First the intensity of one polarization was measured as the phase plates were rotated around the direction of propagation. Since

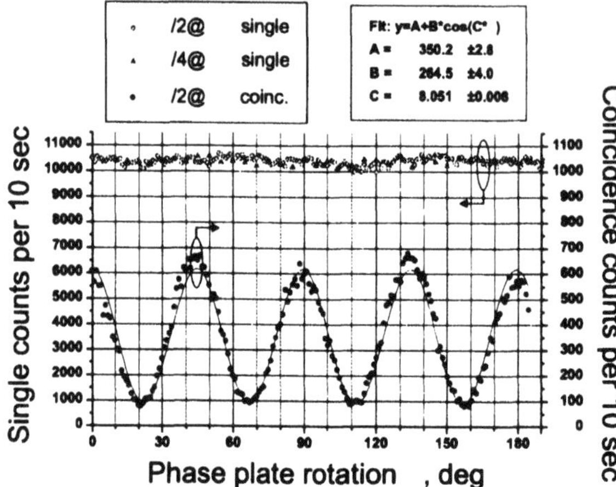

Figure 3 The constant intensities (at the top) found in one linearly polarized mode as the different phase plates were rotated. The coincidence counts (•) as a function of the rotation angle ϑ of the half-wave plate. The sinusoidal curve fit corresponds to a visibility of 76%.

the quantum efficiency of the detectors was only about 5%, the probability of obtaining a click from one detector was linear in the impinging intensity. Therefore we did not have to consider the probability of the presence of two photons in the same polarization.

As seen in Fig. 3, the measured intensity was found to be constant for both the half-wave plate and the quarter-wave plate. Using the Poincaré-sphere representation, the rotations of the phase plates are found to be equivalent to rotations of the Poincaré sphere. According to the classical theory, the first measurement, employing a half-wave plate, is easily seen to imply that the considered state lies on the s_y-axis. Together with the second measurement this then implies that the considered state is located at the origin, and therefore unpolarized. In spite of that, interference fringes were obtained as the half-wave plate was rotated and the coincidences were measured (see Fig. 3). This shows that the state $|1,1\rangle$ has some quantum polarization properties even though it is unpolarized according to the classical theory. This is one result of the fact that the classical theory just put restrictions on the second-order correlation functions for a unpolarized state, whereas quantum theory puts restrictions on higher-order correlation functions too [3].

References

[1] A. Luis and L. L. Sánchez-Soto, "Phase-difference operator," Phys. Rev. A **48**, 4702 (1993).

[2] A. Trifonov *et al.*, "Experimental demonstration of the relative phase operator," J. Opt. B **2**, 105 (2000).

[3] G. S. Agarwal, "On the State of Unpolarized Radiation," Lett. Nuovo Cimento **1**, 53 (1971).

SPIN SQUEEZING AND DECOHERENCE LIMIT IN RAMSEY SPECTROSCOPY

Even sub-optimal entanglement can achieve absolute improvement

Duger Ulam-Orgikh, Masahiro Kitagawa
Graduate School of Engineering Science, Osaka University
1-3 Machikaneyama-cho, Toyonaka, Osaka 560-8531 Japan
CREST, Japan Science and Technology Corporation
uka@laser.ee.es.osaka-u.ac.jp, kit@qc.ee.es.osaka-u.ac.jp

Keywords: spin squeezing, optimal entanglement, Ramsey spectroscopy, sub-quantum-limit phase measurement, quantum circuit

Abstract It is known that partially entangled state gives improved sensitivity of Ramsey spectroscopy in the presence of decoherence, whereas maximally entangled state gives no improvement. However it has been an open question whether the absolute limit in the improvement is attainable by optimal entanglement alone or via measurement optimization. In this Paper, we answer this question by showing that even sub-optimal entanglement generated by the simplest spin squeezing can asymptotically attain the absolute improvement with increasing number of atoms.

Let's suppose we are given a Ramsey spectroscopy with fixed N ions in a trap and do an experiment for a duration of T. The problem on determining the maximum possible precision in frequency is studied here. Several articles have been devoted to the problem and the following results have been obtained . By using maximally entangled input state one can get the possible best precision of $\frac{1}{NT}$ [Bollinger et al., 1996] in the absence of decoherence, but in the presence of decoherence with a rate γ, the uncorrelated and maximally entangled input states do provide the same precision $|\delta\omega_0| = \sqrt{\frac{2\gamma e}{NT}}$. But, as it was numerically shown, a high symmetric partially entangled state gives some relative improvement in precision for a generalized Ramsey spectroscopy [Huelga et al., 1997].

Our contribution is to answer to an open question whether the improvement saturates at theoretical decoherence limit or below for the generalized Ramsey spectroscopy, using a spin squeezed input state , which allows us an analytical solution. We will use a simple spin-squeezing model of one-axis twisting and show that even this sub-optimal entangled state provides saturation at theoretical limit with increasing number of N . The concept of spin squeezing in the sense of quantum correlation was established in [Kitagawa and Ueda, 1993] and squeezing models of one-axis twisting and two-axis twisting, which respectively reduce a quantum noise down to the order of $N^{1/3}$ and $\frac{1}{2}$, were studied.

We will use a spin-representation of two-level atoms for Ramsey spectroscopy. After applying the first Ramsey pulse $R_y(\pi/2)$ to an initial $|0\rangle^{\otimes N}$ state, it will be squeezed via one axis twisting with Hamiltonian $H_{ss} = \chi S_z^2, \quad S_z = \frac{1}{2}\sum_{i=1}^{N}\sigma_{zi}$, where σ_{zi} is a Pauli operator for ith ion and $\hbar = 1$. To get the maximum squeezing in a desired direction we need another $R_x(-\nu)$ rotation pulse and the initial squeezed state becomes of the form

$$|\Psi_{ss}\rangle = R_x(-\nu)\exp(-iH_{ss}t)R_y(\pi/2)|0\rangle^{\otimes N} = \sum_{l=0}^{N} c_l(\mu,\nu)|\varphi_l\rangle,$$

where $\mu = 2\chi t$ is a squeezing parameter, $|\varphi_l\rangle = \sum_{n_1<n_2<\cdots<n_l}|2^{n_1} + 2^{n_2} + \cdots + 2^{n_l}\rangle$, $n_i = 0,1,2,\ldots,N-1$, is a permutation symmetric state with l being the number of 1's and the coefficient is found as

$$\begin{aligned} c_l(\mu,\nu) &= \frac{1}{2^{N/2}}\binom{N}{l}\sum_{M=0}^{N}\sum_{k=0}^{l}\exp\left[i\frac{\mu}{8}(N-2M)^2\right] \\ &\times \binom{N}{M}\binom{M}{k}\binom{N-M}{l-k}\left(\cos\frac{\nu}{2}\right)^{M+l-2k}\left(-i\sin\frac{\nu}{2}\right)^{N-M-l+2k}. \end{aligned}$$

The state $|\Psi_{ss}\rangle$ is also symmetric under exchange of the ground and excited states for each ion, since $c_l(\mu,\nu) = c_{N-l}(\mu,\nu)$. Therefore, a partially entangled state with a high degree of symmetry can be built using a spin squeezing model. But the symmetry property does not guarantee the optimality of the entanglement: it does not related to the quantum noise, i.e., quantum correlation or entanglement, it is only related to the signal strength, i.e., classical correlation among qubits.

During "free evolution" period of Ramsey spectroscopy we include a decoherence governed by master equation $\dot{\rho}(t) = -i[H,\rho] - \gamma(\rho - 1/2)$, i.e., all non-diagonal elements are decayed with a rate γ. Finally, we measure an operator S_y. Under these conditions the relative improvement in the precision becomes of the form

$$P(\tau,\mu) = 1 - \frac{|\delta\omega_0|_{ss}}{|\delta\omega_0|} = 1 - \sqrt{\frac{e^{\tau} + a(\mu) - 1}{eb(\mu)\tau}}, \tag{1}$$

were $\tau = 2\gamma t$, $a(\mu) = 4\langle \Delta S_y(\mu)^2 \rangle / N$, and $b(\mu) = 4\langle S_x(\mu) \rangle^2 / N^2$. The mean $\langle S_x \rangle$ and variance $\langle \Delta S_y^2 \rangle$ are [Kitagawa and Ueda, 1993]:

$$\langle S_x \rangle = \tfrac{N}{2} \cos^{N-1} \tfrac{\mu}{2} \text{ and } \langle \Delta S_y^2 \rangle = \tfrac{N}{4} \left\{ 1 + \tfrac{1}{4}(N-1) \left[A - \sqrt{A^2 + B^2} \right] \right\},$$

where $A = 1 - \cos^{N-2} \mu$, $B = 4 \sin \frac{\mu}{2} \cos^{N-2} \frac{\mu}{2}$, and $\nu_{min} = \frac{\pi}{2} - \frac{1}{2} \arctan \frac{B}{A}$.

The optimal entanglement state for maximum improvement will be found via optimization by only two parameters μ and τ regardless the number of ions, thus drastically simplifying the problem and allowing a solution for any large number of ions. Also it is a good approximation of full optimal solution for small number of ions as seen in Fig. 1a. In Figure 1b one can see a little surprising result that the improvement in the precision asymptotically saturates at the absolute decoherence limit $1 - \frac{1}{\sqrt{e}}$ with increasing number of N for sub-optimal entanglement of a simple spin squeezing. As the fact of the matter, for large number of N we have: $\langle \Delta S_y^2 \rangle \simeq \frac{N}{4} \left(\frac{1}{4\alpha} + \frac{2}{3}\beta^2 \right)$ and $\langle S_x \rangle \simeq \frac{N}{2}(1 - \beta)$, when $\mu \ll 1$, provided $\alpha = \frac{N}{4}\mu \gg 1$ and $\beta = \frac{N}{8}\mu^2 \ll 1$. An estimation says this approximation is valid in less than 1% for $N > 10^5$. When $\tau \ll 1$, we have $P \simeq 1 - \frac{1}{\sqrt{e}} \left[1 + (1/6^{1/6} + 6^{1/3}/2) N^{-1/3} \right]$ or $\lim_{N \to \infty} P = 1 - \frac{1}{\sqrt{e}}$.

For non-squeezed separated state, $a = 1$ and $b = 1$, we get ω_0 at $\tau = 0.5$. Under squeezing a normalized variance, $a_0 \leq a \leq 1$ ($a_0 > 0$ except $N = 2$ for which $a_0 = 0$), as well as the normalized mean, $0 \leq b \leq 1$, are decreased. While the former gives an improvement, the latter decreases the improvement as we can see in Eq. (1) or see Fig. 2. The competition between them justifies the absolute limit. It should also be noted that any small squeezing always gives an improvement.

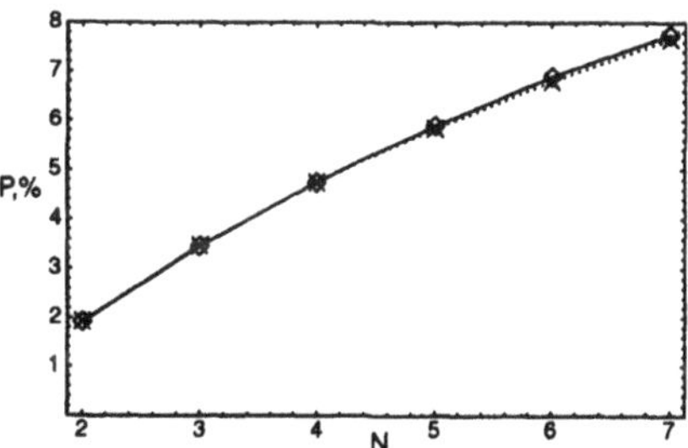

Figure 1a Comparison of maximum improvement for optimal and squeezed input states. ×: one-axis twisting initial state preparation, ◇: full numerical optimization with respect to the initial state preparation [Huelga et al., 1997].

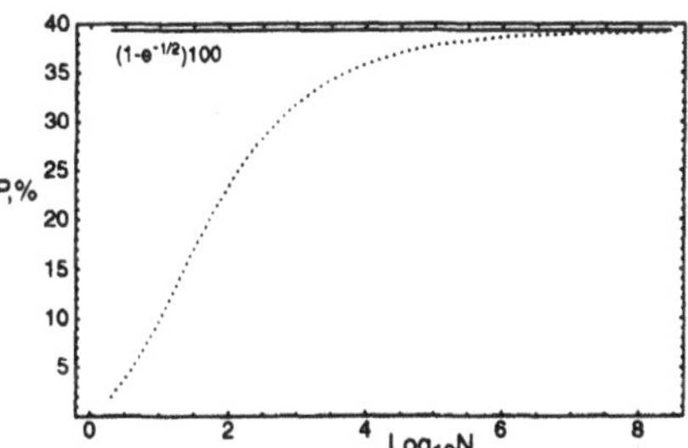

Figure 1b Maximum improvement vs. number of ions N. One-axis twisting improvement in precision of generalized Ramsey spectroscopy (dotted-line) is saturated at the absolute decoherence limit of $1 - 1/\sqrt{e}$ (solid-line).

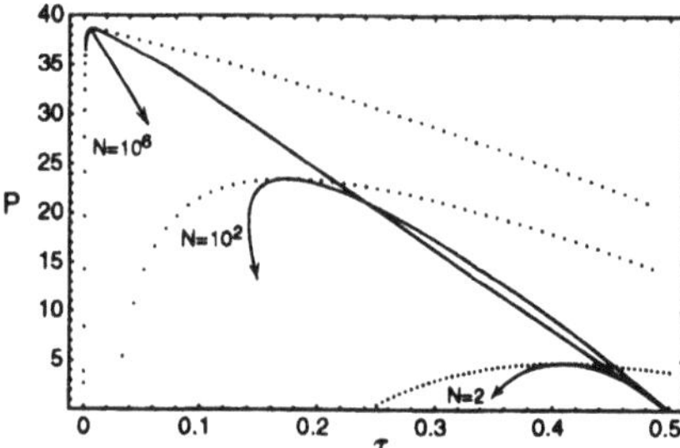

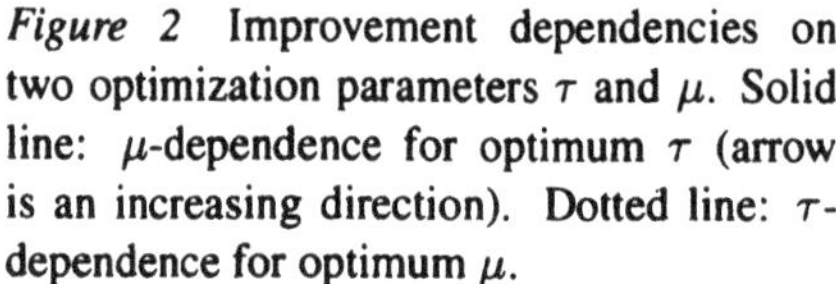
Figure 2 Improvement dependencies on two optimization parameters τ and μ. Solid line: μ-dependence for optimum τ (arrow is an increasing direction). Dotted line: τ-dependence for optimum μ.

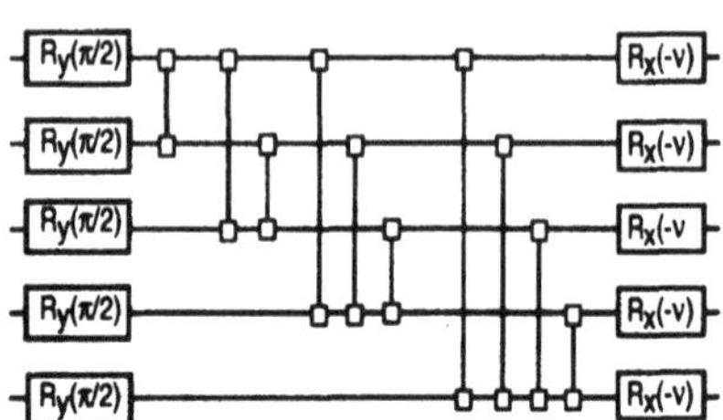

Figure 3 Efficient network for preparation of squeezed initial state. The two-qubit symmetric gate is defined by unitary operator of $\exp[-i\mu\sigma_{zi}\sigma_{zj}/8]$, i.e., a rotation gate of $R_z(\mp\mu/4)$ is applied to the second qubit depending on the value, 0 or 1, of the first qubit.

Finally, an efficient quantum network for preparing the symmetric and sub-optimally entangled initial state can be constructed using $2N$ one-qubit and $N(N-1)/2$ two-qubit gates as shown in Fig. 3. Although our scheme gives a simple and efficient quantum circuit, it does not directly mean that the scheme can be implemented easily, because the actual Hamiltonian is not necessarily the same as our one-axis Hamiltonian. This and measurement optimization will be the subject of future consideration.

In summary, it is shown that even sub-optimal entanglement of spin squeezed input state provides asymptotic saturation at the theoretical improvement limit in the precision of generalized Ramsey spectroscopy with increasing number of atoms. It is also shown that the high symmetric squeezed input state can be prepared via a simple and efficient quantum circuit. The state optimization parameters are reduced from $\lfloor N/2 \rfloor$ for numerical optimization to two for one axis twisting.

References

[Bollinger et al., 1996] Bollinger, J. J., Itano, W. M., Wineland, D. J., and Heinzen, D. J. (1996). Optimal frequency measurements with maximally correlated states. *Phys. Rev. A*, 54:R4649–R4652

[Huelga et al., 1997] Huelga, S. F., Macchiavello, C., Pellizzari, T., Ekert, A. K., Plenio, M. B., and Cirac, J. I. (1997). Improvement of frequency standards with quantum entanglement. *Phys. Rev. Lett*, 79:3865–3868.

[Kitagawa and Ueda, 1993] Kitagawa, M. and Ueda, M. (1993). Squeezed spin states. *Phys. Rev. A*, 47:5138–5143.

TELEPORTATION OF ENTANGLEMENT FOR CONTINUOUS VARIABLES

Anatoly I. Zhiliba
Department of Physics, Tver State University
170002, 35 Sadoviy lane, Tver, Russia
Anatoly.Zhiliba@tversu.ru

Valery N. Gorbachev
vn@vg3025.spb.edu

Andrew I. Trubilko
tai@at3024.spb.edu

Keywords: Entangled states, teleportation, continuous variables

Abstract Teleportation of an unknown two particle pure entangled state using the three-particle quantum channel is considered. Two receivers Bob and Claire can recover the unknown state sent by Alice. But they cannot recover it separately.

Introduction

Teleportation proposed by Bennet et al [1] and demonstrated in optical experiments by several groups [2] is paid great attention in the quantum information processing. It is interesting in its application for quantum computing particularly it can be useful for the problem of a quantum memory for reading of a stored quantum state [3].

In this work, we consider how to teleport an EPR pair or a two particle pure entangled state. The main question is what resource can accomplish the task. In the general case the answer is well known. Due to linearity of quantum mechanics one can teleport any two particles using an one-by-one protocol [1]. For a two qubit state it needs two EPR pairs or the four- particle quantum channel and four bits of information. However, it was shown in [4] that a two-particle entangled state can be teleported by a three-particle quantum channel which is less expensive. A triplet of the Greenberger-Horne-Zeilinger (GHZ) form is useful as the channel. To accomplish the task it needs a three-particle basis for joint measurement we have found for both the discrete and continuous variables to be not maximally entangled [4]. Recently Marinatto et al [5] have shown that a two-qubit pure state of a general form cannot be teleported using the GHZ channel.

1. CONTINUOUS BELL LIKE STATE

Teleportation of continuous variables has been proposed by Vaidman [6] and developed by Braunstein et al [7], [8] and many others [9]. It is convenient to introduced the position and momentum operators $\hat{x}$, $\hat{p}$ and eigenstates of $\hat{x}$: $\hat{x}|x\rangle = x|x\rangle$. For two particles the eigenstates of the total moment $\hat{p}_1 + \hat{p}_2$ and the relative position $\hat{x}_1 - \hat{x}_2$ are continuous Bell-like states $|\Psi(P,Q)\rangle$ with eigenvalues P and Q. For $P = 0$ and $Q = 0$ one can find a maximally entangled state $|\Psi(0,0)\rangle$ or EPR pair.

The continuous entangled states are attractive because of their physical realization. In quantum optics, electromagnetic field is described by the bosonic operators to be associated with canonical momentum and position by the usual way. As well known momentum and position is the quadrature field operator of the form $X(\theta) = \frac{1}{2}(a\exp(i\theta) + h.c.)$ where $p = X(-\pi/2)$ and $x = X(0)$. This quadrature operator or "momentum and position" of light can be measured by a balance detection scheme where the signal and a local field oscillator are mixed by beam splitter. An EPR pair or entangled state of light can be prepared by an optical parametric oscillator well known in quantum optics. Indeed entanglement is the same as squeezing.

2. TELEPORTATION OF CONTINUOUS VARIABLES

Let Alice wishes to teleport an unknown state

$$|A\rangle = \int dx A(x)|x\rangle$$

to Bob. It needs the maximally entangled EPR pair of particles 2 and 3 of the form $|\Psi(0,0)\rangle$, where the particle 2 is in Alice's hands and particle 3 is in Bob's hands. As for the usual protocol Alice performs a joint measurement of particle 1 and 2 on the continuous Bell basis. Then writing the wave function of the combined system

$$|A\rangle_1 \otimes |\Psi(0,0)\rangle_{23} = \frac{1}{\pi}\int dPdQ|\Psi(P,Q)\rangle_{12}|B\rangle_3$$

one finds the state of Bob's particle to be connected with unknown state by a unitary transformation $|B\rangle = U(P,Q)|A\rangle$

$$U(P,Q) = \int dx \exp(2iPx)|x\rangle\langle x - Q|$$

The series expansion on the continuous Bell basis describes a joint measurement of the total momentum $\hat{p}_1 + \hat{p}_2$ and relative position $\hat{x}_1 - \hat{x}_2$. If outcome P and Q is obtained then Bob's particle 3 is projected into a state to be equal an unknown state up to the unitary transformation.

In optical realization the required joint measurement can be performed by four detectors and two local field oscillators.

3. THREE PARTICLE QUANTUM CHANNEL

In the presented scheme instead of an EPR pair the GHZ triplet of particles 3, 4 and 5 is introduced. There are two receivers, Bob and Claire, spatially separated. Alice and the receivers share particles of the triplet GHZ so that a particle 3 is in Alice hand and the other particles are in Bob and Claire site. For continuous variables the GHZ channel has the form

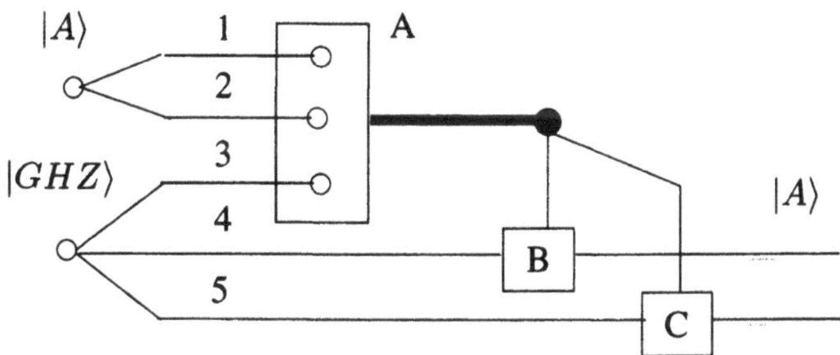

Figure 1 Teleportation of an entangled state of particles 1 and 2 using the GHZ triplet. Alice, Bob and Claire share the qubits 3,4 and 5 of the GHZ triplet. Alice sends outcome of a joint measurement of particle 1, 2 and 3 to Bob and Claire who recover an unknown state.

$$|GHZ\rangle_{345} = \frac{1}{\sqrt{\pi}} \int dx|x\rangle|x\rangle|x\rangle.$$

It can be prepared from squeezed light [8] by mixing of three modes of light by two nonabsorbing beamsplitters.

4. TELEPORTATION OF TWO PARTICLE ENTANGLED STATE

Let now Alice wishes to teleport a two-particle entangled state

$$|A\rangle_{12} = \int dx A(x)|x\rangle|x-q\rangle$$

which is eigenstate of the relative position operator $\hat{x}_1 - \hat{x}_2$, which eigenvalue is q. Then she performs a joint measurement of three particles 1,2 and 3. It would be imagine that the basis of the measurement will consist of a maximally three particle entangled state. However, it does not solve the task because of there is no any unitary transformation connected input and output state. We found the required three particle basis have to be consist of the eigenfunctions $|p\rangle_1$ of the momentum operator of the particle 1 and the Bell state $|\Psi(P,Q)\rangle_{23}$ of particle 2 and 3. In fact, it is the measurement of momentum of the first particle and total momentum and relative position of the particles 2 and 3. Since all operators commute the joint measurement is possible. Using the found basis the combined state can be written in the form

$$|A\rangle_{12} \otimes |GHZ\rangle_{345} = \int \frac{dpdPdQ}{\pi^3} |p\rangle_1 |\Psi(P,Q)\rangle_{23} U_B(P,Q) U_C(p,Q,q) |A\rangle_{45}$$

where two unitary operators

$$U_B(P,Q) = \int dx \exp(2ixP)|x-Q\rangle\langle x|$$

$$U_C(p,Q,q) = \sqrt{\pi} \int dx \langle x+q|p\rangle|x\rangle\langle x-Q|$$

act on particle 4 and 5. It results that unknown state and the state of Bob's and Claire's particles are connected by unitary transformation. Then acting on their particles Bob and Claire can recover an unknown state, however none of them can do it alone.

Acknowledgments

VG wishes to thank Delzell Foundation for financial support.

References

[1] C.H. Bennett, G. Brassard, C. Crepeau, R. Jozsa, A. Peres, W.K. Wootters. Phys.Rev.Lett. **70**, 1895 (1993).

[2] D. Bouwmeester, J-W Pan, M. Mattle, H. Weinfurter, A. Zeilinger. Nature, 390, 575, (1997).

D. Boschi, S. Branca, F. De Martini, L. Hardy, S. Popescu. Phys. Rev. Lett. **80**, 1121, (1998).
A. Furusawa, J.L. Sorensen, S.L. Braunstein, C.A. Fuchs, H.J. Kimble, E.S. Polzik. Science, 282, 706, (1998).

[3] A.E. Kozhekin, K. Molmer and E. Polzik. Quantum Memory for Light. E-print, LANL, quant/ph 9912014.

[4] V.N. Gorbachev, A.I. Trubilko. Quantum teleportation of EPR pair by three particle entanglement. E-print, LANL, quant/ph 9906110. (to be published in Russian JETP, 118, (2000))
V.N. Gorbachev, A.I. Trubilko. Teleportation of entanglement for continuous variables. E-print, LANL, quant/ph 9912061.

[5] L. Marinatto, T. Weber. Which kind of two particle state can be teleported through a three particle quantum channel? E-print, LANL, quant/ph 0004054.

[6] L. Videman. Phys. Rev.A 49, 1473, (1994)

[7] S.L. Braunstein, H.J. Kimble. Phys.Rev.Lett. 80, 869, (1998).
P.van Look, S.L. Braunstein. Unconditional entanglement swapping for continuous variables, e-print, quant/ph 990675.

[8] P.van Look, S.L. Braunstein. Multipatite entanglement for continuous variables. E-print, LANL quant/ph 9906021

[9] R.E.S. Polkinghome, T.C. Ralph. Entanglement swapping using continuous variables. E-print, LANL, quant/ph 9906066.

GENERATION AND DETECTION OF FOCK STATES OF THE RADIATION FIELD

Herbert Walther
Sektion Physik der Universität München and
Max-Planck-Institut für Quantenoptik
85748 Garching, Fed. Rep. of Germany
herbert.walther@mpq.mpg.de

Keywords: Quantum optics, cavity quantum electrodynamics, Fock states

Abstract In this paper we give a survey of our experiments performed with the micromaser on the generation of Fock states. Three methods can be used for this purpose: the trapping states leading to Fock states in a continuous wave operation; state reduction of a pulsed pumping beam and finally using a pulsed pumping beam to produce Fock states on demand where trapping states stabilize the photon number.

1. INTRODUCTION

The quantum treatment of the radiation field uses the number of photons in a particular mode to characterize the quantum states. In the ideal case the modes are defined by the boundary conditions of a cavity giving a discrete set of eigen-frequencies. The ground state of the quantum field is represented by the vacuum state consisting of field fluctuations with no residual energy. The states with fixed photon number are usually called Fock or number states. They are used as a basis in which any general radiation field state can be expressed. Fock states thus represent the most basic quantum states and differ maximally from what one would call a classical field. Although Fock states of vibrational motion are routinely observed in ion traps [1], Fock states of the radiation field are very fragile and very difficult to produce and maintain. They are perfectly number-squeezed, extreme sub-Poissonian states in which intensity fluctuations vanish completely. In order to generate these states it is necessary that the mode considered has minimal losses and the thermal field, always present at finite temperatures, has to be eliminated to a large extent since it causes photon number fluctuations.

The one-atom maser or micromaser [2] is the ideal system to realize Fock states. In the micromaser highly excited Rydberg atoms interact with a single mode of a superconducting cavity which can have a quality factor as high as 4×10^{10}, leading to a photon lifetime in the cavity of 0.3s. The steady-state field generated in the cavity has already been the object of detailed studies of the sub-Poissonian statistical distribution of the field [3], the quantum dynamics of the atom-field photon exchange represented in the collapse and revivals of the Rabi nutation [4], atomic interference [5], bistability and quantum jumps of the field [6], atom-field and atom-atom entanglement [7]. The cavity is operated at a temperature of 0.2 K leading to a thermal field of about 5×10^{-2} photons per mode.

There have been several experiments published in which the strong coupling between atoms and a single cavity mode is exploited (see e.g. Ref. [8]). The setup described here is the only one where maser action can be observed and the maser field investigated. In our setup the threshold for maser action is as small as 1.5 atoms/s. This is a consequence of the high value of the quality factor of the cavity which is three orders of magnitude larger than that of other experiments with Rydberg atoms and cavities [9].

In this paper we present three methods of creating number states in the micromaser. The first is by way of the well known trapping states, which are generated in a c.w. operation of the pumping beam and lead to Fock states with high purity. We also present a second method where the field is prepared by state reduction and the purity of the states generated is investigated by a probing atom. It turns out that the two methods of preparation of Fock states are in fact equivalent and lead to a similar result for the purity of the Fock states. The third method pumps the cavity with a pulsed beam using the trapping condition to stabilize the photon number in our cavity. This method produces Fock states on demand.

2. THE ONE-ATOM MASER AND THE GENERATION OF FOCK-STATES USING TRAPPING STATES

The one-atom maser or micromaser is the experimental realisation of the Jaynes-Cummings model [10], as it allows to study the interaction of a single atom with a single mode of a high Q cavity. The setup used for the experiments is shown in Fig. 1 and has been described in detail previously [11]. Briefly, in this experiment, a $^3He-^4He$ dilution refrigerator houses the microwave cavity which is a closed superconducting niobium cavity. A rubidium oven provides two collimated atomic beams: a central one passing directly into the cryostat and a second one directed to an additional excitation region. The second beam was used as a frequency reference. A frequency doubled dye laser ($\lambda = 294$

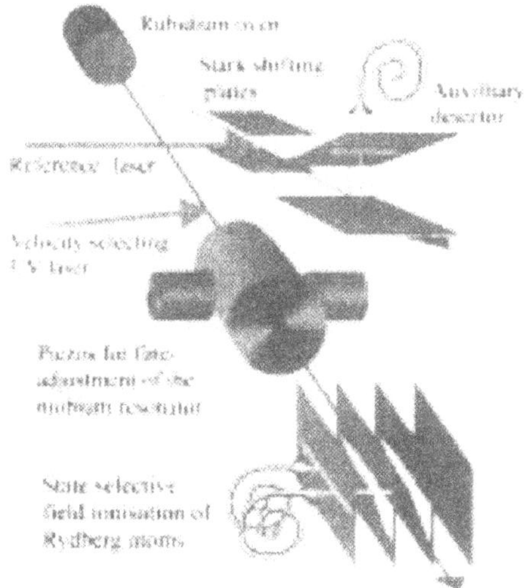

Figure 1 In this figure the micromaser setup is shown. For details see Ref. [9].

nm) was used to excite rubidium (^{85}Rb) atoms to the 63 $P_{3/2}$ Rydberg state from the 5 $S_{1/2}$ (F = 3) ground state.

Velocity selection is provided by angling the excitation laser towards the main atomic beam at about 11° to the normal. The dye laser was locked, using an external computer control, to the 5 $S_{1/2}$ (F = 3)-63 $P_{3/2}$ transition of the reference atomic beam excited under normal incidence. The reference transition was detuned by Stark shifting the resonance frequency using a stabilized power supply. This enabled the laser to be tuned while remaining locked to an atomic transition. The maser frequency corresponds to the transition between 63 $P_{3/2}$ and 61 $D_{5/2}$. The Rydberg atoms are detected by field ionization in two detectors set at different voltages so that the upper and lower states of the maser transition can be investigated separately.

The trapping states are a steady-state feature of the maser field peaked in a single photon number, they occur in the micromaser as a direct consequence of field quantisation. At low cavity temperatures the number of blackbody photons in the cavity mode is reduced and trapping states begin to appear [11, 12]. They occur when the atom field coupling constant given by the Rabi frequency Ω, and the interaction time, $t_{\rm int}$, are chosen such that in a cavity field with n photons each atom undergoes an integer number, k, of Rabi cycles. This is summarised by the condition,

$$\Omega t_{\rm int}\sqrt{n+1} = k\pi. \tag{1}$$

When Eq.1 is fulfilled the cavity photon number is left unchanged after the interaction of an atom and hence the photon number is "trapped". This will

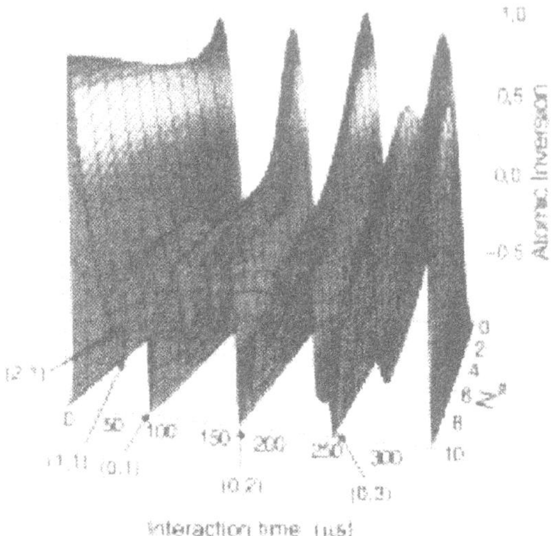

Figure 2 A theoretical plot, in which the trapping states can be seen as valleys in the N_{ex} direction. As the pump rate is increased, the formation of the trapped states from the vacuum can be seen. The positions of trapping states are indicated by arrows with the respective designation (n,k).

occur regardless of the atomic pump rate N_{ex}, where N_{ex} is the rate of pumping atoms in the excited state per decay time of the cavity. The trapping state is therefore characterised by the photon number n and the number of integer multiples of full Rabi cycles k.

The build-up of the cavity field can be seen in Fig. 2, where the emerging atom inversion $\mathrm{I} = \mathrm{P}_g - \mathrm{P}_e$ is plotted against interaction time and pump rate; $\mathrm{P}_{g(e)}$ is the probability of finding a ground (excited) state atom. At low atomic pump rates (low N_{ex}) the maser field cannot build up and the maser exhibits Rabi oscillations due to the interaction with the vacuum field. At the positions of the trapping states, the field increases until it reaches the trapping state condition. This manifests itself as a reduced emission probability and hence as a dip in the atomic inversion. Once in a trapping state the maser will remain there regardless of the pump rate. The trapping states show up therefore as valleys in the N_{ex} direction. Figure 3 shows the photon number distribution as the pump rate is increased for the special condition of the five photon trapping state. The photon distribution develops from a thermal distribution towards higher photon numbers until the pump rate is high enough for the atomic emission to be stabilized by the trapping state condition. As the pump rate is further increased, and in the limit of a low thermal photon number, the field continues to build up to a single trapped photon number and the steady-state distribution approaches a Fock state.

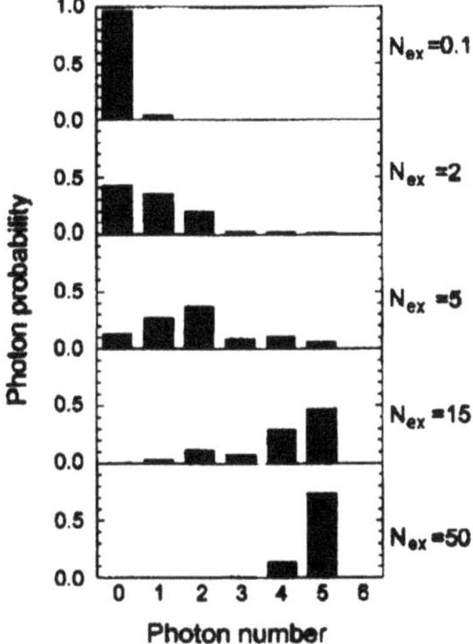

Figure 3 A numerical simulation of the photon number distribution as the atomic pump rate (N_{ex}) is increased until the cavity field is in a Fock state with a high probability. Shown is the example of the n = 5 Fock state. A further increase of N_{ex} beyond 50 does not change the distribution.

Owing to blackbody radiation at finite temperatures, there is always a small probability of having a thermal photon enter the mode. The presence of a thermal photon in the cavity disturbs the trapping state condition and an atom can emit a photon. This causes the field to change around the trapping condition.

Note that under readily achievable experimental conditions, it is possible for the steady-state field in the cavity to approach a Fock state with a high fidelity. Under the present experimental conditions the main deviation from a pure Fock state results from dissipation of the field in the cavity. If a photon disappears it takes a little while until the next incoming excited atom can be used to replace the lost photon. Therefore smaller photon numbers show up besides the considered Fock state. Figure 4 shows micromaser simulations, for achievable experimental conditions, in which Fock states with high purity are created from $n = 0$ to $n = 5$. The experimental realisation requires a pump rate of $N_{\text{ex}} = 50$, a temperature of about 100mK [6], a high selectivity of atomic velocity and very low mechanical noise of the system [11, 13]. The details of the production of Fock states using the trapping condition are described in Ref. [11].

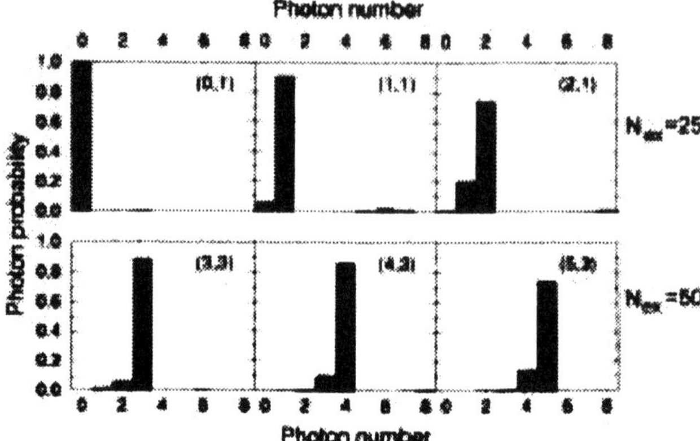

Figure 4 In simulations of the maser operation Fock states from n=0 to n=5 can be readily generated for achievable experimental conditions ($n_{\text{th}} = 10^{-4}$).

3. DYNAMICAL PREPARATION OF NUMBER STATES IN A CAVITY

When the atoms leave the cavity in a micromaser experiment they are in an entangled state with the field. A method of state reduction was suggested by Krause et al. [14] to observe the build up of the cavity field to a known Fock state. State reduction uses the entanglement produced by the interaction of an atom with a cavity field to project the field onto a well defined number state. If the field is in an initial state $|n\rangle$ then an interaction of an atom with the cavity leaves the cavity field in a superposition of the states $|n\rangle$ and $|n+1\rangle$ and the atom in a superposition of the internal atomic states $|e\rangle$ and $|g\rangle$.

$$\Psi = \cos(\phi)|e\rangle|n\rangle - i\sin(\phi)|g\rangle|n+1\rangle \tag{2}$$

where ϕ is an arbitrary phase. The state selective field ionisation measurement of the internal atomic state, reduces the field to one of the states $|n\rangle$ or $|n+1\rangle$. State reduction is independent of the interaction time, hence a ground state atom always projects the field onto the $|n+1\rangle$ state independent of the time spent in the cavity. This results in an *a priori* probability of the maser field being in a specific but unknown number state [14]. If the initial state of the cavity is the vacuum, $|0\rangle$, then a number state created is equal to the number of ground state atoms that were collected within a suitably small fraction of the cavity decay time.

In a system governed by the Jaynes-Cummings Hamiltonian, spontaneous emission is reversible and an atom in the presence of a resonant quantum field undergoes Rabi oscillations. That is the relative populations of the excited and ground states of the atom oscillate at a frequency $\Omega\sqrt{n+1}$, where Ω is the atom field coupling constant. Experimentally we measure the atomic inversion. In the presence of dissipation a fixed photon number n in a particular mode is not observed and the field always evolves into a mixture of such states. Therefore the inversion is generally given by,

$$I(n, t_{\rm int}) = -c \sum_{\rm n} P_{\rm n} \cos(2\Omega\sqrt{n+1}t_{\rm int}) \tag{3}$$

where $P_{\rm n}$ is the probability of finding n photons in the mode, $t_{\rm int}$ is the interaction time of the atoms with the cavity field. The factor c considers the reduction of the signal amplitude as a result of dark counts.

The experimental verification of the presence of Fock states in the cavity corresponds to a pump-probe experiment in which a pump atom prepares a quantum state in the cavity and the Rabi phase of the emerging probe atom measures the quantum state. The signature that the quantum state of interest has been prepared is simply the detection of a defined number of ground state atoms. To verify that the correct quantum state has been projected onto the cavity a probe atom is sent into the cavity with a variable, but well defined, interaction time. As the formation of the quantum state is independent of interaction time we need not to change the relative velocity of the pump and probe atoms, thus reducing the complexity of the experiment. In this sense we are performing a reconstruction of a quantum state in the cavity using a similar method to that described by Bardoff et al. [15]. This experiment reveals the maximum amount of information that can be found relating to the cavity photon number. We have recently used this method to demonstrate the existence of Fock states up to n=2 in the cavity [16].

When the interaction time corresponding to the trapping state condition is met in this experiment, the formation of the cavity field is identical to that which occurs in the steady-state, hence the probe atom should perform an integer number of Rabi cycles. In fact this was observed experimentally [16], which indicates that the pulsed experiment is actually the formation stage of the steady-state experiment. One would therefore expect that the measured photon number distribution, in the dynamical measurement, would be the same as that predicted for the trapping states. State reduction is simply a method of observation that determines the appropriate moment for a measurement. In this sense the observation of a lower emission probability in the steady state is also a field-state measurement as the steady-state inversion measurement occurs for practically the same conditions as for the dynamical measurement described here [16].

4. PREPARATION OF FOCK STATES ON DEMAND

In the following we describe another variant of a dynamical Fock-state preparation with the micromaser [17]. To demonstrate the principle of the

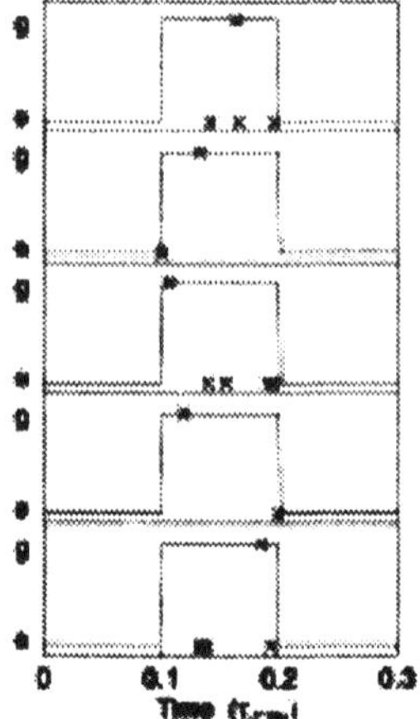

Figure 5 A computer simulation of five sequential excitation pulses for Rydberg atoms. The interaction time is tuned to the (1, 1) trapping state. The cavity is initially in the vacuum state. Rydberg atoms are marked by crosses; those in the upper or lower maser level are marked on the lower and upper row respectively. For high pump rates, every pulse contains at least one lower-state atom with a deviation occurring once in every 50 pulses. The parameters used are: $\tau_{\rm cav}$ = 100 ms, $n_{\rm th} = 0.03$, $\tau_{\rm pulse}$=10 ms, $N_{\rm ex}$ = 40 or $N_{\rm a}$=4 atoms per pulse.

source described here; Fig. 5 shows the simulation of a sequence of five arbitrary atom pulses using a Monte Carlo calculation in which the micromaser is operated in the (1, 1) trapping state. In each pulse there is a single emission event, producing a single lower state atom and leaving a single photon in the cavity. The atom-cavity system is then in the trapping condition, as a consequence the emission probability is reduced to zero and the photon number is stabilized. In steady state operation, the influence of thermal photons and variations in interaction time or cavity tuning complicates this picture, resulting in deviations from Fock states [18]. Pulsed excitation however reduces the influence of such effects and the generated Fock states show a high purity.

Figure 6(a-c) presents three curves obtained from the computer simulation, that illustrate the behaviour of the maser under pulsed excitation as a function of interaction time for more ideal (but achievable) experimental parameters. The simulations show the probability of finding; no lower-state atom per pulse ($P^{(0)}$); finding exactly one lower-state atom per pulse ($P^{(1)}$); and the conditional probability of finding a second lower-state atom in a pulse already containing one ($P^{(>1;1)}$). The latter plot of the conditional probability, $P^{(>1;1)}$, is relatively insensitive to the absolute values of the atomic detection efficiency and therefore has advantages when comparisons with experimental data are performed [17].

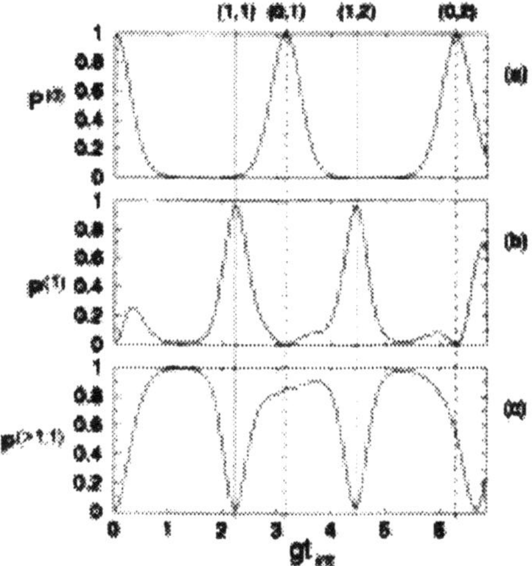

Figure 6 The figure presents the probability of finding; (a) no lower-state atoms per pulse $P^{(0)}$, (b) exactly one lower-state atom per pulse $P^{(1)}$, and (c) of finding a second lower-state atom, if one has already been detected $P^{(>1;1)}$. Parameters for these simulations were τ_{pulse}=0.02τ_{cav}, N_a= 7 atoms and $n_{\text{th}} = 10^{-4}$. The maximum value of $P^{(1)}$ is 98% for the (1, 1) trapping state.

It follows from Fig. 6 that with an interaction time corresponding to the (1, 1) trapping state, both one photon in the cavity and a single atom in the lower-state are produced with a 98% probability. In order to maintain an experimentally verifiable quantity, the simulations presented here relate to the production of lower-state atoms rather than to the Fock state left in the cavity. However, pulse lengths are rather short ($0.01\tau_{\text{cav}} \leq \tau_{\text{pulse}} \leq 0.1\tau_{\text{cav}}$) so there is little dissipation and the probability of finding a one photon state in the cavity following the pulse is very close to the probability of finding an atom in the lower-state. Atomic beam densities must also be chosen with care to avoid short pulses with high atom density that could violate the one-atom-at-a-time condition.

Note that at no time in this process is a detector event required to project the field, the field evolves to the target photon number state, when a suitable interaction time has been chosen so that the trapping condition is fulfilled.

It should be noted that for thermal photon numbers as high as n_{th} = 0.1 or for t_{int} fluctuations of up to 10% (both beyond the current experimental parameters), simulations show that Fock states are still prepared with an 80-90% fidelity. This is considerably better than for steady state trapping states, where highly stable conditions with low thermal photon numbers are required [12, 11, 18].

The present setup of the micromaser was specifically designed for steady state operation. Nevertheless the current apparatus does permit a comparison between theory and experiment in a relatively small parameter range.

A method was used for the comparison with the experiment which is described briefly in the following. During the interaction, strong coupling between pumping atoms and the cavity field creates entanglements between internal atomic levels and the cavity field. Subsequent pumping atoms will therefore also become entangled both with the field and a previous pumping atom. The correlations between subsequent atomic levels are determined by the dynamics of the atom-cavity interaction. The connection between population correlations and the micromaser dynamics, has been studied in detail in previous papers [7, 19]. It is important to note that even in the presence of lost counts, the correlations between subsequently detected atoms are maintained. Thus rather than a measurement of the single-atom-per-pulse probability $P^{(1)}$ which is heavily dependent on exact knowledge of detection efficiency, it is more useful to measure atom pair correlations given by $P^{(>1;1)}$, owing to the insensitivity of the parameter to detector efficiencies. Experimentally the parameter is obtained via,

$$P^{(>1;1)} = \frac{N_{gg}}{N_{gg} + N_{eg} + N_{ge}} \tag{4}$$

where for example N_{eg} is the probability of detecting a pair of atoms containing first an upper state atom (e) and then a lower-state atom (g) within a pulse. Eq. 4 provides both a value appropriate to the existent correlation and is directly related to the total probability of finding one atom per pulse. Although $P^{(>1;1)}$ is insensitive to the absolute detector efficiency it does depend on the relative detector efficiencies (which are nearly equal) and the miscount probability (the probability that a given atomic level is detected in the wrong detector). Each has been measured experimentally.

The source presented here has the significant advantage over our previous method of Fock state creation [16] of being unconditional and therefore significantly faster in preparing a target quantum state. Previously state reduction by detection of a predefined number of lower state atoms was used to prepare the state with 95% fidelity. However, this method has the disadvantage that it is affected by non perfect detectors. In the current experiment however the cavity field is correctly prepared in 83.2% of the pulses and is independent of any detector efficiencies. Improving the experimental parameters we can expect to reach conditions for which 98% of the pulses prepare single photon Fock states and a single atom in the lower-state.

5. CONCLUSION

In this paper we gave a survey of the possibilities for generating Fock states in the micromaser. The generation of Fock states on demand has been experimentally confirmed and will be published elsewhere [17]. The possibility to generate Fock states will allow us to perform the reconstruction of a single photon field or other Fock states in a next step.

References

[1] D. Leibfried, D.M. Meekhof, B.E. King, C. Monroe, W. M. Itano, and D. J. Wineland. Experimental determination of the motional quantum state of a trapped atom. *Phys. Rev. Lett*, 77:4281–4285, 1996.

[2] D. Meschede, H. Walther, and G. Müller. The one-atom-maser. *Phys. Rev. Lett.*, 54:551–554, 1985.

[3] G. Rempe and H. Walther. Sub-Poissonian atomic statistics in a micromaser. *Phys. Rev. A*, 42:1650–1655, 1990.

[4] G. Rempe, H. Walther, and N. Klein. Observation of quantum collapse and revival in a one-atom maser. *Phys. Rev. Lett.*, 58:353–356, 1987.

[5] G. Raithel, O. Benson, and H. Walther. Atomic interferometry with the micromaser. *Phys. Rev. Lett.*, 75:3446–3449, 1995.

[6] O. Benson, G. Raithel, and H. Walther. Quantum jumps of the micromaser field – dynamic behavior close to phase transition points. *Phys. Rev. Lett.*, 72:3506–3509, 1994.

[7] B.-G. Englert, M. Löffler, O. Benson, B. Varcoe, M. Weidinger, and H. Walther. Entangled atoms in micromaser physics. *Fortschr. Phys.*, 46:897–926, 1998.

[8] H. J. Kimble, O. Carnal, N. Georgiades, H. Mabuchi, E. S. Polzik, R. J. Thomson and Q. A. Turchettte. Quantum optics with strong coupling. *Atomic Physics*, 14, D. J. Wineland, C. E. Wieman, and S. J. Smith, eds., AIP Press, 314–335 (1995).

[9] G. Nogues, A. Rauschenbeutel, S. Osnaghi, M. Brune, J. M. Raimond, and S. Haroche. Seeing a single photon without destroying it. *Nature*, 400:239–242, 1999.

[10] E.T. Jaynes and F.W. Cummings. Quantum and semiclassical radiation theories. *Proc. IEEE*, 51:89–109, 1963.

[11] M. Weidinger, B.T.H. Varcoe, R. Heerlein, and H. Walther. Trapping states in the micromaser. *Phys. Rev. Lett.*, 82:3795–3798, 1999.

[12] P. Meystre, G. Rempe, and H. Walther. Very-low temperature behaviour of a micromaser. *Optics Lett.*, 13:1078–1080, 1988.

[13] G. Raithel, *et al.*. The micromaser: a proving ground for quantum physics. In *Advances in Atomic, Molecular and Optical Physics*, Supplement 2, pages 57–121, P. Berman, editor, Academic Press, New York, 1994.

[14] J. Krause, M. O. Scully, and H. Walther. State reduction and |n>-state preparation in a high-Q micromaser. *Phys. Rev. A*, 36:4547–4550, 1987.

[15] P. J. Bardoff, E. Mayr, and W.P. Schleich. Quantum state endoscopy: measurement of the quantum state in a cavity. *Phys. Rev. A,* 51:4963–4966, 1995.

[16] B. T. H. Varcoe, S. Brattke, M. Weidinger, and H. Walther. Preparing pure photon number states of the radiation field. *Nature,* 403:743–746, 2000.

[17] Brattke, S., Varcoe, B. T. H. and Walther, H., publication in preparation.

[18] B. T. H. Varcoe, S. Brattke, and H. Walther. Generation of Fock states in the micromaser. *Journ. of Optics B: Quantum Semiclass. Opt.,* 2:154–157, 2000.

[19] H.-J. Briegel, B.-G. Englert, N. Sterpi, and H. Walther. One-atom maser: statistics of detector clicks. *Phys. Rev. A,* 49:2962–2984, 1994.

Index

Accessible information, 39, 62, 107, 279
Amplifier, 12, 85, 161, 181, 292, 303, 343, 369, 384, 436
Antibunching, 262, 309
Authentication, 286, 299

Bell measurement, 84, 354, 368, 384
Bell state, 270, 300, 345, 360, 368, 376, 385, 412
Bit commitment, 285
Bloch vector, 160
Bose-Einstein condensate, 403
Bunching, 262

Channel capacity, 17, 27, 35, 40
Chaos, 140
Cloning, 13, 82, 84, 160, 296, 343
CN gate, 361
Coherent state, 12, 40, 69, 85, 156, 239, 292, 363, 404
Concurrence, 360, 408, 412
Controlled-NOT gate, 204

Davies, E.B., 63, 101, 107, 108
Decoherence, 92, 115, 121, 151, 171, 231, 235, 421, 453
Dense coding, 384
Depolarizing channel, 8
Detection operator, 27, 40

Eavesdropping, 159, 281, 288, 296, 403
Einstein-Podolsky-Rosen (EPR), 176, 271, 313, 354, 376, 444
Entanglement of formation, 360, 408, 411
Error correction, 176, 195, 199, 231, 280, 296

Fault tolerant, 203, 213, 238, 280
Fidelity, 7, 12, 82, 236, 248, 297, 355, 372, 389
Filtering, 239, 263

Gaussian state, 363
GHZ state, 275, 403, 405
Gram matrix, 36, 41, 80, 361
Group covariant, 82

Hadamard transform, 116, 220, 361
Hash function, 299
Holevo, A.S., 15, 43, 44, 108

Induced transparency, 179, 183, 384
Ion-trap, 235

Jaynes-Cummings model, 72

Kerr effect, 336, 364, 384
Kerr nonlinearity, 336, 365, 384, 443
Key distribution, 11, 60, 254, 277, 285, 290, 299, 446
Key expansion, 286, 290
Kraus theorem, 4, 82, 327

Lindblad, G., 155
Local measurement, 84

Mach-Zender interferometer, 403
Maximum-likelihood estimation, 157
Measurement theory, 97, 100, 101, 151
Mutual entropy, 119, 329
Mutual information, 35, 328

Non locality, 313, 401

Operation, 12, 79, 103, 104
Oracle, 222

Phase-shift-keying (PSK), 40
Polarization state, 61, 255, 376, 385, 449
Positive map, 50, 71, 81, 97, 327
Positive operator valued measure (POVM), 100, 107
Privacy amplification, 289
Projection postulate, 97
Purification, 82, 407

Quantum compression, 207
Quantum cryptography, 278, 307, 354
Quantum non-demolition measurement, 341, 352, 403, 430

Quantum repeater, 280, 373, 425
Quantum state reconstruction, 130
Quantum Tomography, 82
Quasi Bell state, 360

Ramsey pulse, 454
Random coding, 15, 44
Relative entropy, 34, 54, 328
Retrodiction, 144
Rydberg atom, 69, 386, 415, 427

Schumacher, B., 15
Second harmonic generation, 344
Shor, P., 42, 175, 208, 215, 219, 231, 280
Singlet, 315, 345, 359
Soliton, 335, 441, 443
Spin, 27, 129, 160, 164
Square root detection, 28
Square root measurement, 38, 41
Squeezed state, 157, 354, 384, 454
State reduction, 33, 98, 139, 152

Teleportation, 83, 355, 367, 376, 458

Von Neumann entropy, 4, 54, 93, 211, 329, 435

Werner, R., 84, 408, 425
Wigner function, 127, 148, 156, 434

GPSR Compliance
The European Union's (EU) General Product Safety Regulation (GPSR) is a set of rules that requires consumer products to be safe and our obligations to ensure this.

If you have any concerns about our products, you can contact us on

ProductSafety@springernature.com

In case Publisher is established outside the EU, the EU authorized representative is:

Springer Nature Customer Service Center GmbH
Europaplatz 3
69115 Heidelberg, Germany

www.ingramcontent.com/pod-product-compliance
Ingram Content Group UK Ltd.
Pitfield, Milton Keynes, MK11 3LW, UK
UKHW051131260726
13967UKWH00010B/2985

* 9 7 8 1 4 7 5 7 8 6 9 8 9 *